Studies in Classification, Data Analysis, and Knowledge Organization

D1823508

More information about this series at
http://www.springer.com/series/1564

Berthold Lausen • Sabine Krolak-Schwerdt •
Matthias Böhmer

Editors

Data Science, Learning by Latent Structures, and Knowledge Discovery

Fonds National de la
Recherche Luxembourg

Editors

Berthold Lausen
University of Essex
Colchester
United Kingdom

Sabine Krolak-Schwerdt
University of Luxembourg
Walferdange
Luxembourg

Matthias Böhmer
University of Luxembourg
Walferdange
Luxembourg

ISSN 1431-8814
Studies in Classification, Data Analysis, and Knowledge Organization
ISBN 978-3-662-44982-0   ISBN 978-3-662-44983-7   (eBook)
DOI 10.1007/978-3-662-44983-7

Library of Congress Control Number: 2015938905

Springer Heidelberg New York Dordrecht London

Springer-Verlag GmbH Berlin Heidelberg is part of Springer Science+Business Media
(www.springer.com)

Foreword

Dear Scholars,

It is both my privilege and my pleasure to invite you to this volume and to get inspired for further research in the fields of Data Analysis, Learning by Latent Structures, and Knowledge Discovery. After having held the first European Conference on Data Analysis (ECDA) in Luxembourg in the year 2013, this book is the outcome of intensive work of excellent researchers from different disciplines united under the roof of data analysis. When the Gesellschaft für Klassifikation (GfKl) accepted Luxembourg's offer to host the European Conference on Data Analysis, the GfKl was aware to come to the "very data heart of Europe," quoted from the former Prime Minister of the Grand Duchy of Luxembourg, Jean Claude Juncker. With STATEC, the government statistics service, Luxembourg provides an institute serving citizens, companies, and political decision-makers that is one of the country's largest producers and disseminator of information. With EUROSTAT, Luxembourg is hosting the key to European statistics, and with LIS, formerly known as The Luxembourg Income Study, Luxembourg gives home to two databases, the Luxembourg Income Study Database and the Luxembourg Wealth Study Database.

In business "analytics will define the difference between the losers and winners going forward," says Tim McGuire, director at the global management consulting firm McKinsey & Company. Even if this book is dedicated to science and not to business, let's be winners! Thus, I hold high expectations for this volume, and I invite you to read this book with gusto. This book is a work that covers the key challenges of data analysis, so, enjoy the read and gel inspired.

Sincerely yours,
Marc Hansen
Secretary of State for Higher Education and Research

Preface

This volume contains the revised versions of selected papers presented during the European Conference on Data Analysis (ECDA 2013), which was jointly held by the German Classification Society (GfKl) and the French speaking Classification Society (SFC). The conference took place in July 2013 at the CCRN Neumünster Abbey, Luxembourg, under the patronage of the Prime Minister of the Grand Duchy of Luxembourg, and it was hosted by the University of Luxembourg. About 300 participants from 44 countries joined ECDA 2013. At the conference venue, 196 presentations including 11 keynote speeches, 3 invited symposia, and 3 invited talks were given. Thus, the conference provided an international forum for discussions and mutual exchange of knowledge. This international scope was also reflected by the keynote speakers from five European countries and the USA.

The scientific program included sessions from a broad range of topics. Special emphasis was laid on interdisciplinary research and the interaction between theory and practice. The program of ECDA 2013 covered theory, methods, and applications of data analysis from the following seven thematic areas, which were organized by the members of the scientific program committee:

1. Statistics and Data Analysis, organized by Claus Weihs, Eva Ceulemans, Christian Hennig, and Patrice Bertrand;
2. Machine Learning and Knowledge Discovery, organized by Myra Spiliopoulou, Mohamed Nadif, and Eyke Hüllermeier;
3. Data Analysis in Marketing, organized by Daniel Baier, Reinhold Decker, and Vincenzo Esposito Vinzi;
4. Data Analysis in Finance, organized by Krzystof Jajuga and Michael Hanke;
5. Data Analysis in Biostatistics and Bioinformatics, organized by Hans Kestler, Rudi Balling, Rainer Spang, Matthias Schmid, and Amedeo Napoli;
6. Data Analysis in Interdisciplinary Domains, organized by Sabine Krolak-Schwerdt, Claus Weihs, Alexandra Schwarz, Hans-Joachim Mucha, and Pascale Kuntz-Cosperec;
7. The Workshop Library and Information Science (LIS' 2013), organized by Frank Scholze.

This volume with 48 contributions is the result of a post conference paper submission and reviewing process. It comprises papers dedicated to the extraction of knowledge from many types of data such as structural, quantitative, or statistical approaches for the analysis of data; advances in classification, clustering, and pattern recognition methods; strategies for modeling complex data and mining large data sets as well as applications of advanced methods in specific domains of practice. Among the disciplines represented in the contributions are Statistics, Psychology, Biology, Information Retrieval and Library Science, Archeology, Banking and Finance, Computer Science, Economics, Engineering, Geography, Geology, Linguistics and Musicology, Marketing, Mathematics, Medical and Health Sciences, Sociology and Educational Sciences.

Empirical research in these disciplines requires the analysis of multiple data types. Even though underlying research questions and corresponding data emerge from most various areas, they often require similar statistical, structural, or quantitative approaches for the analysis of data. The specific scientific impact of the post conference volume concerns the presentation of methods, which may commonly be used for the analysis of data stemming from different domains and domain-specific research questions with the aim of solving the numerous domain-specific problems of data analysis on a theoretical as well as on a practical level, fostering their effective use for answering specific questions in various areas of application as well as evaluating alternative methods in the framework of applications. Accordingly, the volume is organized into the following sections:

Part I: Invited Papers
Part II: Data Science and Clustering
Part III: Machine Learning and Knowledge Discovery
Part IV: Data Analysis in Marketing
Part V: Data Analysis in Biostatistics and Bioinformatics
Part VI: Data Analysis in Education and Psychology
Part VII: Data Analysis in Musicology
Part VIII: Data Analysis in Communication and Technology
Part IX: Data Analysis in Administration and Spatial Planning
Part X: Data Analysis in Library Science

Organizing the ECDA 2013 conference required the coordination of myriad people and topics; it lived from the effort of dedicated colleagues and the great team of the Faculty of Language and Literature, Humanities, Arts and Education of the University of Luxembourg. We would like to thank the Area Chairs and the LIS Workshop Chair for the organization of the areas during the conference, author recruitment, and the evaluation of submissions.

We are grateful and thank all reviewers:

Daniel Baier, Rudi Balling, Martin Behnisch, Ghazi Bel Mufti, Patrice Bertrand, Matthias Böhmer, Adrian Bruhin, Michael Brusch, Andrea Cerioli, Eva Ceulemans, Miguel Couceiro, Ines Daniel, Reinhold Decker, Christian Dittmar, Florent Domenach, Jozef Dziechciarz, Patrizia Falzetti, Andreas Geyer-Schulz, Giuseppe Giordano, Cynthia-Vera Glodeanu, Erhard Godehardt, Tomasz Górecki, Bettina Gruen,

Andreas Hadjar, Christian Hennig, Thomas Hörstermann, Krzysztof Jajuga, Mehdi Kaytoue, Hans A. Kestler, Florian Klapproth, Wolfgang Konen, Sabine Krolak-Schwerdt, Pascale Kuntz, Berthold Lausen, Ludwig Lausser, Yves Lechevallier, Monika Loesse, Karsten Lübke, Vladimir Makarenkov, Michel Meulders, Henning Meyerhenke, Hans-Joachim Mucha, Mohamed Nadif, Jozef Pociecha, Alexandra Rese, Adam Sagan, Gilbert Saporta, Matthias Schmid, Lars Schmidt-Thieme, Frank Scholze, Alexandra Schwarz, Edlira Shehu, Leo Sögner, Rainer Spang, Myra Spiliopoulou, Christoph Stadtfeld, Winfried J. Steiner, Dirk Thorleuchter, Claus Weihs, and Heidrun Wiesenmüller.

We would also like to express our thanks to the Fonds National de la Recherche of Luxembourg (FNR). Furthermore, we would like to thank Martina Bihn, Springer-Verlag, Heidelberg, for her support and dedication to the production of this volume. Last but not least, we would like to thank all participants of the ECDA 2013 conference for their interest and activities, which made the conference and this volume an interdisciplinary possibility for scientific discussion.

Colchester, UK   Berthold Lausen
Luxembourg, Luxembourg   Sabine Krolak-Schwerdt
Luxembourg, Luxembourg   Matthias Böhmer

Contents

Part IV   Data Analysis in Marketing

Part V   Data Analysis in Biostatistics and Bioinformatics

Contributors

Oguz Akbilgic Department of Statistics, Operations and Management Science, University of Tennessee, Knoxville, TN, USA

Department of Quantitative Methods, Istanbul University School of Business, Istanbul, Turkey

Paul Antony Luxembourg Centre for Systems Biomedicine, University of Luxembourg, Esch-sur-Alzette, Luxembourg

Anthony C. Atkinson Department of Statistics, London School of Economics, London, UK

Athanassios N. Avramidis University of Southampton, Southampton, UK

Daniel Baier Institute of Business Administration and Economics, Brandenburg University of Technology Cottbus-Senftenberg, Cottbus, Germany

Beata Bal-Domańska Wrocław University of Economics, Jelenia Góra, Poland

Antonio Balzanella Department of Political Science "Jean Monnet", Second University of Naples, Caserta, Italy

Hans-Georg Bartel Department of Chemistry at Humboldt University, Berlin, Germany

François Bavaud University of Lausanne, Lausanne, Switzerland

N.Y. Nair Benrekia Orange Labs, Lannion, France

LINA, Nantes, France

Miloud Bessafi LE²P, Université de la Réunion, Réunion, France

Etienne Le Bihan University of Luxembourg, INSIDE, Walferdange, Luxembourg

Maria Biryukov Luxembourg Centre for Systems Biomedicine, University of Luxembourg, Esch-sur-Alzette, Luxembourg

Julien Blanchard Equipe COD-LINA (UMR CNRS 6241), Université de Nantes, Nantes, France

Mindaugas Bloznelis Faculty of Mathematics and Informatics, Vilnius University, Vilnius, Lithuania

Thomas Böttcher Institute of Business Administration and Economics, Brandenburg University of Technology Cottbus-Senftenberg, Cottbus, Germany

Hamparsum Bozdogan Department of Statistics, Operations and Management Science, University of Tennessee, Knoxville, TN, USA

Michael Brusch Chair of Business Administration, Marketing and Corporate Planning, Anhalt University of Applied Sciences, Köthen, Germany

Andre Busche Information Systems and Machine Learning Lab, University of Hildesheim, Hildesheim, Germany

Francisco de A.T. de Carvalho CIn/UFPE, Recife, Brazil

Andrea Cerioli Dipartimento di Economia, Università di Parma, Parma, Italy

Philippe Charton LIM, Université de la Réunion, Réunion, France

Christelle Cocco University of Lausanne, Lausanne, Switzerland

Bruno Daigle Department of Computer Science, Université du Québec à Montréal, Montreal, QC, Canada

Mathieu Delsaut LE²P, Université de la Réunion, Réunion, France

Thierry Despeyroux INRIA, Paris-Rocquencourt, Le Chesnay, France

Abdoulaye Baniré Diallo Department of Computer Science, Université du Québec à Montréal, Montreal, QC, Canada

Florent Domenach Computer Science Department, University of Nicosia, Nicosia, Cyprus

Jörgen Eimecke Institute of Business Administration and Economics, Brandenburg University of Technology Cottbus, Cottbus, Germany

Magda El-Sherbini The Ohio State University Libraries, Columbus, OH, USA

Sebastian Fischer Siemens Professional Education, Siemens Technik Akademie Berlin, Siemens AG, Berlin, Germany

Jörg Fliege University of Southampton, Southampton, UK

Roland Fried Department of Statistics, TU Dortmund University, Dortmund, Germany

Sarah Frost Institute of Business Administration and Economics, Brandenburg University of Technology Cottbus, Cottbus, Germany

Adrian Giurca Brandenburgische Technische Universität Cottbus, Cottbus, Germany

Binarypark, Cottbus, Germany

Erhard Godehardt Clinic of Cardiovascular Surgery, Heinrich Heine University, Düsseldorf, Germany

August Götzfried Eurostat, Luxembourg, Eurostat

Guillaume Guex University of Lausanne, Lausanne, Switzerland

Fabrice Guillet Equipe COD-LINA (UMR CNRS 6241), Université de Nantes, Nantes, France

Philippa A. Hiscock University of Southampton, Southampton, UK

Anke Hofmann Hochschule für Musik und Theater "Felix Mendelssohn Bartholdy" Leipzig, Leipzig, Germany

Thomas Hörstermann University of Luxembourg, LCMI Research Unit, Route de Diekirch, Walferdange, Luxembourg

Carine Hue Orange Labs, Lannion, France

Jerzy Jaworski Faculty of Mathematics and Computer Science, Adam Mickiewicz University, Poznań, Poland

Patrick Jeanty LE²P, Université de la Réunion, Réunion, France

Dieter William Joenssen Ilmenau University of Technology, Ilmenau, Germany

Michael Jungheim Hannover Medical High School, Klinik für Phoniatrie und Pädaudiologie, Hannover, Germany

Hans A. Kestler Medical Systems Biology and Institute of Neural Information Processing, Ulm University, Ulm, Germany

Muhammad Umer Khan Information Systems and Machine Learning Lab, University of Hildesheim, Hildesheim, Germany

Alois Knoll Institute for Informatics, Chair for Robotics and Embedded Systems, Technische Universität München, München, Germany

Patrick Koch Cologne University of Applied Sciences, Köln, Germany

Wolfgang Konen Cologne University of Applied Sciences, Köln, Germany

Daniel Krausche Institute of Business Administration and Economics, Brandenburg University of Technology Cottbus, Cottbus, Germany

Abhimanyu Krishna Luxembourg Centre for Systems Biomedicine, University of Luxembourg, Esch-sur-Alzette, Luxembourg

Anna Król Wrocław University of Economics, Wroclaw, Poland

Sabine Krolak-Schwerdt University of Luxembourg, LCMI Research Unit, Route de Diekirch, Walferdange, Luxembourg

Pascale Kuntz Equipe COD-LINA (UMR CNRS 6241), Université de Nantes, Nantes, France

Valentas Kurauskas Faculty of Mathematics and Informatics, Vilnius University, Vilnius, Lithuania

Jean Daniel Lan-Sun-Luk LE²P, Université de la Réunion, Réunion, France

Ludwig Lausser Medical Systems Biology and Institute of Neural Information Processing, Ulm University, Ulm, Germany

Yves Lechevallier INRIA, Paris-Rocquencourt, Le Chesnay, France

Vincent Lemaire Orange Labs, Lannion, France

Vladimir Makarenkov Department of Computer Science, Université du Québec à Montréal, Montreal, QC, Canada

Małgorzata Markowska Wroclaw University of Economics, Wrocław, Poland

Matthias Mauch Centre for Digital Music, Queen Mary University of London, London, UK

Patrick May Luxembourg Centre for Systems Biomedicine, University of Luxembourg, Esch-sur-Alzette, Luxembourg

Ljubo Mercep Institute for Informatics, Chair for Robotics and Embedded Systems, Technische Universität München, München, Germany

F. Meyer Orange Labs, Lannion, France

Simone Miller Hannover Medical High School, Klinik für Phoniatrie und Pädaudiologie, Hannover, Germany

Gianluca Morelli Dipartimento di Economia, Università di Parma, Parma, Italy

Hans-Joachim Mucha Weierstrass Institute for Applied Analysis and Stochastics (WIAS), Berlin, Germany

Thomas Müllerleile Ilmenau University of Technology, Ilmenau, Germany

Christoph Müssel Medical Systems Biology and Institute of Neural Information Processing, Ulm University, Ulm, Germany

Alexandros Nanopoulos University of Eichstätt, Ingolstadt, Germany

Robert Naundorf Institute of Business Administration and Economics, Brandenburg University of Technology Cottbus-Senftenberg, Cottbus, Germany

Jacques Nicolas IRISA-INRIA, Rennes, France

Marcin Pełka Department of Econometrics and Computer Science, Wroclaw University of Economics, Jelenia Góra, Poland

Lambert Pépin EDF R&D, Clamart, France

Equipe COD-LINA (UMR CNRS 6241), Université de Nantes, Nantes, France

Martin Ptok Hannover Medical High School, Klinik für Phoniatrie und Pädaudiologie, Hannover, Germany

Henri Ralambondrainy LIM, Université de la Réunion, Réunion, France

Ricardo Ramos Cologne University of Applied Sciences, Köln, Germany

John Rayner National Institute for Applied Statistics Research Australia, University of Wollongong, Wollongong, NSW, Australia

School of Mathematical and Physical Sciences, University of Newcastle, Callaghan, NSW, Australia

Marco Riani Dipartimento di Economia, Università di Parma, Parma, Italy

Elvira Romano Department of Political Science "Jean Monnet", Second University of Naples, Caserta, Italy

Günter Rudolph TU Dortmund, Chair of Algorithm Engineering, Dortmund, Germany

Katarzyna Rybarczyk Faculty of Mathematics and Computer Science, Adam Mickiewicz University, Poznań, Poland

Aneta Rybicka Department of Econometrics and Computer Science, Wroclaw University of Economics, Jelenia Góra, Poland

Adam Sagan Cracow University of Economics, Cracow, Poland

Christophe Salperwyck Powerspace, Paris, France

Ronny Scherer Faculty of Educational Sciences, Centre for Educational Measurement (CEMO), University of Oslo, Oslo, Norway

Nicolas Schilling Information Systems and Machine Learning Lab, University of Hildesheim, Hildesheim, Germany

Lars Schmidt-Thieme Information Systems and Machine Learning Lab, University of Hildesheim, Hildesheim, Germany

Ingo Schmitt Brandenburgische Technische Universität Cottbus, Cottbus, Germany

Stefanie Schreiber Institute of Business Administration and Economics, Brandenburg University of Technology Cottbus-Senftenberg, Cottbus, Germany

Daniel Seyfried Technische Universität Braunschweig, Institut für Hochfrequenztechnik, Braunschweig, Germany

Elżbieta Sobczak Wrocław University of Economics, Jelenia Góra, Poland

Marek Sobolewski Rzeszow University of Technology, Rzeszów, Poland

Andrzej Sokołowski Cracow University of Economics, Kraków, Poland

Gernot Spiegelberg Institute for Advanced Study der Technischen Universität München/Siemens AG, München, Germany

Daniel Stoller TU Dortmund, Chair of Algorithm Engineering, Dortmund, Germany

Jörg Stork Cologne University of Applied Sciences, Köln, Germany

Danuta Strahl Wroclaw University of Economics, Wrocław, Poland

Thomas Suesse National Institute for Applied Statistics Research Australia, University of Wollongong, Wollongong, NSW, Australia

Philippe Suignard EDF R&D, Clamart, France

Stephan Szuppa Siemens Professional Education, Siemens Technik Akademie Berlin, Siemens AG, Berlin, Germany

Denis Tagu INRA, UMR 1349 IGEPP, Le Rheu, France

Olivier Thas National Institute for Applied Statistics Research Australia, University of Wollongong, Wollongong, NSW, Australia

Department of Applied Mathematics, Biometrics and Process Control, Ghent University, Gent, Belgium

Christophe Trefois Luxembourg Centre for Systems Biomedicine, University of Luxembourg, Esch-sur-Alzette, Luxembourg

Lionel Trovalet LE²P, Université de la Réunion, Réunion, France

Igor Vatolkin TU Dortmund, Chair of Algorithm Engineering, Dortmund, Germany

Tobias Voigt Department of Statistics, TU Dortmund University, Dortmund, Germany

Peter Voss University of Luxembourg, ECCS, Walferdange, Luxembourg

Claus Weihs TU Dortmund, Chair of Computational Statistics, Dortmund, Germany

Barbara Wiermann Hochschule für Musik und Theater "Felix Mendelssohn Bartholdy" Leipzig, Leipzig, Germany

Valentin Wucher INRA, UMR 1349 IGEPP, Le Rheu, France

IRISA-INRIA, Rennes, France

Part I
Invited Papers

Modernising Official Statistics: A Complex Challenge

August Götzfried

Abstract In Europe, national statistical organisations and Eurostat, the statistical office of the European Union, produce and disseminate official statistics. These organisations come together as partners in the European Statistical System (ESS). This paper describes the ESS, the challenges it faces and the modernisation efforts that have been undertaken based on a redesigned ESS enterprise architecture. It also outlines the probable future direction of the ESS.

1 Introduction

In Europe, national statistical organisations and Eurostat, the statistical office of the European Union,[1] produce and disseminate official statistics. These organisations come together as partners in the European Statistical System (ESS).[2] This paper describes the ESS, the challenges it faces and the modernisation efforts that have been undertaken based on a redesigned ESS enterprise architecture.[3] It also outlines the probable future direction of the ESS.

---

[1] http://ec.europa.eu/eurostat/.

[2] http://epp.eurostat.ec.europa.eu/portal/page/portal/pgp_ess/about_ess.

[3] Enterprise architecture is the organising logic for business processes and IT infrastructure reflecting the integration and standardisation requirements of the operating model of enterprises or institutions. The operating model is the desired state of business process integration and business process standardisation for delivering goods and services to customers.

A. Götzfried (✉)
Head of Unit at Eurostat, European Commission, 2920 Luxembourg, Eurostat
e-mail: august.goetzfried@ec.europa.eu

© Springer-Verlag Berlin Heidelberg 2015
B. Lausen et al. (eds.), *Data Science, Learning by Latent Structures, and Knowledge Discovery*, Studies in Classification, Data Analysis, and Knowledge Organization, DOI 10.1007/978-3-662-44983-7_1

2 Problem Statement

In recent years, national statistical organisations and Eurostat have faced increasing pressure in compiling European statistics because of a fast-changing environment. Some of the stress factors are:

- The advent of new data sources and new data producers using big data sources. These new data sources are accessible through electronic media (e.g. Internet, smart phones) and can be exploited by many new data producers who make non-official statistics available at short notice (e.g. price indices using Amazon sales data).
- The category "European statistics" does not always meet all users' needs for information. Other statistics, therefore, have to complement them in contributing to a better understanding of EU policy development and monitoring and to measuring the impact of different phenomena on our economies or societies.
- Until now, European statistics have been produced in non-integrated production and dissemination lines. This increases the costs and hampers the use of the data outputs.
- Product innovation is not one of the strengths of European or official statistics. More product innovation is needed to respond faster to new user requirements. This also requires integrating the outputs coming from different statistical production lines.
- European statistics do not always need to reach a level of perfection in data quality; in many cases, a "fit-for-purpose" data quality level can suffice.
- Given globalisation, data are not only used at national and European level, but also more and more by countries outside the EU. This also holds true for the cross-border use of data within the European Union.
- Increasing European and international integration undermines and jeopardises some national data sources and therefore also national data production.

The ESS had to react to these previously unseen challenges to maintain the trust placed in it for the collection, compilation, analysis and dissemination of high-quality European statistics.

3 The ESS in a Nutshell

The ESS is the partnership between the European Union statistical authority (i.e. Eurostat), the national statistical institutes (NSIs) and other national authorities responsible in each Member State for the development, production and dissemination of European statistics, i.e. comparable statistics at EU level. This Partnership also includes the European Economic Area (EEA) and European Free Trade Association (EFTA) countries.

Member States collect data and compile statistics for national and EU purposes. The ESS functions as a network in which Eurostat's role is to lead the way in the harmonisation of statistics in close cooperation with the NSIs. ESS work concentrates mainly on EU policy areas—but, with the extension of EU policies, harmonisation has been extended to nearly all statistical fields.

The ESS also coordinates its work with candidate countries, and at European level with other Commission services and agencies as well as the European Central Bank (ECB) and other international organisations such as the Organisation for Economic Cooperation and Development (OECD), the United Nations (UN), the International Monetary Fund (IMF) and the World Bank (WB).

Until recently the business model in force in the ESS was the "stovepipe model" where the production processes are organised by distinct statistical products. This expression is used to illustrate how statistics are traditionally produced within the ESS, i.e. in numerous parallel processes, country by country (even sometimes region by region) and domain per domain. In such a model every single product stovepipe corresponds to a specific domain of statistics, together with the corresponding production system. For each domain, the whole production process from survey design over data collection and processing to dissemination takes place independently of other domains, and each domain has its own data suppliers and user groups. The stovepipe model is also reflected in the way statistical domains are regulated at the European level (i.e. domain-specific legal acts).

The stovepipe model is the outcome of a historic process in which statistics in individual domains developed independently. It has a number of advantages: the production processes are best adapted to the corresponding products; it is flexible in that it can adapt quickly to relatively minor changes in the underlying phenomena that the data describe; it is under the control of the domain manager and it results in a low-risk business architecture, as a problem in one of the production processes should normally not affect the rest of the production. From a European perspective it has the advantage that it can be addressed by limited and specific legal acts.

However, the stovepipe model also has a number of disadvantages. First, it may impose an unnecessary burden on respondents when the collection of data is conducted in an uncoordinated manner and respondents are asked for the same information more than once. Second, the stovepipe model is not well adapted to collect data on phenomena that cover multiple dimensions, such as globalisation, sustainability or climate change. Last but not least, this way of compiling statistics is both inefficient and costly, as it does not make use of standardisation between areas and collaboration between Member States. Redundancies and duplication of work, be it in development, in production or in dissemination processes are unavoidable in the stovepipe model. These inefficiencies and costs for the production of national data are further amplified when it comes to collecting and integrating regional data, which are indispensable for the design, monitoring and evaluation of some EU policies.

In order to produce European statistics, Eurostat compiles the data coming from individual NSIs also area by area. The same stovepipe model thus exists in Eurostat, where the harmonised data in a particular statistical domain are aggregated

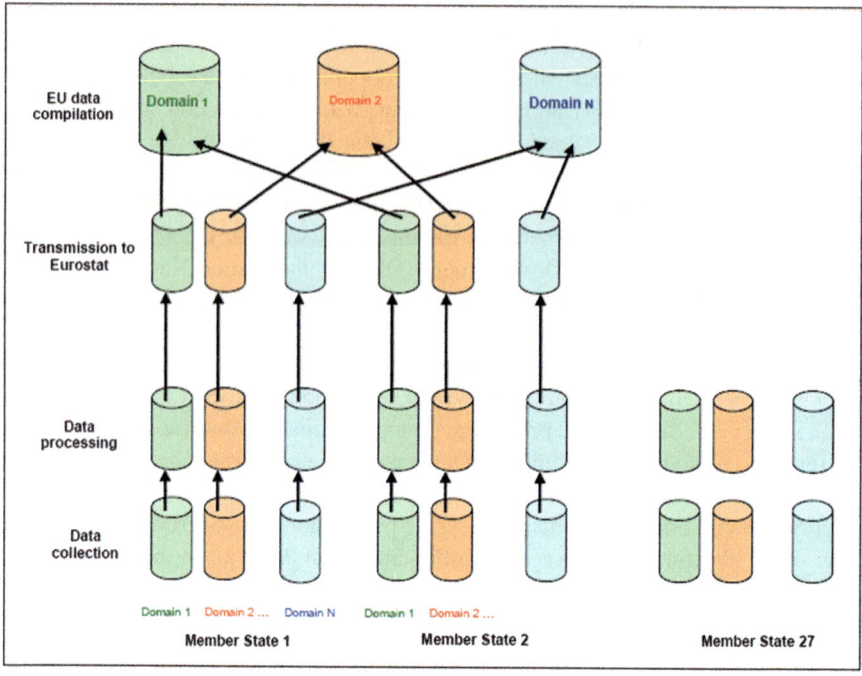

Fig. 1 The statistical world is changing because of big data

to produce European statistics in that domain. The traditional approach for the production of European statistics based on the stovepipe model can thus be labelled as an "augmented" stovepipe model in that the European level is added to the national level (Fig. 1).

4   The Modernisation of the ESS

In recent years, the pressure on official statistical organisations led to global and European modernisation programmes.[4] These programmes were broadly discussed and their focus on statistical processes and product redesign and innovation was widely accepted.

---

[4]These programmes were shaped by the High-Level Group for the Modernisation of Statistical Production and Services working under the UNECE (United Nations Economic Commission for Europe) and by the European Statistical System. In 2009 the ESS also defined its "Vision for the next decade".

4.1 The ESS Modernisation Programme: Some Characteristics

Since 2009, the ESS modernisation programme sets out a medium and longer-term strategy for all ESS statistical organisations. The main characteristics of this programme are:

- Statistical business processes need to be redesigned and better integrated. The integration concerns both statistical domains (e.g. structural business statistics and short-term business statistics) and statistical organisations (e.g. national statistical organisations and Eurostat).
- An increasing number of ESS statistical and technical standards, shared services and IT applications need to be made available and used when statistical business processes are integrated. Examples include the Statistical Data and Metadata eXchange (SDMX)[5] as a technical standard for data and metadata exchange, or metadata standards used for ESS data quality reporting.
- Two international models support the business process integration: the Generic Statistical Information Model (GSIM)[6] and the Generic Statistical Business Process Model (GSBPM).[7] Both models need to be widespread and broadly applied (Fig. 2).
- More diverse data sources need to be exploited for the production of official statistics. There can be different types of administrative data sources or other data sources falling under the big data initiative.[8]
- Greater business process integration also requires the integration of the ESS work structure, its governance and its statistical legislation. This could, for example, lead to more integrated statistical working groups dealing with particular aspects of data processing (e.g. a specific working group on data validation). In terms of legislation, ESS framework regulations for business or social statistics are proposed.

The overall umbrella for these modernisation actions is the enterprise architecture model, which is also applied in the ESS. This model distinguishes four layers: business, data, applications and technology (Fig. 3).

---

[5]http://sdmx.org.

[6]http://www1.unece.org/stat/platform/display/metis/Generic+Statistical+Information+Model.

[7]http://www1.unece.org/stat/platform/display/metis/The+Generic+Statistical+Business+Process+Model.

[8]Many international organisations currently work on the eventual use of "big data" sources (such as scanner data in supermarkets) for official statistics.

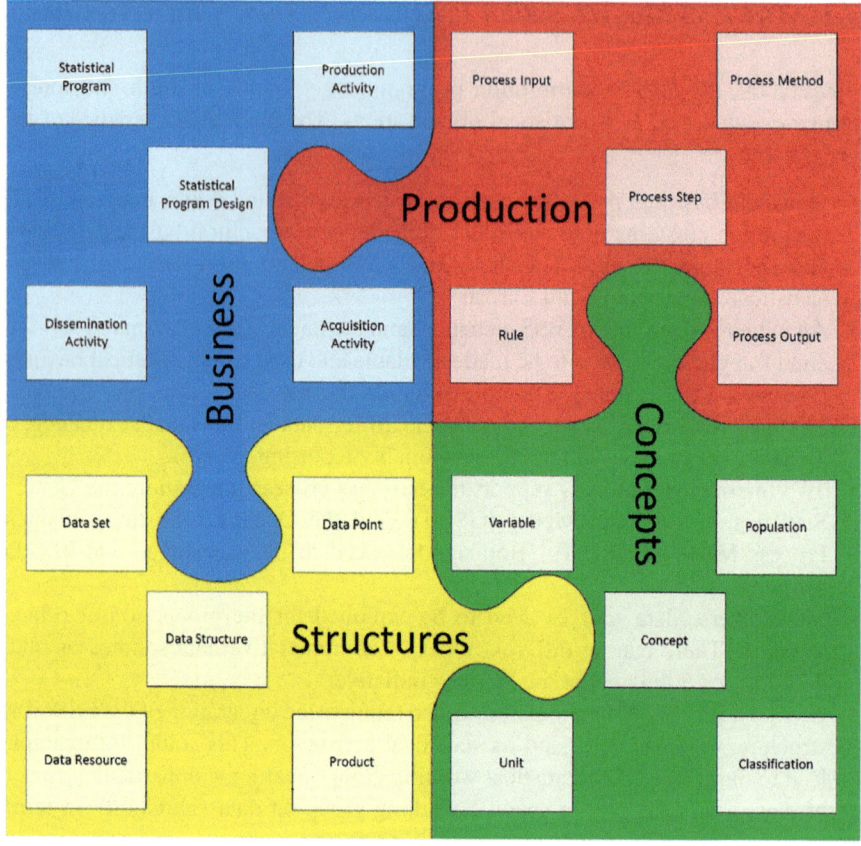

Fig. 2 The generic statistical information model

4.2 From the Current ESS Architecture to a More Integrated ESS Architecture

Implementing such a wide-ranging modernisation programme for the production and dissemination of European statistics will considerably change the way the ESS functions and interacts. The system should develop from the current architecture of relatively separate national and European statistical processes and outputs to a much more integrated system, broadly using ESS standards and shared services (Fig. 4).

This future design of the ESS is characterised by the following:

- National and European statistical processes become much more integrated using shared services, common standards and harmonised metadata. Overall, this should lead to a shortening of processes as duplication of work can be avoided in some cases (e.g. for data validation that only needs to be done once).

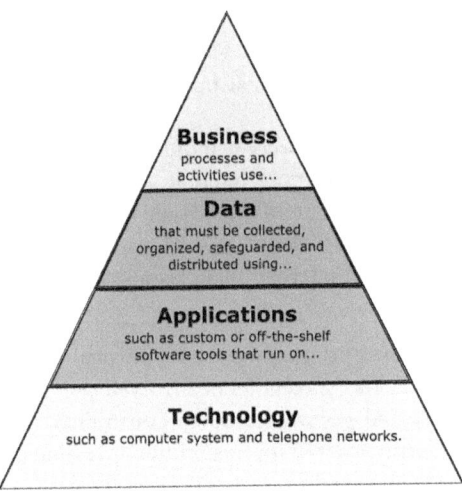

Fig. 3 The four layers of enterprise architecture

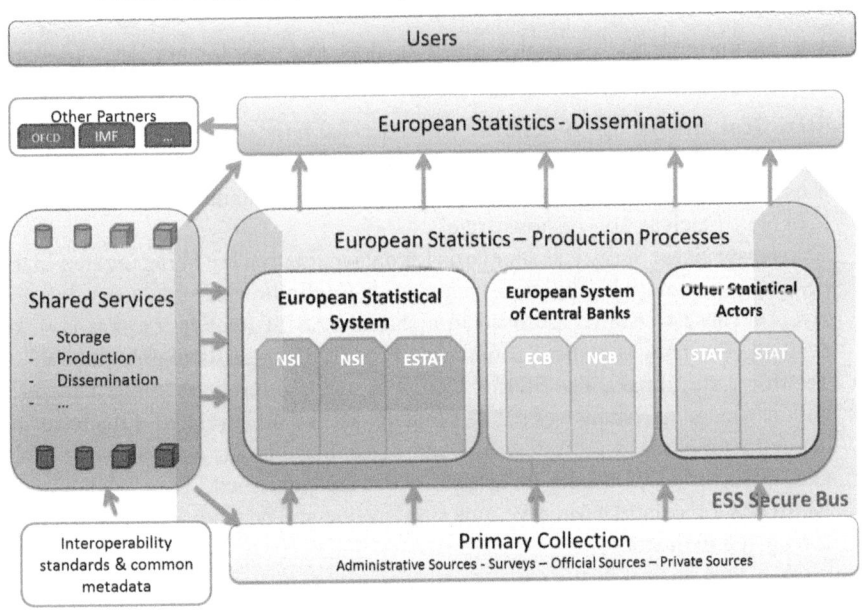

Fig. 4 Future state of the ESS production and dissemination (The European System of Central Banks is made of the European Central Bank (ECB) and the central banks of the 28 European Union (EU) Member States. Interoperability is defined as the ability of two or more systems or components to exchange information and to use the information that has been exchanged; SDMX is an example of such an interoperability standard. Common metadata encompass harmonised concepts, sources and methods underlying the data and therefore helping users to understand and assess the characteristics of the data.)

- An increasing number of shared services, standards and harmonised metadata need to be made available by the ESS. These standards and shared services should be used in more and more statistical business processes and in an increasing number of European countries.
- ESS business processes should become more integrated with related international business processes to enable more international and global data sharing. This would considerably increase international data quality.
- The ESS should appear much more as a corporate system in contrast to today where various national and international organisations are perceived as working together only loosely.

As the ESS modernisation programme is a highly challenging and more medium-term reform programme, the stakeholders involved are adapting only gradually to the new architecture and policy. Resource constraints are one example of the obstacles for this reform programme, as initial investments are needed before resource savings can be expected.

5   Some Modernisation Examples in the ESS

The following concrete examples illustrate how the ESS integration is making progress:

- **Statistical standards:** A standard structure for harmonised ESS data quality reporting was created: the ESS Standard Quality Report Structure (ESQRS). This is used in more and more statistical domains and also contains harmonised quality indicators (such as non-response rate).
- **Shared services**: Statistical standards for data validation are being created in the ESS (e.g. a standard validation language, the definition of standard validation layers). These standards are used in a shared ESS IT application that is made available for many statistical domains and for all ESS statistical organisations.
- **Technical standards**: The SDMX technical standard has been implemented for national accounts, balance of payments statistics and other related statistics using data structures that were integrated and harmonised as far as possible. SDMX Data Structure Definitions for global use were published. These will greatly facilitate global data sharing (i.e. one single value will be made globally available for a given indicator).

6   Summary and Outlook

Confronted with generally increasing pressures, European statistics—and official statistics in general—have to be modernised. Their compilation must become more efficient and they must remain competitive in meeting new information needs. An ambitious modernisation programme for official statistics has been launched at various levels, including for the ESS.

Such ambitious change programmes necessarily require strong and continuous commitment from all stakeholders. These programmes also have an impact on governance and resource distributions. They need to be thoroughly prepared and managed.

The modernisation of official statistics also needs time. This is sometimes not easily at hand when upfront investments are needed before any productivity gains can be harvested.

A New Supervised Classification of Credit Approval Data via the Hybridized RBF Neural Network Model Using Information Complexity

Oguz Akbilgic and Hamparsum Bozdogan

Abstract In this paper, we introduce a new approach for supervised classification to handle mixed-data (i.e., categorical, binary, and continuous) data structures using a hybrid radial basis function neural networks (HRBF-NN). HRBF-NN supervised classification combines regression trees, ridge regression, and the genetic algorithm (GA) with radial basis function (RBF) neural networks (NN) along with information complexity (ICOMP) criterion as the fitness function to carry out both classification and subset selection of best predictors which discriminate between the classes. In this manner, we reduce the dimensionality of the data and at the same time improve classification accuracy of the fitted predictive model. We apply HRBF-NN supervised classification to a real benchmark credit approval mixed-data set to classify the customers into good/bad classes for credit approval. Our results show the excellent performance of HRBF-NN method in supervised classification tasks.

1 Introduction

Credit approval is one of the most critical decisions of banking requiring solid risk analysis. Credit scoring systems are introduced to evaluate the customers' eligibility for credit approval based on historical and current information about the customers. This information can be numeric such as income, age, volume of previous credit history as well as nominal-categorical such as sex, race, type of criminal record, and so on. Although processing such nominal-categorical variables

O. Akbilgic
Department of Business Analytics and Statistics, University of Tennessee, Knoxville,
TN 37996, USA

Department of Quantitative Methods, Istanbul University School of Business, Istanbul, Turkey
e-mail: oguzakbilgic@gmail.com

H. Bozdogan (✉)
Department of Business Analytics and Statistics, University of Tennessee, Knoxville,
TN 37996, USA
e-mail: bozdogan@utk.edu

© Springer-Verlag Berlin Heidelberg 2015
B. Lausen et al. (eds.), *Data Science, Learning by Latent Structures,
and Knowledge Discovery*, Studies in Classification, Data Analysis,
and Knowledge Organization, DOI 10.1007/978-3-662-44983-7_2

can be easy by simple credit scoring systems, it can be difficult to handle them in more sophisticated statistical methods for credit approval decision making.

Traditional techniques such as discriminant analysis and logistics regression suffer in the presence of nominal-categorical data. When the variables are nominal (categorical) definitions of the similarity (dissimilarity) measures become difficult and it requires a new metric. In this paper, our objective is to introduce a new approach for supervised classification using a hybrid radial basis function neural networks (HRBF-NN) with continuity justification on dependent variable so as to handle mixture of nominal-categorical and continuous predictors without using dummy variables for classification. We illustrate the practical utility and the importance of our approach by providing a real example on a benchmark credit approval data from the banking industry to classify good and bad customers. Most of the technical details of this paper can be found in Akbilgic et al. (2013), Akbilgic (2011), Akbilgic and Bozdogan (2011). Here, we only recapitulate the necessary parts from these papers to set up the background of this current paper.

The paper is organized as follows. In Section 2, we briefly explain HRBF-NN and what radial basis function neural network (RBF-NN) model is. In Section 3, we discuss classification trees (CT) and its usage in HRBF-NN model; transforming tree nodes into RBFs. Estimation of the weight parameters is presented in Section 4 using the least-squares method. Later, we explain how to make classification problem look like non-parametric regression by adding a threshold function into output neuron of RBF-NN model. Our threshold function turns out to be a non-linear function of the predictive model. For other threshold selection methods, we refer the readers to Flach et al. (2013) in this volume. In Section 5, for model selection, we develop and use information-theoretic measure of complexity (ICOMP) criterion as our fitness function and show its derived form under both correctly and misspecified HRBF-NN models. We also give the forms of Akaike's information criterion (AIC) (Akaike 1973; Bozdogan 1987) and Rissanen/Schwarz (MDL/SBC) (Rissanen 1978; Schwarz 1978). In Section 6, we briefly explain the background of the genetic algorithm (GA) and the implementation of GA for the subset selection of the best predictors which discriminate between the classes. In Section 7, we give a numerical example to illustrate the performance of the proposed new supervised classification approach via the HRBF-NN model on a real credit approval data set to classify the customers into good/bad credit card customers or classes. Later, in Section 8, we conclude the paper with a discussion.

2 Hybrid Radial Basis Function Neural Networks: HRBF-NN Model

In this section, we briefly introduce the structure of HRBF-NN model as a combination of RBF-NNs, classification trees, ridge regression, information complexity ICOMP, and the genetic algorithm (GA).

2.1 RBF-NN Model

RBF-NNs model is a technique that transforms *non-linearly separable features* to *linearly separable features* using *radial basis functions (RBFs)*. RBF-NN model is a *nonparametric regression* technique (Bishop 1995) defined as

$$y = f(w, x) = \sum_{j=1}^{m} w_j h_j(x) = w_1 h_1 + w_2 h_2 + \cdots + w_m h_m. \qquad (1)$$

In equation (1), y is the dependent variable, $x_1, x_2, \ldots, x_m$ are independent variables, $\{h_j(x)\}_{j=1}^{m}$, and $\{w_j\}_{j=1}^{m}$ are the unknown adaptable coefficients, or weights. Equation (1) is represented in matrix form in equation (2) where H is the $(n \times m)$ *design matrix*, and ε is an $(n \times 1)$ vector of random noise term, such that

$$y = Hw + \epsilon. \qquad (2)$$

2.2 Radial Basis Functions

RBF-NN gains its flexibility from RBFs. We shall consider four most common RBFs in this work although there are many others. These are Gaussian (GS), Cauchy (CH), multi-quadratic (MQ), and inverse multi-quadratic (IMQ) which are given in Table 1.

The RBF-NN non-linearly transforms n-dimensional inputs to m-dimensional space by m basis functions, each characterized by their centers c_j in the (original) input space and a width or radius vector r_j, $j \in \{1, 2, \ldots, m\}$ (Orr 2000).

Table 1 The most common radial basis functions

RBF kernels	Functional form
Gaussian (GS)	$h_j(x) = \exp\left(-\sum_{k=1}^{p} \frac{(x_k - c_{jk})^2}{r_{jk}^2}\right)$
Cauchy (CH)	$h_j(x) = \sqrt[2]{1 + \exp\left(-\sum_{k=1}^{p} \frac{(x_k - c_{jk})^2}{r_{jk}^2}\right)}$
Multiquadric (MQ)	$h_j(x) = \sqrt[2]{1 + \exp\left(-\sum_{k=1}^{p} \frac{(x_k - c_{jk})^2}{r_{jk}^2}\right)}$
Inverse multiquadric (IMQ)	$h_j(x) = \dfrac{1}{\sqrt[2]{1 + \exp\left(-\sum_{k=1}^{p} \frac{(x_k - c_{jk})^2}{r_{jk}^2}\right)}}$

3   Classification Trees and its Use in HRBF-NN Model

3.1   Classification Trees

Classification and regression trees (or CART in short) models are used for both prediction and classification. Classification trees algorithm is based on recursively partitioning of the input space into two parallel hyper-rectangles. The hyper-rectangles with no more splits are called terminal nodes of the tree and a class label is assigned for each terminal nodes. The class assignment rule for a terminal node is simply to correspond the class label having the largest number of members in the terminal node (Sutton 2005).

During the process of recursive partitioning of input space, each split is parallel to one of the axes and can be expressed as an inequality involving of the input components (e.g. $x_k > b$). The input space is divided into hyper-rectangles organized into a binary tree where each branch is determined by the dimension (k) and boundary (b) which together minimize the misclassification error (Orr 2000). The root node of the classification tree is the smallest hyper-rectangle that will include all of the training data $\{x_i\}_{i=1}^{p}$. Its size s_k (half-width) and center c_k in each dimension k are

$$s_k = \frac{1}{2}(\max_{i \in S}(x_{ik}) - \min_{i \in S}(x_{ik}) \tag{3}$$

$$c_k = \frac{1}{2}(\max_{i \in S}(x_{ik}) + \min_{i \in S}(x_{ik})) \tag{4}$$

where $k \in K$ is the set of predictor indices, and $S = \{1, 2, \ldots, p\}$ is the set of training set indices. A split of the root node divides the training samples into left and right subsets, S_L and S_R, on either side of a boundary b in one of the dimensions k such that

$$s_L = \{i : x_{ik} \leq b\}, \tag{5}$$

$$s_R = \{i : x_{ik} > b\}, \tag{6}$$

In classification trees, for a given set of class labels $\{A_1, A_2, , A_3 \ldots\}$, the output values of each side of the bifurcations are

$$\hat{y}_L = A_{\mathrm{argmax}_{i \in s_L}\{a_i\}} \tag{7}$$

$$\hat{y}_R = A_{\mathrm{argmax}_{i \in s_R}\{a_i\}} \tag{8}$$

where the number of members of class label in each subset is defined with the set $a = \{a_1, a_2, a_3 \dots\}$. The misclassification error (MCE) rate is then

$$\text{MCE}(k, b) = \frac{\sum_{i \in S_L} M(y_i, \hat{y}_L) + \sum_{i \in S_R} M(y_i, \hat{y}_L)}{n}, \tag{9}$$

where n is the total sample size, and $M(y_i, \hat{y}_L)$ is a function equal to 0 if $y_i = \hat{y}$, and 1 otherwise.

The split which minimizes MCE (k, b) over all possible choices of k and b is used to create the children of the root node and is found by simple discrete search over m dimensions and p observations. The children of the root node are split recursively in the same manner and the process terminates when every remaining split creates children containing fewer than p_{min} samples, which is another parameter of the method. The children are shifted with respect to their parent nodes and their sizes reduced in the k-th dimension (Akbilgic et al. 2013; Akbilgic 2011; Akbilgic and Bozdogan 2011).

3.2 Transforming Tree Nodes into RBFs

The classification trees contain a root node, some non-terminal nodes (having children) and some terminal nodes (having no children). Each node is associated with a hyper-rectangle of input space having a center c and size s as described above. The node corresponding to the largest hyper-rectangle is the root node and it is divided up into smaller and smaller pieces progressing down the tree (Breiman et al. 1984; Orr 2000). To transform the hyper-rectangle into different basis kernel RBFs we use its center c as the RBF center and its size s, scaled by a parameter α as the RBF radius given by

$$r = \alpha s. \tag{10}$$

The scalar α has the same value for all nodes (Kubat 1998), and it is another parameter of the method. In this study we set $\alpha = \sqrt{2}\alpha_K^{-1}$ where α_K is the Kubat's parameter (Kubat 1998; Orr 2000).

4 Estimation of Weight Parameters

4.1 Least-Squares Estimation

Given a network model in equation (1) consisting of m RBFs with centers $\{c_j\}_{j=1}^m$ and radii $\{r_j\}_{j=1}^m$ and a training set with p patterns, $\{(x_i, y_i)\}_{i=1}^p$, the optimal network weights can be found by minimizing the sum of squared errors:

$$\text{SSE} = \sum_{i=1}^p (f(x_i) - y_i)^2 \tag{11}$$

and is given by

$$\hat{w} = \left(H'H\right)^{-1} H'y \tag{12}$$

the so-called least squares estimation. Here H is the design or model matrix, with its elements $H_{ij} = h_j(x_i)$, and $y = (y_1, y_2, \ldots, y_p)'$ is the p-dimensional vector of training set of output values.

In RBF-NN, one of the most common problems is singularity of the $(H'H)$ matrix. At this point, to overcome possible singularity problem in the model matrix, we use global ridge regression (Tikhonov and Arsenin 1977; Bishop 1991) to regularize HRBF-NN model with the cost function given by

$$C(w, \lambda) = \sum_{i=1}^{p} (f(x_i) - y_i)^2 + \lambda \sum_{i=1}^{m} w_j^2 = \varepsilon'\varepsilon + w'w. \tag{13}$$

$C(w, \lambda)$ is minimized to find a weight vector which is more robust to noise in the training set. The optimal weight vector for global ridge regression is given in equation (14), where I_m is the m dimensional identity matrix, and λ is the regularization parameter.

$$\hat{w} = \left(H'H + \lambda I_m\right)^{-1} H'y. \tag{14}$$

We use *Hoerl, Kennard, and Baldwin (HKB)* (Hoerl et al. 1975) approach to data adaptively determine optimal λ that is given by

$$\hat{\lambda}_{\mathrm{HKB}} = \frac{ms^2}{\hat{\mathbf{w}}'_{LS}\hat{\mathbf{w}}_{LS}}, \tag{15}$$

where $m = k$, the number of predictors not including the intercept term, n is the number of observations, s^2 is the estimated error variance using k predictors so that

$$s^2 = \frac{1}{(n - k + 1)} (y - H\hat{\mathbf{w}}_{LS})' (y - H\hat{\mathbf{w}}_{LS}), \tag{16}$$

where $\hat{\mathbf{w}}_{LS}$ is the estimated coefficient vector obtained from a no-constant model given in matrix form by

$$\hat{\mathbf{w}}_{LS} = \left(H'H\right)^{-1} H'y. \tag{17}$$

4.2 RBF Neural Networks for Classification

The goal of classification is to assign observations into target categories or classes based on their characteristics in some optimal way. Thus, in classification case, outcomes are one of discrete set of possible classes rather than of a continuous function as in non-parametric regression (Bishop 1995). However, we can make the classification problem look like a non-parametric regression by incorporating a threshold function into the output of the neuron of the RBF-NN model.

For a binary dependent variable case, we can assign HRBF-NN predictions to class labels by substituting equation (1) in the threshold function $t\left(f(w, H); t_0\right)$ given by

$$t\left(f(w, H)\right) = \begin{cases} 0 & f(w, x) < t_0 \\ 1 & f(w, x) > t_0 \end{cases} \tag{18}$$

where t_0 is the value separating two classes.

When two clusters have equal number of observations, then $t_0 = 0.5$.

Assuming that the classes are represented with 0, and 1 and having n_1, and n_2, the number of observations in each class, the calculation of threshold value is given by

$$t_0 = \frac{n_1}{n_1 + n_2}. \tag{19}$$

Threshold value can be considered as a prior probability of the first group which is equal to 0.5 when two of the groups have equal number of observations.

5 Information Theoretic Model Selection Criteria

In HRBF-NN, we use ICOMP criterion of Bozdogan (1994, 2000, 2004) and Liu and Bozdogan (2004) as the fitness function to carry out variable selection with GA. The complexity of a nonparametric regression model increases with the number of independent and adjustable parameters, which is also termed effective degrees of freedom in the model. According to the qualitative principle of Occam's Razor, the simplest model that fits the observed data is the best model. Following this principle, we aim to provide a trade-off between how well the model fits the data and the model complexity (Akbilgic et al. 2013).

The derived forms of information criteria are used to evaluate and compare different horizontal and vertical subset selection in the genetic algorithm (GA) for the regularized regression and classification trees and RBF networks model given in equation (1) under the assumption, $\varepsilon \sim N\left(0, \sigma^2 I\right)$ or equivalently $\varepsilon_i \sim N\left(0, \sigma^2\right)$ for $i = 1, 2, \ldots, n$.

General form of ICOMP is an approximation to the sum of two Kullback–Leibler (KL) (Kullback and Leibler 1951) distances. For general multivariate normal linear or nonlinear structural model suppose $C_1\left(\hat{\Sigma}_{model}\right)$ is approximated by the complexity of the IFIM $C_1\left(\hat{\mathscr{F}}^{-1}\left(\hat{\theta}\right)\right)$. Then, we define ICOMP(IFIM) as

$$\text{ICOMP(IFIM)} = -2\log L\left(\hat{\theta}\right) + 2C_1\left(\hat{\mathscr{F}}^{-1}\left(\hat{\theta}\right)\right), \tag{20}$$

where $C_1\left(.\right)$ is a maximal information theoretic measure of complexity of the estimated inverse Fisher information matrix (IFIM) of a multivariate normal distribution given by

$$C_1\left(\hat{\mathscr{F}}^{-1}\left(\hat{\theta}\right)\right) = \frac{s}{2}\log\left(\frac{\text{tr}\left(\hat{\mathscr{F}}^{-1}\left(\hat{\theta}\right)\right)}{s}\right) - \frac{1}{2}\log \mid \hat{\mathscr{F}}^{-1}\left(\hat{\theta}\right) \mid, \tag{21}$$

and where $s = \dim\left(\hat{\mathscr{F}}^{-1}\right) = \text{rank}\left(\hat{\mathscr{F}}^{-1}\right)$. The estimated IFIM for the HRBF-NN model is given by

$$\widehat{\text{Cov}}\left(\hat{w}, \hat{\sigma}^2\right) = \hat{\mathscr{F}}^{-1} = \begin{bmatrix} \hat{\sigma}^2 \left(H'H\right)^{-1} & 0 \\ 0 & \frac{2\hat{\sigma}^4}{4} \end{bmatrix}, \tag{22}$$

where

$$\hat{\sigma}^2 = \frac{(y - H\hat{w})'(y - H\hat{w})}{n}. \tag{23}$$

Then, the definition of ICOMP(IFIM) in equation (20) becomes:

$$\text{ICOMP(IFIM)} = n\log\left(2\pi\right) + n\log\left(\hat{\sigma}^2\right) + n + 2C_1\left(\hat{\mathscr{F}}^{-1}\left(\hat{\theta}\right)\right), \tag{24}$$

where the entropic complexity is

$$C_1\left(\hat{\mathscr{F}}^{-1}\left(\hat{\theta}_m\right)\right) = (m+1)\log\left[\frac{\text{tr}\hat{\sigma}^2\left(H'H\right)^{-1} + \frac{2\hat{\sigma}^4}{4}}{m+1}\right] \tag{25}$$

$$-\frac{1}{2}\log \mid \hat{\sigma}^2\left(H'H\right)^{-1} \mid +\log\left(\frac{2\hat{\sigma}^4}{4}\right).$$

We can also define ICOMP for misspecified models given as follows:

$$\text{ICOMP(IFIM)}_{\text{Misspec}} = -2\log L\left(\hat{\theta}\right) + 2C_1\left(\widehat{\text{Cov}}\left(\hat{\theta}\right)_{\text{Misspec}}\right) \tag{26}$$

$$= n\log\left(2\pi\right) + n\log\left(\hat{\sigma}^2\right) + n + 2C_1\left(\widehat{\text{Cov}}\left(\hat{\theta}\right)_{\text{Misspec}}\right),$$

where

$$\widehat{\mathrm{Cov}}\left(\hat{\theta}\right)_{\mathrm{Misspec}} = \hat{\mathscr{F}}^{-1}\hat{\mathscr{R}}\hat{\mathscr{F}}^{-1} \tag{27}$$

$$\begin{bmatrix} \hat{\sigma}^2(H'H)^{-1} & 0 \\ 0 & \frac{2\hat{\sigma}^4}{n} \end{bmatrix} \begin{bmatrix} \frac{1}{\hat{\sigma}^4}H'D^2H & H'1\frac{Sk}{2\hat{\sigma}^3} \\ \left(H'1\frac{Sk}{2\hat{\sigma}^3}\right)' & \frac{(n-m)(Kt-1)}{4\hat{\sigma}^4} \end{bmatrix} \begin{bmatrix} \hat{\sigma}^2(H'H)^{-1} & 0 \\ 0 & \frac{2\hat{\sigma}^4}{n} \end{bmatrix}$$

is a consistent estimator of the covariance matrix $\mathrm{Cov}(\theta_k^*)$, which is often called the sandwich covariance or robust covariance estimator, since it is a correct covariance regardless whether the assumed model is correct or not. When the model is correct we get $\hat{\mathscr{F}} = \hat{\mathscr{R}}$. Hence, the sandwich covariance reduces to the usual IFIM $\hat{\mathscr{F}}^{-1}$ (White 1982). Note that this covariance matrix takes into account the presence of skewness and kurtosis, which is not possible with AIC (Akaike 1973) and other Akaike-type criteria such as Rissanen/Schwarz (MDL/SBC) (Rissanen 1978; Schwarz 1978). The derived forms of these criteria for the HRBF-NN model are:

$$\mathrm{AIC}(m) = n\log(2\pi) + n\log\left(\frac{(y - H\hat{w})'(y - H\hat{w})}{n}\right) + n + 2(m + 1), \tag{28}$$

$$\mathrm{MDL/SBC}(m) = n\log(2\pi) + n\log\left(\frac{(y - H\hat{w})'(y - H\hat{w})}{n}\right) + n + m\log(n). \tag{29}$$

6 Genetic Algorithm for Subset Selection

There are several standard techniques available for variable selection such as forward selection, backward elimination, a combination of the two, or all possible subset selection. Both forward and backward procedures cannot deal with the collinearity in the predictor variables. Major criticisms on the forward, backward, and stepwise selection are that, little or no theoretical justification exists for the order in which variables enter or exit the algorithm. Stepwise searching rarely finds the overall best model or even the best subsets of a particular size. Stepwise selection, at the very best, can only produce an *"adequate"* model.

All possible subset selection is a fail proof method, but it is not computationally feasible. It takes too much time to compute and it is costly. For 20 predictor variables, for the usual subset regression model, total number of possible models we need to evaluate is: $2^{20} = 1,048,576$. At this point, we use genetic algorithm to carry out variable selection in HRBF-NN with *ICOMP* as the fitness function.

Genetic algorithm is a robust evolutionary optimization search technique with very few restrictions (David and Alice 1996). GA treats information as a series of codes on a binary string, where each string represents a different solution for a given problem. It follows the principles of survival of the fittest, which is introduced by Charles Darwin. The algorithm searches for optimum solution within a defined search space to solve a problem (Eiben and Smith 2010). It has outstanding performance in finding the optimal solution for problems in many different fields (Akbilgic et al. 2013; Akbilgic 2011; Akbilgic and Bozdogan 2011).

7 A Numerical Example: Analysis of Credit Approval Data

In this section, we report our computational results on a credit approval data sets to classify the customers into good/bad classes using our hybrid RBF-NN approach with regularization, GA, and ICOMP(IFIM) as the fitness function.

Our modern world depends upon credit. Entire economies are driven by people's ability to *"buy-now, pay later"* (Anderson 2007).

Therefore, credit approval is one of the most critical decisions of banking industry requiring solid risk analysis.

Credit scoring systems are introduced almost 50 years ago to evaluate the customers' eligibility for credit approval based on historic and current information about the customers.

This information can be numeric such as *income, age, volume of previous credit history* as well as nominal-categorical such as *sex, race, type of criminal record*, and so on with high dimensions.

Our *credit approval data set* is obtained from UCI Machine Learning Repository (2013). Original version of credit approval data set is consisted of 690 observations including fifteen independent variables; six continuous and nine categorical, and one binary dependent variable. However, by excluding the observation with missing attributes, we reduced the data size to 654 representing 296 positive, and 358 negative credit ratings. Because all of the nine categorical independent variables were coded by meaningless letters to protect confidentiality of the data, we transformed them into numbers, $1, 2, 3, \ldots$, based on the number of categories in each variable. The representation of the original data and the usage of them in our study are given in Table 2.

We first analyzed credit approval data via HRBF-NN model separately for four different RBFs: Gaussian, Cauchy, Multi-Quadratic, and Inverse Multi-Quadratic using saturated model. Confusion matrix for different RBFs are reported in the Tables 3, 4, and 5 where ICOMP(IFIM)miss values are reported in the last column of Table 8. For simplicity in text, we will use ICOMP for ICOMP(IFIM)miss in our report in this study. Note that calculation of classification accuracy is carried out using equation (30). The reason we run HRBF-NN model for saturated model is to compare the results after variable selection. The classification accuracy is defined by

$$\text{Classification accuracy} = 100 \frac{\text{number of correctly classified observations}}{\text{total number of observations}}. \quad (30)$$

Table 2 Usage of credit approval data in our analysis

Variables	Original presentation	Usage in our analysis
A1	b, a	1, 2
A2	Continuous	Continuous
A3	Continuous	Continuous
A4	u, y, l, t	1, 2, 3, 4
A5	g, p, gg	1, 2, 3
A6	c, d, cc, i, j, k, m, r, q, w, x, e, aa, ff	1, 2, 3, 4, 5, 6, 7, 8, 9, 10, 11, 12, 13, 14
A7	v, h, bb, j, n, z, dd, ff, o	1, 2, 3, 4, 5, 6, 7, 8, 9
A8	Continuous	Continuous
A9	t, f	1, 2
A10	t, f	1, 2
A11	Continuous	Continuous
A12	t, f	1, 2
A13	g, p, s	1, 2, 3
A14	Continuous	Continuous
A15	Continuous	Continuous
A16	+, − (class attributes)	1, 2

Table 3 Gaussian

Classes	C1	C2	Total	Accuracy (%)
C1	274	22	296	92.57
C2	39	319	358	89.11
Overall			654	90.67

Table 4 Cauchy

Classes	C1	C2	Total	Accuracy (%)
C1	270	26	296	91.22
C2	36	322	358	89.94
Overall			654	90.52

Table 5 MQ

Classes	C1	C2	Total	Accuracy (%)
C1	275	21	296	92.91
C2	52	306	358	85.48
Overall			654	88.84

Tables 3, 4, 5, and 6 show the high performance of HRBF-NN model for classification of credit data which is approximately 90 %. At this point we run variable selection on credit data using GA with ICOMP as the fitness function. Parameter setting of GA is based on our previous studies on HRBF-NN model (Akbilgic et al. 2013). Thus, we set our GA parameters as given in Table 7.

After finishing the first stage of analysis for saturated model and setting the GA parameters, next, we carried out variable selection for credit data using GA separately for four different RBFs. Table 8 shows the selected variable subsets and

Table 6 Inverse MQ RBF

Classes	C1	C2	Total	Accuracy (%)
C1	271	25	296	91.55
C2	33	325	358	91.34
Overall			654	91.44

Table 7 Parameter setting of GA for variable selection

Parameter	Setting
Number of generations	35
Number of populations	20
Mutation probability	0.01
Crossover probability	0.65
Crossover type	Single point
Elitism rule	Yes

Table 8 Variable selection under different RBFs

RBF type	Best subset	ICOMP: Best subset	Saturated model
Gaussian	3-6-9-10-14	191.45	461.56
Cauchy	3-5-6-9-10-11-14	191.28	400.49
Multi-quadratic	1-3-4-7-9-10-11-13-14	248.00	570.77
Inverse multi-quadratic	3-4-5-6-9-10-13-14	214.28	491.45

Table 9 Gaussian RBF

Classes	C1	C2	Accuracy (%)
C1	269	27	90.88
C2	33	325	90.78
Overall			90.83

minimized ICOMP values under selected variable subsets for different RBFs. We also showed the ICOMP values we calculated before for saturated model in Table 8 to give a better comparison.

It is noted from Table 8 that ICOMP values for selected subsets are significantly lower than the ICOMP values calculated for saturated model. At this point, it is important to see if obtained lower ICOMP values correspond to a simple model giving good classification accuracy. To show this, we run HRBF-NN model for all four of the RBFs with corresponding selected best subsets given in Table 8. Confusion matrix and classification accuracy is calculated for each case and the results are reported in Tables 9, 10, 11, 12.

The important results appearing in Tables 9, 10, 11, and 12 show that variable selection within HRBF-NN allows us to reduce dimension of input variables without any loss in classification accuracy. Comparing the classification accuracy results

Table 10 Cauchy RBF

Classes	C1	C2	Accuracy (%)
C1	264	32	89.19
C2	37	321	89.66
Overall			89.45

Table 11 MQ RBF

Classes	C1	C2	Accuracy (%)
C1	271	25	91.55
C2	57	321	89.66
Overall			90.52

Table 12 IMQ RBF

Classes	C1	C2	Accuracy (%)
C1	268	28	90.54
C2	32	326	91.06
Overall			90.83

between saturated model and best subsets shows the similarity of classification performance while the dimensionality is significantly reduced for best subsets. According to Table 8, by carrying out variable selection with Gaussian RBF has resulted in selecting a subset with only five variables out of fifteen where ICOMP value is minimized. Note that, there is even slightly better classification accuracy for best subset selected for Gaussian RBF in comparison with classification accuracy for the saturated model.

Finally, for comparison purposes, we carried out the usual logistic regression analysis, although the assumptions are violated here for this data set, we obtained a classification accuracy of 87.1 % using stepwise variable selection which gave nine predictors as the best predictors including the constant term. These nine predictors are: 0, 4, 5, 7, 8, 9, 10, 11, and 15. Note that this subset docs not include variables 3, 6, 9, 10, and 14 obtained from our results.

8 Conclusions and Discussion

In this paper, we introduced a novel approach for supervised classification using a HRBF-NN model with ICOMP. Our study shows that HRBF-NN model is a highly clever technique to handle hard classification problems even if the data is mixture of continuous and categorical variables. We demonstrated that the GA is a powerful optimization tool for selecting the best subset of predictors that discriminate between the classes or groups. HRBF-NN using ICOMP with GA provides us a flexible variable selection and at the same time a classification tool which gives better results than the full saturated model. With our approach we can now provide a practical method for choosing the best kernel basis RBF for a given data set which was not possible before in the literature of RBF based-methods. In real-world applications, we frequently encounter data sets with 100 and 1,000 of variables. Our results show that HRBF-NN model is a very flexible procedure that can handle dimensionality reduction drastically without losing information in classification accuracy. In our example, we reduced the number of input variables from fifteen to five with even slightly better classification accuracy which is around 91%. As is well known, recently, kernel-based supervised classification techniques such as the support vector machines (SVMs) and multi-class SVMs have become popular. One problem that has not been addressed in the literature is that kernelization and supervised classification takes place in the high dimensional reproducing Hilbert kernel space (RHKS) and not in the original data space. The transformed kernel space mapping is not one-to-one and onto, and not invertible to the original data space due to the dot product operations in using the kernel trick. This makes the practical interpretation of the results difficult even though one can get good classification error rates.

The new HRBF-NN approach proposed in this paper overcomes the difficulties encountered in the RHKS type supervised classification and provides us a flexible technique in the original data space that combines regression trees, regularized regression, and the genetic algorithm (GA) with radial basis function (RBF) neural networks (NN) along with information complexity ICOMP criterion as the fitness function to carry out both classification and at the same time subset selection of best predictors which discriminate between the classes.

Therefore, we believe our approach is a viable means of data mining and knowledge discovery via the HRBF-NN method.

Acknowledgements This paper was invited as a keynote presentation by Prof. Bozdogan at the European Conference on Data Analysis (ECDA-2013) at the University of Luxembourg in Luxembourg during July 10–12, 2013. Prof. Bozdogan extents his gratitude to the conference organizers: Professors Sabine Krolak-Schwerdt, Matthias Bömer, and Berthold Lausen.

References

Akaike, H. (1973). Information theory and an extension of the maximum likelihood principle. In B. H. Petrox & F. Csaki, (Eds.), *Second International Symposium on Information Theory* (pp. 267–281). Budapest: Academiai Kiado.

Akbilgic, O. (2011). *Variable selection and prediction using hybrid radial basis function neural networks: A case study on stock markets.* PhD thesis, Istanbul University.

Akbilgic, O., & Bozdogan, H. (2011). Predictive subset selection using regression trees and rbf neural networks hybridized with the genetic algorithm. *European Journal of Pure and Applied Mathematics, 4*(4), 467–485.

Akbilgic, O., Bozdogan, H., & Balaban, M. E. (2013). A novel hybrid RBF neural network model as a forecaster. *Statistics and Computing.* doi:10.1007/s11222-013-9375-7.

Anderson, R. (2007). *The credit scoring toolkit.* Oxford: Oxford University Press.

Bishop, C. M. (1991). Improving the generalization properties of radial basis function neural networks. *Neural Computation, 3*(4), 579–588.

Bishop, C. M. (1995). *Neural networks for pattern recognition.* Oxford: Oxford University Press.

Bozdogan, H. (1987). Model selection and Akaike's information criterion (AIC): The general theory and it's analytical extension. *Journal of Mathematical Psychology, 52*(3), 345–370.

Bozdogan, H. (1994). Mixture-model cluster analysis using a new informational complexity and model selection criteria. In H. Bozdogan (Ed.), *Multivariate Statistical Modeling, Proceedings of the First US/Japan Conference on the Frontiers of Statistical Modeling: An Informational Approach* (Vol. 2, pp. 69–113). North-Holland: Springer

Bozdogan, H. (2000). Akaike's information criterion and recent developments in informational complexity. *Journal of Mathematical Psychology, 44*, 62–91.

Bozdogan, H. (2004) Intelligent statistical data mining with information complexity and genetic algorithms. In H. Bozdogan (Ed.) *Statistical data mining and knowledge discovery* (pp. 15–56). Boca Raton: Chapman and Hall/CRC

Breiman, L., Freidman, J., Stone, J. C., & Olsen, R. A. (1984). *Classification and regression trees.* Boca Raton: Chapman and Hall.

Credit Approval Data Set by UCI MAchine Learning Repository. http://archive.ics.uci.edu/ml/datasets/Credit+Approval. Cited April 26, 2013

David, W. C., & Alice, E. S. (1996). Reliability optimization of series-parallel systems using a genetic algorithm. *IEEE Transactions on Reliability, 45*(2), 254–266.

Eiben, A. E., & Smith, J. E. (2010). *Introduction to evolutionary computing.* New York: Springer.

Flach, P. A., Hernandez-Orallo, J., & Ferri, C. (2013). *Comparing apples and oranges: Towards commensurate evaluation metrics in classification.* Keynote lecture presented in the European Conference on Data Analysis (ECDA-2013), Luxembourg.

Hoerl, A. E., Kennard, R. W., & Baldwin, K. F. (1975). Ridge regression: Some simulations. *Communications in Statistics, 4*, 105–123.

Kubat, M. (1998). Decision trees can initialize radial basis function networks. *Transactions on Neural Networks, 9*(5), 813–821.

Kullback, A., & Leibler, R. (1951). On information and sufficiency. *Annals of Mathematical Statistics, 22*, 79–86.

Liu, Z., & Bozdogan, H. (2004) Improving the performance of radial basis function classification using information criteria. In H. Bozdogan (Ed.), *Statistical data mining and knowledge discovery* (pp. 193–216). Boca Raton: Chapman and Hall/CRC.

Orr, M. (2000). Combining regression trees and RBFs. *International Journal of Neural Systems, 10*(6), 453–465.

Rissanen, J. (1978). Modeling by shortest data description. *Automatica, 14*(5), 465–471.

Schwarz, G. (1978). Estimating the dimension of model. *Annals of Statistics, 6*, 461–464.

Sutton, C. D. (2005). Classification and regression trees, bagging, and boosting. In *Handbook of statistics* Vol. 24, pp. 303–329. Elsevier B.V. doi: 10.1016/s0169-716(04)24004-4.

Tikhonov, A. H., & Arsenin, V. Y. (1977). *Solutions of ill-posed problems.* New York: Wiley.

White, H. (1982). Maximum likelihood estimation of misspecified models. *Econometrica, 50*, 1–25.

Finding the Number of Disparate Clusters with Background Contamination

Anthony C. Atkinson, Andrea Cerioli, Gianluca Morelli, and Marco Riani

Abstract The Forward Search is used in an exploratory manner, with many random starts, to indicate the number of clusters and their membership in continuous data. The prospective clusters can readily be distinguished from background noise and from other forms of outliers. A confirmatory Forward Search, involving control on the sizes of statistical tests, establishes precise cluster membership. The method performs as well as robust methods such as TCLUST. However, it does not require prior specification of the number of clusters, nor of the level of trimming of outliers. In this way it is "user friendly".

1 Introduction

It is now widely recognized that contamination can strongly affect the results of clustering methods if it is not properly taken into account. The first attempts in this direction mainly aimed at protecting against *contamination from noise*. In fact, the inclusion of noise variables can have dramatic masking consequences on the data structure recovered by distance-based clustering methods. Pioneering studies of such effects were presented by Milligan (1980) and Fowlkes et al. (1988). In a model-based clustering framework, noise can be described through a uniformly distributed component and Fraley and Raftery (2002) suggest a unified approach for dealing with it; see also Coretto and Hennig (2010). However, uniform background noise is not the only type of departure from "ideal" conditions against which one may want to protect. Typical instances that need to be accommodated in practice include both *extreme outliers*, e.g. anomalous observations due to undetected changes in the physical process generating the data or to measurement errors, and *intermediate outliers*, e.g. observations not firmly belonging to any of the

A.C. Atkinson
Department of Statistics, London School of Economics, London WC2A 2AE, UK
e-mail: a.c.atkinson@lse.ac.uk

A. Cerioli (✉) • G. Morelli • M. Riani
Dipartimento di Economia, Università di Parma, Parma, Italy
e-mail: andrea.cerioli@unipr.it; gianluca.morelli@unipr.it; mriani@unipr.it

© Springer-Verlag Berlin Heidelberg 2015
B. Lausen et al. (eds.), *Data Science, Learning by Latent Structures, and Knowledge Discovery*, Studies in Classification, Data Analysis, and Knowledge Organization, DOI 10.1007/978-3-662-44983-7_3

29

established groups (Riani et al. 2014). Robust clustering methods, for a thorough review of which we refer to Gallegos and Ritter (2009) and to García-Escudero et al. (2010), aim at addressing all these issues, by neutralizing the effect of both types of outliers and that of noise.

Robust clustering is often defined in a model-based framework in which observations are assumed to come from distinct multivariate populations. Formally, let $y = \{y_1, \ldots, y_n\}$ denote a sample of v-variate observations $y_i = (y_{i1}, \ldots, y_{iv})'$, for $i = 1, \ldots, n$. In case of no contamination, a popular approach is to search for the partition into K groups that maximizes the "classification likelihood" function

$$\prod_{k=1}^{K} \prod_{i=1}^{n} [\phi(y_i, \theta_k)]^{z_{ik}}, \tag{1}$$

where $\phi(y_i, \theta_k)$ is a v-variate density depending on the multidimensional parameter θ_k, and z_{ik} is the indicator variable taking the value 1 if y_i belongs to group k and 0 otherwise. The value of K, the number of groups, is assumed to be known when fitting the classification likelihood function.

When outliers and noise are present, a trimmed version of (1) has been proposed by García-Escudero et al. (2008). Their robust approach, called TCLUST, is based on maximizing the function

$$\prod_{k=1}^{K} \prod_{i \in R_\alpha} [\pi_k \phi(y_i, \theta_k)]^{z_{ik}}, \tag{2}$$

where $\phi(\cdot)$, z_{ik} and K have the same meaning as in (1), $\pi_k \in [0, 1]$ is an unknown weight taking into account the specific size of group k, $0 \leq \alpha \leq 0.5$ is the *trimming level* and R_α is a subset of the set of indices $\{1, \ldots, n\}$ whose cardinality is $\lfloor n(1 - \alpha) \rfloor + 1$. TCLUST clustering is currently implemented in an R package (Fritz et al. 2012) and in the FSDA toolbox for Matlab (Riani et al. 2012).

The TCLUST methodology has good theoretical properties and has proven to be effective in practical applications (see, e.g., Cerioli and Perrotta 2014). Being a model-based approach, it requires many *a priori* choices on behalf of the user. First, it is necessary to specify the form of the v-variate density $\phi(y_i, \theta_k)$. Although alternative distributions have been recently introduced (see Lee and McLachlan 2013 and the subsequent discussion), the typical choice in (2) is the multivariate normal density function, which is also at the heart of the non-robust proposal (1). Therefore, θ_k corresponds to the mean vector μ_k and to the covariance matrix Σ_k for group k. These parameters are estimated in TCLUST through a complex iterative EM-like procedure, also involving the restrictions

$$\frac{\max_{l=1,\ldots,v} \max_{k=1,\ldots,K} \lambda_l(\hat{\Sigma}_k)}{\min_{l=1,\ldots,v} \min_{k=1,\ldots,K} \lambda_l(\hat{\Sigma}_k)} \leq c. \tag{3}$$

where $\lambda_l(\hat{\Sigma}_k)$ is a generic eigenvalue of the estimated covariance matrix of group k and $c \geq 1$ is a fixed constant which defines constraints on the shape and, implicitly, on the size of the K clusters. For example, the value $c = 1$ corresponds to imposing spherical groups of constant variance, while the limiting case $c \to \infty$ gives unconstrained heterogeneous clusters. The specific value of c is a tuning parameter that must be chosen by the user.

Similarly, both K and α are assumed to be known when fitting the objective function (2). García-Escudero et al. (2011) suggest the use of the so-called classification trimmed likelihood curves (TLC), as useful exploratory tools for selecting K and α from the data. These curves are based on (2) and measure the gain achieved by allowing a unitary increase in the number of groups for a given value of α. However, there might be applications where the evidence provided by TLC is not so clear-cut (Morelli 2013). The curves also depend on the selected value of c. Therefore, there is a crucial interplay among the choices made by the user with respect to the different tuning parameters required for fitting the trimmed likelihood function (2). The development of a unified data-driven framework for assisting these choices is indeed an area of active research.

The goal of this contribution is to compare the performance of TCLUST, with respect both to the selection of the tuning parameters and to cluster recovery, with that obtained through the Forward Search (FS). The FS is a robust diagnostic method which has powerful outlier detection properties (Riani et al. 2009), but also good potential for clustering purposes (Atkinson et al. 2006; Atkinson and Riani 2007). Since it does not aim at finding the maximum of a complex objective function, exploratory use of the FS does not require an explicit choice of a tuning constant like c in (3). Also the density $\phi(y_i, \theta_k)$ is only used for confirmatory purposes, when formal hypothesis testing is required, and not for cluster assignment at an exploratory stage. In fact, the technique adopts some of the classical tools developed for the customary normal-based multivariate model, like Mahalanobis distances, but in its exploratory form the user has the ability for visual detection of departures from such a model. The FS assumes the existence of one or more elliptical populations for the "good" part of the data, which are taken as a reference benchmark against which the observations are compared. Therefore, although computations require specification of a multivariate normal density, as in TCLUST, the diagnostic use of the technique can effectively lead to detection of many alternative structures, such as extreme and intermediate outliers, or skew distributions. Even more importantly, the FS provides simple data-driven rules for the choice of K and α that do not rely on fitting the trimmed likelihood function (2). Therefore, it may be considered as a practical "user-friendly" alternative to complex robust model-based clustering methods. We will see in a difficult example that it can provide powerful information on the data structure, being virtually as illuminating as TCLUST when the latter is appropriately tuned, while requiring only minimal intervention from the user.

The structure of the paper is as follows. Section 2 introduces the FS, with Sect. 2.3 describing the random start FS that we use to identify cluster structure. In Sect. 3 this methodology is applied to a 2,000 observation example from

García-Escudero et al. (2011). Initial cluster identification from the search with random starts is in Sect. 3.1. Section 3.2 uses plots of individual Mahalanobis distances to illustrate the structure of the clusters. Confirmation of this structure, using tests of specified size, is in Sect. 3.3. The paper concludes with a comparison of the analysis of the same data using the robust TCLUST procedure.

2 The Forward Search

2.1 Outlier Detection and Clustering

The main tool in clustering multivariate data with the forward search (FS) is the detection of outliers. For a sample believed to come from a single multivariate normal population, perhaps with outliers, the method described by Riani et al. (2009) has good statistical properties (size and power of the outlier test). As described, it does however require intervention of the data analyst. In this paper we introduce the automatic version of this procedure extended to groups.

The method for a single population starts from a robustly chosen subset of m_0 observations. The subset is increased from size m to $m+1$ by forming the new subset from the observations with the $m + 1$ smallest squared Mahalanobis distances. For each m ($m_0 \leq m \leq n-1$), we test for the presence of outliers, using the observation outside the subset with the smallest Mahalanobis distance.

With data coming from two or more populations, starting with a subset of observations in one of the clusters results in some observations from other clusters being identified as outliers. Atkinson et al. (2004, Sect. 3.4) illustrate this point in an analysis of 200 observations on Swiss banknotes. In that case the banknotes had already been preliminarily classified into two groups, "genuine" and "forgeries". However, in the general clustering problem considered here, the number of clusters is not known, let alone any approximate cluster membership, so that there is no simple robust path to an initial subset. This difficulty was overcome by Atkinson et al. (2006), who suggested a "random-start" forward search in which many initial subsets of size m_0 are selected at random. Monitoring the behaviour of the resulting forward searches leads to an indication of the number and membership of clusters, which can then be refined by the automatic outlier detection procedure. The details are described in Sect. 2.3.

2.2 Mahalanobis Distances

In the forward search we estimate the parameters μ and Σ of the v-dimensional multivariate normal distribution of y by the standard unbiased estimators from a subset of m observations, yielding estimates $\hat{\mu}(m)$ and $\hat{\Sigma}(m)$. From this subset we obtain n squared Mahalanobis distances

$$d_i^2(m) = \{y_i - \hat{\mu}(m)\}' \hat{\Sigma}^{-1}(m)\{y_i - \hat{\mu}(m)\}, \qquad i = 1, \ldots, n. \qquad (4)$$

Let $S^*(m)$ be the subset of size m found by the search. To detect outliers we use the minimum Mahalanobis distance amongst observations not in the subset

$$d_{\min}(m) = \min d_i(m) \quad i \notin S^*(m). \qquad (5)$$

In order to test for outliers we need a reference distribution for $d_i^2(m)$ in (4) and hence for $d_{\min}(m)$ in (5). If we estimated Σ from all n observations, the statistics would have an F distribution. However, in the search we select the central m out of n observations to provide the estimate $\hat{\Sigma}(m)$, so that the variability is underestimated. To allow for estimation from this truncated distribution, Riani et al. (2009) provide a consistency factor to make the estimate approximately unbiased. They also provide an order-statistic argument for the distribution of $d_{\min}(m)$ which obviates the need for simulations in the calculation of the reference distribution. As the search progresses, we perform a series of outlier tests, one for each $m \geq m_0$. To allow for the problem of multiple testing, we use the outlier detection rule of Riani et al. (2009) which depends on the sample size and on the calculated envelopes for the distribution of the test statistic. The results of the FS are conveniently presented graphically through forward plots of quantities of interest as functions of m. We illustrate such plots in Sect. 3.3.

The expected values of the minimum Mahalanobis distances increase rapidly towards the end of the search, since we are looking at the distribution of the largest ordered Mahalanobis distances. In some cases, more informative plots, which we do not use for testing, come from the use of the scaled Mahalanobis distances

$$d_i^{sc}(m) = d_i(m) \times \left(|\hat{\Sigma}(m)| / |\hat{\Sigma}(n)| \right)^{1/2v}, \qquad (6)$$

where $\hat{\Sigma}(n)$ is the estimate of Σ at the end of the search. Examples are in Sect. 3.2 with, again, the details in Riani et al. (2009).

2.3 Random Start Forward Searches

If there are clusters in the data, the robustly chosen initial subset m_0 may lead to a search in which observations from several clusters enter the subset haphazardly in such a way that the clusters are not revealed. Searches from more than one starting point are necessary to reveal the clustering structure. We therefore instead run many forward searches, say R (500 in our example), from randomly selected starting points, monitoring the evolution of the values of $d_{\min}(m, j)$ for each search j, $(j = 1, \ldots, R)$. The criterion used for moving from step m to $m + 1$ is also reminiscent of the Mahalanobis Fixed Point Clusters procedure of Hennig and Christlieb (2002), where, however, a χ_v^2 threshold is used for defining the fitting subset.

At the beginning of the search, a random start produces some very large distances. But, because the search can drop units from the subset as well as adding them, some searches are attracted to cluster centres. As the searches progress, the various random start trajectories converge, with subsets containing the same units. Once trajectories have converged, they cannot diverge again. As we see in Fig. 2, which is typical of those for many data structures, the search is rapidly reduced to only a few trajectories. It is these that provide information on the number and membership of the clusters. Typically, in the last third of the search all trajectories have coalesced into one. Interpreting such figures requires intervention from the data analyst. However, once prospective clusters have been identified, the procedure for cluster confirmation is automatic.

3 An Example of García-Escudero et al.

3.1 Random Start Forward Search

García-Escudero et al. (2011) provide a simulated data test case of 1,800 observations simulated from two-dimensional normal distributions, with 200 outliers generated from a uniform distribution. The data are plotted in Fig. 1. The plot appears to show one clear tight cluster, one moderately clearly defined cluster, a

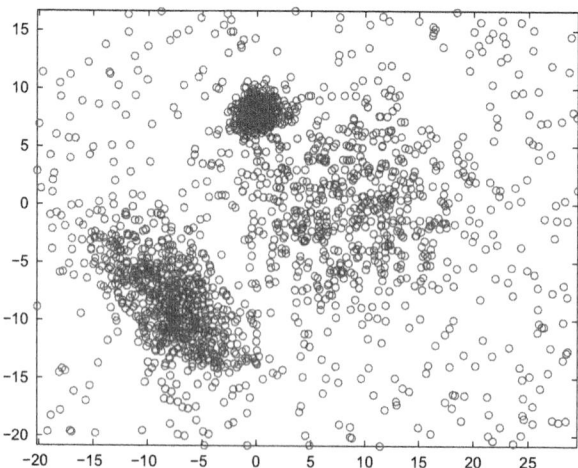

Fig. 1 The 2,000 simulated observations of García-Escudero et al. (2011), which include background contamination

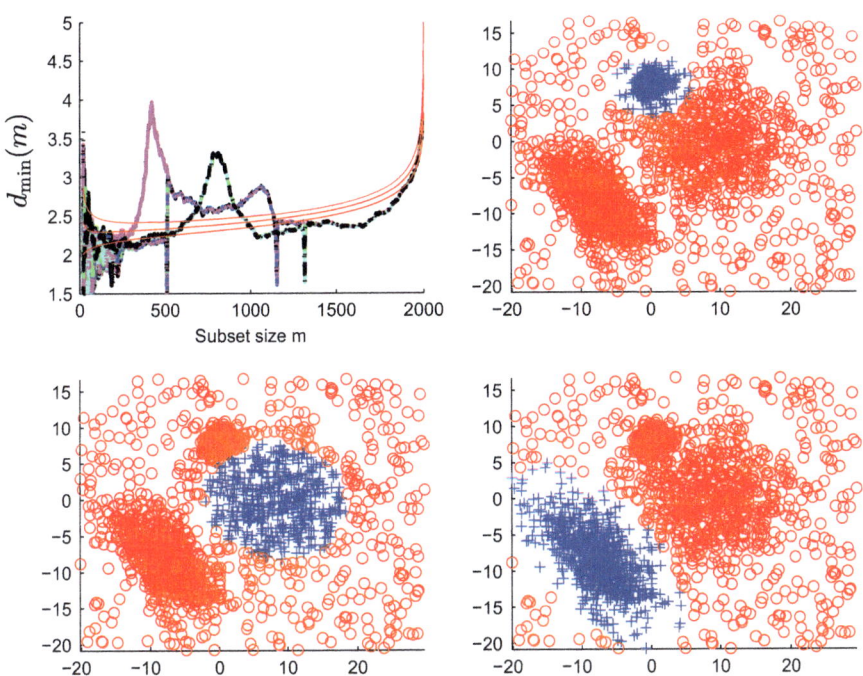

Fig. 2 Preliminary cluster identification. *Top left-hand panel*, the trajectories of minimum Maha-lanobis distances $d_{\min}(m, j)$ from 500 random start forward searches; the vertical trajectories occur when clusters merge. *Top right-hand panel*, scatterplot, with preliminary Group 1 highlighted; the first 420 units from the sharpest peak. *Bottom left-hand panel*, preliminary Group 2; the first 490 units from the lowest trajectory. *Bottom right-hand panel*, preliminary Group 3, the first 780 units from the central peak. Highlighted group in each panel indicated $+$

third more dispersed cluster and a background scatter. The clusters are thus of disparate sizes, orientations and shapes (as measured by the eigenvalue ratio for each cluster) and there is background contamination. A strange feature is that this background contamination seems to contain a hole for large values of y_2 and slightly smaller values of y_1.

A traditional robust clustering method, such as PAM (Partitioning Around Medoids, Kaufman and Rousseeuw 1990) when $K = 3$ finds three large clusters, which include all the observations. A strong advantage of the FS is that we do not have to cluster all observations, but can declare a data-chosen number of them to be outliers.

The top left-hand panel of Fig. 2 shows the forward plot of minimum Maha-lanobis distances from 500 random start forward searches. The structure is exem-plary in its clarity. By the time $m = 300$ there are just three distinct trajectories, which we examine to see if they correspond to clusters. The vertical trajectories occur at values of m when two clusters merge. At this stage in the analysis of cluster structure we are not concerned about using tests of the correct size, but only in the detection of structure. Tests with carefully controlled statistical properties are, however, important in the confirmation of cluster membership (Sect. 3.3).

We start with the first, highest, peak, which we call Group 1. The peak is caused by $d_{\min}(m)$ indicating the remoteness of the nearest outlier to the fitted cluster. Eventually, as more remote observations are included in the subset of size m, the parameter estimates are sufficiently corrupted that the next nearest observation does not seem so far away, and the peak declines. We "interrogate" the search just before or after the peak. Here we choose $m = 420$, which is before the three trajectories collapse into two. The top right-hand panel of Fig. 2 shows a scatterplot of those units in the subset at this value of m for searches with this trajectory. These searches have identified the small, tight cluster. With a small dispersion matrix, even observations which are close to the cluster in the Euclidean norm have appreciable Mahalanobis distances. That we continued a little after the peak in the search is shown in the plot by the slightly more remote sample data points surrounding the central core.

We now repeat the procedure for the other two trajectories. At $m = 490$ the lowest trajectory yields the preliminary clustering shown in the bottom left-hand panel of Fig. 2. This relatively dispersed group does not have a clear boundary, unlike Group 1. We do not get a peak in the trajectory because Groups 1 and 2 merge shortly after $m = 490$. Because Group 2 has a larger dispersion matrix than Group 1, units in Group 1 do not, as the search progresses, seem particularly remote from Group 2. Finally, we look at the third trajectory at $m = 780$. This is a relatively isolated group, the trajectory for which rises to a clear peak before declining. Its membership is shown in the bottom right-hand panel of the figure.

There remains one final feature of interest in Fig. 2. In a forward plot of distances calculated from a sample from a normal population, the curves rise at the end, as the envelopes in Fig. 2 show. However, when all observations are fitted, there is no evidence of any outliers; the sample trajectory lies within the envelope. More importantly, for the last third or more, the trajectory lies appreciably below the envelopes, suggesting that the tails of the single multivariate distribution being fitted are too short; here, this is an indication of uniform outliers over a fixed region.

3.2   Plots of Individual Scaled Distances

Further insight into the structure of the data and the workings of the FS can be obtained from the forward plots of individual Mahalanobis distances, that is the trajectory of the distances for each observation. For this we use the scaled distances (6) which, unlike the distances in the top-left panel of Fig. 2, do not increase appreciably towards the end of the search.

With clustered data, there are many possible series of plots. We divide the observations into four groups based on the classification shown in Fig. 2 and see how the different groups behave during a search starting in tentative Group 1. Figure 3 shows the very different behaviour of these distances. The top left-hand panel is for the members of Group 1. Initially the distances have a distribution ranging upwards

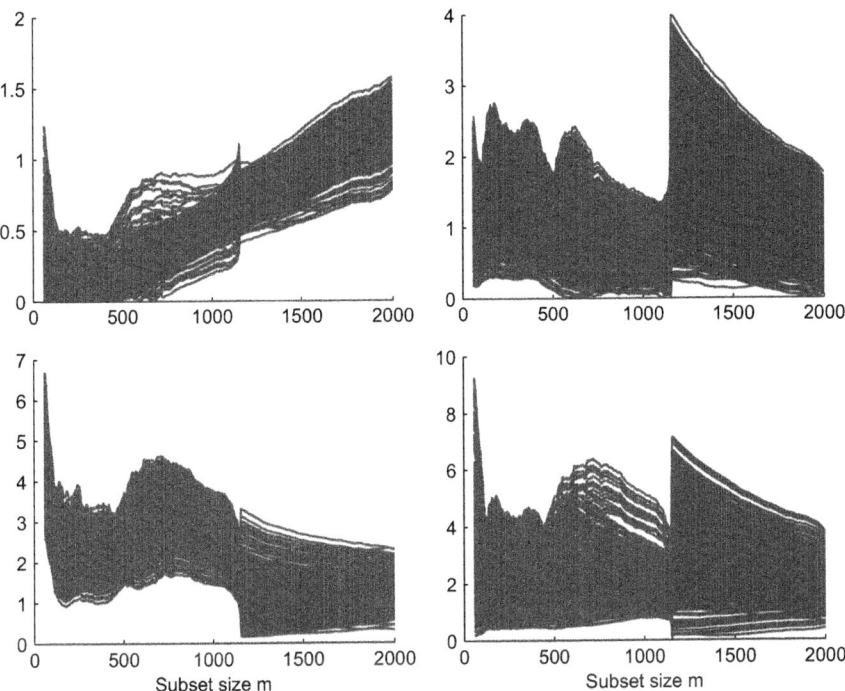

Fig. 3 Forward plots of individual scaled Mahalanobis distances $d_i^{sc}(m)$ from the preliminary classification shown in Fig. 2 when the FS starts in Group 1. Reading across, Groups 1, 2, 3 and zero (the outliers). Note the differing vertical scales in the panels

from zero, whereas the other two groups and the outliers, all remote from the cluster centre, have distances that start away from zero. There is an abrupt change, shown by the virtually vertical line in the top left panel of Fig. 2 around $m = 510$, when the centre of the fitted single population switches near to that of Group 2. The observations in Group 1 become increasingly remote. There is another dramatic change at around $m = 1,150$ when Groups 2 and 3 begin to combine. At this point, the outliers, with distances plotted in the bottom right-hand panel, also become less remote. An interesting feature of the last part of the search is that there are virtually no very small Mahalanobis distances; the cluster centre is lying between groups so that all distances are larger than they should be for a normal population. Only, towards the very end of the search, are there some small values in Group 2. A QQ-plot of the squared distances from fitting all observations might, therefore, indicate that all was not well, without being informative about the precise departure from assumptions.

3.3   Cluster Confirmation and Automatic Outlier Detection

We employ an automatic version of the two-stage procedure of Riani et al. (2009) to calibrate the FS envelopes used to confirm the clusters indicated by the random start FS. In the first stage we run a search on the data, monitoring the bounds for all n observations until we obtain a "signal" indicating that observation $m^\dagger$, and therefore succeeding observations, may be outliers, because the value of the statistic lies beyond our threshold. The hypothesis is that cluster j contains an unknown number n_j of observations. All that we know is that n_j is much less than n. We therefore need to judge the values of the statistics against envelopes from appropriately smaller population sizes. In the second part we accordingly superimpose envelopes for values of n from $m^\dagger - 1$ onwards, until the first time we introduce an observation we recognize as an outlier. The details of the rule are in Riani et al. (2009). Here we use an automatic version that, starting from a set of observations in each tentative cluster, finds $m^\dagger_j$, and then proceeds with the superimposition of envelopes until an outlier is identified and the cluster size n_j is established.

We start with the tentative Group 1. The upper left-hand panel of Fig. 4 shows the first-stage search with the sharp peak we have seen before around $m = 400$. The first outlier, the signal, is identified at $m^\dagger = 244$. The automatic procedure therefore starts with $n = 243$. These new envelopes, for a much smaller sample size, are broader than those for $n = 2,000$ and curve upwards at the end. There is no sign of any outlier for $n = 243$, so the sample size is augmented to 244 and the procedure repeated. Finally, as the bottom left-hand panel of Fig. 4 shows, there is no outlier when $n = 390$, although there is one above the 99.9 % envelope at $n = 397$. The first group therefore contains 396 observations. A similar procedure for Group 3 leads to a cluster of 768 observations.

The analysis for Group 2 needs more care. Because this group is relatively dispersed compared to Group 1, an FS starting from Group 2 will absorb many units from the compact group. We proceed by removing the observations confirmed as belonging to Groups 1 and 3, leaving 836 units. The left-hand panel of Fig. 5 shows the FS for these units. There is a signal at $m = 543$. The automatic procedure accordingly starts at $n = 542$ and proceeds to $n = 661$, as shown in the right-hand panel of Fig. 5. At this point, an outlier is indicated at $m = 597$, the very last part of the data trace lying below the lower threshold, indicating the presence of a few units from a different group. When these four units are removed we have 656 units in the group.

Figure 6 shows the scatterplot of the final clusters, together with the outliers. The histogram in the right-hand panel of the figure shows the outliers with the next-to-darkest colour. Importantly, there are very few outliers in the central part of the plot; virtually all have been classified as being in either Groups 1 or 3. This illustrates the general point that it is not possible to distinguish between clusters of observations and background contamination with the same values of y_1 and y_2.

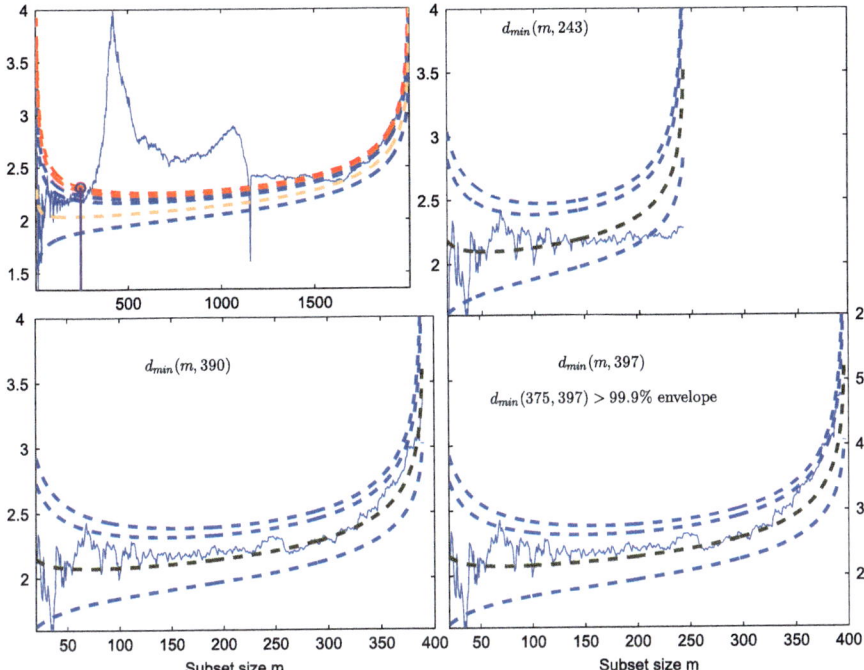

Fig. 4 Confirming Group 1. *Top left-hand panel*, forward plot of minimum Mahalanobis distances $d_{\min}(m)$ starting with units believed to be in Group 1; signal at $m^\dagger = 244$. Succeeding panels, distances for $n = 243, 390$ and 397. 396 units are assigned to the group

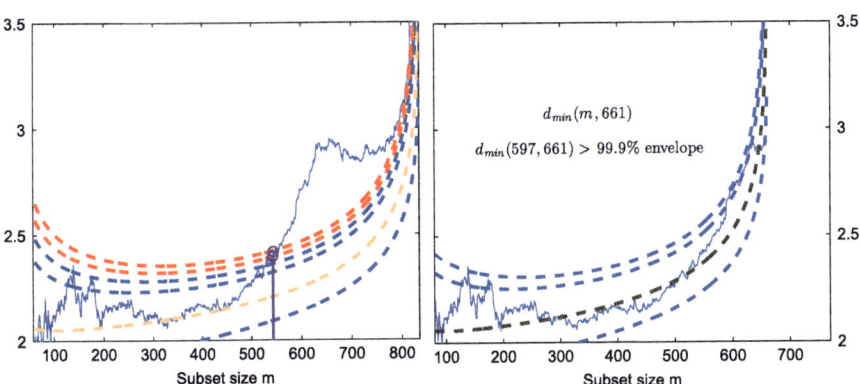

Fig. 5 Confirming Group 2 from 840 unassigned units. *Left-hand panel*, forward plot of minimum Mahalanobis distances $d_{\min}(m)$ starting with units believed to be in Group 2; signal at $m^\dagger = 543$. *Right-hand panel*, distances for $n = 661$. 656 units are ultimately assigned to the group

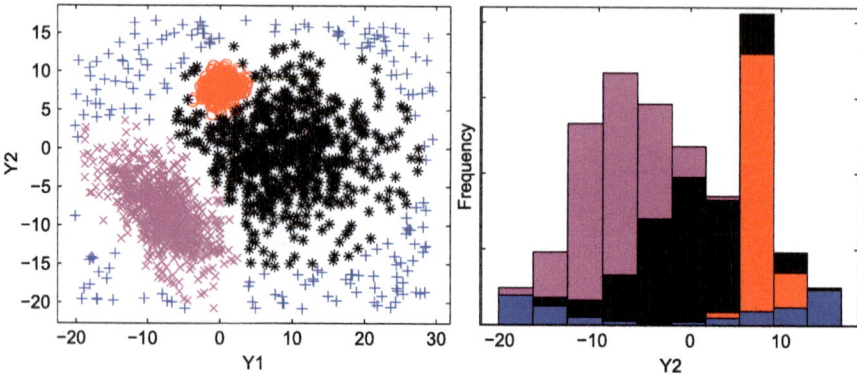

Fig. 6 Final FS clustering. *Left-hand panel*, scatterplot of the three groups and background contamination. *Right-hand panel*, histogram of classification of y_2 values: the contamination is shown in the next-to-darkest colour (*blue* in the pdf version). For central values of y_2, virtually all these observations have been included in one of the groups

4 Comparisons and Discussion

We now compare our clustering results with those of García-Escudero et al. (2011). Unlike our method, theirs requires some input parameters. We take these from the clustering we have found, namely: number of clusters $K = 3$, percentage of background contamination (outliers) 9 % and ratio of maximum to minimum eigenvalue over all clusters 100. With these parameters the agreement between TCLUST and FS clustering is very strong, the modified Rand index being 0.947. Table 1 gives a summary of the clustering performance of the two methods, together with the structure of the simulated data.

Given the choices for TCLUST in the previous paragraph which were derived from FS clustering, it is not perhaps surprising that both procedures produce clusters that differ from the data in a similar way. Clusters 1 and 3, which we established separately, are both larger than they should be, perhaps partly from the absorption of units from the background contamination; Cluster 0 has an identically low number of units for both methods. We were concerned that the relatively dispersed Cluster 2 would include some of the units of Cluster 1. In the event, it is slightly smaller than it should be.

Our results convincingly illustrate the use of the random start FS for discovering the number and identity of clusters. Refinement by the automatic version of the multivariate outlier detection method of Riani et al. (2009) leads to clustering in the presence of disparate clusters and background contamination. The output from the FS procedure can also be used to provide input to other robust clustering algorithms. A full two-step procedure, where confirmation follows exploration along the lines originally described in Atkinson et al. (2004, Sect. 7), could also help to handle overlapping clusters.

Table 1 Comparison of
Forward Search (FS) and
TCLUST (TCL)

Cluster	Numbers			Percentages		
	Data	FS	TCL	Data	FS	TCL
0	200	180	180	10	9.0	9.0
1	360	396	378	18	19.8	18.9
2	720	656	678	36	32.8	33.9
3	720	768	764	36	38.4	38.3

Numbers and percentages of observations in the
groups in the original data and after clustering

Acknowledgements We are very grateful to Berthold Lausen and Matthias Bömher for their
scientific and organizational support during the European Conference on Data Analysis 2013. We
also thank an anonymous reviewer for careful reading of an earlier draft, and for pointing out the
reference to Hennig and Christlieb (2002). Our work on this paper was partly supported by the
project MIUR PRIN *"MISURA—Multivariate Models for Risk Assessment"*.

References

Atkinson, A. C., & Riani, M. (2007). Exploratory tools for clustering multivariate data. *Computational Statistics and Data Analysis, 52*, 272–285.

Atkinson, A. C., Riani, M., & Cerioli, A. (2004). *Exploring multivariate data with the forward search*. New York: Springer.

Atkinson, A. C., Riani, M., & Cerioli, A. (2006). Random start forward searches with envelopes for detecting clusters in multivariate data. In S. Zani, A. Cerioli, M. Riani, & M. Vichi (Eds.), *Data Analysis, Classification and the Forward Search* (pp. 163–171). Berlin: Springer.

Cerioli, A., & Perrotta, D. (2014). Robust clustering around regression lines with high density regions. *Advances in Data Analysis and Classification, 8*, 5–26.

Coretto, P., & Hennig, C. (2010). A simulation study to compare robust clustering methods based on mixtures. *Advances in Data Analysis and Classification, 4*, 111–135.

Fowlkes, E. B., Gnanadesikan, R., & Kettenring, J. R. (1988). Variable selection in clustering. *Journal of Classification, 5*, 205–228.

Fraley, C., & Raftery, A. E. (2002). Model-based clustering, discriminant analysis, and density estimation. *Journal of the American Statistical Association, 97*, 611–631.

Fritz, H., García-Escudero, L. A., & Mayo-Iscar, A. (2012). TCLUST: An R package for a trimming approach to cluster analysis. *Journal of Statistical Software, 47*, 1–26.

Gallegos, M. T., & Ritter, G. (2009). Trimming algorithms for clustering contaminated grouped data and their robustness. *Advances in Data Analysis and Classification, 3*, 135–167.

García-Escudero, L. A., Gordaliza, A., Matrán, C., & Mayo-Iscar, A. (2008). A general trimming approach to robust cluster analysis. *Annals of Statistics, 36*, 1324–1345.

García-Escudero, L. A., Gordaliza, A., Matrán, C., & Mayo-Iscar, A. (2010). A review of robust clustering methods. *Advances in Data Analysis and Classification, 4*, 89–109.

García-Escudero, L. A., Gordaliza, A., Matrán, C., & Mayo-Iscar, A. (2011). Exploring the number of groups in model-based clustering. *Statistics and Computing, 21*, 585–599.

Hennig, C., & Christlieb, N. (2002). Validating visual clusters in large datasets: Fixed point clusters of spectral features. *Computational Statistics and Data Analysis, 40*, 723–739.

Kaufman, L., & Rousseeuw, P. J. (1990). *Finding groups in data. An introduction to cluster analysis*. New York: Wiley.

Lee, S. X., & Mclachlan, G. J. (2013). Model-based clustering and classification with non-normal mixture distributions. *Statistical Methods and Applications, 22*, 427–454.

Milligan, G. W. (1980). An examination of the effect of six types of error perturbation on fifteen clustering algorithms. *Psychometrika, 45*, 325–342.

Morelli, G. (2013). *A comparison of different classification methods*. Ph.D. dissertation, Università di Parma.

Riani, M., Atkinson, A. C., & Cerioli, A. (2009). Finding an unknown number of multivariate outliers. *Journal of the Royal Statistical Society, Series B, 71*, 447–466.

Riani, M., Perrotta, D., & Torti, F. (2012). FSDA: A MATLAB toolbox for robust analysis and interactive data exploration. *Chemometrics and Intelligent Laboratory Systems, 116*, 17–32.

Riani, M., Atkinson, A. C., & Perrotta, D. (2014). A parametric framework for the comparison of methods of very robust regression. *Statistical Science, 29*, 128–143.

Clustering of Solar Irradiance

Miloud Bessafi, Francisco de A.T. de Carvalho, Philippe Charton, Mathieu Delsaut, Thierry Despeyroux, Patrick Jeanty, Jean Daniel Lan-Sun-Luk, Yves Lechevallier, Henri Ralambondrainy, and Lionel Trovalet

Abstract The development of grid-connected photovoltaic power systems leads to new challenges. The short or medium term prediction of the solar irradiance is definitively a solution to reduce the storage capacities and, as a result, authorizes to increase the penetration of the photovoltaic units on the power grid. We present the first results of an interdisciplinary research project which involves researchers in energy, meteorology, and data mining, addressing this real-world problem. In Reunion Island from December 2008 to March 2012, solar radiation measurements have been collected, every minute, using calibrated instruments. Prior to prediction modelling, two clustering strategies have been applied for the analysis of the data base of 951 days. The first approach combines the following proven data-mining methods. principal component analysis (PCA) was used as a pre-process for reduction and denoising and the Ward Hierarchical and K-means methods to find a partition with a good number of classes. The second approach uses a clustering method that operates on a set of dissimilarity matrices. Each cluster is represented by an element or a subset of the set of objects to be classified. The five meaningfully clusters found by the two clustering approaches are compared. The interest and disadvantages of the two approaches for classifying curves are discussed.

M. Bessafi • M. Delsaut • P. Jeanty • J.D. Lan-Sun-Luk • L. Trovalet
LE²P, Université de la Réunion-97490 Sainte-Clotilde, Réunion, France
e-mail: bessafi@univ-reunion.fr; mathieu.delsaut@univ-reunion.fr;
patrick.jeanty@univ-reunion.fr; lanson@univ-reunion.fr; lionel.trovalet@univ-reunion.fr

F.A.T. de Carvalho
CIn/UFPE, Recife-PE, Brazil
e-mail: fatc@cin.ufpe.br

P. Charton • H. Ralambondrainy (✉)
LIM, Université de la Réunion-97490 Sainte-Clotilde, Réunion, France
e-mail: ralambon@univ-reunion.fr; charton@univ-reunion.fr

T. Despeyroux • Y. Lechevallier
INRIA, Paris-Rocquencourt - 78153 Le Chesnay cedex, France
e-mail: Yves.Lechevallier@inria.fr; Thierry.Despeyroux@inria.fr

© Springer-Verlag Berlin Heidelberg 2015
B. Lausen et al. (eds.), *Data Science, Learning by Latent Structures, and Knowledge Discovery*, Studies in Classification, Data Analysis, and Knowledge Organization, DOI 10.1007/978-3-662-44983-7_4

1   Introduction

The energy production by intermittent energy sources (solar production) is significant in Reunion Island. The LIM laboratory and its partners intend to improve the ability to predict the energy production of the photovoltaic power systems through a network of intelligent sensors.

The prediction of the primary energy source (solar irradiance) is a solution to reduce the storage capacities and, as a result, authorizes to increase the penetration of the photovoltaic units on the power grid. This article presents the data treatment that is necessary before prediction modelling. In doing so, we follow the studies on clustering recently performed by Muselli et al. (2000), Soudhan et al. (2009). We defined (Jeanty et al. 2013) k_b like the ratio of the direct radiation over the global radiation. When k_b is close to 1, the direct radiation level is close to the global radiation level, indicating we are in presence of a sunny day. A value close to 0 is the signature of a very cloudy day (see Fig. 1).

A analyzed sample contains 956 daily sequences of k_b measured between 08:00 a.m. and 05:00 p.m. (local time, GMT+4). This restricted time slot is chosen to avoid any effect due to solar shade mask actually present at the beginning and the end of the day.

The methodology proposed is to cluster time series. It combines clustering and data-mining methods to identify patterns into the data. The approach is applied to different solar schemes on Reunion Island, and the resulted clusters are interpreted taking into account meteorological phenomena.

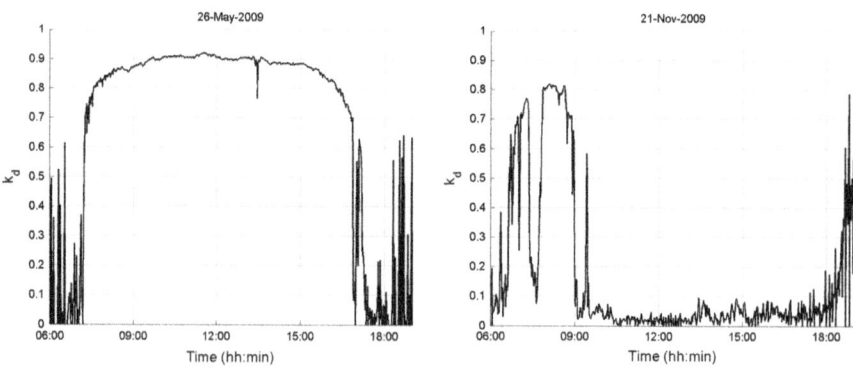

Fig. 1 Two-day examples described by k_b measure

2   Clustering Time Series: The First Approach

A time serie e_i is a sequence of real values representing the measurements of a given attribute at equal time intervals. Let $I = \{1, \ldots, n\}$ and $T = \{t_1, \ldots, t_q\}$. A set of time series $\{e_i | i \in I\}$ can be represented by a matrix $X = (\mathbf{x}^t)_{t \in T} = (x_i^t)_{i \in I, t \in T}$ where x_i^t is the value of the attribute for the time serie e_i at the time t. In our case, a time serie records the different measurements of k_b every minute or hour (average of 60 min) during a day.

The methodology (Jeanty et al. 2013) used to cluster a set of time series consists in combining three data-mining methods: principal component analysis (PCA), Ward and K-means clustering methods. We used the package *FactoMineR* (Lê et al. 2008) that implements this strategy in the R platform.

1. *PCA* is used as a pre-process for hierarchical clustering method for reduction and denoising. The first principal components are a set of new uncorrelated variables which extract the main information contained in the data. This issue is important to analyze time series when measurements are done on a long period, and when the values are correlated. The partitioning obtained by the hierarchical clustering method performed on a set of selected pertinent principal components is more stable than the one obtained from the original data.
2. To determine an optimal number of clusters, the *Ward Hierarchical* method is applied on these PCA principal components. The Ward method organizes the set of days in a sequence of nested partitions forming a hierarchical tree. At each step, the Ward method merges two clusters that minimize the reduction of the between-cluster variance B. A good number of clusters can be picked out by analyzing the decrease of the between-cluster variance B for the partitions of the hierarchical tree.
3. When a number of clusters are selected, the quality of the partition P obtained by cutting the hierarchical tree is improved by applying the K *means* algorithm that minimizes the within-cluster variance W. The percentage of variance explained by the partition $Q(P) = \frac{B}{W+B}$ is the quality of the partition P.

3   Clustering Time Series: The Second Approach

The second approach introduces clustering algorithms that are able to partition objects taking into account simultaneously multiple dissimilarity matrices. These matrices have been generated using different sets of variables and/or dissimilarity functions. These methods build a partition and a prototype for each cluster and learn a weight for each dissimilarity matrix by optimizing an adequacy criterion that measures the fitting between the clusters and their representatives (prototypes and weights). These weights and prototypes change at each iteration. Several clustering algorithms are based on a single dissimilarity matrix:

- HAC Hierarchical Agglomerative Clustering (Ward,...),
- Clustering with Adaptive Distances by Diday and Govaert (1977),
- PAM and CLARA : Partitioning around medoids by Kaufman and Rousseeuw (1990),
- SAHN : Sequential agglomerative hierarchical non-overlapping by Jain et al. (1999) and
- CARD : Clustering and aggregation of relational data by Frigui et al. (2007) that is based on NERFCM (Hathaway and Bezdek 1994) and FANNY (Kaufman and Rousseeuw 1990) and learns a relevance weight for each dissimilarity matrix in each cluster.

Let $E = \{e_1, \ldots, e_n\}$ be a set of n objects. Let $(\mathbf{D}_1, \ldots, \mathbf{D}_j, \ldots, \mathbf{D}_p)$ a vector of p dissimilarity matrices and a weight vector or matrix $\lambda = (\lambda_1, \ldots, \lambda_k, \ldots, \lambda_K)$. Let $\mathbf{D}_j[i, l] = d_j(e_i, e_l)$ where $d_j(e_i, e_l)$ measures the dissimilarity between objects e_i and e_l for the dissimilarity matrix $\mathbf{D}_j$.

The algorithms MRDCA-SWG and MRDCA-SWL extend the dynamic clustering algorithm for relational data (de Carvalho et al. 2009) and for multiple dissimilarity matrices (de Carvalho et al. 2012) where the class prototype, called here *prototype set*, is a subset of E where cardinality q is fixed with $q \leq n$. The set of prototype sets is named $E^{(q)}$ and $(G_1, \ldots, G_K)$ is the vector of K prototype sets where $G_k \in E^{(q)} = \{G \subseteq E / \mid G \mid = q\}$. These algorithms optimize the following *adequacy criterion*

$$W = \sum_{k=1}^{K} W_k = \sum_{k=1}^{K} \sum_{e_i \in C_k} D_{\lambda_k}(e_i, G_k) = \sum_{k=1}^{K} \sum_{e_i \in C_k} \sum_{j=1}^{p} \lambda_{kj} \sum_{e \in G_k} d_j(e_i, e)$$

where $W_k = \sum_{e_i \in C_k} D_{\lambda_k}(e_i, G_k)$ is the homogeneity of the cluster C_k and $D_{\lambda_k}(e_i, G_k) = \sum_{j=1}^{p} \lambda_{kj} \sum_{e \in G_k} d_j(e_i, e)$ measures the matching between an object $e_i \in C_k$ and the prototype set $G_k \in E^{(q)}$.

The MRDCA clustering algorithm starts with a random initial vector of prototype sets $(G_1, \ldots, G_K)$ and $\lambda = (1 \ldots 1)$ and alternates three steps:

- Step 1: *Build the Best Partition.*
- Step 2: *Compute the Relevance Weight.*

  - MRDCA-SWL is the local approach where the weight is a **matrix**;
  - MRDCA-SWG is the global approach where the weight is a **vector**;

- Step 3: *Compute the Prototype Sets.*

until convergence, when the adequacy criterion W reaches a stationary value representing a local minimum.

3.1   Step 1: Build the Best Partition

The fixed elements are a vector of the prototype sets $(G_1, \ldots, G_K)$ and the relevance weight vector or matrix λ.

The partition $P = (C_1, \ldots, C_K)$ is updated by :

$$C_k = \{e_i \in E : D_{\lambda_k}(e_i, G_k) = \sum_{j=1}^{p} \lambda_{kj} \sum_{e \in G_k} d_j(e_i, e)$$

$$\leq D_{\lambda_h}(e_i, G_h) = \sum_{j=1}^{p} \lambda_{hj} \sum_{e \in G_h} d_j(e_i, e) \; \forall h \neq k\}$$

3.2   Step 2: Compute of the Best Relevance Weight Matrix

The fixed elements are the partition $P = (C_1, \ldots, C_K)$ of and the prototype sets $G_1, \ldots, G_K$ of $E^{(q)}$. In this step we propose two approaches:

The first is the local approach **MRDCA-SWL** where the weight is a **matrix**. The weight matrix λ that minimizes the criterion W under the constraints $\lambda_{kj} > 0$ and $\prod_{j=1}^{p} \lambda_{kj} = 1$ is

$$\lambda_{kj} = \frac{\left\{ \prod_{h=1}^{p} \left[\sum_{e_i \in C_k} \sum_{e \in G_k} d_h(e_i, e) \right] \right\}^{\frac{1}{p}}}{\left[\sum_{e_i \in C_k} \sum_{e \in G_k} d_j(e_i, e) \right]}$$

The second is the global approach **MRDCA-SWG** where the weight is a **vector**. The weight vector $\lambda = (\lambda_1, \ldots, \lambda_p)$ that minimizes the criterion W under the constraints $\lambda_j > 0$ and $\prod_{j=1}^{p} \lambda_j = 1$ is

$$\lambda_j = \frac{\left\{ \prod_{h=1}^{p} \left(\sum_{k=1}^{K} \left(\sum_{e_i \in C_k} \sum_{e \in G_k} d_h(e_i, e) \right) \right) \right\}^{\frac{1}{p}}}{\sum_{k=1}^{K} \left(\sum_{e_i \in C_k} \sum_{e \in G_k} d_j(e_i, e) \right)}$$

3.3   Step 3: Find the Best Prototype Sets

The fixed elements are the partition $P = (C_1, \ldots, C_K)$ and the relevance weight vector or matrix λ. The best prototype set $G_k \in E^{(q)}$ is the prototype set that minimizes the following Fréchet function

$$G_k = \arg\min_{G \in E^{(q)}} \sum_{e_i \in C_k} D_{\lambda_k}(e_i, G)$$

$$= \arg\min_{G \in E^{(q)}} \sum_{e_i \in C_k} \sum_{j=1}^{p} \lambda_{kj} \sum_{e \in G} d_j(e_i, e)$$

G_k is a Fréchet mean or Karcher mean of C_k. Unicity of G_k is provided by a priori total order on the finite set $E^{(q)}$.

3.4 Finding the Number of Clusters and Partition and Cluster Interpretation

The choice of the number of clusters k is one of the most difficult problems in the cluster analysis field and no unique solution exists. With a vector of p dissimilarity matrices many tests for the number of clusters can be used, for example the Silhouette coefficient proposed by Kaufman and Rousseeuw (1990). But it's necessary to adapt the pseudo F statistic or all tests based on the decomposition of the inertia (Milligan and Cooper 1996). We propose the decomposition of the total variance or inertia T by using

Global dispersion T of the partition $P = (C_1, \ldots, C_K)$

$$T = \sum_{k=1}^{K} \sum_{e_i \in C_k} D_{\lambda_k}(e_i, G_E) = \sum_{k=1}^{K} \sum_{e_i \in C_k} \sum_{j=1}^{p} \lambda_{kj} \sum_{e \in G_E} d_j(e_i, e)$$

where $G_E \in E^{(q)}$ is the Fréchet mean of the global dispersion T,
Intra-cluster dispersion W of the partition $P = (C_1, \ldots, C_K)$

$$W = \sum_{k=1}^{K} W_k = \sum_{k=1}^{K} \sum_{e_i \in C_k} D_{\lambda_k}(e_i, G_k) = \sum_{k=1}^{K} \sum_{e_i \in C_k} \sum_{j=1}^{p} \lambda_{kj} \sum_{e \in G_k} d_j(e_i, e)$$

where $G_k \in E^{(q)}$ is the Fréchet mean of the intra-cluster dispersion W_k of the cluster C_k. However dispersion between clusters B must be calculated by $B = T - W$.

4   Meteorological Interpretation of the Classes

The five classes partition of first approach (named *a priori* partition) contains five clusters named *Classe 1*, . . ., *Classe 5* and resumes the sunshine regime in Reunion Island. The average of the five clusters are represented in Fig. 2. This figure shows two types of curves :

- T_1 type for which we observe a very overcast sky at the beginning of the day giving anyway relatively low values of k_b for the whole day (*Classe 1* and *Classe 3*); two different situations may be identified: one giving a continued slow degradation during the day (*Classe 1*) and another characterized by a quick and frank improvement for the whole day (*Classe 1*);
- T_2 type that corresponds to an uncluttered sky giving precedence to high values for k_b (*Classe 2*, *Classe 4*, and *Classe 5*). It may be noted that the T_2 category is relative to days which start with a nice weather but showing a degradation initiated at different times later in the day. The atmospheric tide probably plays an important role in the variability of the semi-diurnal k_b value noticed for *Classe 2* and *Classe 4*.

The left part of Figs. 2 and 3 shows the average trend of the index k_b for each class. The right part shows the index $t_{kt} = \frac{\bar{x}^t - \bar{x}_k^t}{\sigma(x^t)}$ associated with the class C_k and the time t where $\bar{x}^t$ and $\sigma(x^t)$ are the mean and standard deviation of a sample $(x_i^t)_{i \in I}$ and $\bar{x}_k^t$ is the mean of a sample $(x_i^t)_{i \in C_k}$ If the absolute value of t_{kt} is greater the time t is discriminating in this class C_k. The sign of this value locates the class C_k relating to the center value.

> *Classe 1* : **Cloudy Days** Size : 146, 15 %. *Classe 1* corresponds to a very low level of sunshine all day. The consequently averaged value of k_b indicates a significant cloud cover. This class presents dominant local phenomena which include, on one hand, the weak trade winds accompanied by a flow of moisture leading to significant effects of orographic clouds, and, on the other hand, the land breeze phenomenon induced by thermal contrasts.

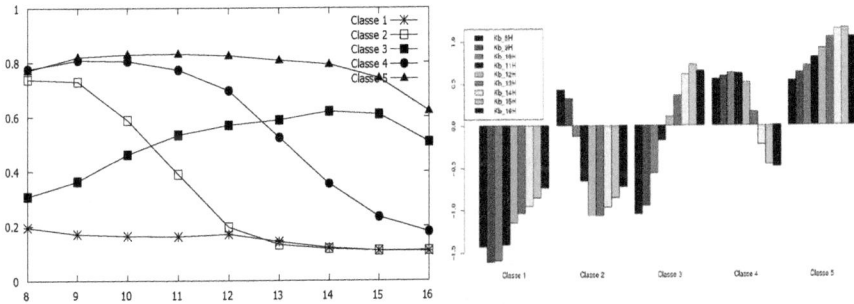

Fig. 2 k_b and t_k parameters for each classes of the first approach

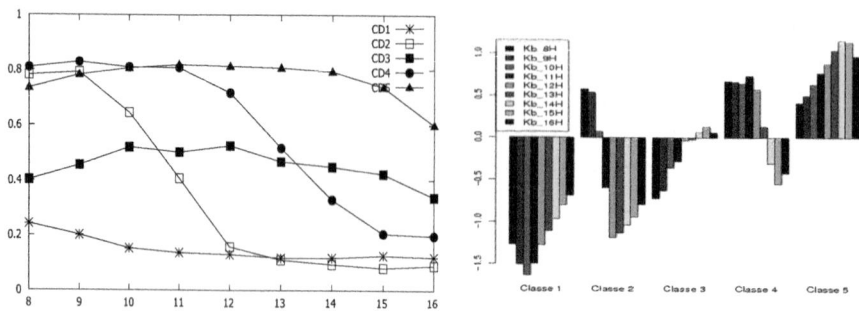

Fig. 3 k_b average and t_k for each class of D'Urso and Vichi partition

Classe 2 : **Intermittent Bad Days** Size 189, 19.7 %. *Classe 2* has a sunny beginning until mid-morning around 09:00–09:30 a.m. and a cloudy afternoon. This class is dominant over the other classes from November to January.

Classe 3 : **Disturbed Days** Size 131, 13.7 %. *Classe 3* corresponds to a day with a variable weather with a high variability of the direct fraction: improvement in the late morning and moderate cloud cover in the afternoon.

Classe 4 : **Intermittent Good Days** Size : 232, 24 %. The behavior of *Classe 4* is similar to that of *Classe 2*, but with a stronger sunny regime during all morning till early afternoon. Diffuse radiation takes place later in *Classe 4* than in *Classe 2*.

Classe 5 : **Clear Sky Days** Size : 258, 26.9 %. *Classe 5* days are characterized by a very little variations of k_b and correspond to a regime of good weather throughout the day. The direct radiation dominates in this class.

5 Interpretation of the Partition Obtained in the Second Approach

In order to compare time series, we consider in this paper a cross-sectional longitudinal dissimilarity function proposed by D'urso and Vichi (1998). The authors propose a compromise dissimilarity, that is a combination of a cross-sectional dissimilarity, which compares the instantaneous position (trend) of each pair of time series, and two longitudinal dissimilarities, based on the concepts of *velocity* and *acceleration* of a time serie.

Let $\mathbf{x}_i = (x_i^t)_{t \in T}$ the i th time serie. The *velocity* of the i th time serie is defined as $\mathbf{v}_i = (v_i^{t_2}, \ldots, v_i^{t_q})$, where $v_i^{t_h} = \frac{x_i^{t_h} - x_i^{t_{h-1}}}{t_h - t_{h-1}}$. The *acceleration* of the i th time serie is defined as $\mathbf{a}_i = (a_i^{t_3}, \ldots, a_i^{t_q})$, where $a_i^{t_h} = \frac{v_i^{t_h} - v_i^{t_{h-1}}}{t_h - t_{h-2}}$.

The compromise dissimilarity between the ith and lth time series is defined as $d^2(i, l) = \alpha_1 \|\mathbf{x}_i - \mathbf{x}_l\|^2 + \alpha_2 \|\mathbf{v}_i - \mathbf{v}_l\|^2 + \alpha_3 \|\mathbf{a}_i - \mathbf{a}_l\|^2$. In D'Urso and Vichi, the weights α of each dissimilarity-component is determined by the global objective criterion. In this paper, they will be learned.

The clustering algorithm has been performed simultaneously on these three data tables (position, velocity, and acceleration) to obtain a partition in five classes. The clustering algorithm is run 100 times and the best result (D'Urso and Vichi partition on Fig. 3) according to the adequacy criterion is selected.

The position dissimilarity matrix has the highest relevant weight for the five clusters of MRDCA-SWL method, also we selected MRDCA-SWG solution where the weight of the position dissimilarity matrix has also the highest weight (Table 1).

The relationship between the classes of the a priori partition and the classes of D'Urso and Vichi partition is performed by minimizing misclassification (MAP). D'Urso and Vichi partition is very close to the a priori partition because the overall classification error (OERC) is 82.95 %. Corresponding to the classification error, the line "Recall" gives a recall very similar for extremes classes (84.04 % for *Classe 1* and 97.29 % for *Classe 5*) and also similar for three intermediate classes where a recall close to 75 % (Table 2).

Table 1 Vector or matrix of relevance weight

		MRDCA-SWL				
Table	MRDCA-SWG	*Classe 1*	*Classe 2*	*Classe 3*	*Classe 4*	*Classe 5*
Position	**2.138767**	**2.103582**	**2.196250**	**1.873207**	**1.976635**	**2.554436**
Velocity	0.785142	0.809051	0.778197	0.815612	0.780199	0.717527
Acceleration	0.595509	0.587577	0.585098	0.654531	0.648438	0.545590

Table 2 Confusion table between a priori partition and D'Urso and Vichi partition

PP	*Classe 1*	*Classe 2*	*Classe 3*	*Classe 4*	*Classe 5*	Sum	%
CD1	**130**	17	7	0	0	154	16.11
CD2	0	**137**	0	1	0	138	14.44
CD3	16	32	**95**	35	3	181	18.93
CD4	0	3	2	**180**	4	189	19.77
CD5	0	0	27	16	**251**	294	30.75
Sum	146	189	131	232	258	956	
%	15.27	19.77	13.70	24.27	26.99		
Recall (%)	**89.04**	72.49	72.52	77.59	**97.29**	**82.95**	

6   Conclusions

This paper deals with the problem of mining daily solar radiation data in Reunion Island. Clustering real-world data is not a trivial task because natural or well-separated classes are rarely present in such data. We proposed some methodologies to cluster time series. Finally, the clustering of time series from the experimental direct fraction allowed us to highlight the various diurnal cycles in an island environment. The establishment of a predictive model for the period 10:00 a.m. to 02:00 p.m. will be based on the identification of the previous day (assigned to a class this day) and the period before 10:00 a.m. in the morning.

The dissimilarity matrices allow to incorporate easily the complex structure of time series and to give a collaborative role to each dissimilarity matrix. The prototypes are time series from our population, giving a simple interpretation of a class. The weights are optimized locally by class and by dissimilarity matrix.

Acknowledgements This work received financial support from Europe, Regional Reunion Island Council and the French government through the ERDF (European Regional Development Fund) and ADEME (French Environment and Energy Management Agency).

References

de Carvalho, F. A. T., Csernel, M., & Lechevallier, Y. (2009). Clustering constrained symbolic data. *Pattern Recognition Letter, 30*, 1037–1045.
de Carvalho, F. A. T., Lechevallier, Y., & De Melo, F. M. (2012). Partitioning hard clustering algorithms based on multiple dissimilarity matrices. *Pattern Recognition, 45*, 447–464.
Diday, E., & Govaert, G. (1977). Classification automatique avec distances adaptatives. *R.A.I.R.O. Informatique Computer Science, 11*(4), 329–349.
D'urso, P., & Vichi, M. (1998). Dissimilarities between trajectories of a three-way longitudinal data set. In A. Rizzi, M. Vichi, & H.-H. Bock (Eds.), *Advances in data science and classification* (pp. 585–592). Berlin: Springer.
Frigui, H., Hwang, C., & Rhee, F. (2007). Clustering and aggregation of relational data with applications to image database categorization. *Pattern Recognition, 40*, 3053–3068.
Hathaway, R. J., & Bezdek, J. C. (1994). Nerf c-means: Non-Euclidean relational fuzzy clustering. *Pattern Recognition, 27*(3), 429–437.
Jain, A. K., Murty, M. N., & Flynn, P. J. (1999). Data clustering: A review. *ACM Computer Survey, 31*(3), 264–323.
Jeanty, P., Delsaut, M., Trovalet, L., Ralambondrainy, H., Lan-sun-luk, J. D., Bessafi, M., Charton, P., & Chabriat, J. P. (2013). Clustering daily solar radiation from reunion island using data analysis methods. In *International Conference on Renewable Energies and Power Quality*. Spain: Bilbao.
Kaufman, L., & Rousseeuw, P. J. (1990). *Finding groups in data*. New York: Wiley.
Lê, S., Josse, J., & Husson, F. (2008). FactoMineR: An R package for multivariate analysis. *Journal of Statistical Software, 25*(1), 1–18.
Milligan, G. W., & Cooper, M. C. (1996). Clustering validation: Results and implications for applied analysis. In P. Arabie, L. Hubert, & G. De Soete (Eds.), *Clustering and classification*. (pp. 341–375). Singapore: World Scientific.

Muselli, M., Poggi, P., Notton, G., & Louche, A. (2000). Classification of typical meteorological days from global irradiation records and comparison between two mediterranean coastal sites in corsica island. *Energy Conversion Management, 41*, 1043–1063.

Soudhan, T., Emilion, R., & Calif, R. (2009). Classification of daily solar radiation distributions using a mixture of Dirichlet distributions. *Solar Energy, 83*, 1056–1063.

Part II
Data Science and Clustering

Factor Analysis of Local Formalism

François Bavaud and Christelle Cocco

Abstract Local formalism deals with weighted unoriented networks, specified by an exchange matrix, determining the selection probabilities of pairs of vertices. It permits to define local inertia and local autocorrelation relatively to arbitrary networks. In particular, free partitioned exchanges amount in defining a categorical variable (hard membership), together with canonical spectral scores, identical to Fisher's discriminant functions. One demonstrates how to extend the construction of the latter to any unoriented network, and how to assess the similarity between canonical and original configurations, as illustrated on four datasets.

1 Introduction

Introducing a *neighbourhood relation* between pairs of observations permits to construct a *local formalism*, and a local variance in particular. Comparing the latter to the ordinary variance defines the Durbin-Watson, Moran or Geary measures of autocorrelation. Local formalisms are central to spatial statistics ever since the 1950s (Moran 1950; Geary 1954); they have also been considered by a few authors in the data analytic community (e.g. Lebart 1969; Le Foll 1982; Meot et al. 1993; Thioulouse et al. 1995).

Section 2 exposes a quite general yet tractable local formalism, based upon *two primitives only*, namely an *exchange matrix E* between observations, determining the selection probabilities of pairs of observations, defining an unoriented weighted network together with the observation weights, and a *dissimilarity matrix D* between observations, chosen as squared Euclidean. This formalism, closely related to reversible Markov chain theory and spectral clustering, defines in particular a *local inertia* and a *relative inertia*, generalising the local variance and Moran's *I*.

Categorical variables, that is groups of observations, emerge as a particular instance of the local formalism, namely under *free partitioned exchange matrices* (Sect. 3.1). This circumstance enables to generalise *for any network* the computation

F. Bavaud (⊠) • C. Cocco
University of Lausanne, 1015 Lausanne, Switzerland
e-mail: francois.bavaud@unil.ch; christelle.cocco@unil.ch

© Springer-Verlag Berlin Heidelberg 2015
B. Lausen et al. (eds.), *Data Science, Learning by Latent Structures, and Knowledge Discovery*, Studies in Classification, Data Analysis, and Knowledge Organization, DOI 10.1007/978-3-662-44983-7_5

of *canonical scores* (Fisher discriminant functions), maximising the between-groups dispersion (Sect. 3.2). Canonical scores and factor scores (the latter maximising the global inertia) are compared by means of two presumably original *weighted configuration similarity* indices (Sect. 3.3).

In the last part (Sect. 4), the theory is illustrated on four datasets. The higher the relative inertia, the more similar are the canonical and factor configurations. Also, the Markovian origin of the formalism allows to consider higher-order exchange matrices, and to construct iterated canonical configurations, in contrast to ordinary discriminant analysis, idempotent in nature.

2 Definitions and Notations

2.1 Exchange Matrix and Local Variance

In spatial or temporal contexts, neighbourhood relations between n observations can be expressed by means of an $n \times n$ *exchange matrix* (Berger and Snell 1957; Bavaud 2008), describing an *unoriented weighted network*:

$$E = (e_{ij}) \qquad e_{ij} \geq 0 \qquad e_{ij} = e_{ji} \qquad f_i := \sum_j e_{ij} > 0 \qquad \sum_{ij} e_{ij} = 1 .$$

Here e_{ij} can be interpreted as the probability to select the pair of individuals (i, j), and f_i as the probability to select individual i, that is the *weight* of the observation. The ordinary (weighted) variance reads as

$$\text{var}(x) = \sum_i f_i (x_i - \bar{x})^2 = \frac{1}{2} \sum_{ij} f_i f_j (x_i - x_j)^2 \qquad \text{where} \quad \bar{x} := \sum_i f_i x_i .$$

The *local variance* (Lebart 1969; see also, e.g., Bavaud 2013) is defined as

$$\text{var}_{\text{loc}}(x) = \frac{1}{2} \sum_{ij} e_{ij} (x_i - x_j)^2 = \text{var}(x) - \text{cov}(x, Wx) \qquad w_{ij} := \frac{e_{ij}}{f_i} \quad (1)$$

where $W = (w_{ij})$ is the transition matrix of a reversible Markov chain. The canonical measure of autocorrelation—*Moran's I*—is defined as

$$I(x) := \frac{\text{var}(x) - \text{var}_{\text{loc}}(x)}{\text{var}(x)} \qquad \text{with} \quad -1 \leq I(x) \leq 1 .$$

Iterated exchange matrices are given by $E^{(r)} := \Pi W^r$, where W is given in (1) and $\Pi = \text{diag}(f)$. As $E^{(0)} = \Pi$ (frozen network) and, at least for regular chains, $E^{(\infty)} = ff'$ (complete network), one gets $\delta^{(r=0)} = 1$ and $\delta^{(r=\infty)} = 0$.

2.2   Eigen-Decomposition

The *standardised exchange matrix* $E^s = (e_{ij}^s)$ and its spectral decomposition

$$e_{ij}^s := \frac{e_{ij} - f_i f_j}{\sqrt{f_i f_j}} \qquad E^s = T \, \Gamma \, T' \qquad T = (t_{i\alpha}) \qquad \Gamma = \text{diag}(\gamma_\alpha) \quad (2)$$

enjoy numerous properties of interest. First, E^s can be iterated: $(E^s)^2$ points to the neighbours of the neighbours, and $(E^s)^r$ generates neighbours of order r. The dependence among nodes of the network can be measured by the chi-square $\chi^2(E) = \text{Tr}((E^s)^2)$, whose minimum 0 is attained with the *complete network* as described by the *free exchange* $e_{ij} = f_i f_j$, and its maximum $n - 1$ for the *frozen network* $e_{ij} = f_i \delta_{ij}$, where δ_{ij} is Kronecker delta.

The standardised exchange matrix is related to the so-called *normalised Laplacian* $\mathcal{L}$ (e.g. Chung 1997) as $E^s = I - \mathcal{L} - \sqrt{f}\sqrt{f}'$, with essentially the same eigenstructure. In particular, the eigenvalues $\gamma_\alpha \in [-1, 1]$ of E^s are those of W with the exception of the trivial Perron-Frobenius unit value, transformed to $\gamma_0 = 0$, with corresponding eigenvector $t_{i0} = \sqrt{f_i}$. The largest eigenvalue obeys $\gamma_1 \leq 1$, with equality iff the network is *reducible*, that is composed of two of more disconnected components. The smallest eigenvalue obeys $\gamma_{n-1} \geq -1$, with equality iff the network is bipartite. The network is *diffusive* if E is positive semi-definite (p.s.d.), that is if $\gamma_\alpha \geq 0$, and *off-diagonal* if $e_{ii} = 0$. Note that off-diagonal networks cannot be diffusive. Proofs of the above elements, exposed in Bavaud (2010, 2013) in the present context, can be found in the standard literature on Markov chains.

2.3   Distances, Kernels, MDS, Covariances and Factor Scores

Squared Euclidean dissimilarities between observations, as obtained from the $n \times p$ (possibly pre-transformed) data matrix $X = (x_{ik})$, are

$$D_{ij} = \sum_{k=1}^{p} (x_{ik} - x_{jk})^2 = \| x_i - x_j \|^2 \ .$$

Scalar products B, and weighted scalar products or *kernels K* are defined as

$$B = -\frac{1}{2}HDH' = X^c(X^c)' \qquad\qquad K = \Pi^{\frac{1}{2}}B\Pi^{\frac{1}{2}}$$

where $H := I - \mathbf{1}f'$ is the centring matrix, centring the data as $X^c = HX$, and $\Pi = \text{diag}(f)$ is the diagonal matrix of object weights. By construction, K possesses a trivial eigenvalue $\lambda_0 = 0$ corresponding to the eigenvector $\sqrt{f}$, and non-negative non-trivial eigenvalues $\lambda_\beta \geq 0$ for $\beta \geq 1$, decreasingly ordered.

Weighted MDS consists in extracting, from K, a set of $\min(n-1, p)$ uncorrelated coordinates $\tilde{X} = (\tilde{x}_{i\beta})$ reproducing the distances as $D_{ij} = \|\tilde{x}_i - \tilde{x}_j\|^2$ and expressing a maximum amount of variance in the first non-trivial spectral dimensions. The solution is

$$\tilde{x}_{i\beta} = \frac{\sqrt{\lambda_\beta}}{\sqrt{f_i}} u_{i\beta} \quad \text{where } K = U\Lambda U', \ U = (u_{i\beta}) \text{ and } \Lambda = \text{diag}(\lambda_\beta). \quad (3)$$

Weighted covariances are given by $S = (X^c)'\Pi X^c$. By the singular value decomposition, the non-zero eigenvalues of S are identical to those of K, that is $S = V\Lambda V'$, where the *loadings V* serve at obtaining the *factor scores* of principal component analysis as $F = (F_{i\beta}) := X^c V$.

2.4 Local Inertia, Relative Inertia and Local Covariance

Moran's I (Sect. 2.1) can be generalised to the multivariate setting by defining the inertia Δ, the *local inertia* Δ_{loc} and the *relative inertia* $\delta(E, D)$ as

$$\Delta := \frac{1}{2}\sum_{ij} f_i f_j D_{ij} = \text{Tr}(K) = \text{Tr}(S) \quad \Delta_{\text{loc}} := \frac{1}{2}\sum_{ij} e_{ij} D_{ij} \quad \delta := \frac{\Delta - \Delta_{\text{loc}}}{\Delta}$$

Here $\Delta_{\text{loc}} = \text{Tr}(S_{\text{loc}})$, where $S_{\text{loc}} := (X^c)'(\Pi - E)X^c$ is the *local covariance*.

Under the null hypothesis of absence of autocorrelation, the expected value of δ is $\delta_0 = -1/(n-1)$ for off-diagonal networks (e.g. Cliff and Ord 1973), and $\delta_0 = (m-1)/(n-1)$ for m isolated complete sub-networks described by the free partitioned exchange matrices of Sect. 3.1 (e.g. Bavaud 2013). Alternative expressions read (cf. (2) and (3)) :

$$\delta = \frac{\text{Tr}(E^s K)}{\text{Tr}(K)} = \frac{\sum_{\alpha,\beta\geq 1} \gamma_\alpha \lambda_\beta C_{\alpha\beta}}{\sum_{\beta\geq 1} \lambda_\beta} \qquad \text{where} \quad C_{\alpha\beta} := (\sum_i t_{i\alpha}u_{i\beta})^2 \leq 1 \ .$$

In particular, $-1 \leq \gamma_{n-1} \leq \delta \leq \gamma_1 \leq 1$, where the maximum of $\delta(E, D)$ for E fixed is attained for $D_{ij} = C (t_{i1}/\sqrt{f_i} - t_{j1}/\sqrt{f_j})^2$ with $C > 0$ (Bavaud 2010).

3 Local Formalism as a Generalised Discriminant Analysis

3.1 Free Partitioned Exchanges and Within-Groups Covariance

An important special case consists of *partitioned complete weighted networks* made of m components, described by *free partitioned exchange matrices*

$$e_{ij} := f_i f_j \sum_{g=1}^{m} \frac{I(i \in g)\, I(j \in g)}{\rho_g} \qquad \rho_g := \sum_{i \in g} f_i \ . \qquad (4)$$

where $I(i \in g)$ is the 0/1 indicator function of the event "observation i belongs to group g". In this set-up, $\Delta_{\mathrm{loc}} = \Delta_W$, the *within-groups inertia* associated with the hard partitioning of the n observations into m groups (e.g. Le Foll 1982; Meot et al. 1993; Lebart 2005). Similarly, S_{loc} is the *within-groups covariance* S_W. It is thus tempting to define, in the general case, a *between-groups covariance* and a *between-groups inertia* as

$$S_B := S - S_{\mathrm{loc}} = (X^c)' E X^c \qquad \Delta_B := \Delta - \Delta_{\mathrm{loc}}$$

with the caveat that, in contrast to S or S_{loc}, S_B is not p.s.d., unless E is diffusive. This is indeed the case for reducible exchanges (4), obeying $(E^s)^2 = E^s$, thus making $\gamma_\alpha = 1$ or $\gamma_\alpha = 0$.

Hence the relative autocorrelation reads as $\delta = \Delta_B / \Delta$, behaving as a kind of F-ratio for any general neighbourhood structure E—reducible or not, diffusive or not. Also, *hard partitioning* and *discriminant analysis* appear as particularly cases of the local formalism.

3.2 Factor Analysis and Canonical Scores for Unoriented Networks

Fisher *canonical scores* consist of mutually orthogonal *linear combinations* of scores maximising the relative between-groups dispersion Δ_B / Δ (or, equivalently, maximising δ under the linear requirement; compare with Sect. 2.4 for the unconstrained maximising configuration). They can be shown to be $F^{\mathrm{can}} := X^c U^{\mathrm{can}}$, where U^{can} contains eigenvectors of $S^{-1} S_B u_\sigma^{\mathrm{can}} = \lambda_\sigma^{\mathrm{can}} u_\sigma^{\mathrm{can}}$, yet to be normalised (e.g. Mardia et al. 1979; Flury 1997; Saporta 2006). Canonical scores are related to the normalised eigenvectors of the symmetric matrix $A := S^{-\frac{1}{2}} S_B S^{-\frac{1}{2}} = \dot{U} \Lambda^{\mathrm{can}} \dot{U}'$

as $U^{\mathrm{can}} = S^{-\frac{1}{2}} \dot{U} \, \varXi$, where the freely adjustable matrix $\varXi = \mathrm{diag}(\xi)$ is diagonal and fixes the normalisation: canonical scores are centred, with covariance

$$S^{\mathrm{can}} = (F^{\mathrm{can}})' \varPi F^{\mathrm{can}} = \varXi^2 = (\delta_{\sigma\tau}\, \xi_\sigma^2) \ . \tag{5}$$

Canonical dissimilarities turn out to be of the form

$$D_{ij}^{\mathrm{can}} = \sum_\sigma (F_{i\sigma}^{\mathrm{can}} - F_{j\sigma}^{\mathrm{can}})^2 = \sum_{kl} m_{kl}(x_{ik} - x_{jk})(x_{il} - x_{jl}) \tag{6}$$

where $M = (m_{kl}) = S^{-\frac{1}{2}} \dot{U} \varXi^2 \dot{U}' S^{-\frac{1}{2}}$. Equivalently, canonical scores F^{can} are given by MDS coordinates resulting of the spectral decomposition of the kernel $K^{\mathrm{can}} = \varPi^{\frac{1}{2}} X^c M (X^c)' \varPi^{\frac{1}{2}}$.

Canonical scores and canonical dissimilarities are, ultimately, fully specified by requiring a normalisation relation of the form $\xi_\sigma^2 = h(\lambda_\sigma^{\mathrm{can}})$, making the canonical and local canonical covariances diagonal:

$$S^{\mathrm{can}} = h(\varLambda^{\mathrm{can}}) \qquad\qquad S_{\mathrm{loc}}^{\mathrm{can}} := (F^{\mathrm{can}})'(\varPi - E) F^{\mathrm{can}} = h(\varLambda^{\mathrm{can}})(I - \varLambda^{\mathrm{can}}) \ .$$

Then

$$M = S^{-\frac{1}{2}} \dot{U} h(\varLambda^{\mathrm{can}}) \dot{U}' S^{-\frac{1}{2}} = S^{-\frac{1}{2}} h(A) S^{-\frac{1}{2}} = S^{-\frac{1}{2}} h(S^{-\frac{1}{2}} S_B S^{-\frac{1}{2}}) S^{-\frac{1}{2}} \ .$$

The choice $h(\lambda) \equiv 1$ yields $M = S^{-1}$, that is canonical dissimilarities (6) identical to Mahalanobis distances. The choice $h(\lambda) = 1/(1-\lambda)$, *which we shall adopt in this paper*, turns out to produce the *local metric* $M = S_{\mathrm{loc}}^{-1}$. Note that $M = S^{-1} S_B S^{-1}$ for $h(\lambda) = \lambda$, a legitimate choice provided E is diffusive (Sect. 3.1), insuring that M is p.s.d. and D^{can} in (6) is Euclidean.

3.3 Similarity Between Two Weighted Configurations

For future use, let us propose two similarity indices between two weighted configurations (f_A, D_A) and (f_B, D_B) on n observations, with associated kernels K_A and K_B, namely

configuration similarity 1 $\quad$ $\mathrm{CS1}_{AB} := \dfrac{\mathrm{Tr}(K_A K_B)}{\sqrt{\mathrm{Tr}(K_A^2)\mathrm{Tr}(K_B^2)}}$

configuration similarity 2 $\quad$ $\mathrm{CS2}_{AB} := \dfrac{\mathrm{Tr}(K_A^{\frac{1}{2}} K_B^{\frac{1}{2}})}{\sqrt{\mathrm{Tr}(K_A)\mathrm{Tr}(K_B)}} = \dfrac{\mathrm{Tr}(K_A^{\frac{1}{2}} K_B^{\frac{1}{2}})}{\sqrt{\Delta_A \Delta_B}} \ .$

The similarity indices constitute a simple alternative to Procrustean analysis, as well as to the *RV-coefficient* (Robert and Escoufier 1976) or to the *distance correlation coefficient* (Székely et al. 2007). By construction, $0 \leq \text{CS1}_{AB} \leq \text{CS2}_{AB} \leq 1$. Also, $\tilde{D}_{AB} := 1 - \text{CS1}_{AB}$ and $\hat{D}_{AB} := 1 - \text{CS2}_{AB}$ constitute squared Euclidean distances between weighted configurations.

4 Illustrations and Cases Studies

Let us illustrate the theory on four datasets, each endowed with a local structure (E, D). After computing δ, the factor scores F and canonical scores F^{can}, together with their correlations with the original features X (correlations circle), are determined and plotted. Also, the similarity coefficients $\text{CS1}(X, F^{\text{can}}) = \text{CS1}(F, F^{\text{can}})$ and $\text{CS2}(X, F^{\text{can}} = \text{CS2}(F, F^{\text{can}})$ are computed.

4.1 Swiss Nineteenth Century Socio-Economic Data

$p = 6$ socio-economic standardised variables are measured on $n = 47$ districts, partitioned into $m = 6$ cantons (source: swiss{datasets} in R). Here E is the free partitioned exchange matrix (Sect. 3.1), and the set-up amounts to ordinary discriminant analysis (Fig. 1).

4.2 Correspondence Analysis of Musical Scores

A simple musical illustration is provided by the *circle of fifths progression*, split into $n = 8$ time intervals of one half-note each (Fig. 2, top left). First, one constructs the 8×7 contingency table $\Psi = (\psi_{ik})$ counting the relative duration of note k ($k = C, D, E, F, G, A, B$) in each interval i, with $\sum_k \psi_{ik} = 1$, on which *chi-square dissimilarities* D_{ij} between the rows of Ψ (that is, between time intervals) are computed. Second, one considers *periodic neighbourhoods of order $r \geq 1$*, in which each column and each row of $E^{[r]}$ has two non-zero entries $1/(2n)$ at distance r (modulo n) from the main diagonal. Note that $f_i = 1/n$ and $E^{[2r]} \neq (E^{[r]})^2$, that is periodic neighbourhoods of order r do not form an iterated family (see Sect. 2.1).

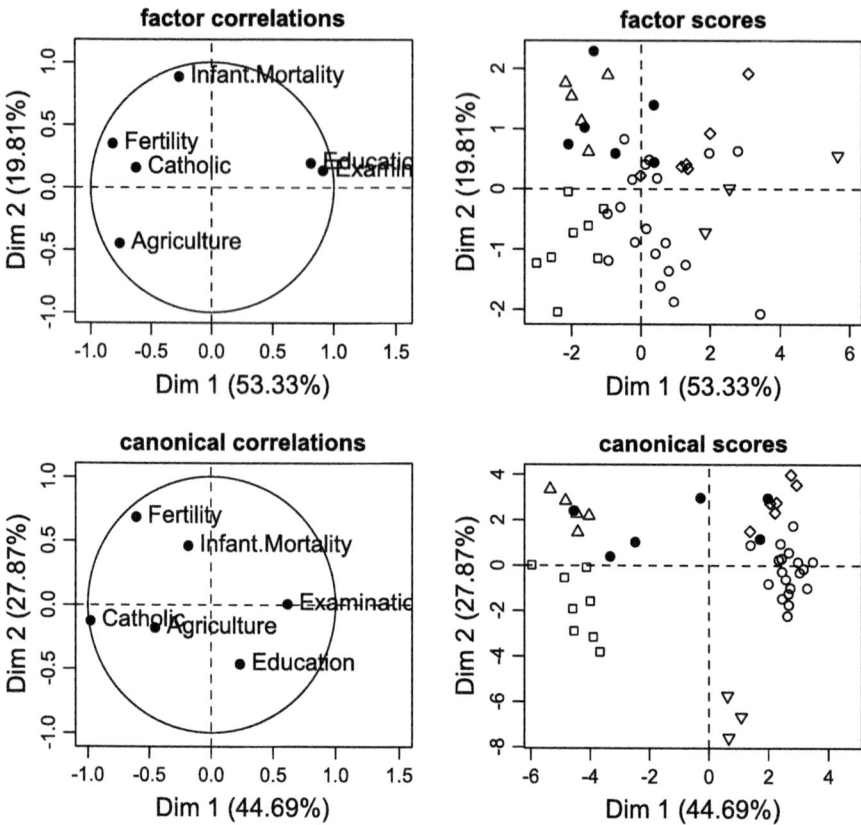

Fig. 1 Factor (*top*) and canonical (*bottom*) correlations circle and spectral scores (Sect. 4.1). Plotting symbols refer to cantons: *open triangle* = FR, *inverted open triangle* = GE, *open circle* = VD, *open square* = VS, *open diamond* = NE, *filled circle* = JU. Groups are fairly homogeneous (δ = 0.63, to be compared to δ_0 = $(6-1)/(47-1)$ = 0.11), resulting in high similarities (CS1 = 0.72 and CS2 = 0.89). Canonical configuration further increases the groups homogeneity, as it must

4.3 Local CA for Text-Document Matrix

The play *Sganarelle ou le Cocu imaginaire* (XVII th Century, by Molière) contains $n = 24$ short scenes (the observations) including $p = 339$ verbs. Analysing the scenes-verbs matrix yields fairly constant chi-square distances D^χ between scenes, in the range [9.9, 25.7], with a non-significant $\chi^2[\mathrm{df} = 7774] = 195.9$, impeding the emergence of a textual structure by low-dimensional CA compression.

Scene profiles x_i are chosen as standardised coordinates resulting from uniform MDS on D^χ. As in the previous section, exchange matrices are periodic neighbourhoods of order r, resulting in uniform weights $f_i = 1/n$. Autocorrelation

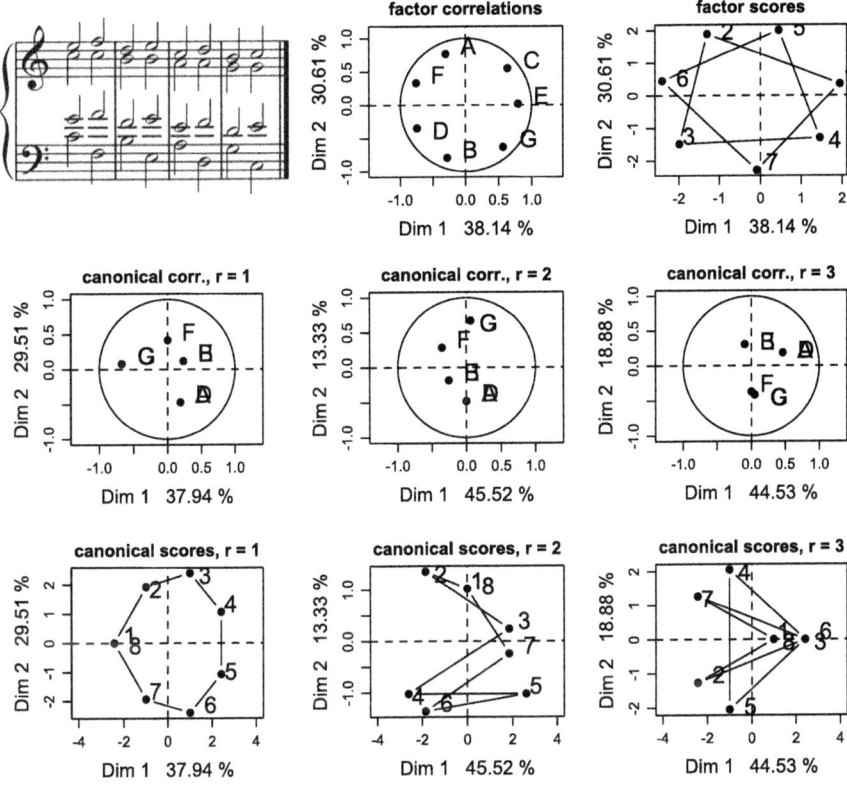

Fig. 2 Factor and canonical analysis of the musical piece, with periodic neighbourhoods of order $r = 1, 2, 3$. For $r = 1$: $\delta = 0.03$, CS1 $= 0.61$, CS2 $= 0.87$. For $r = 2$: $\delta = -0.51$, CS1 $= 0.40$, CS2 $= 0.78$. For $r = 3$: $\delta = -0.12$, CS1 $= 0.54$, CS2 $= 0.84$

is absent, with a δ close, for various r, to its expected value under independence $\delta_0 = -1/23 = -0.043$. Canonical scores, minimising the relative local dispersion, exhibit unexpected patterns (Fig. 3), yet to be elucidated.

4.4 Distances Between World Cities

$n = 313$ world cities over 10^6 inhabitants, with latitudes θ_i, longitudes α_i and relative weights f_i (proportional to the population size) have been extracted from the R file `world.cities{maps}`.

Geodesic or *arc-length* dissimilarities $D_{ij} = \arccos^2(\kappa_{ij})$, where $\kappa_{ij} = \sin\theta_i \sin\theta_j + \cos\theta_i \cos\theta_j \cos(\alpha_i - \alpha_j)$, are squared Euclidean. They may serve at constructing "gravity-like" exchange matrices of the form $e_{ij} = C b_i b_j \exp(-\beta D_{ij})$

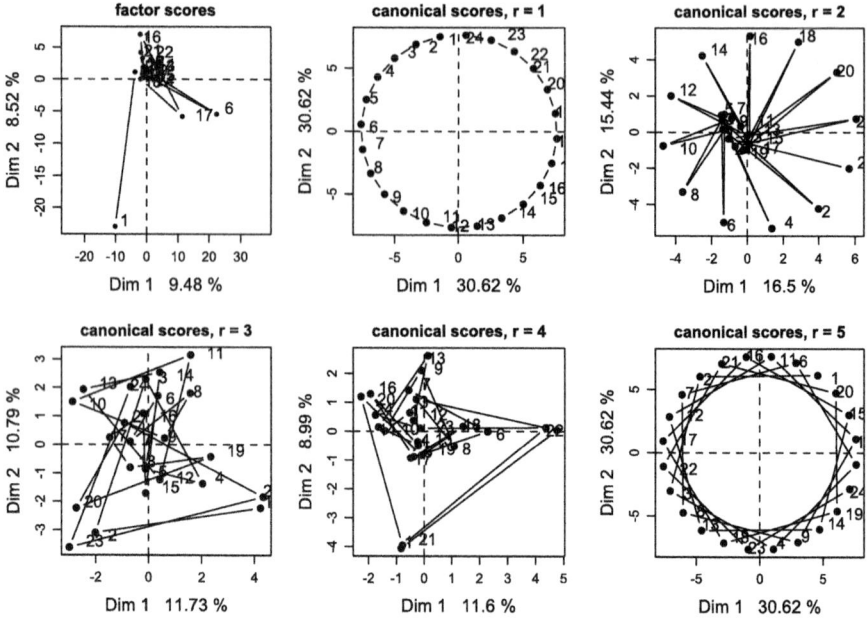

Fig. 3 Factor and canonical scores for the textual example of Sect. 4.2. Autocorrelation is absent $(\delta^{(r)} \cong \delta_0 = -1/(n-1))$, yet similarities between factorial and canonical configurations remain fairly high, despite appearances: CS1 ~ 0.45 and CS2 ~ 0.75 for $r = 1, 5$; CS1 $= 0.61$ and CS2 $= 0.84$ for $r = 2$; CS1 ~ 0.745 and CS2 ~ 0.88 for $r = 3, 4$

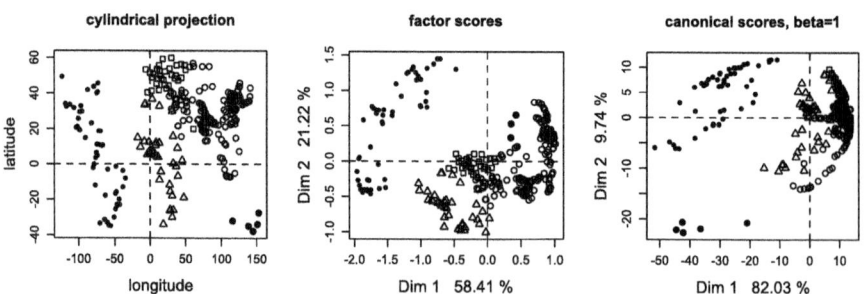

Fig. 4 World cities scatterplots (Sect. 4.4), where plotting symbols refer to the continents: *open triangle* = Africa, *large filled circle* = Oceania, *open circle* = Asia, *open square* = Europa, *small filled circle* = America. The *central figure* results from weighted MDS. The depicted factor scores, whose two first dimensions account for 79.63 % of the inertia, are highly similar to the canonical scores (91.77 % of the inertia): $\delta = 0.98$, CS1 $= 0.73$ and CS2 $= 0.96$

for $i \neq j$, where b is the *eigenvector centrality*, that is the dominant eigenvector of $\exp(-\beta D_{ij})$, and e_{ii} is adjusted so that $\sum_j e_{ij} = f_i$ (see Bavaud 2013 for details). Figure 4 shows the results obtained with $\beta = 1$.

5 Conclusion

Exchange matrices specify weighted unoriented networks. This contribution underlines their role as *soft categorical variables* associated with the neighbourhood relation between observations, permitting to generalise linear discriminant analysis and hard partitioning to any local set-up.

In general, the higher the relative autocorrelation is, the more similar are the factor and canonical configurations accordingly to CS1 and CS2—but not the other way round (see Fig. 3).

Further studies could develop the concepts of *cross-autocorrelation* between two distinct datasets, and of *partial autocorrelation* beyond the time-series framework. *Local clustering*, favouring the grouping of observations with small dissimilarities *and* strong exchanges, is to be further investigated.

References

Bavaud, F. (2008). Local concentrations. *Papers in Regional Science, 87*, 357–370.

Bavaud, F. (2010). Euclidean distances, soft and spectral clustering on weighted graphs. In *Proceedings of the ECML PKDD'10. Lecture notes in computer science* (Vol. 6321, pp. 103–118). Berlin: Springer.

Bavaud, F. (2013). Testing spatial autocorrelation in weighted networks: The modes permutation test. *Journal of Geographical Systems, 15*, 233–247.

Berger, J., & Snell, J. L. (1957). On the concept of equal exchange. *Behavioral Science, 2*, 111–118.

Chung, F. R. K. (1997). Spectral graph theory. In *CBMS Regional Conference Series in Mathematics 92*. Washington: American Mathematical Society.

Cliff, A. D., & Ord, J. K., (1973). *Spatial autocorrelation*. London: Pion.

Flury, B. (1997). *A first course in multivariate statistics*. New York: Springer.

Geary, R. (1954). The contiguity ratio and statistical mapping. *The Incorporated Statistician, 5*, 115–145.

Le Foll, Y. (1982). Pondération des distances en analyse factorielle. *Statistique et Analyse des Données, 7*, 13–31.

Lebart, L. (1969). Analyse statistique de la contiguïté. *Publications de l'Institut de Statistique des Universités de Paris, XVIII*, 81–112.

Lebart, L. (2005). Contiguity analysis and classification. In W. Gaul, O. Opitz, & M. Schader (Eds.), *Data analysis* (pp. 233–244). Berlin: Springer.

Mardia, K. V., Kent, J. T., & Bibby, J. M. (1979). *Multivariate analysis*. London: Academic Press.

Meot, A., Chessel, D., & Sabatier, R. (1993). Opérateurs de voisinage et analyse des données spatio-temporelles. In D. Lebreton & B. Asselain (Eds.), *Biométrie et environnement* (pp. 45–71). Paris: Masson.

Moran, P. A. P. (1950). Notes on continuous stochastic phenomena. *Biometrika, 37*, 17–23.

Robert, P., & Escoufier, Y. (1976). A unifying tool for linear multivariate statistical methods: The RV-coefficient. *Applied Statistics, 25*, 257–265.

Saporta, G. (2006). *Probabilités, analyse des données et statistique*. Paris: Technip.

Székely, G. J., Rizzo, M. L., & Bakirov, N. K. (2007). Measuring and testing independence by correlation of distances. *Annals of Statistics, 35*, 2769–2794.

Thioulouse, J., Chessel, D., & Champely, S. (1995). Multivariate analysis of spatial patterns - A unified approach to local and global structures. *Environmental and Ecological Statistics, 2*, 1–14.

Recent Progress in Complex Network Analysis: Models of Random Intersection Graphs

Mindaugas Bloznelis, Erhard Godehardt, Jerzy Jaworski,
Valentas Kurauskas, and Katarzyna Rybarczyk

Abstract Experimental results show that in large complex networks such as Internet or biological networks, there is a tendency to connect elements which have a common neighbor. This tendency in theoretical random graph models is depicted by the asymptotically constant clustering coefficient. Moreover complex networks have power law degree distribution and small diameter (small world phenomena), thus these are desirable features of random graphs used for modeling real life networks. We survey various variants of random intersection graph models, which are important for networks modeling.

1 Introduction

Given a finite set W and a collection of its subsets $D_1, \ldots, D_n$, the active intersection graph defines an adjacency relation between the subsets by declaring two subsets adjacent whenever they share at least s common elements (in this sense, each subset D_i can be considered as a vertex v_i of a vertex set V). The passive intersection graph defines an adjacency relation between the elements of W. A pair of elements is declared an edge if it is contained in s or more subsets. Here $s \geq 1$ is a model parameter. Both models have reasonable interpretations, for example, in the active graph, two people v_i and v_j with sets of hobbies D_i and D_j establish a communication link whenever they have sufficiently many common hobbies. In the passive graph, students (represented by elements of W)

M. Bloznelis • V. Kurauskas
Faculty of Mathematics and Informatics, Vilnius University, 03225 Vilnius, Lithuania
e-mail: mindaugas.bloznelis@mif.vu.lt; valentas@gmail.com

E. Godehardt (✉)
Clinic of Cardiovascular Surgery, Heinrich Heine University, 40225 Düsseldorf, Germany
e-mail: godehard@uni-duesseldorf.de

J. Jaworski • K. Rybarczyk
Faculty of Mathematics and Computer Science, Adam Mickiewicz University, 60769 Poznań, Poland
e-mail: jaworski@amu.edu.pl; kryba@amu.edu.pl

© Springer-Verlag Berlin Heidelberg 2015
B. Lausen et al. (eds.), *Data Science, Learning by Latent Structures, and Knowledge Discovery*, Studies in Classification, Data Analysis, and Knowledge Organization, DOI 10.1007/978-3-662-44983-7_6

become acquaintances if they participate in sufficiently many joint projects; here, the projects (respectively their managers) form the set V. In order to model active and passive graphs with desired statistical properties, we choose subsets $D_1, \ldots, D_n$ at random and obtain random intersection graphs. Alternatively, a random intersection graph can be obtained from a random bipartite graph with bipartition $V \cup W$, where each vertex *(actor)* v_i from the set $V = \{v_1, \ldots, v_n\}$ selects the set $D_i \subset W$ of its neighbors *(attributes)* in the bipartite graph at random. Now, the active intersection graph defines the adjacency relation on the vertex set V: v_i and v_j are adjacent if they have at least s common neighbors in the bipartite graph. Similarly, vertices $w_i, w_j \in W$ are adjacent in the passive graph whenever they have at least s common neighbors in the bipartite graph.

An attractive property of these models is that they capture important features of real networks, the power-law degree distribution (also called the "scale-free" property), small typical distances between vertices (the "small-world" property, see Strogatz and Watts 1998), and a high statistical dependency of neighboring adjacency relations expressed in terms of clustering and assortativity coefficients (we give a detailed account of these properties in the accompanying paper Bloznelis et al. 2015). In the literature a network possessing these properties is called a *complex network*. Many real life networks such as the Internet network, the world wide web, or many biological networks are believed to be complex networks.

Complex network models based on random intersection graphs help to explain and understand some statistical properties of real networks, like the actor network where actors are linked by an edge when they have acted in the same film, or the collaboration network where authors are declared adjacent when they have co-authored at least s papers. These networks exploit the underlying bipartite graph structure: Actors are linked to films, and authors to papers. Newman et al. (2002) pointed out that the clustering property of those networks could be explained by the presence of such a bipartite graph structure, see also Barbour and Reinert (2011) and Guillaume and Latapy (2004). The bipartite structure does not need to be given explicitly: The members of a social network tend to establish a link if they share some common interests even if the total "set of interests" might be difficult to specify.

In what follows we present several random intersection graph models and describe relations between them. Our analysis shows that random intersection graph models provide remarkably good approximations to some real networks (such as the actor network) as long as the degree and clustering properties are considered. These empirical observations are supported by theoretical findings (see Bloznelis et al. 2015). The study of random intersection graphs has just started and many interesting properties are still unexplored.

2   Models

Let n and m be positive integers. The structure of a random intersection graph results from relations between elements of two disjoint sets $V = \{v_1, \dots, v_n\}$ and $W = \{w_1, \dots, w_m\}$. V is a set of actors and W a set of attributes.

2.1   Binomial Intersection Graph

The first random intersection graph model, denoted by $G(n, m, p)$, was introduced in Karoński et al. (1999). Given $p \in [0; 1]$, in $G(n, m, p)$ each actor v_j adds an attribute w_i to D_j with probability p independently of all other elements of $V \cup W$. This relation may be represented by a bipartite graph with bipartition (V, W) in which each edge joining an element of V with an element of W appears independently with probability p. $G(n, m, p)$ is a graph with vertex set V in which vertices v_i and v_j are adjacent if D_i and D_j intersect on at least one attribute. First results from this model were applied to the gate matrix layout problem that arises in the context of physical layout of "Very Large Scale Integration," see Karoński et al. (1999) and the references therein. The model was generalized in Godehardt and Jaworski (2001, 2003) to an active and passive random intersection graph.

2.2   Active Intersection Graph

Given a set of attributes $W = \{w_1, \dots, w_m\}$, an actor v is identified with the set $D(v)$ of attributes selected by v from W. Let the actors $v_1, \dots, v_n$ choose their attribute sets $D_i = D(v_i)$, $1 \le i \le n$, independently at random, and declare v_i and v_j adjacent ($v_i \sim v_j$) whenever they share at least s common attributes, i.e., $|D_i \cap D_j| \ge s$. Here and below, $s \ge 1$ is the same for all pairs v_i, v_j. The graph on the vertex set $V = \{v_1, \dots, v_n\}$ defined by this adjacency relation is called the *active* random intersection graph, see Godehardt and Jaworski (2003). Subsets of W of size s play a special role; we call them joints. They serve as witnesses of established links: $v_i \sim v_j$ whenever there exists a joint included both in D_i and D_j.

For simplicity, we assume that the random sets $D_1, \dots, D_n$ have the same probability distribution of the form

$$\mathbf{P}(D_i = A) = P(|A|)\binom{m}{|A|}^{-1} \qquad \text{for} \qquad A \subset W. \tag{1}$$

That is, given an integer k, all subsets $A \subset W$ of size $|A| = k$ receive equal chances, proportional to the weight $P(k)$, where P is a probability on $\{0, 1, \ldots, m\}$. The random intersection graph defined in this way is denoted $G_s(n, m, P)$. Note that the active intersection graph $G_s(n, m, P)$ becomes the binomial intersection graph $G_s(n, m, p)$ if we choose $P = \mathbf{Bin}(m, p)$. Also note that $G_1(n, m, p)$ is $G(n, m, p)$ as defined in Karoński et al. (1999).

The active intersection graph $G_s(n, m, \delta_x)$, where δ_x is the probability distribution putting mass 1 on a positive integer x (i.e., all random sets are of the same, non-random size x) has attracted particular attention in the literature (Blackburn and Gerke 2009; Bloznelis and Łuczak 2013; Eschenauer and Gligor 2002; Godehardt and Jaworski 2003; Nikoletseas et al. 2011; Rybarczyk 2011a; Yagan and Makowski 2009) as it provides a convenient model of a secure wireless network. It is called the uniform random intersection graph and denoted $G_s(n, m, x)$.

The sparse active random intersection graph admits power-law degree distribution (asymptotic as $n, m \to \infty$), which has the form of a Poisson mixture (Bloznelis 2008, 2010c; Deijfen and Kets 2009; Jaworski et al. 2006; Rybarczyk 2012; Stark 2004), tunable clustering (Bloznelis 2013; Deijfen and Kets 2009; Yagan and Makowski 2009), and assortativity coefficients (Bloznelis et al. 2013). To get a power-law active intersection graph one chooses $n = O(m)$ and the probability distribution of the size $|D_i|$ of the typical set such that $\sqrt{n/m}|D_i|$ are asymptotically power-law distributed as $n, m \to +\infty$. Detailed results concerning degree distribution, clustering and other properties are given in the accompanying paper (Bloznelis et al. 2015).

The phase transition in the component size of an active random intersection graph has been studied in Behrisch (2007), Bloznelis (2010b), Bloznelis et al. (2009), Godehardt et al. (2007), and Rybarczyk (2011a). The effect of the clustering property on the phase transition in the component size and on the epidemic spread has been studied in Lagerås and Lindholm (2008), Bloznelis (2010c), Britton et al. (2008) respectively. Finally, we would like to mention the intersection graph model of Johnson and Markström (2013), where the sets D_i have a special structure in that they are subcubes of the cube $W = \{0, 1\}^n$.

2.3 Passive Intersection Graph

Let $D_1, \ldots, D_n$ be independent random subsets of W with the same probability distribution (1). Let vertices $w, w' \in W$ be linked by D_j if $w, w' \in D_j$. For example, every $w' \in D_j \setminus \{w\}$ is linked to w by D_j. The links created by $D_1, \ldots, D_n$ define a multigraph on the vertex set W. In the *passive* random intersection graph, two vertices $w, w' \in W$ are defined adjacent whenever there are at least s links between w and w', i.e., the pair $\{w, w'\}$ is contained in at least s subsets of the collection $\{D_1, \ldots, D_n\}$, see Godehardt and Jaworski (2003). We denote the passive random intersection graph $G_s^*(n, m, P)$ with P as the common probability distribution of the random variables $X_1 = |D_1|, \ldots, X_n = |D_n|$. The passive intersection graph $G_s^*(n, m, P)$ becomes the binomial intersection graph $G_s(m, n, p)$ if $P = \mathbf{Bin}(n, p)$.

The sparse passive random intersection graph admits a power-law asymptotic degree distribution which has the form of a compound Poisson distribution, see Bloznelis (2013) and Jaworski and Stark (2008). It has a tunable clustering and assortativity coefficients (Bloznelis et al. 2013; Godehardt et al. 2012; see also Bloznelis 2013).

2.4 Inhomogeneous Intersection Graph

In order to model inhomogeneity of adjacency relations that takes into account the variability of "activity" of actors and "attractiveness" of attributes, the binomial model has been generalized in Nikoletseas et al. (2004, 2008) by introducing inhomogeneous weight sequences, see also Barbour and Reinert (2011), Deijfen and Kets (2009), Rybarczyk (2013), and Shang (2010). Given weight sequences $x = \{x_i\}_{1 \le i \le m}$ and $y = \{y_j\}_{1 \le j \le n}$ with $x_i, y_j \in [0, 1]$, $i, j \ge 1$, consider the random bipartite graph $H_{x,y}$ with bipartition $V = \{v_1, v_2, \ldots, v_n\}$ and $W = \{w_1, w_2, \ldots, w_m\}$, where edges $\{w_i, v_j\}$ are inserted independently and with probabilities $p_{ij} = x_i y_j$. Define the inhomogeneous graph $G_{x,y}$ on the vertex set V by declaring $u, v \in V$ adjacent (denoted $u \sim v$) whenever they have a common neighbor in $H_{x,y}$. Now an attribute w_i is picked with probability proportional to the activity y_j of the actor v_j and the attractiveness x_i of the attribute w_i. Consider, for example, the French actor network G_F, where two actors $v_i, v_j \in V$ are declared adjacent whenever they have acted in the same movie $w_k \in W$ (real network data from the actor network, see http://www.imdb.com). The number of actors $|V| = 43{,}204$ and the number of movies $|W| = 5{,}629$. Let G_F' be the inhomogeneous random intersection graph where the activity of an actor v_j is proportional to the number of movies she or he acted in and the attractiveness of the movie w_i is proportional to the number of actors who acted in this movie. We simulate an instance of G_F' so that in the simulated network actors pick attributes independently at random, but with the probabilities estimated from the true network G_F. In Fig. 1, we plot the clustering coefficients $k \to \mathbf{P}(v_1^* \sim v_2^* | v_1^* \sim v_3^*, v_2^* \sim v_3^*, d(v_3^*) = k)$ of G_F and G_F', and the clustering function $r \to \mathbf{P}(v_1^* \sim v_2^* | d(v_1^*, v_2^*) = r)$ of G_F and G_F' (Bloznelis and Kurauskas 2012). Here (v_1^*, v_2^*, v_3^*) is an ordered triple of vertices sampled at random from V. By $\sim$ we denote the adjacency relation and $d(v)$ counts the number of neighbors of v, $d(v, u)$ counts the number of common neighbors of u and v.

Analysis of the inhomogeneous intersection graph becomes simpler if we impose some regularity conditions on the weight sequences x and y. For example, if we drop the condition $x_i, y_j \in [0, 1]$ and replace p_{ij} by $\tilde{p}_{ij} = \min\{1, x_i y_j / \sqrt{nm}\}$,

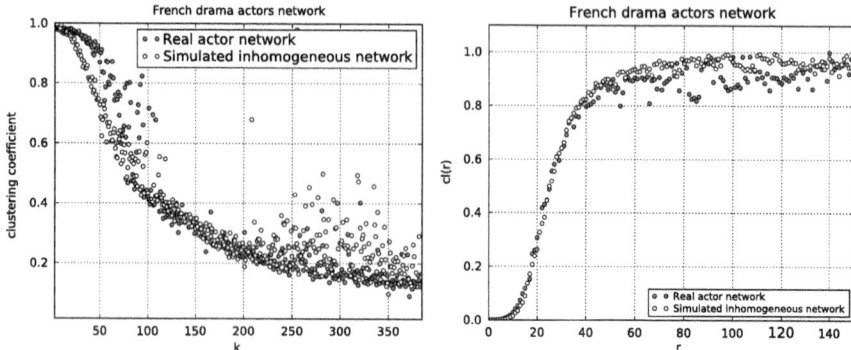

Fig. 1 Clustering coefficient and clustering function

we obtain a modification of $G_{x,y}$, which we denote $\tilde{G}_{x,y}$. In addition a convenient assumption is that x and y are realized values of two independent sequences of i.i.d. random variables $X = \{X_i\}_{1\leq i \leq m}$ and $Y = \{Y_j\}_{1\leq j \leq n}$ respectively. Let P_1 and P_2 denote the probability distributions of the random variables Y_1 and X_1. By $G(n, m, P_1, P_2)$ we denote the random intersection graph $\tilde{G}_{X,Y}$.

Note that $G(n, m, P_1, \delta_x)$ is an active random intersection graph and $G(n, m, \delta_x, P_2)$ a passive one (we recall that δ_x is the probability distribution putting mass 1 on $x > 0$). In particular, the model $G(n, m, P_1, P_2)$ admits a power-law asymptotic degree distribution and tunable clustering and assortativity coefficients (Bloznelis and Damarackas 2013; Bloznelis and Kurauskas 2012; see also Bloznelis and Karoński 2013). Bloznelis and Damarackas (2013) revise an incorrect result of Shang (2010).

Inhomogeneous random intersection graphs reproduce empirically observed clustering properties of a real actor network with remarkable accuracy as shown in Bloznelis and Kurauskas (2012). One drawback of such models is that they do not account for various characteristics of the networks that change over time. This shortage is overcome by the random intersection graph process in Bloznelis and Karoński (2013), aimed at modeling an affiliation network evolving in time, see Martin et al. (2013).

2.5 Intersection Digraph

Often relations between actors of a social network are non-symmetric. Such adjacency relations are usually modeled with the aid of a directed graph (digraph). In order to obtain a random digraph admitting non-vanishing clustering coefficients it is convenient to employ the bipartite structure, similarly as in the case of a random intersection graph.

Given two collections of subsets $S_1, \ldots, S_n$ and $T_1, \ldots, T_n$ of a set $W = \{w_1, \ldots, w_m\}$, define the intersection digraph on a vertex set $V = \{v_1, \ldots, v_n\}$ such that the arc $v_i \to v_j$ is present in the digraph whenever $S_i \cap T_j \neq \emptyset$ for $i \neq j$. Assuming that the sets S_i and T_i, $i = 1, \ldots, n$, are drawn at random, one obtains a random intersection digraph (Bloznelis 2010a). For example, the set S_i may be used to represent the set of papers (co-) authored by the i-th scientist, while T_j may be the set of papers cited by the j-th scientist. Then the corresponding intersection digraph represents scholarly influences. For simplicity, one may consider the class of random intersection digraphs where the pairs of random subsets (S_i, T_i), $i = 1, \ldots, n$, are independent and identically distributed. In addition, we assume that the distributions of S_i and T_i are mixtures of uniform distributions. That is, for every k, conditionally on the event $|S_i| = k$, the random set S_i is uniformly distributed in the class $\mathcal{W}_k$ of all subsets of W of size k. Similarly, conditionally on the event $|T_i| = k$, the random set T_i is uniformly distributed in $\mathcal{W}_k$. In particular, with P_{S*} and P_{T*} denoting the distributions of $|S_i|$ and $|T_i|$, we have that, for every $A \subset W$, $\mathbf{P}(S_i = A) = \binom{m}{|A|}^{-1} P_{S*}(|A|)$ and $\mathbf{P}(T_i = A) = \binom{m}{|A|}^{-1} P_{T*}(|A|)$. By $D(n, m, P_*)$ we denote the random intersection digraph generated by independent and identically distributed pairs of random subsets (S_i, T_i), $1 \leq i \leq n$, where P_* denotes the common distribution of pairs (S_i, T_i). Random intersection digraphs $D(n, m, P_*)$ are flexible enough to model random digraphs with in- and outdegrees having some desired statistical properties as a power-law outdegree distribution and a bounded-support indegree distribution. Assuming, e.g., that S_i and T_i intersect with positive probability, one can obtain a random digraph with a clustering property (Bloznelis 2010a).

3 Relations Between the Models

The theory of random graphs investigates asymptotic properties of random graph models, in particular properties of random intersection graphs. The following sets are examples of properties: $\mathcal{B} = \mathcal{B}_n$ consisting of all connected graphs on n vertices, $\mathcal{C} = \mathcal{C}_n$ consisting of all graphs on n vertices with maximal degree at most 6 or $\mathcal{D} = \mathcal{D}_n$ consisting of all graphs on n vertices which have a given degree distribution. If $G \in \mathcal{B}$, we say that G is connected and so on. Moreover $\mathcal{A}$ is called *increasing (decreasing)* if for every G with property $\mathcal{A}$, G with added (deleted) any edge has property $\mathcal{A}$. A property $\mathcal{A}$ which is either increasing or decreasing is called *monotone*. For instance, $\mathcal{B}$ is increasing, $\mathcal{C}$ is decreasing, and $\mathcal{D}$ is an intersection of one increasing and one decreasing property.

Sometimes it is possible to compare graph models in order to deduce something about asymptotic (usually monotone) properties of one model using known results concerning the other one. The comparison technique is called coupling. It was used in Bloznelis et al. (2009) to determine relationship between $G_s(n, m, \delta_d)$, $G_s(n, m, p)$ and $G_s(n, m, P)$. In particular it was shown that for any fixed $s \geq 1$

and increasing property $\mathcal{A}$, if $\ln n = o(mp)$ then for any integers $0 \le d_- \le mp - t$ and $mp + t \le d_+ \le m$, where $t^2 = 3(mp + \ln n) \ln n$, we have as $n \to \infty$

$$\Pr\big(G_s(n, m, \delta_{d_-}) \in \mathcal{A}\big) - o(1) \le \Pr\big(G_s(n, m, p) \in \mathcal{A}\big)$$

$$\le \Pr\big(G_s(n, m, \delta_{d_+}) \in \mathcal{A}\big) + o(1).$$

Analogue inequalities hold for any decreasing property $\mathcal{B}$. Informally speaking, the inequalities allow to show that for $d \sim mp$ and $\ln n = o(d)$ $G_s(n, m, \delta_d)$ and $G_s(n, m, p)$ have the same monotone properties with probability tending to 1 as $n \to \infty$. A similar theorem concerning more general $G(n, m, P)$ can be found in Bloznelis et al. (2009).

It would be convenient to find any such relation between random intersection graphs and well-studied models such as the Erdős–Rényi random graph $G(n, \hat{p})$, in which each edge appears independently with probability $\hat{p}$. For the inhomogeneous random intersection graph $G_{x,y}$ with $x = (p_1, p_2, ..., p_n)$ and $y = (1, 1, ..., 1)$, $0 \le p_i \le 1$ it is possible to define $\hat{p}$, such that for any increasing property $\mathcal{A}$

$$\liminf_{n \to \infty} \Pr\{G(n, \hat{p}) \in \mathcal{A}\} \le \limsup_{n \to \infty} \Pr\{G_{x,y} \in \mathcal{A})\}.$$

This and other relations between the random graph models are used in Rybarczyk (2011c, 2013) to establish threshold functions in random intersection graphs for many monotone properties such as k-connectivity, Hamiltonicity, and existence of a perfect matching.

Some random graph models have the same asymptotic properties even though they are defined in different ways. We call such models *equivalent*. For example, for np small, the edges of $G(n, m, p)$ appear almost independently. More precisely, in Fill et al. (2000) and Rybarczyk (2011b) it is shown that for $p = o\big((n\sqrt[3]{m})^{-1}\big)$ there is $\hat{p}$ such that for any property $\mathcal{B}$

$$\Pr\{G(n, \hat{p}) \in \mathcal{B}\} \to a \quad \text{if and only if} \quad \Pr\{G(n, m, p) \in \mathcal{B}\} \to a.$$

It is conjectured in Fill et al. (2000) that the condition $p = o\big((n\sqrt[3]{m})^{-1}\big)$ may be replaced by $p = \Omega(n^{-1}m^{-1/3})$ and $p = O(\sqrt{\ln n/m})$, and $m = n^\alpha$ with $\alpha > 3$. The conjecture is still open, however it is shown in Rybarczyk (2011b) to be true for monotone properties.

Acknowledgements The work of M. Bloznelis and V. Kurauskas was supported by the Lithuanian Research Council (grant MIP–067/2013). J. Jaworski and K. Rybarczyk were supported by the National Science Centre—DEC-2011/01/B/ST1/03943. Co-operation between E. Godehardt and J. Jaworski was also supported by Deutsche Forschungsgemeinschaft (grant no. GO 490/17–1).

References

Barbour, A. D., & Reinert, G. (2011). The shortest distance in random multi-type intersection graphs. *Random Structures and Algorithms, 39*, 179–209.

Behrisch, M. (2007). Component evolution in random intersection graphs. *The Electronic Journal of Combinatorics, 14*(1).

Blackburn, S., & Gerke, S. (2009). Connectivity of the uniform random intersection graph. *Discrete Mathematics, 309*, 5130–5140.

Bloznelis, M. (2008). Degree distribution of a typical vertex in a general random intersection graph. *Lithuanian Mathematical Journal, 48*, 38–45.

Bloznelis, M. (2010a). A random intersection digraph: Indegree and outdegree distributions. *Discrete Mathematics, 310*, 2560–2566.

Bloznelis, M. (2010b). Component evolution in general random intersection graphs. *SIAM Journal on Discrete Mathematics, 24*, 639–654.

Bloznelis, M. (2010c). The largest component in an inhomogeneous random intersection graph with clustering. *The Electronic Journal of Combinatorics, 17*(1), R110.

Bloznelis, M. (2013). Degree and clustering coefficient in sparse random intersection graphs. *The Annals of Applied Probability, 23*, 1254–1289.

Bloznelis, M., & Damarackas, J. (2013). Degree distribution of an inhomogeneous random intersection graph. *The Electronic Journal of Combinatorics, 20*(3), R3.

Bloznelis, M., Godehardt, E., Jaworski, J., Kurauskas, V., & Rybarczyk, K. (2015). Recent progress in complex network analysis: Properties of random intersection graphs. In B. Lausen, S. Krolak-Schwerdt, & M. Boehmer (Eds.), *European Conference on Data Analysis*. Berlin/Heidelberg/New York: Springer.

Bloznelis, M., Jaworski, J., & Kurauskas, V. (2013). Assortativity and clustering of sparse random intersection graphs. *Electronic Journal of Probability, 18*, N-38.

Bloznelis, M., Jaworski, J., & Rybarczyk, K. (2009). Component evolution in a secure wireless sensor network. *Networks, 53*(1), 19–26.

Bloznelis, M., & Karoński, M. (2013). Random intersection graph process. In A. Bonato, M. Mitzenmacher, & P. Pralat (Eds.), *Algorithms and models for the web graph. WAW 2013. Lecture notes in computer science* (Vol. 8305, pp. 93–105). Switzerland: Springer International Publishing.

Bloznelis, M., & Kurauskas, V. (2012). Clustering function: A measure of social influence. http://www.arxiv.org/abs/1207.4941.

Bloznelis, M., & Łuczak, T. (2013). Perfect matchings in random intersection graphs. *Acta Mathematica Hungarica, 138*, 15–33.

Britton, T., Deijfen, M., Lindholm, M., & Lagerås, N. A. (2008). Epidemics on random graphs with tunable clustering. *Journal of Applied Probability, 45*, 743–756.

Deijfen, M., & Kets, W. (2009). Random intersection graphs with tunable degree distribution and clustering. *Probability in the Engineering and Informational Sciences, 23*, 661–674.

Eschenauer, L., & Gligor, V. D. (2002). A key-management scheme for distributed sensor networks. In *Proceedings of the 9th ACM Conference on Computer and Communications Security* (pp. 41–47).

Fill, J. A., Scheinerman, E. R., & Singer-Cohen, K. B. (2000). Random intersection graphs when $m = \omega(n)$: an equivalence theorem relating the evolution of the $G(n, m, p)$ and $G(n, p)$ models. *Random Structures and Algorithms, 16*, 156–176.

Godehardt, E., & Jaworski, J. (2001). Two models of random intersection graphs and their applications. *Electronic Notes in Discrete Mathematics, 10*, 129–132.

Godehardt, E., & Jaworski, J. (2003). Two models of random intersection graphs for classification. In M. Schwaiger & O. Opitz (Eds.), *Exploratory data analysis in empirical research* (pp. 67–81). Berlin/Heidelberg/New York: Springer.

Godehardt, E., Jaworski, J., & Rybarczyk, K. (2007). Random intersection graphs and clas-
  sification. In R. Decker & H.-J. Lenz (Eds.), *Advances in data analysis* (pp. 67–74).
  Berlin/Heidelberg/New York: Springer.
Godehardt, E., Jaworski, J., & Rybarczyk, K. (2012). Clustering coefficients of random inter-
  section graphs. In W. Gaul, A. Geier-Schulz, L. Schmidt-Thieme, & J. Kunze (Eds.),
  Challenges at the interface of data analysis, computer science, and optimization (pp. 243–253).
  Berlin/Heidelberg/New York: Springer.
Guillaume, J. L., & Latapy, M. (2004). Bipartite structure of all complex networks. *Information
  Processing Letters, 90,* 215–221.
Jaworski, J., Karoński, M., & Stark, D. (2006). The degree of a typical vertex in generalized random
  intersection graph models. *Discrete Mathematics, 306,* 2152–2165.
Jaworski, J., & Stark, D. (2008). The vertex degree distribution of passive random intersection
  graph models. *Combinatorics, Probability and Computing, 17,* 549–558.
Johnson, J. R., & Markström, K. (2013). Turán and Ramsey properties of subcube intersection
  graphs. *Combinatorics, Probability and Computing, 22*(1), 55–70.
Karoński, M., Scheinerman, E. R., & Singer-Cohen, K. B. (1999). On random intersection graphs:
  The subgraph problem. *Combinatorics, Probability and Computing, 8,* 131–159.
Lagerås, A. N., & Lindholm, M. (2008). A note on the component structure in random intersection
  graphs with tunable clustering. *Electronic Journal of Combinatorics, 15*(1).
Martin, T., Ball, B., Karrer, B., & Newman, M. E. J. (2013). Coauthorship and citation patterns in
  the Physical Review. Phys. Rev. E 88, 012814.
Newman, M. E. J., Watts, D. J., & Strogatz, S. H. (2002). Random graph models of social networks.
  Proceedings of the National Academy of Sciences of the USA, 99(Suppl. 1), 2566–2572.
Nikoletseas, S., Raptopoulos, C., & Spirakis, P. (2004). The existence and efficient construction of
  large independent sets in general random intersection graphs. In J. Daz, J. Karhumki, A. Lepist,
  & D. Sannella (Eds.), *ICALP. Lecture notes in computer science* (Vol. 3142, pp. 1029–1040).
  Berlin: Springer.
Nikoletseas, S., Raptopoulos, C., & Spirakis, P. (2008). Large independent sets in general random
  intersection graphs. *Theoretical Computer Science, 406,* 215–224.
Nikoletseas, S., Raptopoulos, C., & Spirakis, P. G. (2011). On the independence number and
  Hamiltonicity of uniform random intersection graphs. *Theoretical Computer Science, 412,*
  6750–6760.
Rybarczyk, K. (2011a). Diameter, connectivity, and phase transition of the uniform random
  intersection graph. *Discrete Mathematics, 311,* 1998–2019.
Rybarczyk, K. (2011b). Equivalence of the random intersection graph and $G(n, p)$. *Random
  Structures and Algorithms, 38,* 205–234.
Rybarczyk, K. (2011c). Sharp threshold functions for random intersection graphs via a coupling
  method. *The Electronic Journal of Combinatorics, 18*(1), P36.
Rybarczyk, K. (2012). The degree distribution in random intersection graphs. In W. Gaul, A. Geier-
  Schulz, L. Schmidt-Thieme, & J. Kunze (Eds.), *Challenges at the interface of data analysis,
  computer science, and optimization* (pp. 291–299). Berlin/Heidelberg/New York: Springer.
Rybarczyk, K. (2013). The coupling method for inhomogeneous random intersection graphs.
  ArXiv:1301.0466.
Shang, Y. (2010). Degree distributions in general random intersection graphs. *The Electronical
  Journal of Combinatorics, 17,* #R23.
Stark, D. (2004). The vertex degree distribution of random intersection graphs. *Random Structures
  and Algorithms, 24,* 249–258.
Strogatz, S. H., & Watts, D. J. (1998). Collective dynamics of small-world networks. *Nature, 393,*
  440–442.
Yagan, O., & Makowski, A. M. (2009). Random key graphs – Can they be small worlds? In *2009
  First International Conference on Networks & Communications* (pp. 313–318).

Recent Progress in Complex Network Analysis: Properties of Random Intersection Graphs

Mindaugas Bloznelis, Erhard Godehardt, Jerzy Jaworski, Valentas Kurauskas, and Katarzyna Rybarczyk

Abstract Experimental results show that in large complex networks (such as internet, social or biological networks) there exists a tendency to connect elements which have a common neighbor. In theoretical random graph models, this tendency is described by the clustering coefficient being bounded away from zero. Complex networks also have power-law degree distributions and short average distances (small world phenomena). These are desirable features of random graphs used for modeling real life networks. We survey recent results concerning various random intersection graph models showing that they have tunable clustering coefficient, a rich class of degree distributions including power-laws, and short average distances.

1 Introduction

In this paper we present a selection of recent results showing the properties of random intersection graphs desired for applications in network modeling. In Sect. 2 we show how to pick parameters to obtain a random intersection graph with a power-law degree distribution. In Sect. 3 we show how the real-world network tendency to cluster is reproduced in random intersection graphs. Section 4 deals with the existence of cliques, which is closely related to clustering properties. In Sect. 5 we mention results on connectivity and phase transition in random intersection graphs.

This article is intended to extend and supplement the accompanying paper Bloznelis et al. (2015), where various models of random intersection graphs are

M. Bloznelis • V. Kurauskas
Faculty of Mathematics and Informatics, Vilnius University, 03225 Vilnius, Lithuania
e-mail: mindaugas.bloznelis@mif.vu.lt; valentas@gmail.com

E. Godehardt (✉)
Clinic of Cardiovascular Surgery, Heinrich Heine University, 40225 Düsseldorf, Germany
e-mail: godehard@uni-duesseldorf.de

J. Jaworski • K. Rybarczyk
Faculty of Mathematics and Computer Science, Adam Mickiewicz University, 60769 Poznań, Poland
e-mail: jaworski@amu.edu.pl; kryba@amu.edu.pl

© Springer-Verlag Berlin Heidelberg 2015
B. Lausen et al. (eds.), *Data Science, Learning by Latent Structures, and Knowledge Discovery*, Studies in Classification, Data Analysis, and Knowledge Organization, DOI 10.1007/978-3-662-44983-7_7

presented. Let us briefly remind that a random intersection graph is obtained from the bipartite graph with the bipartition $V \cup W$, where each vertex v_i from $V = \{v_1, \ldots, v_n\}$ selects a set $D_i \subset W$ of its neighbors in the bipartite graph at random. Elements of $W = \{w_1, \ldots, w_m\}$ are called attributes. Assuming that the sets $D_1, \ldots, D_n$ are selected independently and according to the same probability distribution putting mass $P(k)\binom{n}{k}^{-1}$ on every set $A \subset W$ of size $|A| = k$, $k = 0, 1, \ldots, m$, we obtain the active random intersection graph $G_s(n, m, P)$ on the vertex set V by declaring v_i and v_j adjacent whenever $|D_i \cap D_j| \geq s$. Here $P = \{P(0), \ldots, P(m)\}$ is a given probability distribution of the sizes $X_i = |D_i|$ of the random sets. The passive graph $G_s^*(n, m, P)$ defines the adjacency relation on the set W: two attributes w_i, w_j are declared adjacent if the pair $\{w_i, w_j\}$ belong to s or more random subsets from the list $D_1, \ldots, D_n$. We use special notation for the binomial graph $G_s(n, m, p)$ and the uniform graph $G_s(n, m, d)$, which are active intersection graphs with $P = \mathbf{Bin}(m, p)$ and $P = \delta_d$ (probability putting mass 1 on an integer $d > 0$), respectively. The inhomogeneous random intersection graph $G(n, m, P_1, P_2)$ on the vertex set V is obtained from the bipartite graph, where attributes $w_i \in W$ are assigned random weights $X_1, \ldots, X_m$ distributed according to the probability distribution P_2, and vertices $v_j \in V$ are assigned random weights $Y_1, \ldots, Y_n$ distributed according to P_1. Given the realized values of weights, links between w_i and v_j in the bipartite graph are inserted independently and with probabilities $\min\{1, X_i Y_j (mn)^{-1/2}\}$. Vertices v_j, v_r of $G(n, m, P_1, P_2)$ are declared adjacent whenever there is an attribute linked to both of them.

All presented results are asymptotic. We use Landau's notation and stochastic symbols consistently with Janson et al. (2010b).

2 Degree Distribution

The number of neighbors of a vertex v_1 of the active graph $G_1(n, m, P)$

$$d(v_1) = \sum_{2 \leq j \leq n} \mathbb{I}_{D_1 \cap D_j}$$

counts random subsets that intersect with D_1. Hence, given D_1 the random variable $d(v_1)$ has the binomial distribution $\mathbf{Bin}(n - 1, q)$, where the success probability $q = \mathbf{P}(D_1 \cap D_2 | X_1)$ depends on the size $X_1 = |D_1|$ of D_1. Observing that $q \approx m^{-1} X_1 \mathbf{E} X_2$, for $X_1, X_2 = o_P(\sqrt{m})$, and assuming that $X_1 \sqrt{n/m}$ converges in L_1 to an integrable random variable Z, we can approximate the distribution of $d(v_1)$ by a mixed Poisson distribution,

$$\forall k \qquad \mathbf{P}(d(v_1) = k) \approx \mathbf{E}\frac{\lambda_X^k}{k!}e^{-\lambda_X} \to \mathbf{E}\frac{(\lambda_Z)^k}{k!}e^{-\lambda_Z}, \quad \text{as} \quad n, m \to \infty. \tag{1}$$

Here $\lambda_X = \frac{n}{m} X_1 \mathbf{E} X_1$ converges to $\lambda_Z = Z \mathbf{E} Z$ (see Bloznelis 2008, 2010c; Deijfen and Kets 2009; Jaworski et al. 2006; Rybarczyk 2012; Stark 2004).

More generally, for any given $s \geq 1$ the degree $d(v_1)$ of a vertex v_1 of $G_s(n, m, P)$ has a mixed Poisson asymptotic distribution defined by the right-hand side of (1), provided that $\binom{X_1}{s} \sqrt{n / \binom{m}{s}}$ converges to a random variable Z in L_1 (Bloznelis 2013). An extension of (1) to in- and out-degrees of a random intersection digraph is shown in Bloznelis (2010a).

The degree distribution of passive random intersection graphs has been studied in Jaworski et al. (2006), Jaworski and Stark (2008) and Bloznelis (2013). We say that vertices $w, w' \in W$ of a passive graph $G_1^*(n, m, P)$ are linked by the set D_j if $w, w' \in D_j$. For example, every $w' \in D_j \setminus \{w\}$ is linked to $w \in D_j$ by D_j. The links created by $D_1, \ldots, D_n$ define a multigraph on the vertex set W. Let $d(w_1)$ denote the degree of w_1 in $G_1^*(n, m, P)$, and let $\mathcal{L}_1 = \mathcal{L}_1(w_1)$ denote the number of links incident to w_1. We first approximate $d(w_1) = \mathcal{L}_1 + o_P(1)$ and then establish the asymptotic distribution of $\mathcal{L}_1$. We write

$$\mathcal{L}_1 = \sum_{1 \leq j \leq n} (X_j - 1) \mathbb{I}_{w_1 \in D_j} \tag{2}$$

and note that links contributing to the sum (2) come in bunches (the size $X_j - 1$ of the j-th bunch counts neighboring elements to w_1 from the set D_j that covers w_1). Since the conditional probability $\mathbf{P}(w_1 \in D_j | X_j) = X_j / m$ that D_j contributes $X_j - 1$ to sum (2) is $O_P(m^{-1})$, we may conclude that the distribution of $\mathcal{L}_1$ is asymptotically a compound Poisson distribution in the case where m and n are of the same order of magnitude. In order to give a rigorous result we first assume that $m/n \to \beta$ for some $\beta \in (0, \infty)$. Secondly, we assume that X_1 converges in distribution to a random variable Z and the moment $\mathbf{E} X_1^{4/3}$ converges to $\mathbf{E} Z^{4/3}$ as $n, m \to \infty$. Then $\mathcal{L}_1$ together with $d(w_1)$ converge in distribution to the random variable $\tilde{Z}_1 + \tilde{Z}_2 + \cdots + \tilde{Z}_A$, where $\tilde{Z}_1, \tilde{Z}_2, \ldots$ are independent random variables with common probability distribution

$$\mathbf{P}(\tilde{Z}_1 = k) = (k + 1) \mathbf{P}(Z = k + 1) / \mathbf{E} Z, \qquad k = 0, 1, 2, \ldots.$$

The random variable A is independent of the sequence $\tilde{Z}_1, \tilde{Z}_2, \ldots$ and Poisson distributed with mean $\mathbf{E} A = \beta^{-1} \mathbf{E} Z$.

The inhomogeneous graph $G(n, m, P_1, P_2)$ retains some properties of the active and passive models: It becomes an active graph for $P_2 = \delta_d$, and it becomes a passive graph for $P_1 = \delta_d$. Therefore, it is not surprising that its asymptotic degree distribution retains the structural components of active and passive graphs, namely the mixed and compound Poisson. Thus, e.g., for $m/n \to \beta \in (0, \infty)$ the degree $d(v_1)$ of a vertex v_1 of $G(n, m, P_1, P_2)$ converges in distribution to the random variable $\tau_1 + \tau_2 + \cdots + \tau_{A_1}$, where $\tau_1, \tau_2, \ldots$ are independent and identically distributed random variables independent of the random variable Λ_1. They are distributed as follows. For $r = 0, 1, 2, \ldots$

$$\mathbf{P}(\tau_1 = r) = \frac{r+1}{\mathbf{E}\Lambda_2}\mathbf{P}(\Lambda_2 = r+1) \quad \text{and} \quad \mathbf{P}(\Lambda_i = r) = \mathbf{E}\,e^{-\lambda_i}\frac{\lambda_i^r}{r!}, \quad i = 1, 2.$$

Here $\lambda_1 = Y_1\beta^{1/2}\mathbf{E}X_1$, $\lambda_2 = X_1\beta^{-1/2}\mathbf{E}Y_1$ (Bloznelis and Damarackas 2013). Bloznelis and Karoński (2013) extended this result to a graph process.

Let us mention that all three models admit power-law asymptotic degree distributions: For active and passive graphs this is the case, when Z has a power-law distribution; for inhomogeneous graph this happens, when the heavier of the tails of X_1 and Y_1 obeys a power-law.

3   Clustering

Often relations between members of a real network look as if they were generated by statistically dependent events: while exploring the network the chances of a link $u \sim v$ seem to increase as we learn about common neighbors of u and v. A convenient (theoretical) measure of such a dependence is the conditional probability

$$\alpha = \mathbf{P}(u^* \sim v^* \mid u^* \sim t^*, v^* \sim t^*).$$

Here the triple of distinct vertices $\{u^*, v^*, t^*\}$ is sampled uniformly at random. We say that a (sparse) random graph model has the clustering property whenever this conditional probability (called clustering coefficient) is bounded away from zero as the number of vertices tends to infinity. We refer to Barrat and Weigt (2000), Newman et al. (2001), Newman (2003) and Strogatz and Watts (1998) for local and global empirical characteristics of real networks related to the clustering property. We also consider a related conditional probability

$$\alpha_k = \mathbf{P}(u^* \sim v^* \mid u^* \sim t^*, v^* \sim t^*, d(t^*) = k),$$

with the extra condition that the common neighbor t^* has degree k (Foudalis et al. 2011; Jackson and Rogers 2007; Ravasz and Barabási 2003). Another interesting characteristic is the sequence of conditional probabilities

$$Cl(r) = \mathbf{P}(u^* \sim v^* \mid d(u^*, v^*) = r), \qquad r = 0, 1, 2, \ldots,$$

where $d(u, v)$ denotes the number of common neighbors of vertices u, v, (Bloznelis and Kurauskas 2012). The function $r \to Cl(r)$ (called clustering function) is a measure of the influence exercised by the common neighbors on u and v to establish a relation. An attractive property of random intersection graph models is that such conditional probabilities (at least their first order asymptotics as $n, m \to \infty$) can be calculated exactly.

Assuming that $\beta_n = \binom{m}{s} n^{-1}$ converges to some $\beta \in (0, \infty)$, and that $\binom{X_1}{s} \beta_n^{-1/2}$ converges in distribution to a random variable Z and $\mathbf{E}\binom{X_1}{s}^2 \beta_n^{-1} \to \mathbf{E}Z^2$, where $\mathbf{E}Z^2 < \infty$, we obtain a first order asymptotics of the clustering coefficient of $G_s(n, m, P)$,

$$\alpha = \frac{1}{\sqrt{\beta}} \frac{\mathbf{E}Z}{\mathbf{E}Z^2} + o(1) = \frac{1}{\sqrt{\beta}} \frac{(\mathbf{E}d_*)^{3/2}}{\mathbf{E}d_*^2 - \mathbf{E}d_*} + o(1). \tag{3}$$

Here d_* is a random variable with the asymptotic degree distribution (1), see Bloznelis (2013), Deijfen and Kets (2009). We remark that (3) revises an incorrect result of Godehardt et al. (2012).

For the conditional probability α_k of $G_s(n, m, P)$ we have (Bloznelis 2013),

$$\alpha_k = \frac{1}{k} \frac{\mathbf{E}Z}{\sqrt{\beta}} \frac{\mathbf{P}(d_* = k-1)}{\mathbf{P}(d_* = k)} + o(1) = \frac{1}{k} \frac{\sqrt{\mathbf{E}d_*}}{\sqrt{\beta}} \frac{\mathbf{P}(d_* = k-1)}{\mathbf{P}(d_* = k)} + o(1).$$

In the case of a power-law asymptotic degree distribution we obtain from these relations that $\alpha_k \sim ck^{-1}$, for large k. The scaling factor k^{-1} of the conditional probability α_k in fact has been empirically observed in real networks, see Ravasz and Barabási (2003) and the references therein.

A first order asymptotic of the clustering coefficient of the passive random graph $G_1^*(n, m, P)$ has been shown in Godehardt et al. (2012):

$$\alpha = \frac{\beta_*^2 m^{-1} (\mathbf{E}(X_1)_2)^3 + \mathbf{E}(X_1)_3}{\beta_* (\mathbf{E}(X_1)_2)^2 + \mathbf{E}(X_1)_3} + o(1), \qquad \beta_* := nm^{-1}. \tag{4}$$

Here it is assumed that $\mathbf{E}(X_1)_2 > 0$ and $\mathbf{E}(X_1)_2 = o(m^2 n^{-1})$ as $m, n \to \infty$. The conditional probability α_k is evaluated in Bloznelis (2013).

The clustering function $r \to Cl(r)$ of active, passive and inhomogeneous models is studied in Bloznelis and Kurauskas (2012). The latter paper also gives an asymptotic expression of the clustering coefficient of the inhomogeneous graph. Clustering coefficients of an intersection digraph are calculated in Bloznelis (2010a). Clustering coefficients of an evolving random intersection graph are given in Bloznelis and Karoński (2013). The relation between the clustering coefficient and the assortativity coefficient (Pearson's correlation coefficient of degrees of the endpoints of a randomly chosen edge) is discussed in Bloznelis et al. (2013).

4 Cliques

Karoński et al. (1999) observed that in random intersection graphs cliques are "born" at much smaller edge densities than, for example, in the Erdős–Rényi random graph with independent edges. In particular, in $G_1(n, m, p)$ with $m = m(n) = \lfloor n^a \rfloor$ and constant $a > 0$, the birth threshold for K_h (a clique of size h, h a constant) is of the form

$$\tau = \tau(K_h, n) = \begin{cases} n^{-1} m^{-1/h} & \text{for } 0 < a < 2h/(h-1); \\ n^{-1/(h-1)} m^{-1/2} & \text{for } a \geq 2h/(h-1). \end{cases} \tag{5}$$

This means that with probability tending to 1 as $n \to \infty$, $G_1(n, m, p)$ contains (respectively, does not contain) a copy of K_h if $p/\tau \to \infty$ (respectively, $p/\tau \to 0$). Rybarczyk and Stark (2010) showed that on the threshold (for $p = c\tau$, where τ is given by (5)) the number of copies of K_h in $G(n, m, p)$ is approximately Poisson distributed with parameter λ, where $\lambda = c^h/h!$ if $a < 2h/(h-1)$, (i.e. almost all K_h arise from one attribute) and $\lambda = c^{h(h-1)}/h!$ if $a > 2h/(h-1)$ (i.e. almost all K_h arise from various attributes), whereas λ is the sum of the two, if $a = 2h/(h-1)$. The diversity of the answers is explained by the fact that a clique of an intersection graph may correspond to several different subgraphs of the underlying bipartite graph. For a similar reason the description of the birth threshold of a complete subgraph of an intersection digraph is a bit more complex, see Kurauskas (2013).

In random intersection graphs the clustering property affects not only the number of small cliques, but also the size of the largest clique (the clique number). The largest clique of $G_1(n, m, p)$, with $m = \lfloor n^a \rfloor$, $0 < a < 1$ and $mp^2 = O(1)$ was considered by Nikoletseas et al. (2012). Motivated by a seemingly different subject (the Erdős–Ko–Rado theorem), Balogh et al. (2009) considered the random k-uniform hypergraph where each of the $\binom{n}{k}$ hyperedges is included independently with the same probability. Their result yields the clique number in $G_1(n, m, \delta_d)$ for a wide range of parameters n, m and d.

Bloznelis and Kurauskas (2013) established an exact order of the clique number of sparse active graphs $G_1(n, m, P)$ and examined its relationship with the clustering property and tail-fatness of the degree distribution. The clique number is shown to diverge polynomially in n in the case of a power-law asymptotic degree distribution having infinite second moment. A similar result was obtained earlier by Janson et al. (2010a) for another random graph model without the clustering property. In the case of a square integrable asymptotic degree distribution the clique number of $G_1(n, m, P)$ may still diverge to infinity as $n, m \to \infty$, but at a rate not faster than $O\left(\frac{\ln n}{\ln \ln n}\right)$, see Bloznelis and Kurauskas (2013) for exact asymptotics. Here a comparison with the sparse random Erdős–Rényi graph having a clique number at most three or with the corresponding random graph of Janson et al. (2010a) having a clique number at most four with high probability (w.h.p.) shows that the clustering property indeed affects the clique number of $G_1(n, m, P)$.

Finally, Bloznelis and Kurauskas (2013) provide simple polynomial algorithms which find a clique of optimal order in $G_1(n, m, P)$ with high probability.

We note that Balogh et al. (2009), Nikoletseas et al. (2012), Bloznelis and Kurauskas (2013) discover the same phenomenon: For a wide range of parameters the largest clique in an intersection graph is formed by a single attribute.

5 Connectivity and Related Properties

A key issue considered in network analysis is information transmission. This includes routing protocols and questions related to them such as network connectivity, the size of the largest component, the diameter, and average distances between vertices. As an example we state the problem of how to pick the parameters of a secure wireless sensor network with random key predistribution to ensure that all (or at least 95 %) of sensors are able to transmit information fast among each other (see Eschenauer and Gligor 2002). This problem comes down to the following question: given n, how to pick $m(n)$ and $d(m)$ so that $G_1(n, m, \delta_d)$ is w.h.p. connected (has a giant component of size $0.95\,n$) and has diameter of size $O(\ln n)$ (see Blackburn and Gerke 2009; Bloznelis et al. 2009; Rybarczyk 2011a).

A nice property of the classical Erdős–Rényi random graph G, where edges are inserted independently and with the same probability denoted $\hat{p}(G)$, is that

1. G exhibits phase transition near $\hat{p}(G) = 1/n$, i.e. for $\hat{p}(G) = c/n$

   - if $c < 1$, then w.h.p. all connected components of G are of size $O(\ln n)$;
   - if $c > 1$, then w.h.p. in G there is exactly one connected component of size $\Theta(n)$ and all other connected components are of size $O(\ln n)$.

2. For $\hat{p}(G) = (\ln n + c_n)/n$

   - if $c_n \to -\infty$, then w.h.p. is disconnected;
   - if $c_n \to \infty$, then w.h.p. is connected.

An interesting question is whether and how these properties may extend to random intersection graphs. The answer for $G_1(n, m, \delta_d)$ was given in Blackburn and Gerke (2009), Rybarczyk (2011a), Yagan and Makowski (2012). For $G(n, m, p)$ the phase transition is shown in Behrisch (2007), Lagerås and Lindholm (2008). The connectivity threshold is shown in Rybarczyk (2011c, 2013), Singer (1995).

For the general $G_s(n, m, P)$ model we have knowledge about the phase transition and the size of the giant component of $G_s(n, m, p)$ and $G_s(n, m, \delta_d)$ for $s \geq 1$ (Bloznelis et al. 2009), the connectivity and perfect matching threshold of $G_s(n, m, \delta_d)$ for $s \geq 1$ (Bloznelis and Łuczak 2013), the phase transition and the size of the giant component of $G_1(n, m, P)$ (Bloznelis 2010b,c), the connectivity threshold and the phase transition of $G_1(n, m, P)$ and $G_1^*(n, m, P)$ (Godehardt et al. 2007). For inhomogeneous intersection graphs we refer to Rybarczyk (2013) and Bradonjic et al. (2010).

Distances between vertices of a random intersection graph (distance distribution, typical distances, diameter) have been considered in Barbour and Reinert (2011), Bloznelis (2009), Rybarczyk (2011a).

We note that generally random intersection graphs may tend to act similarly to an Erdős–Rényi random graph for m large comparing to n and differently otherwise. However what "m large" means depends on the studied property. This conjecture is supported by results concerning Hamilton cycles (see Nikoletseas et al. 2011; Bloznelis and Radavičius 2011; Rybarczyk 2013). Also greedy algorithms perform differently on random intersection graphs. For small m, unlike in Erdős–Rényi graphs, greedy algorithms construct an asymptotically optimal independent set in $G_1(n, m, p)$ and $G_1(n, m, \delta_d)$ (see Rybarczyk 2014), an almost optimal coloring in $G_1(n, m, p)$ (Behrisch et al. 2009) and $G_1(n, m, \delta_d)$ (Kurauskas and Rybarczyk 2013).

Finally let us mention two other areas where models and results for random intersection graphs were recently used. In Ball et al. (2014), Britton et al. (2008) epidemic processes were studied while Blackburn et al. (2012) applied random intersection graphs results in cryptology.

Acknowledgements The work of M. Bloznelis and V. Kurauskas was supported by the Lithuanian Research Council (grant MIP-067/2013). J. Jaworski and K. Rybarczyk acknowledge the support by the National Science Centre (NCN)—DEC-2011/01/B/ST1/03943. Co-operation between E. Godehardt and J. Jaworski was also supported by Deutsche Forschungsgemeinschaft (grant no. GO 490/17-1).

References

Ball, F., Sirl, D., & Trapman, P. (2014). Epidemics on random intersection graphs. *The Annals of Applied Probability, 24*, 1081–1128.

Balogh, J., Bohman, T., & Mubayi, D. (2009). Erdős–Ko–Rado in random hypergraphs. *Combinatorics, Probability and Computing, 18*, 629–646.

Barbour, A. D., & Reinert, G. (2011). The shortest distance in random multi-type intersection graphs. *Random Structures and Algorithms, 39*, 179–209.

Barrat, A., & Weigt, M. (2000). On the properties of small-world networks. *The European Physical Journal B, 13*, 547–560.

Behrisch, M. (2007). Component evolution in random intersection graphs. *The Electronic Journal of Combinatorics, 14*(1), R17

Behrisch, M., Taraz, A., & Ueckerdt, M. (2009). Colouring random intersection graphs and complex networks. *SIAM Journal on Discrete Mathematics, 23*, 288–299.

Blackburn, S., & Gerke, S. (2009). Connectivity of the uniform random intersection graph. *Discrete Mathematics, 309*, 5130–5140.

Blackburn, S., Stinson, D., & Upadhyay, J. (2012). On the complexity of the herding attack and some related attacks on hash functions. *Designs, Codes and Cryptography, 64*, 171–193.

Bloznelis, M. (2008). Degree distribution of a typical vertex in a general random intersection graph. *Lithuanian Mathematical Journal, 48*, 38–45.

Bloznelis, M. (2009). Loglog distances in a power law random intersection graphs. Preprint 09059, CRC701. http://www.math.uni-bielefeld.de/sfb701.

Bloznelis, M. (2010a). A random intersection digraph: Indegree and outdegree distributions. *Discrete Mathematics, 310*, 2560–2566.

Bloznelis, M. (2010b). Component evolution in general random intersection graphs. *SIAM Journal on Discrete Mathematics, 24*, 639–654.

Bloznelis, M. (2010c). The largest component in an inhomogeneous random intersection graph with clustering. *The Electronic Journal of Combinatorics, 17*(1), R110.

Bloznelis, M. (2013). Degree and clustering coefficient in sparse random intersection graphs. *The Annals of Applied Probability, 23*, 1254–1289.

Bloznelis, M., & Damarackas, J. (2013). Degree distribution of an inhomogeneous random intersection graph. *The Electronic Journal of Combinatorics, 20*(3), R3.

Bloznelis, M., Godehardt, E., Jaworski, J., Kurauskas, V., & Rybarczyk, K. (2015). Recent progress in complex network analysis—Models of random intersection graphs. In B. Lausen, S. Krolak-Schwerdt, & M. Boehmer (Eds.), *European Conference on Data Analysis.* Berlin/Heidelberg/New York: Springer (in this volume).

Bloznelis, M., Jaworski, J., & Kurauskas, V. (2013). Assortativity and clustering of sparse random intersection graphs. *Electronic Journal of Probability, 18*, N-38.

Bloznelis, M., Jaworski, J., & Rybarczyk, K. (2009). Component evolution in a secure wireless sensor network. *Networks, 53*(1), 19–26.

Bloznelis, M., & Karoński, M. (2013). Random intersection graph process. In A. Bonato, M. Mitzenmacher, & P. Pralat (Eds.), *WAW 2013. Lecture notes in computer science* (Vol. 8305, pp. 93–105). Switzerland: Springer International Publishing.

Bloznelis, M., & Kurauskas, V. (2012). Clustering function: A measure of social influence. http://arxiv.org/abs/1207.4941.

Bloznelis, M., & Kurauskas, V. (2013). Large cliques in sparse random intersection graphs. arXiv:1302.4627 [math.CO].

Bloznelis, M., & Łuczak, T. (2013). Perfect matchings in random intersection graphs. *Acta Mathematica Hungarica, 138*, 15–33.

Bloznelis, M., & Radavičius, I. (2011). A note on Hamiltonicity of uniform random intersection graphs. *Lithuanian Mathematical Journal, 51*(2), 155–161.

Bradonjic, M., Hagberg, A., Hengartner, N. W., & Percus, A. G. (2010). Component evolution in general random intersection graphs. In R. Kumar & D. Sivakumar (Eds.), *WAW 2010. Lecture notes in computer science* (Vol. 6516, pp. 36–49). Berlin/Heidelberg: Springer.

Britton, T., Deijfen, M., Lindholm, M., & Lagerås, N. A. (2008). Epidemics on random graphs with tunable clustering. *Journal of Applied Probability, 45*, 743–756.

Deijfen, M., & Kets, W. (2009). Random intersection graphs with tunable degree distribution and clustering. *Probability in the Engineering and Informational Sciences, 23*, 661–674.

Eschenauer, L., & Gligor, V. D. (2002). A key-management scheme for distributed sensor networks. In *Proceedings of the 9th ACM Conference on Computer and Communications Security* (pp. 41–47).

Foudalis, I., Jain, K., Papadimitriou, C., & Sideri, M. (2011). Modeling social networks through user background and behavior. In A. Frieze, P. Horn, & P. Pralat (Eds.), *WAW 2011. Lecture notes in computer science* (Vol. 6732, pp. 85–102). Berlin/Heidelberg: Springer.

Godehardt, E., Jaworski, J., & Rybarczyk, K. (2007). Random intersection graphs and classification. In R. Decker & H.-J. Lenz (Eds.), *Advances in data analysis* (pp. 67–74). Berlin/Heidelberg/New York: Springer.

Godehardt, E., Jaworski, J., & Rybarczyk, K. (2012). Clustering coefficients of random intersection graphs. In W. Gaul, A. Geyer-Schulz, L. Schmidt-Thieme, & J. Kunze (Eds.), *Challenges at the interface of data analysis, computer science, and optimization* (pp. 243–253). Berlin/Heidelberg/New York: Springer.

Jackson, O. M., & Rogers, B. W. (2007). Meeting strangers and friends of friends: How random are social networks? *American Economic Review, 97*, 890–915.

Janson, S., Łuczak, T., & Norros, I. (2010a). Large cliques in a power-law random graph. *Journal of Applied Probability, 47*, 1124–1135.

Janson, S., Łuczak, T., & Ruciński, A. (2010b). *Random graphs*. New York: Wiley.

Jaworski, J., Karoński, M., & Stark, D. (2006). The degree of a typical vertex in generalized random intersection graph models. *Discrete Mathematics, 306*, 2152–2165.

Jaworski, J., & Stark, D. (2008). The vertex degree distribution of passive random intersection graph models. *Combinatorics, Probability and Computing, 17*, 549–558.

Karoński, M., Scheinerman, E. R., & Singer-Cohen, K. B. (1999). On random intersection graphs: The subgraph problem. *Combinatorics, Probability and Computing, 8*, 131–159.

Kurauskas, V. (2013). On small subgraphs in a random intersection digraph. *Discrete Mathematics, 313*, 872–885.

Kurauskas, V., & Rybarczyk, K. (2013). On the chromatic index of random uniform hypergraphs *SIAM Journal on Discrete Mathematics*. To appear.

Lagerås, A. N., & Lindholm, M. (2008). A note on the component structure in random intersection graphs with tunable clustering. *Electronic Journal of Combinatorics, 15*(1), N10.

Newman, M. E. J. (2003). Properties of highly clustered networks. *Physical Review E, 68*, 026121.

Newman, M. E. J., Strogatz, S. H., & Watts, D. J. (2001). Random graphs with arbitrary degree distributions and their applications. *Physical Review E, 64*, 026118.

Nikoletseas, S., Raptopoulos, C., & Spirakis, P. G. (2011). On the independence number and hamiltonicity of uniform random intersection graphs. *Theoretical Computer Science, 412*, 6750–6760.

Nikoletseas, S., Raptopoulos, C., & Spirakis, P. (2012). Maximum cliques in graphs with small intersection number and random intersection graphs. In *Mathematical foundations of computer science* (pp. 728–739). Berlin/Heidelberg: Springer.

Ravasz, E., & Barabási, A. L. (2003). Hierarchical organization in complex networks. *Physical Review E, 67*, 026112.

Rybarczyk, K. (2011a). Diameter, connectivity, and phase transition of the uniform random intersection graph. *Discrete Mathematics, 311*, 1998–2019.

Rybarczyk, K. (2011c). Sharp threshold functions for random intersection graphs via a coupling method. *The Electronic Journal of Combinatorics, 18*(1), P36.

Rybarczyk, K. (2012). The degree distribution in random intersection graphs. In W. Gaul, A. Geyer-Schulz, L. Schmidt-Thieme, & J. Kunze (Eds.), *Challenges at the interface of data analysis, computer science, and optimization* (pp. 291–299). Berlin/Heidelberg/New York: Springer.

Rybarczyk, K. (2013). The coupling method for inhomogeneous random intersection graphs. arXiv:1301.0466.

Rybarczyk, K. (2014). Constructions of independent sets in random intersection graphs. *Theoretical Computer Science, 524*, 103–125.

Rybarczyk, K., & Stark, D. (2010). Poisson approximation of the number of cliques in random intersection graphs. *Journal of Applied Probability, 47*, 826–840.

Singer, K. (1995). *Random intersection graphs*. Ph.D. thesis, The Johns Hopkins University.

Stark, D. (2004). The vertex degree distribution of random intersection graphs. *Random Structures and Algorithms, 24*, 249–258.

Strogatz, S. H., & Watts, D. J. (1998). Collective dynamics of small-world networks. *Nature, 393*, 440–442.

Yagan, O., & Makowski, A. M. (2012). Zero-one laws for connectivity in random key graphs, IEEE Transactions on Information Theory, 58, 2983–2999

Similarity Measures of Concept Lattices

Florent Domenach

Abstract Concept lattices fulfil one of the aims of classification by providing a description by attributes of each class of objects. We introduce here two new similarity/dissimilarity measures: a similarity measure between concepts (elements) of a lattice and a dissimilarity measure between concept lattices defined on the same set of objects and attributes. Both measures are based on the overhanging relation previously introduced by the author, which are a cryptomorphism of lattices.

1 Introduction

Concept lattices, which constitute the basic frame of Formal Concept Analysis (FCA), provide classes (the extents) of objects sharing similar characters (the intents), a description by attributes is associated with each class. In practice however the number of concepts generated is too big, and the lattices are too large to be of direct use. The main contribution of this paper is two new measures in and between lattices: a similarity measure between concepts, and a dissimilarity measure between lattices. Each of these measure represents a normalization compared to existing measures by taking into account the width of the lattice. To our knowledge few similarity indices exist between concepts, most of them stemming from ontology design, and a dissimilarity measure between concepts lattices is new.

This paper is organized as follows: we recall in Sect. 2.1 the fundamentals of FCA, followed in Sect. 2.2 by the definition and some properties of the overhanging relation that will be the core of the new similarity measures. Section 3.1 presents existing similarities between concepts before introducing a new similarity measure in Sect. 3.2. A new dissimilarity measure between lattices is given in Sect. 4.1, and a preliminary study using this measure is introduced in Sect. 4.2.

F. Domenach (✉)
Computer Science Department, University of Nicosia, 46 Makedonitissas Av., 1700 Nicosia, Cyprus
e-mail: domenach.f@unic.ac.cy

© Springer-Verlag Berlin Heidelberg 2015
B. Lausen et al. (eds.), *Data Science, Learning by Latent Structures, and Knowledge Discovery*, Studies in Classification, Data Analysis, and Knowledge Organization, DOI 10.1007/978-3-662-44983-7_8

2 Formal Concept Analysis

2.1 Introduction to FCA

FCA takes its roots in lattice theory (Birkhoff 1967; Barbut and Monjardet 1970) and was subsequently developed in Darmstadt as a mathematical theory for the modelling of the notion of "concept" (Ganter and Wille 1996), i.e. a pair of intent and extent of a set. Following classical definition of set theory, the intent is the set of all common properties shared by the elements of the set, and the extent is the set of all elements having all the properties of the set.

More precisely, a *formal context* (G, M, I) where G is a set of objects, M is a set of attributes (properties) and I is a cross table $I \subseteq G \times M$. We read $(g, m) \in I$ as "object g has attribute m". Table 1 shows an example of a cross table of numbers less than 11 together with the properties of being composite, prime, square, even or odd. For example, we have $(1, odd) \in I$ as 1 has the property of being odd.

The notion of concepts associated with the formal context (G, M, I) arises through a Galois connection between $\mathcal{P}(G)$ and $\mathcal{P}(M)$, the power sets of G and M, which associates with any set of objects its intent, and conversely, with any set of attributes its extent. Formally, it is defined as follows:

$$\forall A \subseteq G, A' = \{m \in M : \forall g \in A, (g, m) \in I\}$$

$$\forall B \subseteq M, B' = \{g \in G : \forall m \in B, (g, m) \in I\}$$

The compositions of these two mappings form dual closure operators on $\mathcal{P}(G)$ and $\mathcal{P}(M)$, respectively. Continuing with our example, the intent of $\{3, 5\}$ is $\{Odd, Prime\}$, and the extent of $\{Odd, Prime\}$ is $\{3, 5, 7\}$.

A formal context, together with its associated Galois connection, leads to the notion of concept, i.e. of pair of maximal sets of objects and attributes. Formally, a pair $(A, B), A \subseteq G, B \subseteq M$, is a *concept* if $A' = B$ et $B' = A$, or,

Table 1 Example of a binary table on numbers less than 11 with their properties

	Composite	Even	Odd	Prime	Square
1			X		X
2		X		X	
3			X	X	
4	X	X			X
5			X	X	
6	X	X			
7			X	X	
8	X	X			
9	X		X		X
10	X	X			

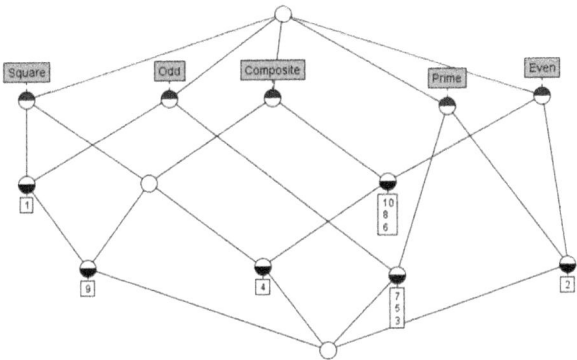

Fig. 1 Galois lattice associated with the formal context of Table 1.

equivalently, (A'', A') is a concept. Continuing with our example of Table 1, $(\{3, 5, 7\}, \{Odd, Prime\})$ and $(\{1, 9\}, \{Odd, Square\})$ are concepts.

The sets of all concepts of a formal context, ordered by inclusion on their extents, form a complete lattice (Wille 1982; Davey and Priestley 2002; Caspard et al. 2012), which generates and visualizes hierarchies of concepts. The *Galois lattice (concept lattice)* associated with a formal context is the set of all concepts (A, B), with the partial order

$$(A_1, B_1) \le (A_2, B_2) \iff A_1 \subseteq A_2 \; (\iff B_1 \supseteq B_2)$$

Figure 1 shows the Galois lattice associated with the context of Table 1 as a Hasse diagram with minimal labelling: every vertex is a concept which inherits attributes that are above it and objects that are under it, and the edges represent the cover relation.

2.2 Overhanging Relation

An interesting aspect of Galois lattices is the many existing equivalent cryptomorphisms[1] (Caspard and Monjardet 2003). One of them, called overhanging relation (Domenach and Leclerc 2004), is defined on the power set of the set of objects as follows:

---

[1] Birkhoff (1967) defines two objects, especially systems of axioms or semantics for them, as *cryptomorphic* if they are equivalent but not obviously equivalent.

Definition 1 A binary relation $\mathcal{O}$ on $\mathcal{P}(G)$ is an *overhanging relation* if for every $A, B, C \subseteq G$ it satisfies the following conditions:

- $(A, B) \in \mathcal{O}$ implies $A \subseteq B$
- $A \subset B \subset C$ imply $(A, C) \in \mathcal{O} \iff ((A, B) \in \mathcal{O}$ or $(B, C) \in \mathcal{O})$
- $(A, A \cup B) \in \mathcal{O}$ implies $(A \cap B, B) \in \mathcal{O}$

An alternative definition that links overhanging relation and concept lattice is the following: the overhanging relation associated with the concept lattice is a binary relation $\mathcal{O}$ on $\mathcal{P}(G)$ such that:

$$(A, B) \in \mathcal{O} \iff A \subset B \text{ and } A'' \subset B''$$

In other words, two sets are overhanged if one is a subset of the other and they have a different closure. Continuing with the example of the context of Table 1, we have $(\{4, 6\}, \{2, 4, 6\}) \in \mathcal{O}$.

The overhanging relation associated with a formal context allows us to define a ternary index expressing the similarity between two objects r and s relatively to other objects of G:

Definition 2 Consider a formal context (G, M, I), and the associated Galois lattice $\mathbb{L}$. We define the similarity $\varepsilon_{\mathbb{L}}$ by, for all $r, s \in G$:

$$\varepsilon_{\mathbb{L}}(r, s) = |\{k \in G : (\{r, s\}, \{r, s, k\}) \in \mathcal{O}\}|$$

We showed in Domenach and Leclerc (2004) that when the lattice $\mathbb{L}$ is restricted to a hierarchy (a poset containing all singletons and for which the intersection of two concepts is either empty or one of the concepts), then $\varepsilon(rs)$ is maximal when r and s belong to the same concept.

Lemma 1 *Let $\mathbb{H}$ be a hierarchy on S, r and s two distinct elements of S such that $\varepsilon_{\mathbb{H}}(rs)$ is maximal. Then r and s belong to a non-trivial class of $\mathbb{H}$ that is minimal for inclusion. If $\mathbb{H}$ is a binary hierarchy, then $\{r, s\} \in \mathbb{H}$.*

3 Similarity Measure Between Concepts

3.1 Existing Similarity Measures

There exist many different similarity measures that can be applied on concepts. They can be roughly divided into two main categories: one, based on Tversky (1977) model, only considers concepts as sets of objects. The other is rooted in ontology theory and uses the fact that concepts are ordered in the lattice to evaluate the similarity between two concepts.

Set based similarity measures can be expressed using Tversky similarity model. Given 2 concepts $C_1 = (A_1, B_1)$ and $C_2 = (A_2, B_2)$, we define *Tversky similarity* (Tversky 1977) on the extents of C_1 and C_2 as follows:

$$S_{\alpha\beta}(C_1, C_2) = \frac{|A_1 \cap A_2|}{|A_1 \cap A_2| + \alpha|A_1 - A_2| + \beta|A_2 - A_1|} \quad \alpha, \beta \geq 0$$

Depending on the values of α and β, the Tversky index can be seen as a generalization of Jaccard index and Dice's coefficient.

When taking $\alpha = 1$ and $\beta = 1$, we have the Jaccard similarity:

$$S_{\text{Jaccard}}(C_1, C_2) = \frac{|A_1 \cap A_2|}{|A_1 \cup A_2|}$$

Taking $\alpha = \beta = 1/2$ in the Tversky similarity model, we obtain the Dice's coefficient:

$$S_{\text{dice}}(C_1, C_2) = \frac{2|A_1 \cap A_2|}{|A_1| + |A_2|}$$

With $\alpha = 0, \beta = 1$, we have the inclusion measure:

$$S_{\text{inclusion}}(C_1, C_2) = \frac{|A_1 \cap A_2|}{|A_1|}$$

The following measures of similarities use the order structure of the lattice to calculate the similarity between two concepts C_1 and C_2. They are adapted from ontology-based similarities that use only taxonomic links, by considering the lattice $\mathbb{L}$ as a generalization of a tree. In order to define those similarities, we need the notion of length between C_1 and C_2, denoted by $\text{length}(C_1, C_2)$, which is the topological distance in the covering graph of the lattice, and the depth of concept C_1, i.e. the distance between C_1 and top concept of $\mathbb{L}$.

The first one, due to Rada et al. (1989), is the inverse of the topological distance between C_1 and C_2:

$$S_{\text{rada}}(C_1, C_2) = \frac{1}{|\text{length}(C_1, C_2) + 1|}$$

In Leacock and Chodorow (1998) similarity, the shortest path between two concepts is normalized by dividing by twice the maximum depth of the lattice:

$$S_{lc}(C_1, C_2) = -\log\left(\frac{\text{length}(C_1, C_2)}{2 \times \max_{x \in L}(\text{depth}(x))}\right)$$

The Wu and Palmer (1994) measure is based on the idea that the deeper the concepts are, the more relevant they become:

$$S_{\text{wup}}(C_1, C_2) = \frac{2 * \text{depth}(C_1 \vee C_2)}{\text{depth}(C_1) + \text{depth}(C_2)}$$

3.2 Overhanging-Based Similarity

Let C be a concept of $\mathbb{L}$. Similar to Definition 2, let $o(C)$ be the set of objects that is overhanged with C, that is:

$$o(C) = \{k \in S : (C, C \cup \{k\}) \in \mathcal{O}\}$$

In other words, $o(C)$ is the set of objects that, when added to C, have a closure different than the closure of C. For example, from our recurring example, we have $o(\{4, 8\}) = \{1, 2, 3, 5, 7, 9\}$.

From this mapping we can define a similarity measure, inspired from that of Wu and Palmer, as follows: for any two concepts C_1 and C_2 of $\mathbb{L}$,

$$S_o(C_1, C_2) = \frac{|o(C_1 \vee C_2)|}{|o(C_1) \cup o(C_2)|}$$

The idea behind this measure is to take into account the width of the lattice. Two concepts will be more similar if they are close in the lattice and are not sharing objects with other concepts. In other words, it indicates the similarity of two concepts in relation with all the other concepts.

We have the following properties for S_o, $\forall C_1, C_2 \in \mathbb{L}$:

- S_o is normalized, i.e. it takes values in $[0, 1]$;
- $S_o(C_1, C_2) = 0$ if and only if $C_1 \vee C_2 = 1_{\mathbb{L}}$;
- $S_o(C_1, C_2) = 1$ if and only if $C_1 = C_2$;
- if $C_1 \leq C_2$, then $S_o(C_1, C_2) = \frac{|o(C_2)|}{|o(C_1)|}$ as $C_1 \vee C_2 = C_2$ and o is an antitone mapping.

To illustrate, consider the lattice of Fig. 2 defined on $G = \{a, b, c, d, e, f\}$. Let $C_1 = \{a\}$, $C_2 = \{b\}$, $C_3 = \{c\}$ and $C_4 = \{d\}$. In this example, we have all the Tversky similarities between C_1 and C_2 on the one hand, and C_3 and C_4 on the other hand, equal to zero as the concepts don't share elements. We also have $S_{\text{rada}}(C_1, C_2) = S_{\text{rada}}(C_3, C_4) = \frac{1}{3}$, $S_{lc}(C_1, C_2) = S_{lc}(C_3, C_4) = \log(3)$ and $S_{\text{wup}}(C_1, C_2) = S_{\text{wup}}(C_3, C_4) = \frac{1}{4}$. None of those existing similarity measures allow us to discriminate between (C_1, C_2) and (C_3, C_4). However, we have $S_o(C_1, C_2) = \frac{4}{6}$ and $S_o(C_3, C_4) = \frac{2}{6}$. So S_o capture the fact that C_1 and C_2 "represent" all of $C_1 \vee C_2$, while C_3 and C_4 don't capture all the information of $C_3 \vee C_4$, i.e. $\{e\}$ and $\{f\}$.

Fig. 2 Example of lattice on
$G = \{a, b, c, d, e, f\}$

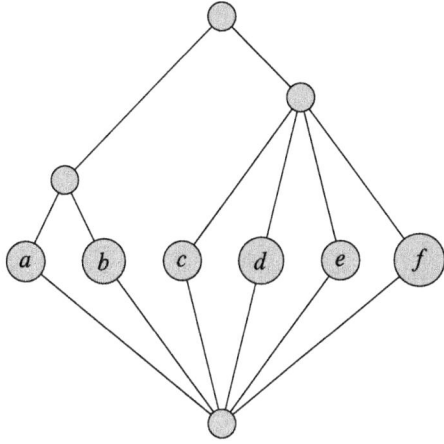

4 Similarity Measure Between Lattices

4.1 A Dissimilarity Measure

Suppose that we have two Galois lattices $\mathbb{L}$ ($\mathbb{L} = \{\mathbb{L}_1, \mathbb{L}_2\}$) on the same set of objects, n. Using an approach similar to the previous similarity measure between concepts, we can define a similarity measure between Galois lattices $\mathbb{L}_1$ and $\mathbb{L}_2$ defined, respectively, on the formal contexts (G, M, I_1) and (G, M, I_2).

Let the two lattices $\mathbb{L} = \{\mathbb{L}_1, \mathbb{L}_2\}$, and consider the $N = \frac{n(n-1)}{2}$ pairs of different objects $\{r, s\}, r, s \in G$. We can define a binary variable $m_{ik\mathbb{L}}$ with

- $m_{ik\mathbb{L}} = 1$ if objects in $i = \{r, s\}$ are overhanged in $i \cup \{k\}$ in lattice $\mathbb{L}$
- $m_{ik\mathbb{L}} = 0$ otherwise

A global matrix $\mathbf{M}_{\mathbb{L}}$ can be derived for each lattice $\mathbb{L}$:

- columns are binary vectors $\mathbf{m}_{k\mathbb{L}} = [m_{1k\mathbb{L}}, \ldots, m_{Nk\mathbb{L}}]^T$
- rows are $\mathbf{m}_{i\mathbb{L}} = [m_{i1\mathbb{L}}, \ldots, m_{in\mathbb{L}}]$

A global measure of dissimilarity that considers all overhanged pairs simultaneously may be defined as follows:

$$D = \frac{\|\mathbf{M}_1 - \mathbf{M}_2\|}{\|\mathbf{M}_1\| + \|\mathbf{M}_2\|} \tag{1}$$

with $\|\mathbf{M}\| = \Sigma_i \Sigma_k |m_{ik}|$, with $\|\mathbf{M}\|$ is the L_1 norm of matrix $\mathbf{M}$. In expression (1), since the matrices involved take only binary values, the L_1 norm is the square of the L_2 norm.

Table 2 Table 1 on the left, and same table with the relation ({2}, {composite}) added on the right

	Comp.	Even	Odd	Prime	Square
1			X		X
2		X		X	
3			X	X	
4	X	X			X
5			X	X	
6	X	X			
7			X	X	
8	X	X			
9	X			X	X
10	X	X			

	Comp.	Even	Odd	Prime	Square
1			X		X
2	X	X		X	
3			X	X	
4	X	X			X
5			X	X	
6	X	X			
7			X	X	
8	X	X			
9	X			X	X
10	X	X			

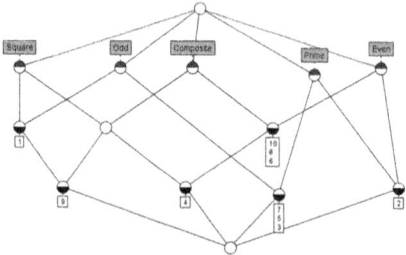 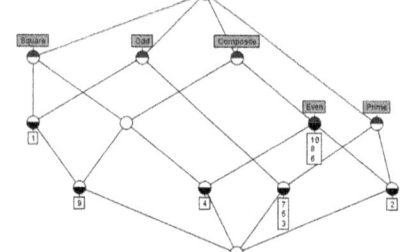

Fig. 3 Corresponding lattices of Table 2

The measure D has the following properties:

- It is bounded in $[0, 1]$
- $D = 0$ if and only if the two lattices are identical
- $D = 1$ if and only if no concept of $\mathbb{L}_1$ intersect a concept of $\mathbb{L}_2$, i.e. $\forall C_1 \in \mathbb{L}_1, C_2 \in \mathbb{L}_2, C_1 \cap C_2 = \emptyset$.
- D is a distance, since it satisfies the conditions of non-negativity, identity, symmetry and triangular inequality.
- The complement to 1 of D is a similarity measure since it satisfies the conditions of non-negativity, normalization and symmetry.
- It does not depend on object labels since it refers to pairs of objects only.

Consider the example of Table 2 in order to illustrate this dissimilarity measure. On the left is the original Table 1, and on the left is the same table with the ({2}, {composite}) entry added. Figure 3 shows the corresponding lattices. As we can see, there is only a small change in the lattice: the concepts ({2, 4, 6, 8, 10}, {even}) and ({4, 6, 8, 10}, {even, composite}) are merged in the new lattice. The dissimilarity value between these two lattices is $D = 0.022452504$.

Fig. 4 Boxplots of S for
5,000 pairs of unrelated
matrices

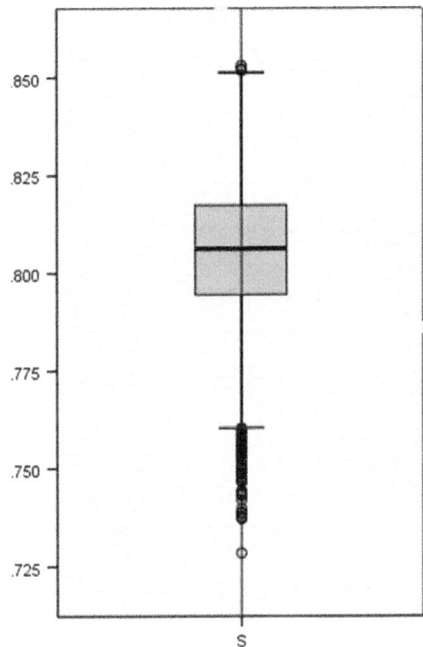

We created some simulation experiments in order to do some random testing. All simulations have been implemented in the programming language C♯. We generated a random data set according to the following steps:

- We generated independently pairs of $n \times m$ Boolean matrices;
- The sample size is $n = 50$ and $m = 20$;
- We repeat previous steps 5,000 times and each time we compute the associated overhanging relation on all possible pairs of objects in order to create the matrices $\mathbf{M}_{\mathbb{L}}$;
- We compute the dissimilarity measure D each time.

Figure 4 shows the boxplot of the values of D. The range of D is $[0, 1]$ and thus the median value cannot be one. However the average value is approximately 0.81 and it indicates an absence of similarity between the lattices.

4.2 Example on Climate Data

We investigated in a preliminary study the interest of the dissimilarity value presented in the previous section. We used climate data obtained from the National Climatic Data Center (NCDC) of the National Oceanic and Atmospheric Administration in the USA. The NCDC makes available 1981–2010 climate

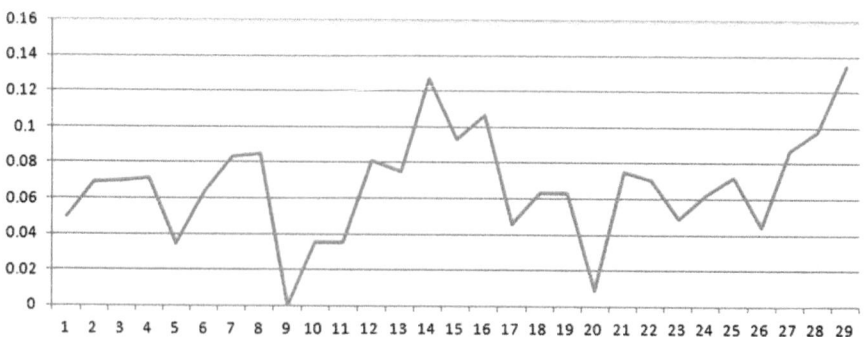

Fig. 5 Dissimilarity measure indicating the changes in weather between years

normals containing averages of different climatological variables, including temperature and precipitation over 9,800 stations across the USA, with monthly, daily, annual/seasonal, or hourly normals. It is available for free at www.ncdc.noaa. gov.

We pre-processed the data as follows:

- Amongst the 9,800 stations available, we chose 10 weather stations randomly
- Then we calculated the monthly average of each of these station's temperature for each month, over a period of 30 years (1981–2010)

Then we created for each year, from 1981 to 2010, a binary matrix where the objects are the stations and the attributes are the months. In that matrix, we put true for the entry (station$_i$, month$_i$) if the temperature at station$_i$ at month$_i$ of this year is greater than the previously calculated average temperature for this month and this station. Figure 5 shows the dissimilarity values between matrices from one year to the next. It can be seen as an indicator of variability of the climate on a specific period.

5   Conclusion

In this paper we introduced two new measures to compare concept lattices: one between concepts and one between lattices. They summarize the closeness of two structures taking into account their size, thus can be seen as a normalization. Our preliminary study on climate data looks promising—more experiments remain to be done, comparing them to established methods, as well as a thorough study of the various pattern exhibited thanks to the dissimilarity measure. We are also planning to investigate whether there is an explicit relationship between those two measures and what is their computational complexity.

References

Barbut, M., & Monjardet, B. (1970). *Ordres et classification: Algèbre et combinatoire (tome II)*. Paris: Hachette.

Birkhoff, G. (1967). *Lattice theory*, (3rd ed.). Providence: American Mathematical Society.

Caspard, N., Leclerc, B., & Monjardet, B. (2012). *Finite ordered sets: Concepts, results and uses*. Cambridge: Cambridge University Press.

Caspard, N., & Monjardet, B. (2003). The lattices of Moore families and closure operators on a finite set: A survey. *Discrete Applied Mathematics, 127*, 241–269.

Davey, B. A., & Priestley, H. A. (2002). *Introduction to lattices and order* (2nd ed.). Cambridge: Cambridge University Press.

Domenach, F., & Leclerc, B. (2004). Closure systems, implicational systems, overhanging relations and the case of hierarchical classification. *Mathematical Social Sciences, 47(3)*, 349–366.

Ganter, B., & Wille, R. (1996). *Formal concept analysis : Mathematical foundations*. New York: Springer.

Leacock, C., & Chodorow, M. (1998). Combining local context and wordnet similarity for word sense identification. In C. Fellbaum (Ed.), *WordNet: An electronic lexical database* (pp. 265–283). Cambridge: MIT.

Rada, R., Mili, H., Bicknell, E., & Blettner, M. (1989). Development and application of a metric on semantic nets. *IEEE Transactions SMC, 19*, 17–30.

Tversky, A. (1977). Features of similarity. *Psychological Review, 84(4)*, 327–352.

Wille, R. (1982). Restructuring lattice theory: An approach based on hierarchies of concepts. In I. Rival (Ed.), *Ordered sets* (pp. 314–339).

Wu, Z., & Palmer, M. (1994). Verb semantics and lexical selection. In *Proceedings of the 32nd Annual Meeting of the Associations for Computational Linguistics* (pp. 33–138).

Flow-Based Dissimilarities: Shortest Path, Commute Time, Max-Flow and Free Energy

Guillaume Guex and François Bavaud

Abstract Random-walk based dissimilarities on weighted networks have demonstrated their efficiency in clustering algorithms. This contribution considers a few alternative network dissimilarities, among which a new max-flow dissimilarity, and more general *flow-based dissimilarities*, freely mixing shortest paths and random walks in function of a free parameter—the temperature. Their geometrical properties and in particular their squared Euclidean nature are investigated through their power indices and multidimensional scaling properties. In particular, formal and numerical studies demonstrate the existence of critical temperatures, where flow-based dissimilarities cease to be squared Euclidean. The clustering potential of medium range temperatures is emphasized.

1 Introduction

The last decade has witnessed an increasing interest in centrality indices, community detection algorithms, or network clustering assisted by random walks. Most approaches imply network dissimilarities, among which the shortest path and the commute time, closely linked to the minimization of path functionals, namely a resistance or energy functional, respectively a relative entropy functional. The maximum flow and p-resistances (Alamgir and Von Luxburg 2011) constitute alternative path functionals. Optimal flows minimizing mixtures of path functionals characterize the global properties of the network, beyond the limited local view provided by binary or weighted adjacency matrices. Optimal flows also generate network dissimilarities, such as the presumably original maximum-flow dissimilarity (Sect. 2.2).

In particular, the free energy path functional (Saerens et al. 2009) generates optimal flows interpolating between shortest paths and random walks (Sect. 2.4), where the edge resistances and transition matrix can be fixed independently (Bavaud and Guex 2012). After reviewing the main definitions involved in the taxonomy of

G. Guex • F. Bavaud (✉)
University of Lausanne, 1015 Lausanne, Switzerland
e-mail: guillaume.guex@unil.ch; francois.bavaud@unil.ch

© Springer-Verlag Berlin Heidelberg 2015
B. Lausen et al. (eds.), *Data Science, Learning by Latent Structures,
and Knowledge Discovery*, Studies in Classification, Data Analysis,
and Knowledge Organization, DOI 10.1007/978-3-662-44983-7_9

dissimilarities (Sect. 2.1), the geometric properties of the energy and free energy path functional dissimilarities (Sect. 2.5) are investigated. In particular, the question of their squared Euclidean character is studied through the well-known Torgeson criterium, as well as through the less-known power index criterium of Joly and Le Calvé (1986). Numerical examples and applications demonstrate the existence of phase transitions, where path functional dissimilarities become non-Euclidean below some critical temperatures (Sect. 3.1). Section 2.5 addresses the issue of multidimensional scaling network reconstruction by path functional dissimilarities. Their better efficiency in clustering and classification of categorical data, as compared to chi-square dissimilarities, is illustrated in Sect. 3.3 for intermediate temperature ranges.

2 Dissimilarities

2.1 A Few Definitions and Properties

Let us recall a few standard definitions (e.g., Joly and Le Calvé 1986; Critchley and Fichet 1994): a dissimilarity on a set S of n objects is an $n \times n$ symmetric non-negative matrix $D = (d_{ij})$ with a null diagonal. The dissimilarity is *separable* if $d_{ij} = 0$ iff $i = j$, and *metric* if $d_{ij} \leq d_{ik} + d_{kj}$ (for all triples of S). A *distance* is a metric dissimilarity. One further distinguishes between

- **ultrametric** distances ($D \in \mathcal{D}_U$) for which $d_{ij} \leq \max(d_{ik}, d_{jk})$.
- **Minkowski**(q) distances ($D \in \mathcal{D}_q$, where $q \geq 1$) if one can find n vectors $x_{ik} \in \mathbb{R}^p$ such that $d_{ij} = \left(\sum_{l=1}^{p} |x_{il} - x_{jl}|^q\right)^{\frac{1}{q}}$. $\mathcal{D}_2$ corresponds to the **Euclidean distance**.
- **squared Euclidean** dissimilarities ($D \in \mathcal{D}_2^2$) if there exists an embedding of the form $d_{ij} = \sum_{l=1}^{p} (x_{il} - x_{jl})^2$.
- **Chebychev** or **Frechet** distances ($D \in \mathcal{D}_\infty$) if there exists an embedding of the form $d_{ij} = \max_{l=1}^{p} |x_{il} - x_{jl}|$.

$\mathcal{D}_\infty$ is the set of all distances, and $\mathcal{D}_U \subset \mathcal{D}_2 \subset \mathcal{D}_1 \subset (\mathcal{D}_\infty \cap \mathcal{D}_2^2)$ holds.

Our study of the squared Euclidean character of the graph dissimilarities (Sects. 2.5 and 3.2) mainly relies upon the following results, the first often attributed to Torgeson (1958) (with many precursors e.g. mentioned by Lew 1978), and the second due to Joly and Le Calvé (1986). Here $f_i > 0$ is the relative weight of object i, normalized to unity, and δ_{ij} is Kronecker's delta:

Proposition 1 *A dissimilarity D on finite set S is $\mathcal{D}_2^2$ iff the matrix of scalar products $B := -\frac{1}{2}HDH'$ is positive semi-definite. Here, $H = (h_{ij})$ is the centering matrix where $h_{ij} = \delta_{ij} - f_j$, for any fixed normalized distribution f.*

Proposition 2 *For any dissimilarity D, there is a number $a \geq 0$ such that the elementwise power D^a is a squared Euclidean dissimilarity. Also, D^b is $\mathcal{D}_2^2$ as well for $0 \leq b \leq a$.*

*Define $\mathrm{pow}(D)$, the **power** of D, as the maximum value of a making D^a squared Euclidean. Then $\mathrm{pow}(D) \geq 1$ iff $D \in \mathcal{D}_2^2$, $\mathrm{pow}(D) \geq 2$ iff $d \in \mathcal{D}_2$ and $\mathrm{pow}(D) = \infty$ iff $d \in \mathcal{D}_U$.*

2.2 Commute-Time, Max-Flow Graph, and Chi-Square Distances

A binary graph $G = (V, E)$ on $|V| = n$ nodes is specified by an $n \times n$ symmetric adjacency matrix $A = (a_{ij})$ taking on values 0 or 1. The associated random walk is defined by the Markov transition matrix $W = (w_{ij})$ with $w_{ij} = a_{ij}/a_{i\bullet}$ (here "$\bullet$" denotes the summation over the replaced index).

Conversely, any regular Markov chain $W = (w_{ij})$ with stationary distribution f defines a *weighted graph* with associated node weights f_i, and *edge weights* or *exchange matrix* (Berger and Snell 1957) $e_{ij} := f_i w_{ij}$, giving the probability to select the pair of nodes ij. For unoriented graphs, $e_{ij} = e_{ji}$, that is W is *reversible*. By construction, $e_{i\bullet} = e_{\bullet i} = f_i$ and $e_{\bullet\bullet} = 1$.

Let $\mathcal{X}_{st}$ denote the set of *unit st-flows* from the source node $s \in V$ to the target node $t \in V$, as specified by the edge transitions counts $X = (x_{ij})$ of the trajectories or paths. By construction

$$x_{ij} \geq 0, \qquad x_{i\bullet} - x_{\bullet i} = \delta_{is} - \delta_{it} \qquad \text{and} \qquad x_{t\bullet} = 0 \ . \tag{1}$$

The *commute-time distance* d_{st}^{ct} associated with W is the average time to go from s to t and back to i. That is $d_{st}^{\mathrm{ct}} = x_{\bullet\bullet}^{st} + x_{\bullet\bullet}^{ts}$, where x_{ij}^{st} is the random walk flow, obeying (1) and $x_{ij}^{st} = x_{i\bullet}^{st} w_{ij}$. One knows that $D^{\mathrm{ct}} \in (\mathcal{D}_\infty \cap \mathcal{D}_2^2)$, even in the oriented case (e.g., Boley et al. 2011). Also, D^{ct} is *graph-geodetic* (Klein and Zhu 1998; Chebotarev 2010), that is obeys $d_{ik}^{\mathrm{ct}} + d_{kj}^{\mathrm{ct}} = d_{ij}^{\mathrm{ct}}$ whenever all ij-paths and ji-paths moving over edges with non-zero weights pass through k.

Let us introduce a presumably new distance on unoriented weighted graphs, the ultrametric *max-flow distance* D^{mf}. In this setup, e_{ij} represents the *edge capacity*, controlling for the flow of maximum value v_{st} between s and t, solution of the problem

$$v_{st} := \max v \quad \text{suchthat} \quad 0 \leq x_{ij} \leq e_{ij}, \quad x_{i\bullet} - x_{\bullet i} = v(\delta_{is} - \delta_{it}), \quad x_{t\bullet} = 0.$$

By construction, $v_{ij} \geq 0$, $v_{ij} = v_{ji}$, $v_{ii} = \infty$ and $v_{ij} \geq \min(v_{ik}, v_{kj})$ for all triples in V^3. For $e_{ij} = e_{ji}$, define the max-flow distance as $d_{ij}^{\mathrm{mf}} := 1/v_{ij}$. Then d_{ij}^{mf} is a dissimilarity obeying $d_{ij}^{\mathrm{mf}} \leq \max(d_{ik}^{\mathrm{mf}}, d_{jk}^{\mathrm{mf}})$, that is $D^{\mathrm{mf}} \in \mathcal{D}_U$.

Categorial data analysis can also be cast in the above setup: let $N = (n_{il})$ be an $n \times m$ contingency table. Define a pair selection scheme by first choosing a row i, then a category l present in i, then another row j containing l. The resulting edge weight, node weight, and transition matrix read (e.g., Bavaud and Xanthos 2005)

$$e_{ij} = \sum_{l=1}^{m} \frac{n_{il} n_{jl}}{n_{\bullet\bullet} n_{\bullet l}} \qquad f_i = \frac{n_{i\bullet}}{n_{\bullet\bullet}} \qquad w_{ij} = \sum_{l=1}^{m} \frac{n_{il} n_{jl}}{n_{i\bullet} n_{\bullet l}} \tag{2}$$

On the other hand, the chi-square dissimilarity $D^\chi = (d_{ij}^\chi)$ between rows reads

$$d_{ij}^\chi := n_{\bullet\bullet} \sum_l \frac{1}{n_{\bullet l}} \left(\frac{n_{il}}{n_{i\bullet}} - \frac{n_{jl}}{n_{j\bullet}} \right)^2 \tag{3}$$

$D^\chi \in \mathcal{D}_2^2$, but $D^\chi \notin \mathcal{D}_\infty$. Neither D^χ nor D^{mf} are graph-geodetic.

2.3   Shortest-Path Distances

Let $r_{ij} \geq 0$ denote the length, cost, travel time or *resistance* of edge ij. The *shortest-path* length from s to t is

$$d_{st}^{\mathrm{sp}} := \min_{X \in \mathcal{X}_{st}} U(X) \qquad \text{where} \quad U(X) := \sum_{ij} r_{ij} x_{ij}$$

Then $d_{ii}^{\mathrm{sp}} = 0$, $d_{ik}^{\mathrm{sp}} + d_{jk}^{\mathrm{sp}} \geq d_{ij}^{\mathrm{sp}}$ and $d_{ij}^{\mathrm{sp}} = d_{ji}^{\mathrm{sp}}$ for symmetric $R = (R_{ij})$. In general, $D^{\mathrm{sp}} \notin \mathcal{D}_2^2$ (e.g., Deza and Laurent 1997 or Bavaud 2010).

An important body of literature considers the *plain setup* $r_{ij} = c/a_{ij}$, where $c > 0$ is a normalization constant, as, e.g., in Yen et al. (2008) and the references therein, or as in the celebrated Doyle and Snell monograph (1984); see also the illustrations of Sect. 3. For $c = 1$, the seemingly counterintuitive inequality $d_{ij}^{\mathrm{ct}} \leq d_{ij}^{\mathrm{sp}}$ holds, with equality iff the graph is a tree (e.g., Chandra et al. 1989; Deza and Deza 2009; Bavaud 2010). Also, D^{sp} is graph-geodetic.

2.4   Interpolating Random Walks and Shortest Paths

Measuring the navigation effort within a weighted network depends on the nature of the moving agents (people, goods, money, information, etc.), either knowledgeable of all networks characteristics, or only aware of their immediate neighborhood: d^{sp} models agents moving directly to their target, while d^{ct} models agents just wandering randomly until the target is reached. The former is more sensitive to short-cuts and to the length of paths in the network, while the latter is more sensitive to the degree and the number of paths between two nodes. Both capture information on the network structure, although of different kind.

This section presents a flow formalism aimed at continuously interpolating between the shortest path and the random walk, already detailed in Bavaud and Guex (2012); see also Yen et al. (2008) and Saerens et al. (2009) for a close yet independent proposal, distinct in its implementation.

First, consider the general *twofold setup* endowed with two distinct edges valuations, namely the transition matrix $W = (w_{ij})$ of Sect. 2.2, and the resistances $R = (r_{ij})$ of Sect. 2.3. Both can be chosen independently, except for the consistency condition $r_{ij} = \infty$ iff $w_{ij} = 0$.

Second, define for each path $X = (x_{ij})$ the **flow energy** as

$$U(X) := \sum_{ij} r_{ij}\, \varphi(x_{ij})$$

where $\varphi(x)$ is a smooth non-decreasing function with $\varphi(0) = 0$. Flows of $\mathcal{X}_{st}$ minimizing $U(X)$ yield st-shortest paths for the choice $\varphi(x) = x$ and st-electric currents for the choice $\varphi(x) = x^2/2$ (Alamgir and Von Luxburg 2011; Li et al. 2011). Define also the **flow entropy**:

$$G(X) := \sum_{ij} x_{ij} \ln \frac{x_{ij}}{x_{i\bullet} w_{ij}} = \sum_i x_{i\bullet} K_i(X\|W) = x_{\bullet\bullet} \sum_i \frac{x_{i\bullet}}{x_{\bullet\bullet}} K_i(X\|W)$$

where $K_i(X\|W) := \sum_j \frac{x_{ij}}{x_{i\bullet}} \ln \frac{x_{ij}}{x_{i\bullet} w_{ij}} \geq 0$ is the Kullback–Leibler divergence between the empirical transitions $x_{ij}/x_{i\bullet}$ and the theoretical transitions w_{ij}. The entropy $G(X)$ takes on its minimum value zero iff $x_{ij}/x_{i\bullet} = w_{ij}$. Note that the multiplicative factor $x_{\bullet\bullet}$ aims at making $G(X)$ *homogeneous*, that is $G(vX) = vG(X)$ for $v > 0$.

Third, the aforementioned interpolation is implemented by considering the minimizing solution, noted $\tilde{X}^{st}$ or simply $\tilde{X}$, of the **free energy**

$$\tilde{X}^{st} := \arg\min_{X \in \mathcal{X}_{st}} F(X) \qquad\qquad F(X) := U(X) + T\, G(X)$$

where $T > 0$ is a free parameter, the **temperature**, which arbiters between the conflicting objectives: st-flows $\tilde{X}^{sp}$ minimizing $F(X)$ realize shortest paths when $T \to 0$ (and $\varphi(x) = x$), and random walks $\tilde{X}^{rw}$ when $T \to \infty$. We will also use $\beta = 1/T$, the **inverse temperature**.

On the one hand, the feasible set $\mathcal{X}_{st}$ defined by (1) is convex; on the other hand, $F(X)$ is convex iff $\varphi(x)$ is convex, in which case the solution $\tilde{X}^{st}$ is unique and given by Bavaud and Guex (2012)

$$\tilde{x}_{ij} = \tilde{x}_{i\bullet} w_{ij} \exp(-\beta[r_{ij}\varphi'(\tilde{x}_{ij}) + \lambda_i - \lambda_j]) \qquad (4)$$

where λ_i are the Lagrange multipliers associated with (1). Equivalently, defining $v_{ij} := w_{ij}\exp(-\beta r_{ij}\varphi'(\tilde{x}_{ij}))$ as well as $V := (v_{ij})$ (where $i \neq t$ and $j \neq t$), $M := (I - V)^{-1}$, $q := (v_{it})_{i \neq t}$ and $z := Mq$, the solution reads:

$$\tilde{x}_{ij} = m_{si}\, v_{ij}\, \frac{z_j}{z_s} \quad (j \neq t) \qquad\qquad \tilde{x}_{it} = m_{si}\, \frac{q_i}{z_s}\ .$$

The optimal flow $\tilde{X} = (\tilde{x}_{ij})$ can be interpreted as the expected number of passages on edge ij when starting from s until eventually reaching t. The minimum free energy simply expresses as $F(\tilde{X}) = -T \ln z_s$. In what follows, we assume $\varphi(x) = x$. Then (4) is solved in a single step, instead of iteratively.

2.5 Energy and Free Energy Dissimilarities

Define, for any pair st of nodes, the **energy dissimilarity** D^U and the **free energy dissimilarity** D^F as

$$d_{st}^U := \frac{1}{2}(U(\tilde{X}^{st}) + U(\tilde{X}^{ts})) \qquad d_{st}^F := \frac{1}{2}(F(\tilde{X}^{st}) + F(\tilde{X}^{ts}))\ . \quad (5)$$

d_{st}^U is the expected resistance for going from s to t and coming back. In general, $D^U \notin \mathcal{D}_\infty$; however, $D^F \in \mathcal{D}_\infty$, and is graph-geodetic as well (see Kivimäki et al. 2012 and the references therein). Furthermore, one can prove that

$$\lim_{\beta\to\infty} d_{st}^{\mathrm{F}} = \lim_{\beta\to\infty} d_{st}^{\mathrm{U}} = \frac{1}{2}(d_{st}^{\mathrm{sp}} + d_{ts}^{\mathrm{sp}})$$

$$\lim_{\beta\to 0} d_{st}^{\mathrm{F}} = \lim_{\beta\to 0} d_{st}^{\mathrm{U}} = \frac{1}{2}\sum_{ij} r_{ij}\,(\tilde{x}_{ij}^{st\mathrm{rw}} + \tilde{x}_{ij}^{ts\mathrm{rw}}) := d_{st}^{\mathrm{wct}}$$

where d_{st}^{wct} is the commute cost or *weighted commute time*, proportional to d_{st}^{ct} (Françoisse et al. 2013; Kivimäki et al. 2012). The above suggests the possibility of an *Euclidean phase transition*, with dissimilarities in $\mathcal{D}_2^2$ for $T \geq T_c$, but not anymore for $T < T_c$ whenever $D^{\mathrm{sp}} \notin \mathcal{D}_2^2$.

3 Numerical Examples and Applications

3.1 Experiments with Small Graphs

Both D^U and D^F capture network information related to D^{sp} as well as to D^{ct}. Let us investigate their squared Euclidean nature by using the two criteria of Sect. 2.1.

Figure 1 depicts the behavior of the smallest eigenvalue of B and the power index, in the plain setup $r_{ij} := 1/a_{ij}$ and $w_{ij} := a_{ij}/a_{i\bullet}$.

The first example, K_{23}, demonstrates the existence of *critical temperatures* T_c in the sense of Sect. 2.5. The second example, C_{15}, shows D^F to be $\mathcal{D}_2^2$ in the whole temperature range, contrarily to D^U which is not $\mathcal{D}_2^2$ for intermediate values of β. In the third example, both D^F and D^U are $\mathcal{D}_2^2$ over the whole temperature range, with a contrasted behavior between the last eigenvalue of B, monotonously decreasing, and the power, maximum around $\beta = 0.5$. On all examples, D^F stays in $\mathcal{D}_2^2$ longer than D^U when β is raised, a potentially interesting property since squared Euclidean dissimilarities are often needed for statistical applications. Lacunary as they are, those results underline the wide behavioral range of simple graphs, as expected from statistical mechanical entities.

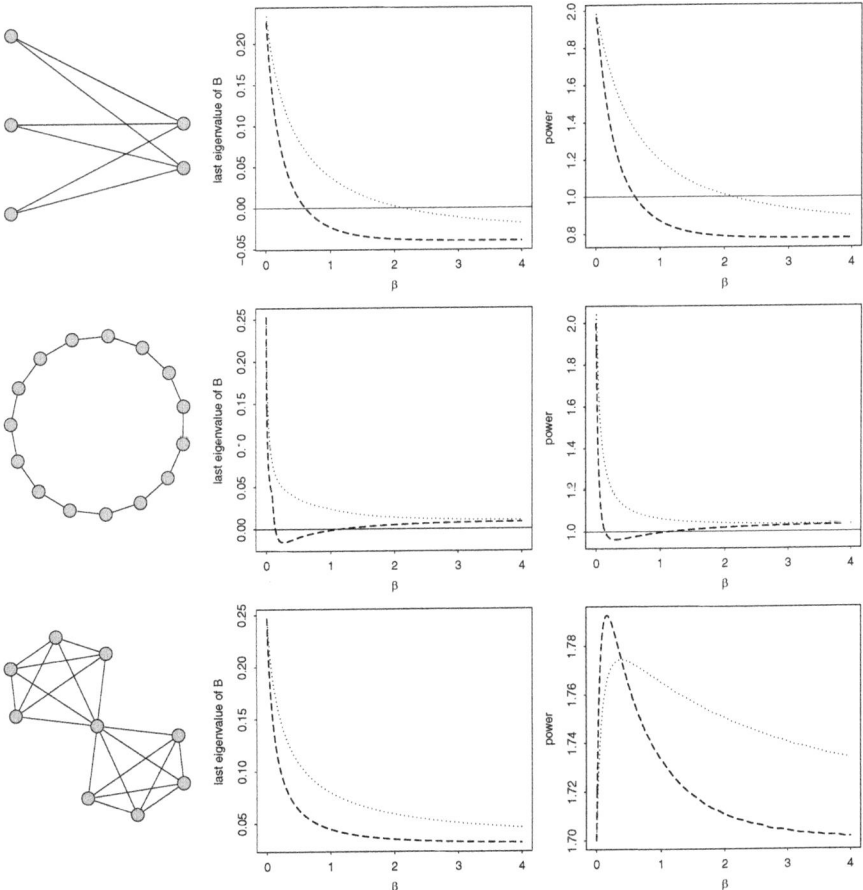

Fig. 1 Three toy graph examples, scanning the squared Euclidean character of D^U (*dashed line*) and D^F (*dotted line*). The *first plot* depicts the evolution of the last eigenvalue of B versus β. The *second plot* exhibits the power index versus β

3.2 Multidimensional Scaling

A planar graph with $n = 50$ nodes, aimed at imitating a realistic road network, is generated following a variant of an algorithm due to Gastner and Newman (2006) (Fig. 2). We define r_{ij} as the Euclidean distances between the pairs of nodes, and apply the simple setup $e_{ij} = c/r_{ij}$. After computing D^U or D^F for various β, we extract the MDS eigen-coordinates in the two first dimensions (regardless of the possible negativity of the last eigenvalue of B^U).

In addition, we evaluate the similarity between the original dissimilarities and the path functional dissimilarities by means of a presumably original *configuration similarity* index $CS_{ab} := \dfrac{\text{Tr}(B^a B^b)}{\sqrt{\text{Tr}((B^a)^2)\text{Tr}((B^b)^2)}} \in [0, 1]$, where B^a and B^b are the scalar products corresponding to configurations D^a and D^b. The maximum similarity of $CS_{\text{planar, U}} = 0.86$ for which D^U is still squared Euclidean obtains for $\beta = 0.3$ (Fig. 2).

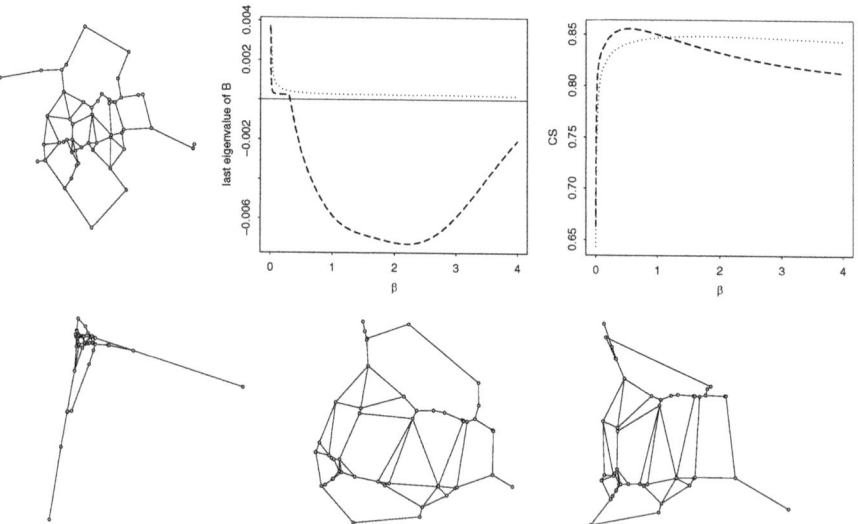

Fig. 2 *Top*: the original planar graph (*left*) and the behavior, regarding β, of the last eigenvalue of B (*middle*) and the configuration similarity CS (*right*) between the original configuration and the graph dissimilarities D^U (*dashed line*) and D^F (*dotted line*). *Bottom*: MDS graph reconstruction from D^U, whose first two dimensions, respectively, explain 42.6 % of the inertia (*left*, for $\beta = 0.001$), 53.8 % (*middle*, for $\beta = 0.3$) and 48.1 % (*right*, for $\beta = 4$). In the last case, the dissimilarity is not squared Euclidean and negative eigenvalues have been removed from the inertia

3.3   Clustering

A standard approach to clustering categorical data consists in computing chi-square dissimilarities D^χ (3) between objects, and then applying a k-means procedure. Alternatives based upon D^U and D^F might reveal more efficient, as demonstrated here in a supervised context with groups known a priori.

Specifically, one considers the document-terms contingency table $N = (n_{il})$ of the $n = 160$ documents of the `Reuters21578` corpus, belonging to $m = 8$ different groups (20 documents in each group). Exchanges e_{ij} obtain as in (2), and resistances as $r_{ij} = 1/e_{ij}$ (plain setup). Document eigen-coordinates are obtained from *weighted MDS* on D^U and D^F, that is by considering the spectral decomposition of $K = (k_{ij})$ with $k_{ij} = \sqrt{f_i f_j}\, b_{ij}$ instead of $B = (b_{ij})$ (e.g., Bavaud 2010); also, negative eigenvalues of K^U or K^F are set to zero. A k-means procedure with 8 clusters is then applied on the eigen-coordinates, and the resulting partition C is compared to the true partition C^{true} by means of the *variation of information dissimilarity* $d_{\text{VI}}(C, C^{\text{true}}) := H(C) + H(C^{\text{true}}) - 2I(C, C^{\text{true}})$, where H is the entropy of the partition and I the mutual information (Meila 2003). Figure 3 shows that both D^U and D^F yield noticeably better results than D^χ for intermediate value of β, but quickly cease to be squared Euclidean. D^U gives for $\beta = 5 \cdot 10^{-6}$ the clustering most similar to the true classification, with a rate of correct classification (under optimal group permutation) of 65.63 %. See Kivimäki et al. (2012) for further clustering experiments involving D^U and D^F, and, e.g., Liu et al. (2013) for random walk clustering.

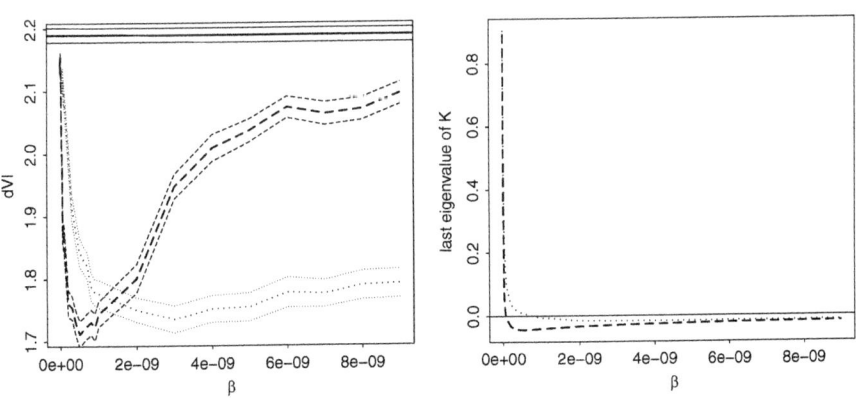

Fig. 3 *Left*: average information dissimilarity and 95 % CI, comparing the true classification and the clustering obtained from D^U (*dashed line*), D^F (*dotted line*) and D^χ (*baseline, top*). *Right*: last eigenvalue of K^U (*dashed line*) and K^F (*dotted line*) as a function of β

4 Conclusion

Transforming a graph into a dissimilarity matrix most facilitates the discussion of clustering and visualization issues. It permits to open graph problems to a large body of statistical and mathematical methods: classical data analysis, machine learning, spectral graph theory and operations research. To that extent, enriching the family of flow-based dissimilarities by considering further sensible yet tractable functionals appears as a research priority. In particular, squared Euclidean dissimilarities immediately allow for MDS visualization and Ward hierarchical clustering.

Various dissimilarities capture (and hide) various aspects of the graph. Results show that D^U generally gives a better representation of the graph structure than D^F for a precise value of the temperature T^{opt}, while the latter is more stable over temperature changes. This suggests that D^U should be used when T^{opt} is known, whereas D^F is preferable otherwise.

Despite the above results, the questions of knowing how to determine, even approximatively, T^{opt}, which aspect of the graph structure is best enlightened by which dissimilarity, and which dissimilarity is most efficient for a specific clustering or visualization task, remain largely open.

References

Alamgir, M., & Von Luxburg, U. (2011). Phase transition in the family of p-resistances. In *Neural Information Processing Systems (NIPS 2011)* (pp. 379–387).

Bavaud, F. (2010). Euclidean distances, soft and spectral clustering on weighted graphs. In *Proceedings of ECML-PKDD 2010. Lecture notes in computer science* (Vol. 6321, pp. 103–118).

Bavaud, F., & Guex, G. (2012). Interpolating between random walks and shortest paths: A path functional approach. In *Proceedings of SocInfo 2012. Lecture notes in computer science* (Vol. 7710, pp. 68–81).

Bavaud, F., & Xanthos, A. (2005). Markov associativities. *Journal of Quantitative Linguistics, 12,* 123–137.

Berger, J., & Snell, J. L. (1957). On the concept of equal exchange. *Behavioral Science, 2,* 111–118.

Boley, D., Ranjan, G., & Zhang, Z.-L. (2011). Commute times for a directed graph using an asymmetric Laplacian. *Linear Algebra and its Applications, 435,* 224–242.

Chandra, A. K., Raghavan, P., Ruzzo, W. L., Smolensky, R., & Tiwari, P. (1989). The electrical resistance of a graph captures its commute and cover times. In *Proceedings of the Twenty-First Annual ACM Symposium on Theory of Computing (STOC '89)* (pp. 574–586).

Chebotarev, P. (2010). A class of graph-geodetic distances generalizing the shortest-path and the resistance distances. *Discrete Applied Mathematics, 159,* 295–302.

Critchley, F., & Fichet (1994) The partial order by inclusion of the principal classes of dissimilarity on a finite set, and some of their basic properties. In B. van Cutsem (Ed.), *Classification and dissimilarity analysis. LNS* (Vol. 93, pp. 5–65).

Deza, M., & Deza, E. (2009). *Encyclopedia of distances.* New York: Springer.

Deza, M., & Laurent, M. (1997). *Geometry of cuts and metrics.* New York: Springer.

Doyle, P., & Snell, J. (1984). *Random walks and electric networks.* Washington, DC: Mathematical Association of America.

Françoisse, K., Kivimäki, I., Mantrach, A. Rossi, F., & Saerens, M. (2013). A bag-of-paths framework for network data analysis. arXiv:1302.6766.

Gastner, M. T., & Newman, M. E. J. (2006). The spatial structure of networks. *The European Physical Journal B, 49,* 247–252.

Joly, S., & Le Calvé, G. (1986). Etude des puissances d'une distance. *Statistique et analyse des donnes, 11,* 30–50.

Kivimäki, I., Shimbo, M., & Saerens, M. (2012). Developments in the theory of randomized shortest paths with a comparison of graph node distances. arXiv:1212.1666.

Klein, D. J., & Zhu, H. Y. (1998). Distances and volumina for graphs. *Journal of Mathematical Chemistry, 23,* 179–195.

Lew, J. S. (1978). Some counterexamples in multidimensional scaling. *Journal of Mathematical Psychology, 17,* 247–254.

Li, Y., Zhang, Z.-L., & Boley, D. (2011). The routing continuum from shortest-path to all-path: A unifying theory. *31st International Conference on Distributed Computing Systems (ICDCS)* (pp. 847–856).

Liu, S., Matzavinos, A., & Sethuraman, S. (2013). Random walk distances in data clustering and applications. *Advances in Data Analysis and Classification, 7,* 83–108.

Meila, M. (2003). Comparing clusterings by the variation of information. In *Proceedings of the Sixteenth Annual Conference of Computational Learning Theory (COLT).* New York: Springer.

Saerens, M., Achbany, Y., Fouss, F., & Yen, L. (2009). Randomized shortest-path problems: Two related models. *Neural Computation, 21,* 2363–2404.

Torgeson, W. S. (1958). *Theory and methods of scaling.* New York: Wiley.

Yen, L., Saerens, M., Mantrach, A., & Shimbo, M. (2008). A family of dissimilarity measures between nodes generalizing both the shortest-path and the commute-time distances. In *Proceedings of the 14th SIGKDD International Conference on Knowledge Discovery and Data Mining* (pp. 785–793).

Resampling Techniques in Cluster Analysis: Is Subsampling Better Than Bootstrapping?

Hans-Joachim Mucha and Hans-Georg Bartel

Abstract In the case of two small toy data sets, we found out that subsampling has a much weaker behavior in the finding of the true number of clusters K than bootstrapping (Mucha and Bartel, Soft bootstrapping in cluster analysis and its comparison with other resampling methods. In: M. Spiliopoulou, L. Schmidt-Thieme, R. Janning (eds.) Data analysis, machine learning and knowledge discovery. Springer, Cham, 2014). In contradiction, Möller and Dörte (Intell Data Anal 10:139–162, 2006) pointed out that "subsampling ... clearly outperformed the bootstrapping technique in the detection of correct clustering consensus results." Obviously, there is a need for further investigations. Therefore here we compare these two resampling techniques based on real and artificial data sets by means of different indices: ARI or Jaccard. We consider hierarchical cluster analysis methods because they find all partitions into $K = 2, 3, \ldots$ clusters in one run only, and, moreover, these results are (usually) unique (Spaeth, Cluster analysis algorithms for data reduction and classification of objects. Ellis Horwood, Chichester, 1982). The methods are tested on two synthetic data sets and two real data sets. Obviously, bootstrapping is better than subsampling in finding the true number of clusters.

1 Introduction

Efron (1981) investigated nonparametric estimates of standard error using different resampling methods, and his conclusion was "The bootstrap performs notably best." Originally, nonparametric bootstrapping is a statistical method for estimating the sampling distribution of an estimator by sampling with replacement from the

H.-J. Mucha (✉)
Weierstrass Institute for Applied Analysis and Stochastics (WIAS), Mohrenstraße 39, 10117
Berlin, Germany
e-mail: mucha@wias-berlin.de

H.-G. Bartel
Department of Chemistry at Humboldt University, Brook-Taylor-Straße 2, 12489 Berlin,
Germany
e-mail: hg.bartel@yahoo.de

© Springer-Verlag Berlin Heidelberg 2015
B. Lausen et al. (eds.), *Data Science, Learning by Latent Structures,
and Knowledge Discovery*, Studies in Classification, Data Analysis,
and Knowledge Organization, DOI 10.1007/978-3-662-44983-7_10

original sample. This very simple technique allows estimation of the sampling distribution of almost any statistic. Bootstrapping falls in the broader class of resampling methods and simulation schemes. Some alternative resampling methods are subsampling (draw a subsample to a smaller size without replacement) and jittering (add noise to every single observation), and a combination of both simulation schemes.

We found out that bootstrapping performs best for finding the number of clusters K in hierarchical cluster analysis (HCA) of two small toy data sets (Mucha and Bartel 2014). This is in contradiction to Möller and Dörte (2006) The value of K which provides the most stable partitions is the estimate of the number of clusters in a given original data set.

2 Resampling Techniques in Cluster Analysis

Usually, the starting point of clustering is a data matrix $\mathbf{X} = (x_{ij})$, $i = 1, 2, \ldots, I$, $j = 1, 2, \ldots, J$ of I observations and J variables. Cluster analysis means finding a partition of the set of I observations into K non-empty clusters (subsets) C_k, $k = 1, 2, \ldots, K$.

Why is validation of clustering so important? That's because cluster analysis presents clusters in almost any case. Real clusters should be stable, i.e., they should be confirmed and reproduced to a high degree if the data set is changed in a non-essential way. Thus, clustering of a randomly drawn sample of the data should lead to similar results.

For instance, a resampling method can be used to investigate the variations of the centroids of the clusters. As a reminder: The nonparametric bootstrap approach is resampling taken with replacement from the original data, whereas subsampling is resampling taken without replacement. Figure 1 introduces a toy data set consisting of at least three classes $C_1 = \{1, 2, \ldots, 17\}$, $C_2 = \{18, 19, \ldots, 23\}$, and $C_3 = \{24, 25, \ldots, 32\}$. In addition, Fig. 2 shows estimates of the location parameters that are the result of hierarchical Ward's clustering of 250 nonparametric subsamples of this toy data into three clusters. Here, each subsample clustering provides estimates of three cluster centers. So, the plot represents 750 ($= 250 * 3$) estimates. Obviously, almost all estimates of the cluster centers reflect the true classes. The estimated expected values of class C_1 look most homogeneous. But, some estimates of expected values of class C_3 (in the center of the plot) present clearly a mixture of classes C_1 and C_3. Of course, such an estimation is strictly restricted to metric data.

In clustering, the estimation of parameters such as expected values is not the main task. The final aim of clustering is the formation of groups either as a partition or a hierarchy of a given set of observations. Therefore, here the focus is on general investigation of stability based on partitions. This covers also hierarchies because they can be considered as a set of partitions (Mucha 2007). To assess the stability of a cluster in a most general way, resampling techniques can be used. For example, bootstrapping the Jaccard coefficient between sets (clusters) is

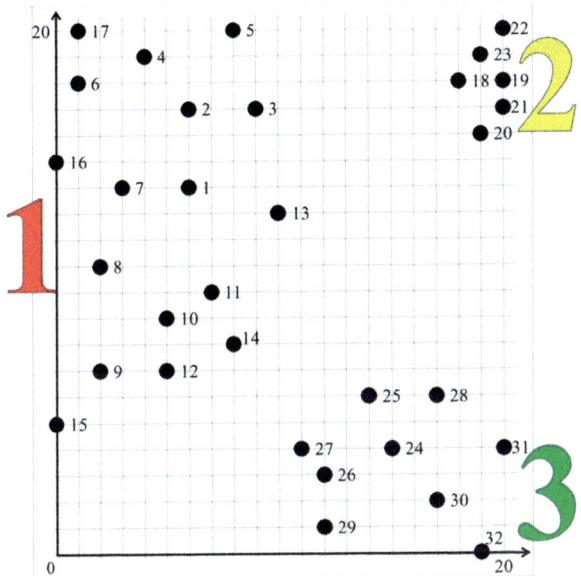

Fig. 1 Plot of the two-dimensional toy data set. The data values are integers. They can be taken directly from the plot. The observations are numbered

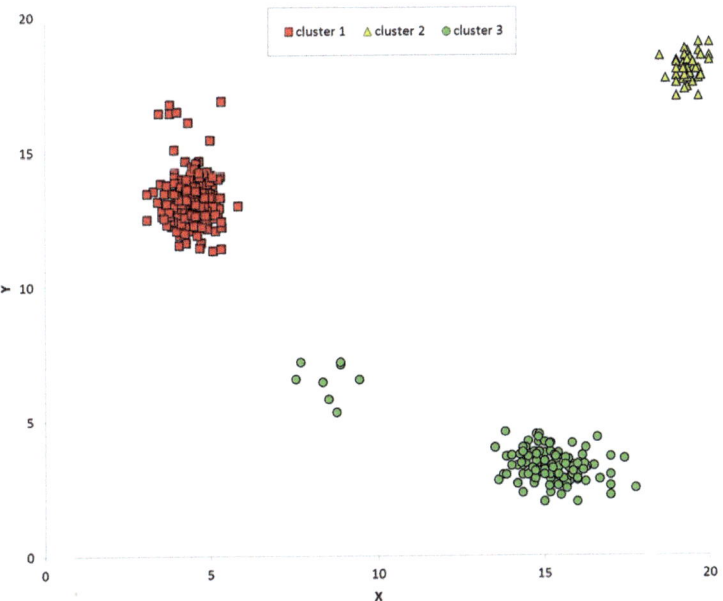

Fig. 2 Plot of the estimates of the location parameter of clusters. They are the result of Ward's HCA of 250 subsamples (75 % resampling rate) into three clusters

recommended (Hennig 2007). Both the adjusted Rand index (ARI) and the averaged Jaccard measure are recommended for an investigation of the stability of a partition.

Nonparametric bootstrapping is resampling taken with replacement from the original data. It generates multiple observations. In clustering of small data sets, this can cause problems. Subsampling is resampling taken without replacement from the original data. A parameter $p < I$ is needed which causes an additional problem: the cardinality of the drawn sample. This is different from bootstrapping where no parameter is needed because here the cardinality of the drawn sample is always I. The way out would be to discard multiple points in a bootstrap scheme (named "Boot2Sub" in the figures below). A subsampling rate of 90 % (i.e., $p = 0.9$: this is related to tenfold-cross-validation in some sense) or greater performs very bad in HCA. Is bootstrapping really the best choice for stability investigations in cluster analysis? To answer this question we will look at other toy and real data.

3   Bootstrapping in Cluster Analysis

Suppose, there is the (unique) cluster analysis result of a data set. Let us call this an original clustering. Furthermore, suppose, the same cluster analysis method is applied to a Bootstrap sample of the data set. A cluster $\mathcal{C}_k$ from this original clustering is compared to a corresponding (i.e., most similar) bootstrap cluster $\mathcal{B}_j$ using the Jaccard index γ

$$\gamma(\mathcal{C}_k, \mathcal{B}_j) = \frac{|\mathcal{C}_k \cap \mathcal{B}_j|}{|\mathcal{C}_k \cup \mathcal{B}_j|}, \tag{1}$$

which is a symmetric similarity measure of the proportion of elements belonging to both clusters of all elements. This bootstrap cluster is the result of clustering of a random sample bootstrapping the Jaccard coefficient works as follows (for details, see Hennig 2007):

1. Draw many new data sets from the original one
2. Apply the clustering method to them
3. For every original cluster find the most similar bootstrap cluster $\mathcal{B}_j$ and record the similarity value $\gamma(\mathcal{C}_k, \mathcal{B}_j)$.

Finally, assess the stability of every original cluster $\mathcal{C}_k$ by the mean Jaccard γ_k taken over all resampled data sets.

Figure 3 shows an informative dendrogram of Average Linkage cluster analysis of the toy data containing the Jaccard values (1) as stability assessment of clusters. The three cluster solution ($\gamma_1 = 0.98659$, $\gamma_2 = 0.98498$, and $\gamma_3 = 1$) is very stable.

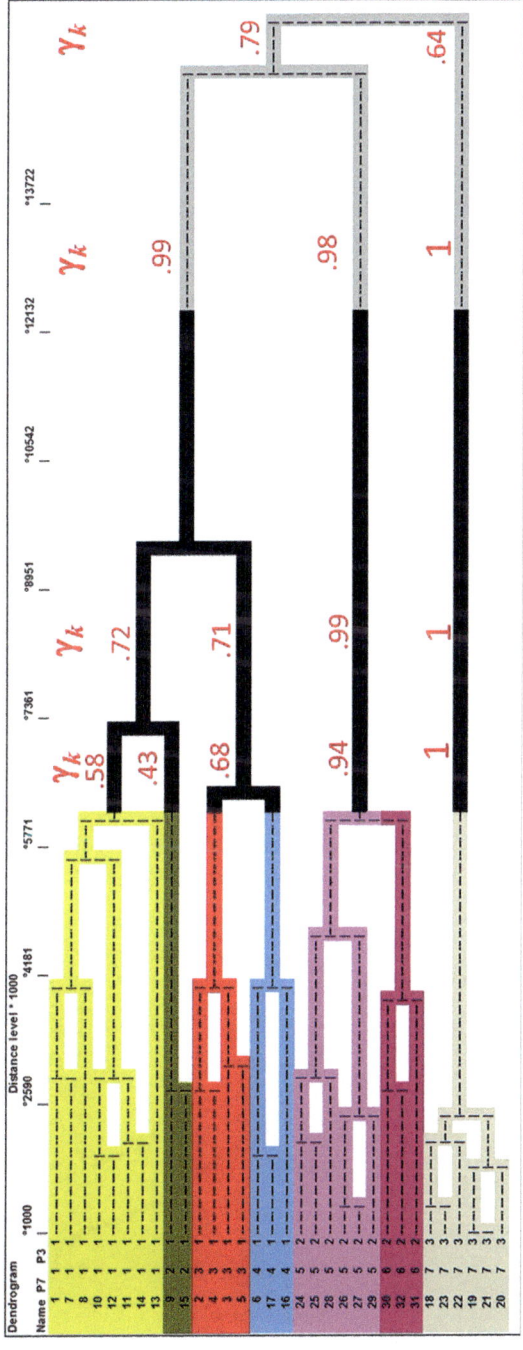

Fig. 3 Dendrogram of average linkage cluster analysis of the toy data with Jaccard measures as stability assessment of the individual clusters

For a decision about the number of clusters K, as it is usual using the ARI (Hubert and Arabie 1985), an averaged Jaccard value γ_K regarding all γ_k of individual clusters $C_k, k = 1, 2, \ldots, K$, of a partition is recommended:

$$\gamma_K = \frac{1}{I} \sum_{k=1}^{K} n_k \gamma_k, \tag{2}$$

where n_k is the cardinality of the cluster C_k. Both the ARI (denoted as R below) and the averaged Jaccard index γ_K are most appropriate to decide about the number of clusters K, i.e. to assess the stability of a partition consisting of all clusters $C_k, k = 1, 2, \ldots, K$.

4 Bootstrapping Versus Subsampling

In Fig. 4, different resampling techniques are compared based on the averaged Jaccard measure (2) for validation of results of Average Linkage cluster analysis of the toy data. Bootstrapping performs best in finding the three classes (see also Fig. 3) because it has

1. its maximum value at $K = 3$,
2. the most steeply rising when going from $K = 2$ to $K = 3$, and
3. the most steeply sloping when going from $K = 3$ to $K = 4$.

Figure 5 shows the same as Fig. 4 but based on the ARI R. The ARI "outperforms" Jaccard with respect to the most steeply rising when going from $K = 2$ to $K = 3$. In general, both present similar results.

Similar to Figs. 4 and 5, Figs. 6 and 7 show validation results of Ward's HCA of the toy data based on the averaged Jaccard measure and the ARI, respectively.

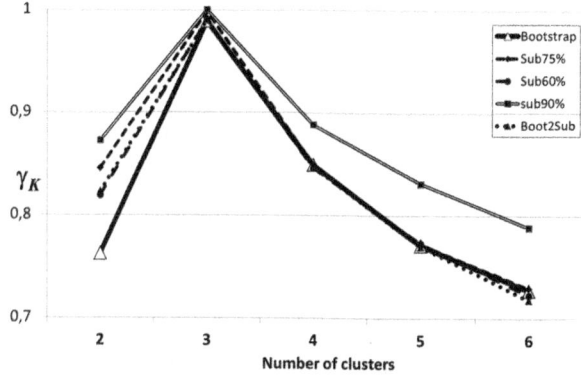

Fig. 4 Jaccard's measures of the average linkage cluster analysis of the toy data

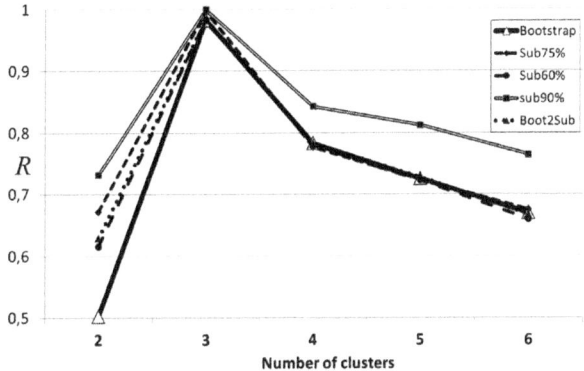

Fig. 5 ARI measures of the average linkage cluster analysis of the toy data

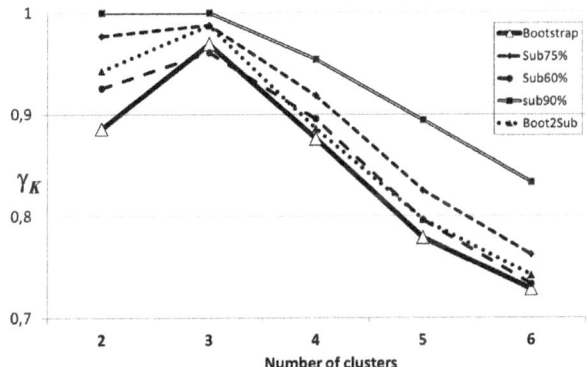

Fig. 6 Jaccard's measures of Ward's HCA of the toy data

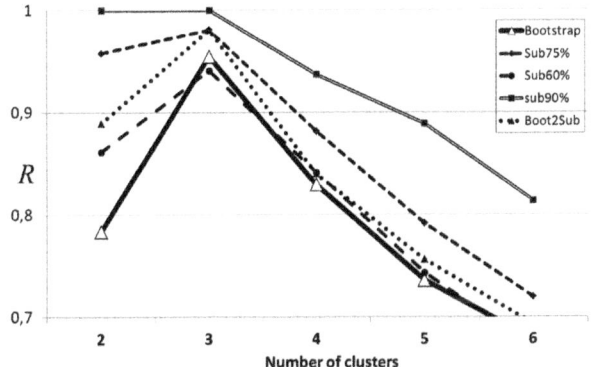

Fig. 7 ARI measures of Ward's HCA of the toy data

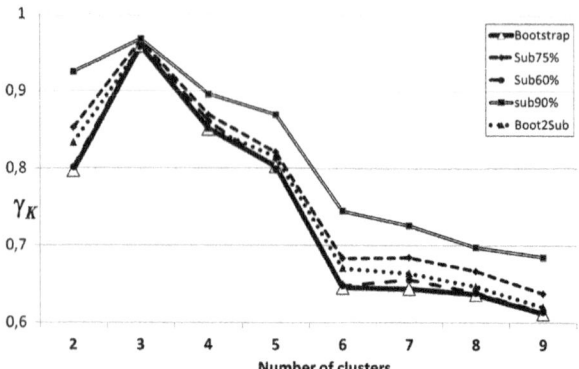

Fig. 8 Jaccard's measures of Ward's HCA of the Gaussian data

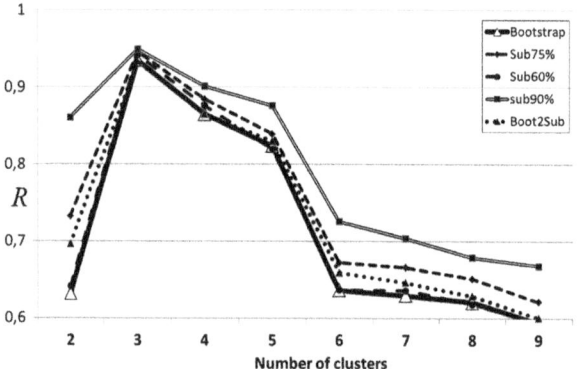

Fig. 9 ARI measures of Ward's HCA of the Gaussian data

Similar to Figs. 4 and 5, Figs. 8 and 9 show validation results of Ward's HCA of the randomly generated two-dimensional three class data based on the averaged Jaccard measure and the ARI, respectively. The three Gaussian sub-populations were generated with the following different parameters: cardinalities 80, 130, and 90, mean values $(-3, 3)$, $(0, 0)$, and $(3, 3)$, and standard deviations $(1, 1)$, $(0.7, 0.7)$, and $(1.2, 1.2)$. Ward's method is successful in dividing (decomposing) the data into three subsets: four errors are counted only.

Figure 10 shows an informative dendrogram of Ward's clustering of the Swiss banknotes data (Flury and Riedwyl 1988) containing Jaccard values as stability assessment of clusters. All together 200 Swiss bank notes are characterized by

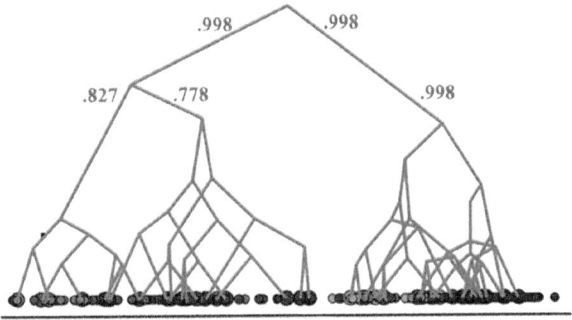

Fig. 10 Dendrogram of Ward's cluster analysis of the Swiss bank notes data with Jaccard measures as stability assessment of the individual clusters

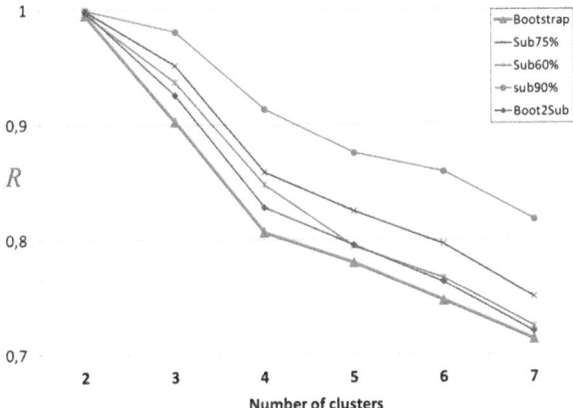

Fig. 11 ARI measures of Ward's HCA of the Swiss bank notes data

six measurements. The genuine bank notes on the right-hand side look more homogeneous than the forged ones. Figure 11 shows validation results of Ward's HCA based on the ARI.

Figure 12 shows validation results of Ward's HCA of the well-known Iris data (Fisher 1936) based on the ARI. All resampling techniques vote for $K = 2$ clusters. Obviously, Ward's method is not the appropriate one to discover the true three species (classes).

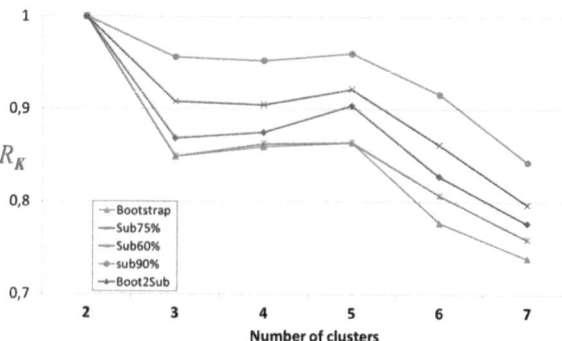

Fig. 12 ARI measures of Ward's HCA of the Iris data

5   Summary

Bootstrapping seems to be the first choice for a decision about the number of clusters. In all cases investigated so far (toy and real data), it outperforms subsampling. But, we investigated here only two popular HCA methods: Average Linkage and Ward (Spaeth 1982). The latter is the simplest model-based HCA. In subsampling, the choice of the parameter "resampling rate" p causes an additional problem. A subsampling rate of 90 % or greater seems to be very problematic. A way out can be recommended: Take the usual bootstrap scheme but discard multiple observations (named "Boot2Sub" in the figures). Here no parameter p is necessary. In fact, asymptotically, 63.2 % of the observations will be presented in such a subsample.

References

Efron, B. (1981). Nonparametric estimates of standard error: The jackknife, the bootstrap and other methods. *Biometrika, 68*, 589–599.

Fisher, R. A. (1936). The use of multiple measurements in taxonomic problems. *Annals of Eugenics, 7*, 179–188.

Flury, B., & Riedwyl, H. (1988). *Multivariate statistics: A practical approach*. London: Chapman and Hall.

Hennig, C. (2007). Cluster-wise assessment of cluster stability. *Computational Statistics and Data Analysis, 52*, 258–271.

Hubert, L. J., & Arabie, P. (1985). Comparing partitions. *Journal of Classification, 2*, 193–218.

Möller, U., & Dörte, R. (2006). Performance of data resampling methods for robust class discovery based on clustering. *Intelligent Data Analysis, 10*, 139–162.

Mucha, H.-J. (2007). On validation of hierarchical clustering. In: R. Decker & H.-J. Lenz (Eds.), *Advances in data analysis* (pp. 115–122). Berlin: Springer.

Mucha, H.-J., & Bartel, H.-G. (2014). Soft bootstrapping in cluster analysis and its comparison with other resampling methods. In: M. Spiliopoulou, L. Schmidt-Thieme, & R. Janning (Eds.), *Data analysis, machine learning and knowledge discovery*. Cham: Springer (pp. 97–104).

Spaeth, H. (1982). *Cluster analysis algorithms for data reduction and classification of objects*. Chichester: Ellis Horwood.

On-Line Clustering of Functional Boxplots for Monitoring Multiple Streaming Time Series

Elvira Romano and Antonio Balzanella

Abstract In this paper we introduce a micro-clustering strategy for functional boxplots. The aim is to summarize a set of streaming time series split in non-overlapping windows. It is a two-step strategy which performs at first, an on-line summarization by means of functional data structures, named Functional Boxplot micro-clusters; then, it reveals the final summarization by processing, off-line, the functional data structures. Our main contribute consists in providing a new definition of micro-cluster based on Functional Boxplots and in defining a proximity measure which allows to compare and update them. This allows to get a finer graphical summarization of the streaming time series by five functional basic statistics of data. The obtained synthesis will be able to keep track of the dynamic evolution of the multiple streams.

1 Introduction

Data stream mining has gained a lot of attention due to the development of applications where sensor networks are used for monitoring physical quantities such as electricity consumptions, environmental variables, computer network traffic. In these applications it is necessary to analyze potentially infinite flows of temporally ordered observations which cannot be stored and which have to be processed using reduced computational resources. The on-line nature of these data streams requires the development of incremental learning methods which update the knowledge about the monitored phenomenon every time a new observation is collected.

Among the exploratory tools for data stream processing, clustering methods are widely used knowledge extraction tools. Some of the main proposals are Balzanella et al. (2011), Bi-Ru Dai et al. (2006), Guha et al. (2003), Aggarwal et al. (2003), and Romano et al. (2011).

E. Romano (✉) • A. Balzanella
Department of Political Science "Jean Monnet", Second University of Naples, Caserta, Italy
e-mail: elvira.romano@unina2.it; antonio.balzanella@unina2.it

© Springer-Verlag Berlin Heidelberg 2015
B. Lausen et al. (eds.), *Data Science, Learning by Latent Structures, and Knowledge Discovery*, Studies in Classification, Data Analysis, and Knowledge Organization, DOI 10.1007/978-3-662-44983-7_11

Usually clustering methods also perform a summarization of the observed data. This is accomplished by identifying a set of centroids which provides the synthesis of each homogeneous group of observations. Since the on-line arriving observations are deleted after being processed, the type of adopted synthesis is a key point in the development of methodologies for data stream clustering. With this aim, the CluStream algorithm proposed in Aggarwal et al. (2003) provides a two-step strategy. The first is an on-line step, named micro-clustering, that performs a first on-line summarization of the streams keeping updated a specific set of data structures (micro-clusters). The second is an off-line step named macro-clustering, which reveals the final summarization by processing the micro-clusters with an appropriate clustering algorithm. The CluStream provides only a basic summarization of the data coming from sensors since it only records the average and the variance of groups of similar multidimensional items. In this paper we extend this algorithm in order to use the Functional Boxplot introduced in Sun and Genton (2011) as tool for gaining knowledge from multiple streaming time series. This will allow to get a finer summarization of the streaming time series that keeps into account five basic statistics (first and third quartile, median, maximum and minimum value) of data and which can be graphically represented.

2 CluStream of Functional Boxplots

Let $y_i(t)$, $i = 1, \ldots, n$, $t \in [1, \infty]$ a set of streaming time series made by real valued temporally ordered observations of a variable $Y(t)$ in n sites, on a discrete time grid. This work proposes an incremental clustering algorithm with the aim to supply a set of data descriptions or synopsis to reduce dimensionality and to keep track of the dynamic evolution of the streams. It processes each example in constant time and memory and is incremental in the sense that data synopsis are incrementally maintained as more and more data are received.

It is a Clustream algorithm on Functional boxplots obtained by a set of n streaming time series split in non-overlapping windows and opportunely approximated by functional data. The method can be summarized by the following steps:

- On-line phase (FBP-micro-clustering)

  - Splitting of incoming data streams into non-overlapping windows;
  - Detection of the Functional Boxplot associated with each window;
  - Updating of appropriate data synopsis called Functional Boxplot micro-clusters.

- Off-line phase (FBP-macro-clustering)

  - Clustering algorithm performed on the Functional Boxplot micro-clusters.

2.1 On-Line Phase

The first step of the on-line phase consists in splitting the incoming parallel streaming time series into a set of non-overlapping windows W_j, $j = 1, \ldots, \infty$, that are compact subsets of T having size $w \in \Re$ and such that $W_j \cap W_{j+1} = \emptyset$. The defined windows frame for each $y_i(t)$ a subset $y_i^{w_j}(t)$ $t \in W_j$ of ordered values of $y_i(t)$ called subsequence.

Following the Functional Data Analysis approach (Ramsay and Silverman 2005), we consider each subsequence $y_i^{w_j}(t)$ of $y_i(t)$ the raw data which includes noise information. Then we determinate a true functional form $f_i^{w_j}(t)$, we call functional subsequence, which describes the trend of the flowing data. For each W_j we have that all the subsequences $y_i^{w_j}(t)$ $i = 1, \ldots, n$ follow the model:

$$y_i^{w_j}(t) = f_i^{w_j}(t) + \epsilon_i^{w_j}(t), \ t \in W_j \ i = 1, \ldots, n \tag{1}$$

where $\epsilon_i^{w_j}(t)$ are residuals with independent zero mean and $f_i^{w_j}(\cdot)$ is the mean function.

The second step of the on-line phase aims at detecting a summary of the set $f_i^{w_j}(t)$ (with $i = 1, \ldots, n$) of the batched streaming time series by means of a functional boxplot variables FBP_j, $j = 1, \ldots, \infty$, defined as follows:

Definition 1 (Functional Boxplot) Let W_j be a window which frames the subsequences $f_1^{w_j}(t), \ldots, f_i^{w_j}(t), \ldots, f_n^{w_j}(t)$ (with $t \in W_j$). A Functional Boxplot FBP_j is a compound of five functions $\left\{ f_{[u]}^{w_j}(t), f_{[l]}^{w_j}(t), f_{[1]}^{w_j}(t), f_{[b_{\min}]}^{w_j}(t), f_{[b_{\max}]}^{w_j}(t) \right\}$ where:

$f_{[u]}^{w_j}(t)$ is the upper bound of the central region;
$f_{[l]}^{w_j}(t)$ is the lower bound of the central region;
$f_{[1]}^{w_j}(t)$ is the median curve
$f_{[b_{\min}]}^{w_j}(t)$ is the upper bound of the subsequences
$f_{[b_{\max}]}^{w_j}(t)$ is the lower bound of the subsequences

A Functional Boxplot is the analog of classical boxplot for functional data (Sun and Genton 2011). The only difference consists in the data ordering criterion. In particular, since functions varies over a continuum, data ordering is based on the notion of band depth or modified band depth (Lopez-Pintado and Romo 2009).

Based on the center outward ordering induced by band depth for functional data, the descriptive statistics of a functional boxplot are: the envelope of the 50 % central region, the median curve, and the maximum non-outlying envelope. The 50 % central region is the analog to the "interquartile range" (IQR), it is defined by the band delimited by the 50 % of deepest, or the most central observations. The border of the 50 % central region is defined as the envelope representing the box in a classical boxplot. The median is the most central observation in the box. The maximum envelope of the dataset identified by the vertical lines of the plot are the "whiskers" of the boxplot. Formally, let $f_{[i]}^{w_j}(t)$ denote the sample of functional subsequence associated with the ith largest band depth value. The set $f_{[1]}^{w_j}(t) \ldots, f_{[n]}^{w_j}(t)$ are order statistics, with $f_{[1]}^{w_j}(t)$ the median curve, that is the most central curve (the deepest), and $f_{[n]}^{w_j}(t)$ is the most outlying curve.

Moreover, let the ordered set $f_{[1]}^{w_j}(t), \ldots, f_{[n]}^{w_j}(t)$ where $f_{[1]}^{w_j}(t)$ is the minimum value of the n curves on t and $f_{[n]}^{w_j}(t)$ is the maximum value. The subsequence i belongs to the central region on t if:

$$f_{[n/4]}^{w_j}(t) \leq f_i^{w_j}(t) \leq f_{[3n/4]}^{w_j}(t) \tag{2}$$

In the third step of the on-line phase, the FBP$_j$ variables concur to update a set of specific data structures FBP$_{C_k}$, $k = 1, \ldots, K$ we name FBP-micro-clusters, defined as:

Definition 2 (Functional Boxplot Micro-cluster) An FBP-micro-cluster FBP$_{C_k}$, $k = 1, \ldots, K$, for a set of FBP$_j$ (with $j = 1, \ldots, n^k$) of functional boxplots is the tuple $(\overline{\text{FBP}}_k, n^k, tl^k)$ where:

- $\overline{\text{FBP}}_k$ is the functional boxplot which assumes the role of centroid;
- n^k is the number of allocated functional boxplots;
- tl^k is the time stamp of the last update;

The Functional Boxplot micro-cluster is an extension of the micro-cluster introduced in Aggarwal et al. (2003). In our method, its task is to summarize very similar Functional boxplots, through a set of statistics which are updatable on-line and able to adapt to the change of data.

In order to achieve the desired space saving, we keep a set of FBP$_{C_k}$ with $k = 1, \ldots, K$ where K is chosen to keep a high representativeness of data. Thus K is much higher than the clusters in data but much lower than the number of processed windows.

In the on-line step, every time the data of a new window W_j becomes available, an FBP$_j$ is constructed and then allocated to an FBP$_{C_k}$. The allocation is obtained evaluating the distance between the FBP$_j$ and the centroid $\overline{\text{FBP}}_k$ so that if the minimum value of distance is lower than the threshold value th stored in the micro-cluster, the allocation is performed to the corresponding FBP$_{C_k}$, otherwise a new one is started setting the functional boxplot of the window as centroid and $n^k = 1$.

The allocation is based on the definition of an appropriate distance measure for comparing FBP$_j$. It is computed by considering that each couple of correspondent functions is compared on the same time interval by means of an alignment of the FBP$_j$.

Let us consider two functional boxplots FBP$_j$, FBP$_{j'}$ defined on two windows W_j, $W_{j'}$. Each of them is characterized by the set of five functions $f^{W_j}(t) : W_j \longrightarrow \Re$, $f^{W_{j'}}(t) : W_{j'} \longrightarrow \Re$.

Aligning FBP$_{j'}$ to FBP$_j$ means finding a function $g(t) : W_{j'} \longrightarrow W_j$ such that $f^{W_j}(t)$ and $g \circ f^{W_{j'}}(t) = h^{W_j}(t)$ are defined on the same interval W_j, with the function $g(t)$ expressed by $g(t) = a + bt$.

The function g must be a bijective function and replace $g \circ f^{W_{j'}}(t) = h^{W_j}(t)$ by $h^{W_j}(g(t)) = h^{W_{j'}}(t)$ with $t \in W_{j'}$ or by $h^{W_j}(t) = h^{W_{j'}}(g^{-1}(t))$ with $t \in W_j$ because the function g is defined on the index t.

We consider $a \in \Re$ and $b = 1$, that is an alignment. If $b \neq 1$, that is for not only misaligned but also warped functions, the function $g(t)$ can be considered a warping function as in Sangalli et al. (2010), Adelfio et al. (2012).

Thus, formally, the distance between a pair of functional boxplots FBP_j, $\text{FBP}_{j'}$ is defined as follows:

Definition 3 (Distance) Let $\text{FBP}_j = \left\{ f_{[u]}^{w_j}(t), f_{[l]}^{w_j}(t), f_{[1]}^{w_j}(t), f_{[b_{\min}]}^{w_j}(t), f_{[b_{\max}]}^{w_j}(t) \right\}$ and $\text{FBP}_{j'} = \left\{ f_{[u]}^{w_{j'}}(t), f_{[l]}^{w_{j'}}(t), f_{[1]}^{w_{j'}}(t), f_{[b_{\min}]}^{w_{j'}}(t), f_{[b_{\max}]}^{w_{j'}}(t) \right\}$ be two functional boxplots defined, respectively, on W_j and $W_{j'}$, $g(t)$ be the alignment function so that $g \circ f_{\cdot}^{W_{j'}}(t) = h_{\cdot}^{W_j}(t)$, the distance between FBP_j and $\text{FBP}_{j'}$ is:

$$
d(\text{FBP}_j, \text{FBP}_{j'}) = \sqrt{\int_{t \in W} (f_{[u]}^{w_j}(t) - h_{[u]}^{w_j}(t))^2 dt} + \sqrt{\int_{t \in W} (f_{[l]}^{w_j}(t) - h_{[l]}^{w_j}(t))^2 dt} +
$$

$$
+ \sqrt{\int_{t \in W} (f_{[1]}^{w_j}(t) - h_{[1]}^{w_j}(t)^2 dt)} + \sqrt{\int_{t \in W} (f_{[b_{\min}]}^{w_j}(t) - h_{[b_{\min}]}^{w_j}(t))^2 dt} +
$$

$$
+ \sqrt{\int_{t \in W} (f_{[b_{\max}]}^{w_j}(t) - h_{[b_{\max}]}^{w_j}(t))^2 dt}
$$

The consequences of an allocation are the unitary increment of n^k, the setting of the current time stamp for the parameter W^k and the computation of the FPB-micro-cluster centroid. The latter is performed so that for each of the five functions which define the Functional Boxplot, the average is kept. This can be obtained starting from the information stored in the FBP-micro-cluster self and from the just allocated Functional Boxplot.

In our method, the size K of the set of FPB-micro-cluster is not defined a-priori but it adapts to the structure of data, however it strongly depends on the choice of the threshold th. A too high value involves that only few FBP-micro-clusters are generated; on the contrary, a too low value brings to generate too many FBP-micro-clusters. To deal with this issue we introduce an heuristic to set the value of the threshold and a criterion to keep the number of functional boxplot micro-clusters under a value $K_{\max}$ (this allows to keep a constant upper bound of the used memory space).

Particularly, we propose to compute the threshold th as follows:

$$
th = \min d(\overline{\text{FBP}}_j, \overline{\text{FBP}}_k) \quad \forall j, k = 1, \dots, K \text{ with } k \neq j \tag{3}
$$

thus, th is set to the minimum distance between the FBP-micro-cluster centroids.

If the number of FBP-micro-clusters grows too much so to exceed the available memory resources, we propose, alternatively, to discard the micro-clusters recording concepts no longer present in the data or to merge the two nearest FBP-micro-clusters into one. The choice is made by evaluating the time stamp of the last updating stored in the parameter tl^k of each FBP-micro-cluster:

$$\begin{cases} \text{If } (t_{\text{now}} - tl^k) > t^* \quad \forall k = 1, \dots, K \quad \Rightarrow \quad \text{Discard FBP}_{C_k}, \\ \text{Else } \operatorname{argmin}_{j,k} d(\overline{\text{FBP}}_j, \overline{\text{FBP}}_k) \quad \forall j, k = 1, \dots, K \text{ with } k \neq j \quad \Rightarrow \quad \text{Merge FBP}_j, \text{FBP}_k \end{cases}$$

where t_{now} is the time stamp of the current window and t^* indicates the age over which an FBP-micro-cluster has to be considered no longer useful.

2.2 Off-Line Phase

In order to reveal the final summarization of the streams, the off-line phase analyzes the FBP-micro-clusters computed on-line. We provide a method to get the summarization of data behavior over user-defined time slots. It is based on storing, at predefined time instants, a snapshot of the set of FBP-micro-cluster. Each snapshot will collect the state of updating of each FBP_{C_k} in that time instant.

In order to get the summarization of the user-defined time slot, the procedure identifies the snapshot that is temporally closer to the lower end of the time interval (lower snapshot) and the one which is temporally closer to the upper end (upper snapshot). The next step is to remove from the state of the functional boxplot micro-clusters the effects of the updates that occurred before the beginning of the lower snapshot. Since the centroid $\overline{\text{FBP}}_k$ of each FBP-micro-cluster is the average of the allocated functional boxplots, it is possible to recover the state of each FBP_{C_k} removing what has happened before the beginning of the time slot, by computing a component by component weighted difference between the centroid $\overline{\text{FBP}}_k$ as available from the upper snapshot and the corresponding $\overline{\text{FBP}}_k$, obtained from the lower snapshot (the weights are the number of allocations stored in the parameter n^k).

From the output of the previous step, the obtained centroids $\overline{\text{FBP}}_k$, together with the number of allocated items n_k (which assumes the role of weight), become the data to be processed by a k-means like algorithm which provides, as output, a partition of the FBP-micro-clusters centroids into a set $\text{FBP}_{C_1}, \dots, \text{FBP}_{C_c}, \dots, \text{FBP}_{C_C}$ (with $C < K$) of macro-clusters and a new set $\overline{\overline{\text{FBP}}}_c$, (with $c = 1, \dots, C$) of functional boxplots which are the final summaries of the required time interval.

Similarly to the k-means, this algorithm minimizes an internal heterogeneity measure:

$$\Delta = \sum_{c=1}^{C} \sum_{\overline{FBP_k} \in MC_c} d(\overline{FBP_k}; \overline{\overline{FBP_c}}) n_k \tag{4}$$

where $d(\overline{FBP_k}; \overline{\overline{FBP_c}})$ is computed according to the Definition 3.

In order to optimize the criterion Δ, our macro-clustering algorithm iterates, until the convergence, an allocation and a centroid computation step. In the allocation step, each $\overline{FBP_k}$ is attributed to the macro cluster whose distance is minimal. In the centroid computation step, the representation of each macro-cluster MC_c is obtained by means of a component by component weighted average where n_k is the weight for the corresponding $\overline{FBP_k}$.

3   Daily Rainfall Monitoring by Clustering of FBP

This section shows the results on real data of the proposed method. We have analyzed a dataset provided by the Australian Government Bureau of Meteorology, available on-line at http://www.bom.gov.au/climate/data/, which records the daily rainfall in Australia from 1/4/1961 to 30/4/2012. We have downloaded 77 time series, each one made by 15,139 observations and corresponding to a weather station located in the Australia region. The choice of the observation period and the selection of the weather stations has been carried out in such a way to have no missing data. Precipitation is most often rain, but also includes other forms such as snow. Observations of daily rainfall are nominally made at 9 a.m. local clock time and record the total precipitation for the preceding 24 h. If, for some reason, an observation is unable to be made, the next observation is recorded as an accumulation, since the rainfall has been accumulating in the rain gauge since the last reading. As can be seen from Fig. 1 , daily rainfall is characterized by intense variations. The highest values of the mean precipitation reached in the 15 days of the first window could seem comparable with the maxima of the 25 days of the second window. However it is not the same daily rainfall stream but a stream related to different stations. In this sense, the overall trend of the phenomenon cannot be detected. In the following we show as our method can help to catch the main rainfall behaviors along the whole observation period and to describe and graphically represent them by means of a set of Functional boxplots.

The assessment of the method requires to set two input parameters: the size of each window w_j and the maximum number K_{max} of generated FBP_{C_k} micro-clusters. We set the first one to $w = 30$ in order to get on-line computed functional boxplots summarizing 30 days of observations. The second parameter has been set to $K_{max} = 50$, which represents a good compromise between the detail of summarization and the memory usage.

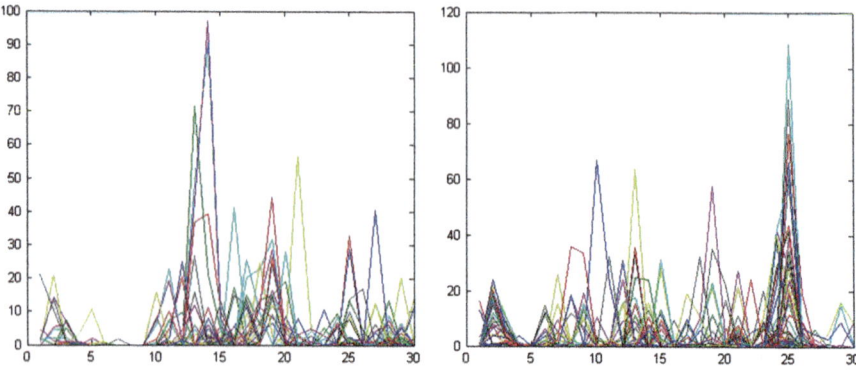

Fig. 1 Daily rainfall in two different time windows made by 30 observations

Table 1 The number of on-line computed FBP allocated to each FBP_{C_k}

FBP_{C_k}	n^k	FBP_{C_k}	n^k	FBP_{C_k}	n^k	FBP_{C_k}	n^k	FBP_{C_k}	n^k
FBP_{C_1}	45	$\text{FBP}_{C_{11}}$	1	$\text{FBP}_{C_{21}}$	1	$\text{FBP}_{C_{31}}$	1	$\text{FBP}_{C_{41}}$	15
FBP_{C_2}	2	$\text{FBP}_{C_{12}}$	1	$\text{FBP}_{C_{22}}$	1	$\text{FBP}_{C_{32}}$	1	$\text{FBP}_{C_{42}}$	1
FBP_{C_3}	2	$\text{FBP}_{C_{13}}$	4	$\text{FBP}_{C_{23}}$	1	$\text{FBP}_{C_{33}}$	4	$\text{FBP}_{C_{43}}$	1
FBP_{C_4}	380	$\text{FBP}_{C_{14}}$	4	$\text{FBP}_{C_{24}}$	4	$\text{FBP}_{C_{34}}$	1	$\text{FBP}_{C_{44}}$	1
FBP_{C_5}	26	$\text{FBP}_{C_{15}}$	1	$\text{FBP}_{C_{25}}$	1	$\text{FBP}_{C_{35}}$	1	$\text{FBP}_{C_{45}}$	1
FBP_{C_6}	1	$\text{FBP}_{C_{16}}$	1	$\text{FBP}_{C_{26}}$	1	$\text{FBP}_{C_{36}}$	28	$\text{FBP}_{C_{46}}$	1
FBP_{C_7}	5	$\text{FBP}_{C_{17}}$	2	$\text{FBP}_{C_{27}}$	1	$\text{FBP}_{C_{37}}$	1	$\text{FBP}_{C_{47}}$	2
FBP_{C_8}	24	$\text{FBP}_{C_{18}}$	45	$\text{FBP}_{C_{28}}$	2	$\text{FBP}_{C_{38}}$	38	$\text{FBP}_{C_{48}}$	1
FBP_{C_9}	2	$\text{FBP}_{C_{19}}$	4	$\text{FBP}_{C_{29}}$	2	$\text{FBP}_{C_{39}}$	1	$\text{FBP}_{C_{49}}$	1
$\text{FBP}_{C_{10}}$	4	$\text{FBP}_{C_{20}}$	1	$\text{FBP}_{C_{30}}$	1	$\text{FBP}_{C_{40}}$	1	$\text{FBP}_{C_{50}}$	1

From the results, see Table 1, we can observe that eight FBP_{C_k} collect more than five on-line computed functional boxplots so these are the ones that record the main concepts in the data. The remaining FBP-micro-clusters summarize the anomalous or residual rainfall behaviors.

The off-line procedure, which is performed taking as input the whole set of FBP_{C_k}, provides a final summarization of the data. We are interested in discovering how the whole trend changes over the days and if there are dominant structure in the data behaviors. Thus, we choose to get four final functional boxplots summarization (Fig. 2).

Comparing the original curves to the four functional boxplots, we see that the latter are very informative to underline the main changes in the data. In all the four cases, the curve distributions are asymmetric and positively skewed. The four functional boxplots differ mostly for: the median curve, that can be interpreted as the most representative observed patterns of rainfall data; the central region, that gives a less biased visualization of the curves' spread.

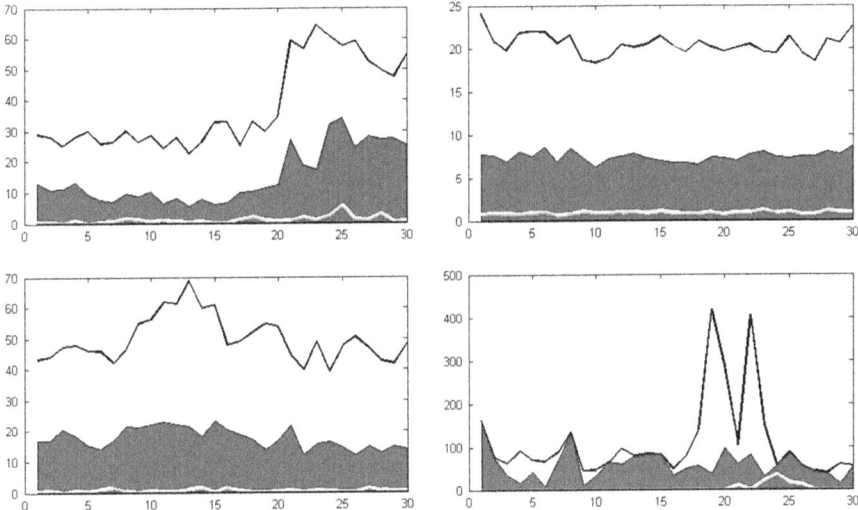

Fig. 2 Functional boxplots summarization for daily rainfall data with four micro-cluster. The *black curves* denote the envelope. The *dark gray* area delimits the 50% central region. The *white curve* represents the median

In the first Functional Boxplot, the median curve is characterized by low and oscillating rainfall trend around 1.5 mm with higher values between the 23th and 27th day. In this case, more information is detected by observing the box. It highlights that the 23rd and 25th rainfall are high in the last 10 days.

At the opposite the second Boxplot depicts a quasi-constant rainfall trend around the 3 mm (the median curve) with a similar shape of the box but with higher values of the rainfall. This indicates that the trend rainfall vary with a constant trend among 5 and 7.5 mm. The third Functional Boxplot instead shows lower values of the rainfall median curve with a highest values of the box bounds (the values vary among 18 and 20 mm). Finally the fourth Functional Boxplot highlights a median rainfall near to zero except for the 20th and 28th and a box with a concentration of rainfall curves in the third quartile with high variability.

All the four Functional Boxplots have an envelope bounded by the blue curve which has a minimum value corresponding to an absence of rainfall. Thus, the lower curve shall be the same with the x-axis. The upper curve limit, indicating the maximum value of the fall of rain, is characterized by four different behaviors linked evidently to different period of summarization. In the first Functional Boxplot it can be observed a curve with an almost constant trend with a value oscillating around 30 mm for the first 20 days and up to 80 mm in the other 10 days. In the second and third boxplot on the contrary, the trends vary around the value of 20 and 55 mm. Finally the fourth boxplot shows a narrow central region with high peaks in the interval 19th and 24th day.

4 Concluding Remarks

In this paper we have introduced a new Clustream strategy for multiple streaming time series. It is based on a two-step process to handle incremental time series. In a first step (the on-line step) graphical summarizing structures, named Functional Boxplots, continuously updated are detected. In the second step (the off-line step) a final graphical summarization of the flow data is obtained.

Unlike the existent CluStream strategy in streaming time series literature, we have introduced a tool able also to provide a graphic synthesis.

References

Adelfio, G., Chiodi, M., D'alessandro, A., Luzio, D., D'anna, G., & Mangano, G. (2012). Simultaneous seismic wave clustering and registration. *Computers Geosciences, 44*, 60–69. ISSN: 0098-3004. doi: 10.1016/j.cageo.2012.02.017.

Aggarwal, C. C., Han, J., Wang, J., & Yup, S. (2003). A framework for clustering evolving data stream. In *Proceedings of the 29th VLDB Conference*.

Balzanella, A., Lechevallier, Y., & Verde, R. (2011). Clustering multiple data streams. In *New perspectives in statistical modeling and data analysis*. Heidelberg: Springer. ISBN: 978-3-642-11362-8. doi: 10.1007/978-3-642-11363-5-28.

Dai, B. R., Huang, J. W., Yeh M. Y., & Chen, M. S. (2006). Adaptive clustering for multiple evolving streams. *IEEE Transactions on Knowledge and Data Engineering, 18*(9), 1166–1180.

Guha, S., Meyerson, A., Mishra, N., & Motwani, R. (2003). Clustering data streams: Theory and practice. *IEEE Transactions on Knowledge and Data Engineering, 15*(3), 515–528.

Lopez-Pintado, S., & Romo, J. (2009). On the concept of depth for functional data. *Journal of the American Statistical Association, 104*, 718–734.

Ramsay, J. E., & Silverman, B. W. (2005). *Functional data analysis*, 2nd ed. New York: Springer.

Romano, E., Balzanella, A., & Rivoli, L. (2011). Functional boxplots for summarizing and detecting changes in environmental data coming from sensors. In *Electronic Proceedings of Spatial 2, Spatial Data Methods for Environmental and Ecological Processes 2nd Edition*. Foggia, 1–3 Settembre.

Sangalli, L. M., Secchi, P., Vantini, S., & Vitelli, V. (2010). K-mean alignment for curve clustering. *Computational Statistics and Data Analysis, 54*(5), 1219–1233. ISSN 0167-9473. 10.1016/j.csda.2009.12.008.

Sun, Y., & Genton, M. G. (2011). Functional boxplots. *Journal of Computational and Graphical Statistics, 20*, 316–334.

Smooth Tests of Fit for Gaussian Mixtures

Thomas Suesse, John Rayner, and Olivier Thas

Abstract Model based clustering and classification are often based on a finite mixture distribution. The most popular choice for the mixture component distribution is the Gaussian distribution (Fraley and Raftery, J Stat Softw 18(6):1–13, 2007). Many tests, for example those based on goodness of fit measures, focus on detecting the order of the mixture. However what is often neglected are diagnostic tests to confirm the distributional assumptions. This may lead to the cluster analysis having invalid conclusions.

Smooth tests (Rayner et al., Smooth tests of goodness of fit: using R, 2nd edn. Wiley, Singapore, 2009) can be used to test the distributional assumptions against the so-called general smooth alternatives in the sense of Neyman (Skandinavisk Aktuarietidskr 20:150–99, 1937). To test for a mixture distribution we present smooth tests that have the additional advantage that they permit the testing of sub-hypotheses using components. These test statistics are asymptotically chi-squared distributed. Results of the simulation study show that bootstrapping needs

T. Suesse (✉)
National Institute for Applied Statistics Research Australia, University of Wollongong, Wollongong, NSW 2522, Australia
e-mail: tsuesse@uow.edu.au

J. Rayner
National Institute for Applied Statistics Research Australia, University of Wollongong, Wollongong, NSW 2522, Australia

School of Mathematical and Physical Sciences, University of Newcastle, Callaghan, NSW 2308, Australia
e-mail: John.Rayner@newcastle.edu.au

O. Thas
National Institute for Applied Statistics Research Australia, University of Wollongong, Wollongong, NSW 2522, Australia

Department of Applied Mathematics, Biometrics and Process Control, Ghent University, 9000 Gent, Belgium
e-mail: olivier.thas@Ugent.be

© Springer-Verlag Berlin Heidelberg 2015
B. Lausen et al. (eds.), *Data Science, Learning by Latent Structures, and Knowledge Discovery*, Studies in Classification, Data Analysis, and Knowledge Organization, DOI 10.1007/978-3-662-44983-7_12

to be applied for small to medium sample sizes to maintain the P(type I error) at the nominal level and that the proposed tests have high power against various alternatives. Lastly the tests are illustrated on a data set on the average amount of precipitation in inches for each of 70 United States and Puerto Rico cities (Mcneil, Interactive data analysis. Wiley, New York, 1977).

1 Introduction

Mixture distributions are often used to model multi-modal distributions, for example populations containing sub-populations, and are particularly popular in model-based clustering. A mixture distribution is characterised by the density

$$f(x) = f(x, \boldsymbol{\eta}_1, \ldots, \boldsymbol{\eta}_m, p_1, \ldots, p_m) = \sum_{i=1}^{m} p_i f_i(x, \boldsymbol{\eta}_i)$$

where p_i are mixing probabilities and f_i is the component density for the ith sub-population (governed by nuisance parameters $\boldsymbol{\eta}_i$). Often the normal distribution $N(\mu, \sigma^2)$ is used as the component distribution (Fraley and Raftery 2007):

$$f_i(x, \mu_i, \sigma_i^2) = \frac{1}{\sigma_i \sqrt{2\pi}} \exp\left(-\frac{(x - \mu_i)^2}{2\sigma_i^2}\right)$$

where $\boldsymbol{\eta}_i = (\mu_i, \sigma_i^2)'$, as it will be throughout this paper.

Similarly, for multivariate observations, the multivariate normal is a common choice. In cluster analysis, it is assumed that the ith cluster is represented by f_i and often the means μ_i and variances σ_i^2 are of interest. Estimates of such mixture models are usually obtained by applying maximum likelihood (ML) and using the EM algorithm. The literature on finite mixture models focuses on the detection of the order of the mixture, i.e. m; see Lo et al. (2001) and Li and Chen (2010). Related to testing for the order of a mixture is testing for homogeneity, i.e. testing whether two normal component distributions are the same (Li et al. 2009; Chen and Li 2009).

However, to our best knowledge, more general questions, as to whether the distribution f is correctly specified, or more detailed questions, as whether a particular component distribution f_i is correctly specified, have not been addressed in the literature. Mis-specification of the mixture distribution is an important topic, because inference depends on the correct specification of the component densities f_i. For the normal distribution, the mean μ and variance σ^2 will not be "mis-specified", because they are essentially estimated from the data, but the skewness and kurtosis of the data may not agree with those of the normal distribution.

In this paper, we introduce smooth tests for the overall null hypothesis that f is correctly specified and formulate two so-called "smooth" alternative distributions, which will be called alternatives 1 and 2. Smooth tests based on the alternative 1 model will test departure from the null in the moments of f, whereas the alternative 2 model is used to test the correct specification of the component distributions. The next section reviews the methodology of score tests and introduces the two alternatives. Our smooth tests are score tests with respect to smooth alternatives in the sense of Neyman (1937). A simulation then investigates and compares their performance with standard goodness of fit (GOF) tests. The tests are illustrated in Sect. 4 using the average amount of rainfall for each of 70 United States and Puerto Rico cities (Mcneil 1977). The last section gives a brief conclusion.

2   Smooth Tests

A smooth test is essentially a score test with respect to a smooth alternative. To start we review the score test in its general form. Based on a random sample $x_1, \ldots, x_n$ we test

$$H_0: \text{distribution is specified by } f, \text{ i.e. } \boldsymbol{\theta} = \mathbf{0}$$

versus

$$H_1: \text{distribution is specified by } g, \text{ where } f \text{ is nested within } g$$

Here we assume $g(\boldsymbol{\eta}, \boldsymbol{\theta})$ is characterised by $\boldsymbol{\eta}$ and $\boldsymbol{\theta}$ and $g(\boldsymbol{\eta}, \mathbf{0}) = f(\boldsymbol{\eta})$, where $\boldsymbol{\eta} = (\eta_1', \ldots, \eta_m', p_1, \ldots, p_{m-1})'$.

Let $\mathbf{u}(\boldsymbol{\zeta}) := \partial L(\boldsymbol{\zeta})/\partial \boldsymbol{\zeta}$ be the score vector, where $\boldsymbol{\zeta} = (\boldsymbol{\eta}', \boldsymbol{\theta}')'$ and $L = \sum_{j=1}^{n} \log g(x_j)$ is the log-likelihood based on the sample $x_1, \ldots, x_n$, and let $\mathbf{I}(\boldsymbol{\zeta}) = -E(\partial^2 L/\partial \boldsymbol{\zeta} \partial \boldsymbol{\zeta}')$ be the information matrix.

Then the general score test statistic has the following form

$$S = \mathbf{u}'(\hat{\boldsymbol{\zeta}}_0)\mathbf{I}^{-1}(\hat{\boldsymbol{\zeta}}_0)\mathbf{u}(\hat{\boldsymbol{\zeta}}_0) \tag{1}$$

where $\hat{\boldsymbol{\zeta}}_0 = (\mathbf{0}', \hat{\boldsymbol{\eta}}')'$ is an estimate of $\boldsymbol{\zeta}_0 = (\mathbf{0}', \boldsymbol{\eta}')'$. The statistic S is asymptotically chi squared distributed with degrees of freedom d.f.$=|\boldsymbol{\theta}|$, the number of restrictions imposed by the full model over and above those imposed by H_0. $\hat{\boldsymbol{\eta}}'$ is the ML estimate under H_0. The advantages of the score test are: the test is asymptotically optimal and estimates are only needed for the null model and not under the alternative.

Next we present two alternative models, one with respect to the whole mixture density (*Alternative 1*) and one with respect to each component density (*Alternative 2*)

2.1 Alternative 1

Following the general approach of Rayner et al. (2009) when testing for an arbitrary density, the smooth alternative is defined as

$$g(x; \boldsymbol{\theta}, \boldsymbol{\eta}) = C(\boldsymbol{\theta}, \boldsymbol{\eta}) \exp\left(\boldsymbol{\theta}' \mathbf{h}(x, \boldsymbol{\eta})\right) f(x, \boldsymbol{\eta}) \tag{A1}$$

where now $\mathbf{h} = \mathbf{h}(x) = (h_3(x), \ldots, h_K(x))'$ are orthonormal polynomials with respect to f, i.e. $E_f(h_r(X)h_s(X)) = 0$ for $r \neq s$ and $E_f(h_r(X)^2) = 1$. For the normal mixture distribution the elements $h_1(x)$ and $h_2(x)$ are zero, essentially because the first two moments of the fitted normal mixture agree with the moments of the data, and are consequently removed from $\mathbf{h}$; see Rayner et al. (2009) for non-mixture distributions. If they are not removed, the information matrix is singular, and the first two elements u_1 and u_2 of the score vector $\mathbf{u}$ with $u_r = \frac{1}{\sqrt{n}} \sum_{j=1}^{n} h_r(x_j)$ are zero.

A small K gives a focused test, whereas a large K gives an omnibus test. The choice of a moderate K, such as $K = 4$, is often a desirable compromise. For example for the standard normal distribution, the first four orthonormal polynomials are

$$h_1 = x, h_2 = \frac{1}{\sqrt{2}}(x^2 - 1), h_3 = \frac{1}{\sqrt{6}}(x^3 - 3x), h_4 = \frac{\sqrt{6}}{12}(x^4 - 6x^2 + 3).$$

By definition the zeroth order polynomial is $h_0 = 1$, so that $E_f h_0 = E_f h_0^2 = 1$. The orthonormal polynomials with respect to f need to be constructed, i.e. the coefficients need to be determined given the moments of f, which can be calculated from the moments of the component densities by $E_f(X^k) = \sum_{i=1}^{m} p_i E_{f_i}(X^k)$ where $E_{f_i} X^k = E_Z(\sigma_i Z + \mu_i)^k$ and Z is standard normal. Note that $EZ^k = 0$ if k is odd and if $k = 2i$ is even, then $EZ^{2i} = (2i - 1)!!$ Given $\eta_i = (\mu_i, \sigma_i^2)$, the moments w.r.t. f can be calculated and consequently the coefficients of the orthonormal polynomials. The R-package `orth` available on request from the third author can be used to calculate these orthonormal polynomials. It requires the moments of the mixture and gives the coefficients of the orthonormal polynomials.

Consider the term $\boldsymbol{\theta}' \mathbf{h}(x, \boldsymbol{\eta}) = \sum_{r=3}^{K} \theta_r h_r(x, \boldsymbol{\eta})$. If the coefficient θ_r is significant, referring to the hypothesis $H_{0r} : \theta_r = 0$ versus $H_{1r} : \theta_r \neq 0$, then roughly speaking the hypothesised density f does not agree with the data in the rth moment. To test H_{0r} versus H_{1r} the score vector u only contains one element u_r. The score statistic, denoted by S_r, is based on u_r only, and is asymptotically chi-squared distributed with d.f.=1. Now $S_r = T_r^2$ with $T_r = u_r / \sqrt{[\mathbf{I}^{-1}(\hat{\xi}_0)]_{rr}}$, where $[\cdot]_{rr}$ denotes the rth diagonal element of the inverse of the information matrix. T_r is asymptotically standard normal under H_0 and for smaller sample sizes a t-distribution may be preferred.

In general, interpretation-wise, K is usually limited to 4, because deviations from the third and fourth moment can be interpreted as different skewness and kurtosis of the data compared to f. Interpretation of higher moments is often more difficult and therefore omitted here.

2.2   Alternative 2

Instead of formulating the smooth alternative with respect to f, we now formulate an alternative g_i against each component density f_i by

$$g_i(x; \boldsymbol{\theta}_i, \boldsymbol{\eta}_i) = C_i(\boldsymbol{\theta}_i, \boldsymbol{\eta}_i) \exp\left\{\boldsymbol{\theta}_i' \mathbf{h}_i(x, \boldsymbol{\eta}_i)\right\} f_i(x; \boldsymbol{\eta}_i), i = 1, \ldots, m,$$

where $\mathbf{h}_i = (h_{i3}, \ldots, h_{iK_i})'$ is a vector of orthonormal polynomials for density f_i. Since f_i is normal, $\{h_{ir}\}$ is the set of Hermite polynomials mentioned above. To obtain the Hermite polynomials for the general normal, we note that $(X - \mu_i)/\sigma_i$ is standard normal, so we apply the Hermite polynomials of the standard normal, but evaluated at $(X - \mu_i)/\sigma_i$.

Overall the alternative model g versus f is

$$g(x; \boldsymbol{\theta}, \boldsymbol{\eta}) = \sum_{i=1}^{m} p_i g_i(x; \boldsymbol{\theta}_i, \boldsymbol{\eta}_i), \tag{A2}$$

which we call the order $\mathbf{K} = (K_1, \ldots, K_m)'$ alternative 2 to f.

Define H_{0ir}: $\theta_{ir} = 0$ versus H_{1ir}: $\theta_{ir} \neq 0$. If all $\theta_{ir} = 0$, then clearly H_0 is true. The advantage of defining multiple hypothesis is that it allows us to access the correctness of each component i and its moments, e.g. $\theta_{i3} \neq 0$, would indicate that the third moment (skewness) of the ith component density is mis-specified. In a similar way, we can define the null $H_{0i} : \boldsymbol{\theta}_i = \mathbf{0}$ and the alternative H_{1i}: $\boldsymbol{\theta}_i \neq \mathbf{0}$, to check whether the ith component density as a whole is correctly specified. For example, to check whether the data agree with the kurtosis and skewness of ith component density, we check whether θ_{i3} and θ_{i4} are significant.

Instead of using the score statistic S one may use $T_{ir} = u_{ir}/\sqrt{[\mathbf{I}^{-1}(\hat{\boldsymbol{\zeta}}_0)]_{ir,ir}}$, which is asymptotically standard normal. If θ_{i3} or θ_{i4} is significantly different from zero, then f_i is not normal. The proposed tests allow testing a variety of different hypotheses, such as H_{i0} and H_{ir} (alternative 2), and H_r (alternative 1) or more generally H_0, and are very flexible.

3   Simulation Study

To investigate the proposed smooth tests, we conduct a simulation study. The general set-up is to generate 2,000 data sets under H_0, a mixture of two normal distributions, and under H_1, where one or both components follow a scaled and shifted t-distribution. We chose the t-distribution, because we expect it to be difficult for standard GOF tests to detect the difference between a t and a normal distribution.

We considered sample sizes of $n = 20, 100, 200$. The mixing probabilities are set to $p_1 = 0.4$ and $p_2 = 0.6$, the means are $\mu_1 = 1$, $\mu_2 = 5$ and common variance $\sigma_1^2 = \sigma_2^2 = 2$. The choice of the mixing probabilities aims at a balanced scenario and the means and variances have been chosen to avoid a strong overlap and separation of the two component densities.

The three basic scenarios are

(i)   two normals with parameters μ_i and σ^2
(ii) t scaled (by σ_1^2) and shifted t-distribution (by μ_1) with d.f. $= 5$ and normal distribution with parameters μ_2 and σ_2^2
(iii) t two scaled (by σ_i^2) and shifted t-distributions (by μ_i), both with d.f. $= 5$

We include the parametric bootstrap method to calculate bootstrap p-values, because the asymptotic distribution only serves as a good approximation if sample sizes are large. For comparison other standard GOF tests are included: the Pearson–Fisher (PF) test statistic with equiprobable ten groups, the Anderson–Darling (AD) and the Kolmogorov–Smirnov (KS) test statistics, see, for example, Thas (2010) and Rayner et al. (2009).

Instead of only considering the overall hypothesis H_0 versus H_1 of the introduced smooth tests, we test for sub-hypotheses, i.e. H_{0r} versus H_{1r} for (A1) and H_{0ir} versus H_{1ir} for (A2) using the test statistics T_r and T_{ir}.

The estimated P(type I error) of the tests can be found in Table 1, and the power under scenarios (ii) and (iii) can be found in Tables 2 and 3, all for $n = 200$, which are representative results. These results are based on the asymptotic distributions. For $n = 20$ and $n = 100$, the type I errors based on asymptotic distributions are not successful at maintaining the type I error rate, but those using the parametric bootstrap method are similar to the results presented for $n = 200$.

Table 1 P(Type I error) of smooth tests for case (i)

Component r	Alt 1	Alt 2 ($i = 1$)	Alt 2 ($i = 2$)
3	0.040	0.041	0.038
4	0.038	0.044	0.046
5	0.045	0.045	0.042
6	0.044	0.044	0.050
7	0.040	0.044	0.047
8	0.044	0.038	0.048
	0.056	0.039	0.047
	PF	AD	KS

Table 2 Power of smooth tests for case (ii)

Component r	Alt 1	Alt 2 ($i = 1$)	Alt 2 ($i = 2$)
3	0.351	0.461	0.157
4	0.391	0.450	0.074
5	0.389	0.330	0.098
6	0.384	0.322	0.081
7	0.316	0.261	0.081
8	0.318	0.273	0.073
	0.133	0.236	0.144
	PF	AD	KS

Table 3 P(Type I error) of smooth tests for case (i)

Component r	Alt 1	Alt 2 ($i = 1$)	Alt 2 ($i = 2$)
3	0.040	0.041	0.038
4	0.038	0.044	0.046
5	0.045	0.045	0.042
6	0.044	0.044	0.050
7	0.040	0.044	0.047
8	0.044	0.038	0.048
	0.056	0.039	0.047
	PF	AD	KS

The column "component" refers to the index $r = 3, 4, 5, 6$ of T_r and T_{ir}. We consider each "component" as a separate test, because if H_{0r} or H_{0ir} (whichever is applicable) is rejected, then H_0 is rejected.

The results show that the P(type I error) is around $\alpha = 0.05$. Under scenario (ii), where the first component distribution is non-normal, the smooth tests based on alternative 2 and referring to the first mixture component are most powerful, whereas those for the second component are not powerful. It confirms that the proposed tests based on alternative 2 are useful in detecting mis-specification in the mixture component densities. Scenario (iii), on the other hand, clearly shows that the smooth tests based on the more general non-component based alternatives are more powerful if the mixture is mis-specified in more than one mixture component.

The results also show that the standard GOF tests have relatively low power in detecting minor mis-specification of the normal distribution, such as the t-distribution.

For cases (ii) and (iii), Fig. 1 gives a plot of the true mixture distribution ("true density") and a plot of a normal mixture with means and variances equal to the averaged ML estimates over all 2,000 simulated data sets ("estimated normal").

In case (ii) the components corresponding to the normal population with the alternative 2 model are quite correctly small, because the hypothesised component distribution is correct. In this case the mode to the right appears to be well placed. However the mode to the left is misplaced, being somewhat to the left and smaller than the data suggest. Not surprisingly the components corresponding to the t_5 population are all large. They decrease, mainly identifying skewness and kurtosis

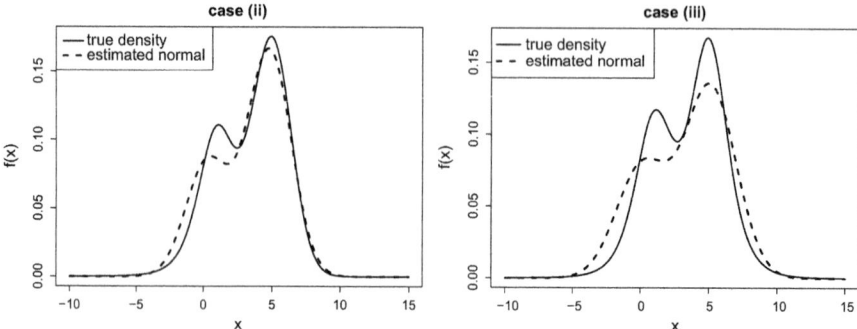

Fig. 1 *Left*: One scaled t and one normal distribution, case (ii), *right*: two scaled t-distributions, case (iii), *dashed line*: normal mixture with "average" estimated parameter estimates, *solid line*: true density

deviations from the hypothesised mixture. With the tests based on the alternative 2 model the fifth and sixth components are more powerful than the third. These tests are quite correctly seeing the overall model as more complex than simply skewness and kurtosis differences from the null.

In case (iii) the component tests based on the alternative 2 model are seeing non-agreement everywhere and are unable to single out particular difficulties. The component tests based on the alternative 1 model identify the even components. They are assessing the agreement of the data with the overall model. The true plot has both modes higher than those for the null mixture, and we would therefore expect the kurtosis component to have good power. The substantial sixth and eighth components may be seen as aliases of the kurtosis component.

4   Example

Let us now consider the data set of the average amount of rainfall in inches for each of 70 United States and Puerto Rico cities (Mcneil 1977). We fitted a mixture of two normal distributions and also plotted the estimated density using non-parametric kernel density estimates. The mean estimates of the two normal component densities are $\hat{\mu}_1 = 12.77$ in. and $\hat{\mu}_2 = 39.76$ in. of rainfall, the estimated variances are $\hat{\sigma}_1^2 = 16.68$ and $\hat{\sigma}_2^2 = 90.74$ and the mixing probabilities $\hat{p}_1 = 0.181$ and $\hat{p}_2 = 0.819$.

The parametric bootstrap p-values of the tests can be found in Table 4. It shows that all tests have a non-significant p-value at the 5 % significance level. However at the 10 % level, the statistic t_{23} ($i = 2, r = 3$) is significant. We can conclude that using a mixture of two normal distributions to model the data is appropriate overall. However in regard to skewness there is some doubt about the appropriateness of the normal distribution for the 2nd component density (Fig. 2).

Table 4 p-Values for testing for a mixture of two normals

Component r	Alt 1	Alt 2 ($i = 1$)	Alt 2 ($i = 2$)
3	0.083	0.931	0.086
4	0.142	0.392	0.228
5	0.829	0.453	0.447
6	0.128	0.268	0.173
7	0.724	0.166	0.940
8	0.666	0.279	0.515
	0.75	0.173	0.234
	PF	AD	KS

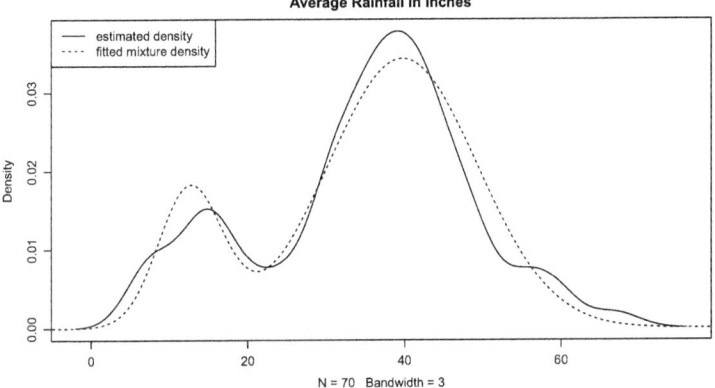

Fig. 2 Fitted normal mixture distribution and estimated density for the rainfall data set

5 Conclusion

We proposed two types of smooth tests based on two alternatives to test whether data can be modelled by a normal mixture distribution. Alternative 1 seems extremely useful because it can detect violations in each component density and in each moment of each component density. Overall the proposed tests are more powerful than existing standard tests, such as the Fisher–Pearson. Smooth tests based on alternative 2 are most powerful among the considered tests if one component density is not normal. In contrast the smooth tests based on alternative 1 are most powerful if both component densities are not normal.

An important caveat on the conclusions here is that significant effects at a particular order (using orthonormal polynomials of order r say) may affect conclusions at higher orders. This effect is well explored in testing GOF. See, for example, Rayner et al. (2009, section 5.3.3 and p. 196). In that context we argue that a significant component of order r may affect the significance or not of components up to order $2r$, but most attention should focus on components up to order r. The situation here requires both theoretical and empirical exploration that we defer to another time. We suggest that from a data analytic perspective, effects using orthonormal polynomials of order r be interpreted as reflecting moment effects of that order.

If the null hypothesis is rejected, other distributions have to be considered. Future research will aim at developing smooth tests for arbitrary densities.

References

Chen, J., & Li, P. (2009). Hypothesis test for normal mixture models: The EM approach. *The Annals of Statistics, 37*, 2523–2542.

Fraley, C., & Raftery, A. E. (2007). Model-based methods of classification: Using the mclust software in chemometrics. *Journal of Statistical Software, 18*(6), 1–13. http://www.jstatsoft. org/.

Li, P., & Chen, J. (2010). Testing the order of a finite mixture. *Journal of the American Statistical Association, 105*(491), 1084–1092.

Li, P., Chen, J., & Marriott, P. (2009). Non-finite fisher information and homogeneity: An EM approach. *Biometrika, 96*(2), 411–426.

Lo, Y., Mendell, N. R., & Rubin, D. B. (2001). Testing the number of components in a normal mixture. *Biometrika, 88*(3), 767–778.

Mcneil, D. R. (1977). *Interactive data analysis.* New York: Wiley.

Neyman, J. (1937). Smooth test for goodness of fit. *Skandinavisk Aktuarietidskr, 20*, 150–99.

Rayner, J. C. W., Thas, O., & Best, D. J. (2009). *Smooth tests of goodness of fit: Using R* (2nd ed.). Singapore: Wiley.

Thas, O. (2010). *Comparing distributions.* New York: Springer.

Part III
Machine Learning and Knowledge Discovery

P2P RVM for Distributed Classification

Muhammad Umer Khan, Alexandros Nanopoulos, and Lars Schmidt-Thieme

Abstract In recent years there is an increasing interest for analytical methods that learn patterns over large-scale data distributed over Peer-to-Peer (P2P) networks and support applications. Mining patterns in such distributed and dynamic environment is a challenging task, because centralization of data is not feasible. In this paper, we have proposed a distributed classification technique based on relevance vector machines (RVM) and local model exchange among neighboring peers in a P2P network. In such networks, the evaluation criteria for an efficient distributed classification algorithm is based on the size of resulting local models (communication efficiency) and their prediction accuracy. RVM utilizes dramatically fewer kernel functions than a state-of-the-art "support vector machine" (SVM), while demonstrating comparable generalization performance. This makes RVM a suitable choice to learn compact and accurate local models at each peer in a P2P network. Our model propagation approach, exchange resulting models with peers in a local neighborhood to produce more accurate network wide global model, while keeping the communication cost low throughout the network. Through extensive experimental evaluations, we demonstrate that by using more relevant and compact models, our approach outperforms the baseline model propagation approaches in terms of accuracy and communication cost.

1 Introduction

In recent years there is an increasing interest for analytical methods that learn patterns over large-scale data distributed over Peer-to-Peer (P2P) networks and support applications. For example, distributed classification of large amount of

M.U. Khan (✉) • L. Schmidt-Thieme
Information Systems and Machine Learning Lab, University of Hildesheim, Hildesheim, Germany
e-mail: khan@ismll.uni-hildesheim.de; schmidt-thieme@ismll.uni-hildesheim.de

A. Nanopoulos
University of Eichstätt, Ingolstadt, Germany
e-mail: alexandros.nanopoulos@ku.de

© Springer-Verlag Berlin Heidelberg 2015
B. Lausen et al. (eds.), *Data Science, Learning by Latent Structures, and Knowledge Discovery*, Studies in Classification, Data Analysis, and Knowledge Organization, DOI 10.1007/978-3-662-44983-7_13

tagged text and image data stored in online newspapers, digital libraries, and blogs. P2P matchmaking analyzes user profiles to recommend more appropriate profiles to connect with. Clustering content with respect to user's interest in media sharing P2P networks (e.g., BitTorrent, Shareaza, LimeWire, etc). Other applications include collaborative and distributed spam classification (Caruana et al. 2012) and outlier detection and scene segmentation in sensor networks.

1.1 Motivation

Mining patterns from such large-scale distributed P2P networks is a challenging task, because centralization of data is not feasible due to prohibitive communication cost and user's privacy concerns. In P2P networks, computing devices might be connected to the network temporarily, communication is unreliable and perhaps with limited bandwidth, resources of data and computation can be distributed sparsely, and the data collections are evolving dynamically. A scheme which centralizes the data stored all over a P2P network is not feasible, because any change must be reported to the central peer, since it might very well alter the result. Therefore, the goal is to develop distributed mining algorithms that are communication efficient, scalable, asynchronous, and robust to peer dynamism, which achieve accuracy as close as possible to centralized but in-feasible ones. Recently, researchers have proposed model propagation approaches based on support vector machine (SVM; Papapetrou et al. 2011) and its variants (Hock et al. 2008) which tend to reduce model size by random sub-sampling techniques, for distributed classification in P2P networks. An inherent problem with these approaches is that, being based on SVM based classifiers and their variance, the size of resulting model (number of support vectors) typically grows linearly with the size of the training set. Therefore, such schemes incur a high communication cost required to exchange models among peers, a fact that negatively impacts the efficiency of the distributed data mining approach.

1.2 Contribution

In this paper, we have presented a distributed classification approach (P2P-RVM) for P2P networks, which is based on relevance vector machines (RVM) (Tipping 2001) and model exchange between peers within a local neighborhood. The key feature of RVM is that while capable of generalization performance equivalent to a regular SVM, it utilizes significantly fewer kernel functions. This sparsity is achieved since posterior distributions of many of the kernel weights tend to get *zero*, during learning process. This makes it extremely effective to keep only those kernel functions which are more prototypical or relevant vectors of the local data, for making accurate predictions using compact models. Each peer in the P2P network learns an RVM model locally, and exchange this model in a synchronized way with its directly connected neighbors in the local neighborhood of network.

To perform extensive experimental evaluation of the proposed method, we have developed a simulation test-bed and compared our approach to baseline methods, which deploy variants of SVM based on random sub-sampling for model propagation in P2P networks. Experimental results demonstrate that P2P-RVM exhibits high classification accuracy and significantly reduces communication cost outperforming the bench-marked methods, and comparable in accuracy to any *state-of-the-art* centralized classifier. We also show that the proposed method is scalable i.e. independent of the size of P2P network.

The rest of this paper is organized as follows: Section 2 describes related work. Section 3 introduces our proposed distributed classification approach. Our simulation framework and experimental evaluations are described in Sect. 4, and the last section concludes the paper.

2 Related Work

Current *state-of-the-art* research in P2P data mining focuses on developing local classification or clustering algorithms which in turn make use of primitive operations such as distributed averaging, majority voting, and other aggregate functions. Most representative work in this regard is distributed association rule mining by Wolff and Schuster (2004), distributed decision tree induction by Bhaduri et al. (2008), distributed K-Means clustering by Datta et al. (2009), and distributed classification by Ping Luo et al. (2007). Most of these locally synchronized algorithms are reactive in a sense that they tend to create a consensus in the local neighborhood of peers, by monitoring every single change in data and keeping track of data statistics, which also require extra polling messages for coordination.

Based on model propagation, an important work for distributed classification in P2P networks is by Hock et al. (2008), in which they build an RSVM (Lee and Mangasarian 2001) model using each peer's local data, then iteratively propagate and merge the models to create an improved model. Papapetrou et al. (2011) use SVM model exchange for this purpose. These approaches tend to rely on random sub-sampling of local data, to control the size of resulting model and then optimizing it for reduced errors. Since with SVM based classifiers the size of model grows linearly with the size of local data and many redundant support vectors also get through, these methods incur high communication cost without any significant gain in classification accuracy.

In our study, we focus on the aforementioned category of classification in P2P networks based on model propagation. Our approach is based on the idea that instead of learning models from random perturbations of data, consider the significance of each instance in data, and keep only those which are most prototypical or relevant to local data set. By using RVM as a base-learner in our model propagation, we intend to optimize both size (communication cost)and quality (accuracy) of the resulting model.

3 Approach

In this section, we present our proposed method P2P-RVM illustrating learning base classifiers locally at each peer, and iterative model propagation and update by the peers in the local neighborhood. More generally, P2P-RVM creates a "cascade" of base-classifiers, i.e. instead of analyzing whole data in one optimization step, the data is partitioned into subsets and optimized separately with multiple base-classifiers. The partial models are combined and re-learned iteratively, until the globally optimal model is obtained. We adapt this general approach for classification in a P2P network. We learn *light weight* local models on presumably naturally distributed data sets, iteratively propagate models to or receive from neighbors and update (relearn) to obtain more accurate global models.

3.1 Building Local Classifier

Classification first builds a model (denoted as classifier) based on labeled training data and then predicts class labels for new (unseen) data instances. In P2P networks, each peer contains its own training data set that is not directly available to the rest of peers. More formally, we consider an ad-hoc P2P network comprising of a set of such k autonomous peers $P = \{1, 2, \ldots, k\}$. The topology of the P2P network is represented by a (connected) graph $G(P, E)$, in which each peer $p \in P$ is represented by a vertex and an edge $\{p, q\} \in E$, where $E \subseteq \{\{p, q\} : p, q \in P\}$, whenever peer p is connected to peer q. The local training data set on a peer p is denoted as $X_p \subseteq \mathbb{R}^d$, where d is the number of data features. Finally, with $\mathcal{X} = \bigcup_{p=1}^{k} X_p$ we denote the global training data set of the entire P2P network. Please notice that $\mathcal{X}$ is not feasible to be centralized (i.e., be collected in a single peer).

Based on the local training data set X_p, each peer p can first build its local classification model m_p. However, when X_p is small, and thus not representative, the accuracy of the local model m_p is reduced. To overcome this problem, a possible solution is to learn models in a collaborative fashion, where each peer p shares its local model m_p with its immediate neighbors.

Since propagating classification models in large-scale ad-hoc P2P networks results in prohibitive communication cost, therefore, it is required to build models that are both accurate and compact, i.e. they can be represented with the *least*, as well as the most *prototypical* information, needed to be exchanged between neighboring peers.

Based on these requirements, we employ RVM, a probabilistic kernel model based on the theory of sparse Bayesian learning. The key feature of this approach is that it utilizes significantly fewer kernel functions while offering good generalization performance. This is because, inferred models are exceedingly sparse in that posterior distributions of majority of kernel weights are found to have maximum values around *zero*. Training instances associated with remaining very few non-zero weights are termed *relevant vectors*. Below we briefly describe RVM formulation derived from Tipping (2001).

3.1.1   RVM: Formulation

At each peer p, given is a training set of instance-label pairs $\{(x_j, y_j)\}_{j=1}^{|X_p|}$, where $x_j \in \mathbb{R}^d$ is an input vector and $y_j \in \{-1, 1\}$ is the corresponding class label. Since we denote the size of local training data at peer p as $|X_p|$, in the following notation $|X_p|$ denotes total number of instances. The aim of the classification task is to predict the posterior probability of class membership of x_j. Considering a generalized additive model $y(x)$ having the similar form as that of SVM prediction function,

$$y(x) = \sum_{i=1}^{|X_p|} w_i K(x, x_i) + w_0 \tag{1}$$

where w_i are model weights and $K(x, x_i)$ is a kernel function. Applying sigmoid logistic function $\sigma(y(x)) = 1/1 + e^{-y(x)}$, we can write the likelihood function for Bernoulli distribution of $P(y|x)$ as:

$$P(\hat{y}|\hat{w}) = \prod_{i=1}^{|X_p|} \sigma(y(x_i))^{y_i} [1 - \sigma(y(x_i))]^{1-y_i} \tag{2}$$

RVM uses the basis function $\phi(x_i) \equiv K(x, x_i)$ based on the kernel function in (1). Using this, we can re-write (2) as:

$$P(\hat{y}|\hat{w}) = \prod_{i=1}^{|X_p|} \sigma(\phi(x_i)w_i)^{y_i} [1 - \sigma(\phi(x_i)w_i)]^{1-y_i} \tag{3}$$

where $\hat{y} = (y_1, \ldots, y_{|X_p|})^T$, $\hat{w} = (w_1, w_2, \ldots, w_{|X_p|})^T$, and $\phi(x_i) = [1, K(x_i, x_1), K(x_i, x_2), \ldots, K(x_i, x_{|X_p|})]^T$.

RVM utilizes an $[|X_p| \times |X_p * | - 1]$ basis matrix $\Phi = [\phi(x_1), \phi(x_2), \ldots, \phi(x_{|X_p|})]$, for filtering out the most relevant basis vectors, as described next.

Estimating maximum likelihood for (3), with as many parameters w as the training examples, would lead to severe over-fitting. To avoid this, RVM puts a constraint on parameters by explicitly defining the following *prior* probability distribution over them, using the principle of *automatic relevance determination (ARD)* proposed by MacKay (1996).

$$p(\hat{\mathbf{w}}|\hat{\boldsymbol{\alpha}}) = \prod_{i=0}^{|X_p|} \mathcal{N}\left(w_i|0, \alpha_i^{-1}\right) \tag{4}$$

with α a vector of $|X_p + 1|$ hyper-parameters, i.e. each hyper-parameter moderates the strength of the weight with which it is associated. Hyper-parameters α are estimated from the training data using Gamma distribution with uniform scales.

$$p(\hat{\boldsymbol{\alpha}}) = \prod_{i=0}^{|X_p|} \text{Gamma}\left(\alpha_i|a, b\right) \tag{5}$$

Using a broad prior over the hyper-parameters α, posterior probability of the associated weights approaches to zero, thus considering those inputs as *irrelevant*. This key feature of RVM is ultimately responsible for significantly reducing the number of basis functions (and corresponding support vectors) to most *relevant* ones. These support vectors are considered to be the most *prototypical* representatives of data set. Ultimately, RVM maximizes the posterior probability of class labels parameterized by hyper-parameters α, which is known as maximizing *marginal likelihood*.

For detailed explanation of inference procedure and hyper-parameter optimization, we refer the reader to original paper by Tipping (2001), for reasons of space limitation.

3.2 Model Exchange and Update

Based on the local training data set X_p, each peer p can first build its local classification model m_p, as described in previous section. Let N_p denote the set of immediate neighbors of peer p, i.e., $N_p = \{q \in P | q \neq p, \{p, q\} \in E\}$. After learning the local model, a peer p uses its neighbor list N_p to propagate m_p to all directly connected neighbors. Moreover for receiving models, each peer p waits for time t until m_q from all the $q \in N_p$ have been received. Once all the neighboring models have been received, each peer updates its local model with the support vectors in the received ones. The resulting global model built through this collaborative process is more accurate and helps improving the classification performance of the whole P2P network. Algorithm 1 describes the working of P2P-RVM at a local peer p.

Algorithm 1 P2P-RVM algorithm for peer p

Input: X_p = Local training data set, t = Time to wait for receiving models before updating, N_p
= List of neighbors
Output: Updated model M
    Train local classifier model m_p using RVM on X_p
    foreach $q \in N_p$ **do**
        Propagate the support vectors of m_p to q ▷ Exchange with direct neighbors
    end for
    $RECEIVED_p := \varnothing$ ▷ Initialize an empty set to keep received models
    while *waiting_time* $< t$ **do**
        if *receive_request* **then**   ▷ Handle receive requests from neighbors
            if $m_q \notin RECEIVED_p$ **then**
                Send ACK
                $RECEIVED_p := RECEIVED_p \cup \{m_q\}$
            end if
        end if
    end while
    if $RECEIVED_p \neq \varnothing$ **then**
        foreach $m_q \in RECEIVED_p$ **do**   ▷ Merge all models
      $m_p = m_p \cup m_q$
        end for
    end if
    M = RVM model trained using updated m_p
    return M

Since P2P networks are highly dynamic, i.e. peers usually leave and join the network in an ad-hoc manner. Model propagation approach implicitly deals with such *peer dynamism*, because even if a peer leaves the network, its local knowledge remains in the network in the form of its model, it had shared with other peers. Moreover, as new data keeps arriving in a P2P network, our simulation considers this data as a new peer, and executes Algorithm-1 for it, consequently dealing with *data dynamism*.

4 Experiments and Results

In this section, we present our simulation setup for P2P network, experiments with P2P-RVM and the baseline methods, and finally the evaluations to compare their performance. We have performed evaluations based on two most significant criteria for the problem of learning in P2P networks, i.e. classification accuracy and communication cost. The communication cost is measured as the sum of size of all propagated models, whereas the size of each model is measured as the number of support vectors it contains. We have compared P2P-RVM with state-of-the-art model propagation technique for distributed classification called Cascade Reduced-SVM proposed by Hock et al. (2008). The performance of two methods is also compared with standard SVM, especially to analyze how better they perform relative to any state-of-the-art centralized classifier. Finally, to demonstrate the effectiveness of model exchange, we also consider a baseline that performs classification only locally, without any model exchange.

4.1 Experimental Setup

Our evaluation needs to determine the network topology with edge delays and local computations at each peer with message exchange. For this purpose we used the BRITE topology generator for P2P networks, with ASWaxman model. Other BRITE parameters we used are $HS = 1,000$, $LS = 100$ (size of plane) and constant bandwidth distribution with $MaxBW = 1,024$ and $MinBW = 10$ (please refer to BRITE documentation for more details: www.cs.bu.edu/brite). We evaluated our experiments with varied number of peers ranging from 20 to 200. We did not experiment with more peers, since it would result in unrealistically small sizes of local data at peers and could adversely affect the performance of P2P classification systems. For local computation at each peer and monitoring of message exchange, we have built our own simulator (using Java) that simulates distributed dynamic P2P overlay trees. Experiments were conducted on a cluster of 41 machines, each with ten Intel-Xeon 2.4 GHz processors, 20 GB of Ram, and connected by gigabit Ethernet.

For learning regular SVM as a centralized baseline, we used C-SVC implementation provided by LibSVM (Chih-Chung et al. 2011). We used RBF kernel for learning RVM, RSVM, and SVM classifiers. Optimal values of hyper-parameters such as kernel width for RVM, C and γ for RSVM and SVM, were found using tenfold cross validation.

4.2 Data Sets

We have used two standard benchmark classification data sets for our experiments. These are,

- *covertype* (581012×54, 7 classes) data set from UCI repository, and
- *cod-rna* (488565×8, 2 classes) data set from LIBSVM repository.

Both data sets are among the largest in these widely used repositories. In recent literature of P2P classification, *covertype* data has been used by several researchers to evaluate their models. Data from the abovementioned repositories, is already partitioned into training and test sets. Data in both training and test sets is distributed uniformly among the peers of the network, before performing classification task.

4.3 Results

Algorithms were compared with respect to quality/cost ratio, i.e. what accuracy can be achieved with a given communication cost. Communication cost is given relative to the upper bound of central scheme, which is cost of centralizing the network's whole data to some server. Figure 1 illustrates average prediction performance in correlation to communication cost for the whole network, for a complete execution of each algorithm on a network of 50 peers.

The results clearly show that P2P-RVM outperforms the baseline Cascade-RSVM (denoted as C-RSVM) by achieving average accuracy which is quite close to that of centralized SVM (denoted as Central), by utilizing only 20 % of the communication cost of C-RSVM and in-feasible Central approach. Table 1 compares the Central, C-RSVM and P2P-RVM in terms of average accuracy and the latter two in terms of average number of support vectors (nSV) used per peer, for a network of 50 peers.

Figure 2 shows the performance of P2P-RVM in terms of scalability, i.e. influence of network size (no. of peers). Secondly, it also illustrates the effectiveness of model propagation approach by comparing its accuracy with that of local models (denoted by L-RVM and L-RSVM), i.e. models learned without any exchange.

Model exchange for collaborative classification significantly improves the network wide prediction accuracy. Moreover, model exchange seems to get more beneficial, as the network size increases (especially in case of P2P networks, there is a majority of free riders with very little amount of data to perform any meaningful classification). Figure 2 also depicts the scalability of P2P-RVM as compared to C-RSVM. P2P-RVM has shown high resilience to performance degradation as the

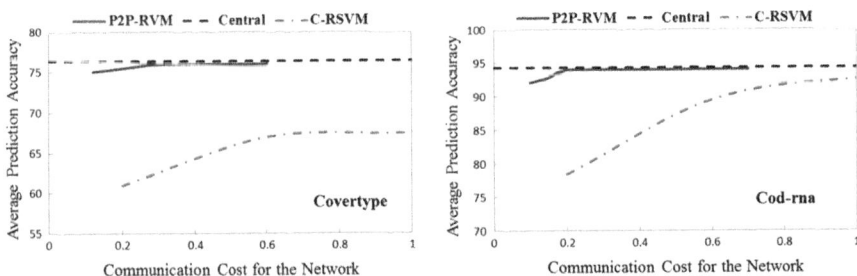

Fig. 1 Classification quality and communication cost

Table 1 Comparison of average accuracy for whole network and average model size per peer (nSV)

Data sets	Accuracy (%)			Average nSV per peer	
	Central	C-RSVM	P2P-RVM	C-RSVM	P2P-RVM
Covertype	76.5	69.5	75.9	174 ± 0	70 ± 5
Cod-rna	94.3	93.4	94.1	195 ± 0	10 ± 1

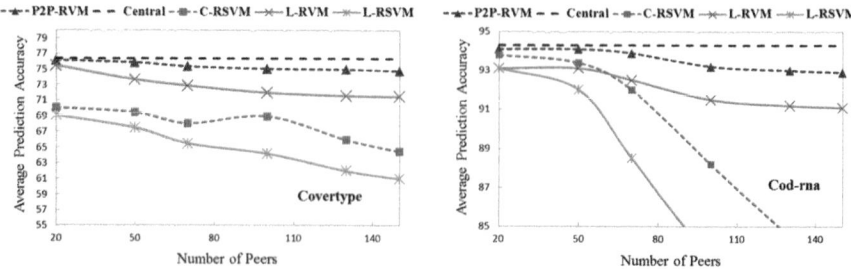

Fig. 2 Scalability: effect of number of peers in the network on average accuracy

network size increases. On the other hand, C-RSVM exhibits a decline in accuracy, as local data sets get smaller in size. The reason for this performance is that RSVM algorithm significantly depends on the size of data as it uses a random subset (n_u percent of data) to be considered as support vectors while learning the model whereas RVM approach considers only the most prototypical vectors following the principle of automatic relevance detection.

5   Conclusion

In this paper, we proposed P2P-RVM, a collaborative distributed classification approach which utilizes RVM to learn local models, and exchange them among peers in the local neighborhood of a P2P network through cascade model propagation and updates. P2P-RVM has shown a strong performance benefits in terms of classification accuracy, communication cost, and scalability.

In our future work we will investigate the problem of classification in dynamic distributed networks, where nodes are allowed to physically move in space, such as in case of vehicular ad hoc networks (VANET). We expect that approaches like RVM can be useful in scenarios where communication resources are scarce but distributed learning applications still need to be highly accurate.

Acknowledgements This work is funded by the Seventh Framework Program of European Commission, through the project REDUCTION (No. 288254). www.reduction-project.eu.

References

Ang, H.-H., Gopalkrishnan, V., Hoi, S. C., & Ng, W. W. (2008). Cascade RSVM in Peer-to-Peer Networks. In *European Conference on Machine Learning and Knowledge Discovery in Databases*.

Bhaduri, K., Wolff, R., Giannella, C., & Kargupta, H. (2008). Distributed decision-tree induction in peer-to-peer systems. *Statistical Analysis and Data Mining, 1*(2), 85–103.

Caruana, G., & Li, M. (2012). A survey of emerging approaches to spam filtering. *ACM Computing Surveys, 44*(2), Article 9, 27.

Chang, C.-C., & Lin, C.-J., LIBSVM. (2011). A library for support vector machines. *ACM Transactions on Intelligent Systems and Technology, 2,* 27:1–27:27.

Datta, S., Giannella, C., & Kargupta, H. (2009). Approximate distributed k-means clustering over a peer-to-peer network. *Transactions on Knowledge and Data Engineering, 21*(10), 1372–1388.

Lee, Y.-J., & Mangasarian, O. L.(2001). RSVM: Reduced support vector machines. In *First SIAM International Conference on Data Mining*, 5–7.

Luo, P., Xiong, H., Kevin, L., & Shi, Z. (2007). Distributed classification in peer-to-peer networks. In *13th ACM SIGKDD International Conference on Knowledge Discovery and Data Mining (KDD'07)*

MacKay, D. J. (1996). *Bayesian methods for back propagation networks. Models of neural networks III* (pp. 211–254). New York: Springer

Odysseas, P., Siberski, W., & Siersdorfer, S. (2011). Collaborative classification over P2P networks. In *20th International Conference Companion on World Wide Web (WWW '11)*

Tipping, M. E. (2001). Sparse bayesian learning and the relevance vector machine. *Journal of Machine Learning Research, 1,* 211–244.

Wolff, R., & Schuster, A. (2004). Association rule mining in peer-to-peer systems. *Transactions on Systems, Man, and Cybernetics, Part B, 34*(6), 2426–2438.

Selecting a Multi-Label Classification Method for an Interactive System

Noureddine-Yassine NAIR-BENREKIA, Pascale Kuntz, and Frank Meyer

Abstract Interactive classification-based systems engage users to coach learning algorithms to take into account their own individual preferences. However most of the recent interactive systems limit the users to a single-label classification, which may be not expressive enough in some organization tasks such as film classification, where a multi-label scheme is required. The objective of this paper is to compare the behaviors of 12 multi-label classification methods in an interactive framework where "good" predictions must be produced in a very short time from a very small set of multi-label training examples. Experimentations highlight important performance differences for four complementary evaluation measures (Log-Loss, Ranking-Loss, Learning and Prediction Times). The best results are obtained for Multi-label k Nearest Neighbors (ML-kNN), ensemble of classifier chains (ECC), and ensemble of binary relevance (EBR).

1 Introduction

The usual classification systems do not allow users to directly interact with the learning models. Consequently, in practice, their results may deviate from their preferences. Modeling human preferences remains a difficult task, especially for personalized systems where classical interviews are out of reach and large-scale behavioral logs are not available. An alternative is to embed the user into the learning process via an interactive visual support (Ware et al. 2001). The user plays

N.-Y. NAIR-BENREKIA (✉)
Orange Labs, AV. Pierre Marzin, 22307 Lannion cedex, France

LINA, la Chantrerie-BP 50609, 44360 NANTES cedex, France
e-mail: yacinenoureddine.nairbenrekia@orange.com

P. Kuntz
LINA, la Chantrerie-BP 50609, 44360 NANTES cedex, France
e-mail: pascale.kuntz@univ-nantes.fr

F. Meyer
Orange Labs, AV. Pierre Marzin, 22307 Lannion cedex, France
e-mail: franck.meyer@orange.com

© Springer-Verlag Berlin Heidelberg 2015
B. Lausen et al. (eds.), *Data Science, Learning by Latent Structures, and Knowledge Discovery*, Studies in Classification, Data Analysis, and Knowledge Organization, DOI 10.1007/978-3-662-44983-7_14

the role of a trainer for an automatic classification algorithm and steers it towards his/her desired concepts. More precisely, in a dynamical process, he/she can define a set of subjective labels $\mathcal{L}_t$ on a set of training examples $\mathcal{T}_t$ described by a set of features $\mathcal{F}$ and correct, if and/or when necessary, the labels predicted by an automatic classifier on a set of unlabeled examples $\mathcal{S}_t$.

Such interactive machine learning process has recently received increasing attention and found applications in several domains. For instance, for document organization, *iCluster* (Drucker et al. 2011) is an interactive, mono-label system that assists users with item-to-group and group-to-item recommendations. It only requires few user-classified examples. With a similar objective, *Smart selection* (Ritter and Basu 2009) interactively helps users achieve complex file selections. Based on the restricted set of files selected by the user, it automatically generalizes the selection to the rest of the files. For social recommendation, *ReGroup* (Amershi et al. 2012) assists users in the creation of personalized groups in social networks on the basis of a small list of friends provided by the user. s However, most of these interactive systems limit the user to a single-label classification which may not be sufficiently meaningful in some organization tasks where examples may belong to different labels. Interactive multi-label classification is the generalization of the interactive single-label classification where examples can be associated with more than one subjective label ($\mathcal{L} > 1$). As an illustration, let us consider a very simplified example (Table 1) from the applicative context of Video-on-Demand (*VoD*) in which we are notably interested. We suppose that, after a short interaction t, a user has created a set $\mathcal{L}_t = \{\text{Funny, I like, Good music, Sad}\}$ of four preferred labels and associated a set $\mathcal{T}_t = \{\text{Twilight, Ice age, Titanic, Kill Bill}\}$ of four films described by a set of features $\mathcal{F} = \{\text{Year, Actor}_1, \ldots \}$ with their most relevant labels in $\mathcal{L}_t$. For instance, Titanic $\in \mathcal{T}_t$ is associated with the following three labels: I like, Good music and Sad. The learning model suggests personalized predictions for selected unlabeled examples (e.g., Sparkle $\in \mathcal{S}_t$ and Man of Steel $\in \mathcal{S}_t$) to him (the user).

Motivated by the growing number of recent interactive systems, our research intends to develop an interactive multi-label classification-based system for film recommendation in a *VoD* application. Its efficiency depends on the quality of the classifier, the visual restitutions and the interaction modalities. Its evaluation is consequently a difficult open question which requires competencies from different scientific communities.

Table 1 Motivating example

Film	Year	Actor$_1$	...	Funny	I like	Good music	Sad
Twilight	2008	R. Pattinson	...	0	1	1	0
Ice age	2002	R. Romano	...	1	1	0	0
Titanic	1997	L. DiCaprio	...	0	1	1	1
Kill bill	2003	U. Thurman	...	1	1	0	0
Sparkle	2013	J. Sparks	...	?	?	?	?
Man of steel	2013	H. Cavill	...	?	?	?	?

In this paper, we restrict ourselves to the learning component: which are the classifiers that simultaneously withstand the interactivity and the multi-label related constraints? We here propose an experimental comparison of 12 multi-label classification methods adapted to interaction. Tests have been made on five classical benchmarks of increasing difficulties. We evaluate the predictive quality with two multi-label measures from the literature (Log-Loss and Ranking-Loss), and the efficiency is roughly assessed by the learning and prediction times. Our results highlight a variety of behaviors, and help us better understand the evolution of the performances while the training set grows. Let us note that, in this paper, we restrict ourselves to learning concepts of each individual or each family independently and we do not consider the collaborative multi-users framework as in the *Smart Selection* system (Ritter and Basu 2009).

This paper is organized as follows. Section 2 precisely defines the objectives. Section 3 briefly recalls the main principles of the 12 multi-label classifiers selected for the comparison. The benchmarks and the experimental protocol are described in Sect. 4. Finally, the obtained results are discussed in Sect. 5.

2 Problem Statement

Throughout the paper, we consider a $\mathcal{F}$-dimensional feature space $\mathcal{F} = \{f_1, \ldots, f_j, \ldots, f_{|\mathcal{F}|}\}$ such that $\text{dom}(f_j) \in \mathcal{R}$, and a $\mathcal{L}$-dimensional label space $\mathcal{L} = \{\lambda_1, \ldots, \lambda_k, \ldots, \lambda_{|\mathcal{L}|}\}$ such that $\text{dom}(\lambda_k) \in \{0, 1\}$ (0: irrelevant, 1: relevant). Let $\mathcal{D} = \{(x_i, y_i) \mid i = 1..|\mathcal{D}|\}$ be a multi-label dataset of $|\mathcal{D}|$ multi-label examples. Each example $x_i = (x_i^1, \ldots, x_i^j, \ldots, x_i^{|\mathcal{F}|})$ is associated with a set of labels $y_i = (y_i^1, \ldots, y_i^k, \ldots, y_i^{|\mathcal{L}|})$ with $\text{dom}(x_i) \in \mathcal{R}^{|\mathcal{F}|}$, $\text{dom}(y_i) \in \{0, 1\}^{|\mathcal{L}|}$ and $|y_i| \leq |\mathcal{L}|$ where $|y_i|$ and $|\overline{y_i}|$ are, respectively, the number of relevant and irrelevant labels of x_i. A multi-label classifier h aims at producing labels of unlabeled examples $x_i \in \mathcal{S}$ from a very small training set $\mathcal{T} \subset \mathcal{D}$, where $\mathcal{S}$ is a large set case s.t. $|\mathcal{T}| << |\mathcal{S}|$. More precisely, $\hat{y}_i = h(x_i) = (\hat{y}_i^1, \ldots, \hat{y}_i^k, \ldots, \hat{y}_i^{|\mathcal{L}|})$ with $\text{dom}(\hat{y}_i) \in [0..1]^{|\mathcal{L}|})$.

Madjarov et al. (2012) recently proposed an extensive experimental comparison of methods for multi-label learning. Their study aimed at evaluating the predictive performances alongside the efficiency of 12 well-known multi-label classifiers with 16 evaluation measures. This thorough comparison led to the recommendation of RF-PCT, HOMER, BR, and CC for multi-label classification. Here, we add the interactivity constraints to the multi-label learning problem. Thus, an efficient classifier should be able to produce "good" predictions from a very small set of multi-label training examples, in a very short time. In the following, we consider three criteria which appeared to be the most important for the problem we are facing in the *VoD* framework:

1. **Preservation of the label ordering**. The user is more likely to be interested in a label-ranking than in a label-classification which hides the prediction confidence of each label,
2. **Maximization of the similarity between the predicted scores and the ground-truth labels**. Each label-prediction must be as close as possible to the true label,
3. **Minimization of learning and prediction times**. Time is an inescapable factor in our interactive framework.

For the first two criteria, we have selected extensions of two classical measures (ranking-loss and log-loss) for the multi-label case. The label ordering preservation is evaluated by the classical Ranking-Loss (RL) (Schapire and Singer 2000) which indicates the number of times that irrelevant labels are ranked higher than relevant labels. More precisely, let us consider a ranking function r_i that sorts the labels of each example x_i with respect to their prediction precision. We suppose that r_i is an increasing function with the quality of the prediction: the highest rank (i.e., $r_i = |\mathcal{L}|$) is given to the most relevant label and conversely (i.e., $r_i = 1$). The Ranking-Loss function $\mathrm{RL}(h, \mathcal{S})$ of the classifier h on the test set $\mathcal{S}$ is defined by

$$\mathrm{RL}(h, \mathcal{S}) = \frac{1}{|\mathcal{S}|} \sum_{i=1}^{|\mathcal{S}|} \frac{1}{|y_i| \times |\overline{y_i}|} |(\lambda_a, \lambda_b) \in y_i \times \overline{y_i} : r_i(\hat{y}_i^a) < r_i(\hat{y}_i^b)|$$

It is defined on $[0..1]$ and, as a loss measure, the lowest values indicate the best performances.

The ability of maximizing the similarity between the predicted scores and the ground-truth labels is measured by the Log-Loss measure (LL) (Read et al. 2009):

$$\mathrm{LL}(h, \mathcal{S}) = \frac{1}{|\mathcal{S}| \times |\mathcal{L}|} \sum_{i=1}^{|\mathcal{S}|} \sum_{j=1}^{|\mathcal{L}|} \min(-(\ln(\hat{y}_i^j) \times y_i^j + \ln(1 - \hat{y}_i^j) \times (1 - y_i^j)), \ln(|\mathcal{S}|))$$

It is defined on $[0.. \ln |\mathcal{S}|]$ and the lowest values are associated with good performances. Its upper artificial bound (i.e., $\ln |\mathcal{S}|$) limits magnitudes of the penalty. The LL measure provides large margins of contrasts between competing multi-label methods: worse label-errors are more harshly penalized.

For the classifier speed evaluation, we only consider the learning and prediction times (in seconds) independently. We are aware that they are closely linked to the implementation of the classifiers, and more accurate measures will be proposed in the next future. However, as all the algorithms have been implemented in the same framework, they provide first interesting tendencies of the complexity.

3 Multi-Label Methods

The multi-label learning approaches can be organized in three main families:

1. the ***problem transformation*** methods transform the multi-label learning problem into one or several single-label classification or regression problems,
2. the ***algorithm adaptation*** methods extend single-label learning algorithms for the multi-label data,
3. the ***ensemble*** methods use ensembles of classifiers either from the problem transformation or the algorithm adaptation approaches.

Our numerical comparisons are based on 12 frequently used multi-label classifiers whose implementation is available on MEKA[1] or MULAN[2] multi-label learning libraries. These classifiers are listed below. Default parameters were always used except for ML-kNN: due to the small training subsets, the number of neighbors k was set to 1.

3.1 Problem Transformation Methods

We consider five problem transformation methods. (1) The Most Frequent label set is our Baseline. For a new instance, it returns the most frequent label set in the training set. (2) Binary Relevance (BR) is probably the most popular transformation method. It learns $|\mathcal{L}|$ binary classifiers, one for each label (Schapire and Singer 2000). (3) Classifier Chain (CC) is an extension of BR that not only trains one classifier per label but also extends the dimensionality of each classifier's training data with labels of the previous classifiers, in a chain, as new features (Read et al. 2009). (4) Label Powerset (*LP*) considers each unique label set in the training data as one of the classes of a multi-class classification task (Tsoumakas and Katakis 2007). And, (5) Calibrated Label Ranking (*CLR*) extends the Ranking by Pairwise Comparison (RPC) (Hüllermeier et al. 2008) by introducing an additional virtual label to separate the relevant labels from the irrelevant ones (Fürnkranz et al. 2008). Let us note that, for the problem transformation methods, two base-learners are commonly used (Madjarov et al. 2012; Read 2010): support vector machine (SVM) (Hearst et al. 1998) and C4.5 decision tree (Quinlan 1993). We here preferred C4.5 decision tree for its low computational complexity: unlike SVM, it only requires a selected number of features for constructing a model. This choice is very important in our interactive learning framework and even more when classifiers are trained on sets with a large number of features (e.g., our VoD data).

---

[1] http://meka.sourceforge.net/.

[2] http://mulan.sourceforge.net/.

3.2 Algorithm Adaptation Methods

We consider two adaptation methods. (1) AdaBoost.MH is an extension of the famous AdaBoost which was implemented into the BoosTexter classification system for multi-label data (Schapire and Singer 2000). It was designed to minimize the Hamming loss. (2) ML-kNN is a binary relevance method which extends the lazy learning algorithm kNN by using a Bayesian approach (Zhang and Zhou 2007). It retrieves the k nearest examples of each new instance, and then determines its label set from the maximum a posteriori principle (MAP).

3.3 Ensemble Methods

We consider three ensemble methods. (1) The RAndom k labEL sets (RAkEL) constructs an ensemble of m LP classifiers. Each LP classifier is trained with a different random subset of a small size k (Tsoumakas and Vlahavas 2007), (2) Hierarchy Of Multi-label classifiERs (HOMER) constructs a hierarchy of LP classifiers such that each classifier deals with a much smaller label set compared to $\mathcal{L}$ (Tsoumakas et al. 2008). (3) Ensemble of classifier chains (ECC) and ensemble of binary relevance (EBR) are ensemble methods whose base learners are CC and BR, respectively (Read et al. 2009).

4 Experimental Setting

For the experimental comparison, we used five open datasets of various complexities (Table 2). These datasets are very small compared to the huge feature space of a VoD catalogue. Yet, they can provide first insights of the behavior properties of our selected classifiers. *Emotions* is a small dataset where each piece of music can be labeled with six emotions (e.g., sad-lonely, angry-aggressive, amazed-surprised). *Yeast* is a widely used biological dataset where genes can be associated with 14

Table 2 Basic statistics of the selected multi-label benchmarks with DL: number of distinct label sets in each dataset, $Lcard$: average number of labels associated with examples in each dataset, n: numeric and b: binary

Dataset	$|\mathcal{F}|$	$|\mathcal{D}|$	$|T|$	$|\mathcal{S}|$	$|\mathcal{L}|$	#DL	#Lcard
Emotions	72n	592	118	474	6	27	1.87
Yeast	103n	2,417	483	1,934	14	198	4.24
Scene	294n	2,407	481	1,926	6	15	1.07
Slashdot	1079b	3,782	756	3,026	22	156	1.18
Imdb	1001b	7,500	1,500	6,000	28	1,021	2.00

biological functions. *Scene* is a dataset where images can be annotated with up to six concepts (e.g., Beach, Sunset, Mountain). *Slashdot* is a dataset where documents can be associated with 22 subject categories (e.g., linux, technology, science). And, a sample of *IMDB* dataset where movies can be labeled with 17 genres (e.g., Romance, Comedy, Drama).

In order to evaluate both the predictive performance and the efficiency of each classifier, we have designed a simple experimental protocol that simulates a user that progressively classifies examples and that expects good predictions as a reward for his effort. The classifiers are tested on relatively large test sets using very small nested training sets (from 2 until 64 examples). The objective is to highlight the classifiers able to learn from restricted training sets in a short time and to measure their prediction improvement when the training sets double in size. Precisely, the experimental protocol is the following:

1. Divide each dataset $\mathcal{D}$ into five distinct folds. Use one fold for training ($\mathcal{T} = 20\%$ of $\mathcal{D}$) and the four remaining folds for test ($\mathcal{S} = 80\%$ of $\mathcal{D}$), and carry out a five cross-validation.
2. From each training set $\mathcal{T}$, extract m sets of p nested training subsets of size 2^1, $2^2, \ldots, 2^p$.
3. For each measure, evaluate the average performance of each classifier on the five test sets of each dataset. Each classifier is trained with the nested subsets of increasing size (2^1, then 2^2, $\ldots$; until 2^p). Then, for each classifier, compute the average performance on the five datasets for each training subset.

For all experiments, m was fixed to 20 (100 tested sets for 5 cross-validations) and p to 64. The number of tested sets ($m = 50$) is sufficient to evaluate the global performance of each classifier. Larger values of m did not significantly improve the results. And, the threshold ($p = 64$) is consistent with real-life experiments where users do not annotate more than 64 examples by themselves without any assistance of a learning algorithm.

5 Results

Let us note that the presented results are averages on all datasets, and that the differences between the datasets are evaluated by the standard deviations. In Figs. 1 2, 3, 4, the classifiers are ordered according to their overall performance when considering together their average performances obtained per training subset size (from 2 to 64 examples).

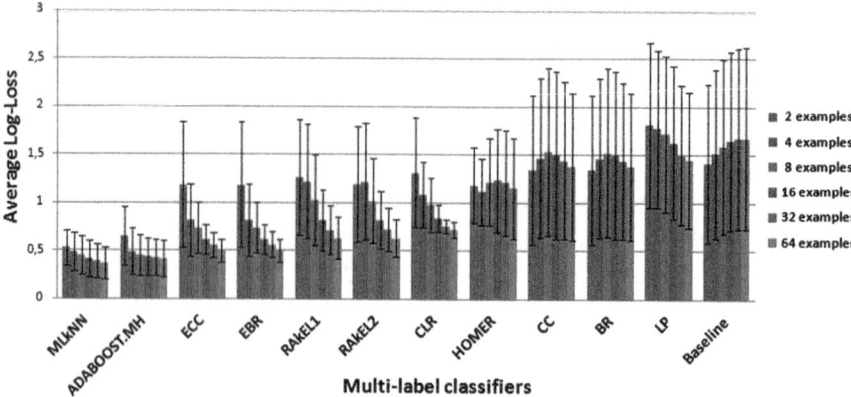

Fig. 1 Performances of the multi-label classifiers in terms of Log-Loss measure. Classifiers are ordered from left to right according to their average performances

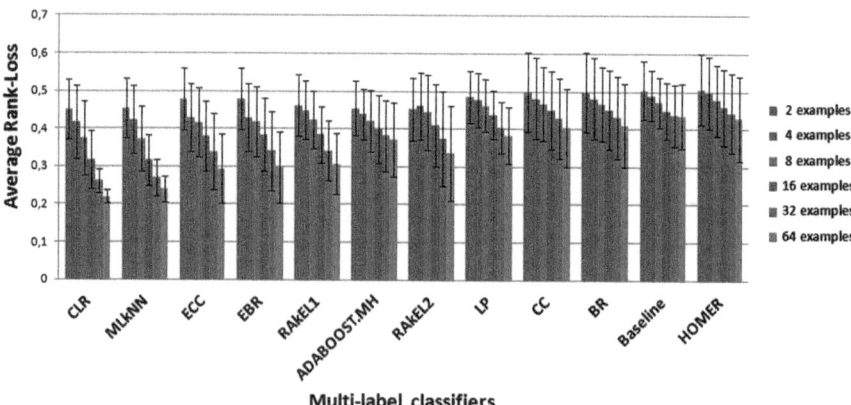

Fig. 2 Performances of the multi-label classifiers in terms of Rank-Loss measure. Classifiers are ordered from left to right according to their average performances

5.1 Log-Loss

The Log-Loss measure intensifies the discrimination between competing methods. Figure 1 clearly shows that the Multi-label k Nearest Neighbors (ML-kNN) and AdaBoost.MH outperform the other approaches with a slight advantage for ML-kNN. ML-kNN, Ensemble of BR (EBR) and ECC are known to get good performances for this measure (Read 2010) whereas, to the best of our knowledge, AdaBoost.MH has not been yet evaluated with this later. Furthermore, the worst results are obtained for Label Power-set (LP), Binary Relevance (BR), Classifier Chain (CC), and our Baseline (the most frequent label-set). The following couples of methods share similar performances: (BR, CC), (EBR, ECC) and (Random k

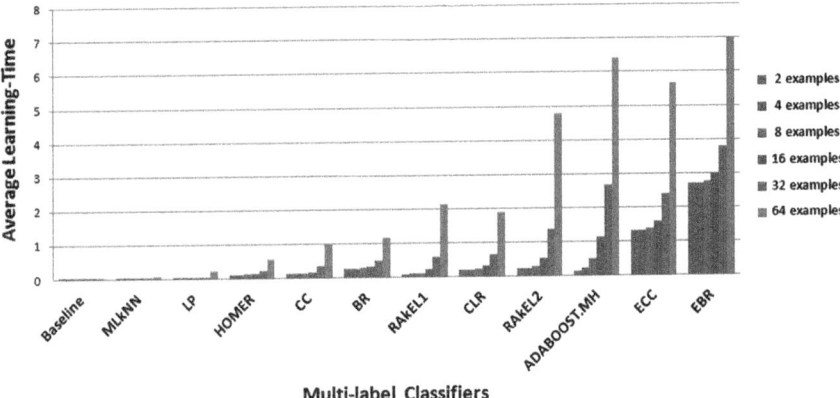

Fig. 3 Performances of the multi-label classifiers in terms of Learning-Time. Classifiers are ordered from left to right according to their average performances

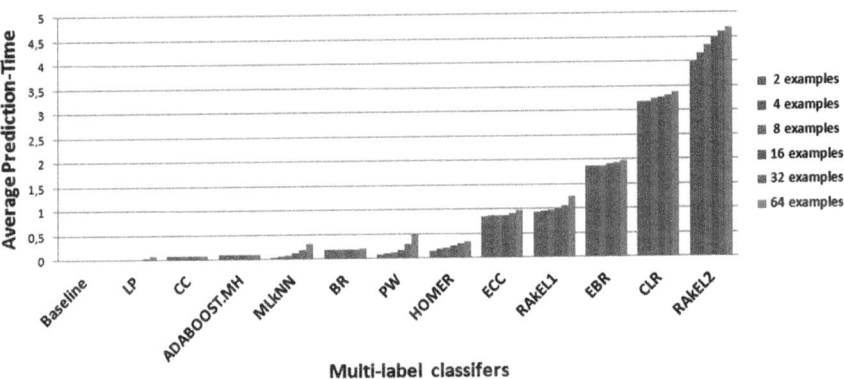

Fig. 4 Performances of the multi-label classifiers in terms of Prediction-Time. Classifiers are ordered from left to right according to their average performances

labEL-set 1 & 2). This is not surprising because CC is closely related to BR and (RAkEL$_1$, RAkEL$_2$) are variations of the original method RAkEL. However, when the cardinality of the training subsets increases, only eight algorithms are able to significantly improve their predictions. This is especially true for EBR and ECC.

5.2 *Rank-Loss*

Roughly speaking, the orderings of the classifiers with the Log-Loss and the Rank-Loss measures are quite similar (Fig. 2). This was expected since these measures are intuitively correlated. Nevertheless, a different behavior is observed for CLR and AdaboostMH; CLR is the best approach for the rank-loss and AdaBoost.MH moves from the second place to the sixth. It is not surprising to obtain the best results with CLR because it was specially designed to improve the quality of the predicted ranking (Fürnkranz et al. 2008). Similarly, ML-kNN aims at minimizing this measure. And, the ensemble methods preserve their ranks. The worst classifiers remain the same as with the Log-Loss measure. Moreover, when the training subset cardinalities increase CLR and ML-kNN are the fastest to improve the quality of the predicted ranking.

5.3 Computation Time

As all our experiments were conducted with implementations of MEKA and MULAN libraries that are not necessarily optimized, the observed learning/prediction times only provide a tendency of the algorithmic complexity (Figs. 3 and 4). For instance, EBR is expected to be computationally faster than ECC but our results show the opposite. Among the three best classifiers (ML-kNN, ECC, EBR) for the previous measures, ML-kNN seems the fastest. It does not learn a model but it requires some milliseconds for the estimation of the prior and posterior probabilities from the training subset. And, it requires less than a $\frac{1}{2}$ second for the prediction because the number of neighbors is here small ($k = 1$). EBR and ECC seem to be more expensive but they are fast enough to finish learning and predicting within the first few seconds.

6 Conclusion and Future Works

This paper studies the behaviors of different multi-label strategies in an interactive setting where training data and computation time are limited resources. The final objective, whose description goes beyond this communication, is notably the development of an interactive classification system for film recommendation in a *VoD* application. Our comparison of 12 multi-label methods from the three main popular families (problem transformation, adaptations, and ensemble methods) on five benchmarks of various complexities has shown that the best performances are reached by Multi-label k Nearest neighbors (ML-kNN), ECC and EBR. However, the limitations of ML-kNN are well known: its scaling is not assured because, in addition to a training stage, $|\mathcal{T}|$ operations are needed to predict labels of one instance only. That is quite expensive in both time and memory space. In addition, a metric learning stage may be required to bring out relevant features from the pool of irrelevant features. In spite of its limitations, ML-kNN is more efficient than ECC for larger training sets (Madjarov et al. 2012). Since implementations of ECC and EBR are closely related, ML-kNN is here considered as the "best" classifier for both small and large training data.

In the near future, we plan to extend the experiments to analyze the behavior of the classifiers on very large datasets where the number of features is very large ($|\mathcal{F}| > 5000|$) (e.g., our *VoD* data). Furthermore, we are willing to check whether the experimental learning and prediction times are consistent with the theoretical complexity.

References

Amershi, S., Fogarty, J., & Weld, D. (2012). Regroup: Interactive machine learning for on-demand group creation in social networks. In *Proceeding of the SIGCHI Conference on Human Factors in Computing Systems* (pp. 21–30). New York: ACM.

Drucker, S. M., Fisher, D., & Basu, S. (2011). Helping users sort faster with adaptive machine learning recommendations. In *Human-Computer Interaction-INTERACT* (pp. 187–203). Berlin/Heidelberg: Springer.

Fürnkranz, J., Hüllermeier, E., Mencía, E. L., & Brinker, K. (2008). Multilabel classification via calibrated label ranking. *Machine Learning, 73*(2), 133–153.

Hearst, M. A., Dumais, S. T., Osman, E., Platt, J., & Scholkopf, B. (1998). Support vector machines. *IEEE Intelligent Systems and Their Applications, 13*(4), 18–28.

Hüllermeier, E., Fürnkranz, J., Cheng, W., & Brinker, K. (2008). Label ranking by learning pairwise preferences. *Artificial Intelligence, 172*(16), 1897–1916.

Madjarov, G., Kocev, D., Gjorgjevikj, D., & Džeroski, S. (2012). An extensive experimental comparison of methods for multi-label learning. *Pattern Recognition, 45*(9), 3084–3104.

Quinlan, J. R. (1993). *C4. 5: Programs for machine learning.* San Mateo: Morgan kaufmann.

Read, J. (2010). *Scalable multi-label classification.* Doctoral dissertation, University of Waikato.

Read, J., Pfahringer, B., Holmes, G., & Frank, E. (2009). Classifier chains for multi-label classification. *Machine Learning, 85*(3), 333–359.

Ritter, A., & Basu, S. (2009). Learning to generalize for complex selection tasks. In *Proceeding of the 14th International Conference on Intelligent User Interfaces* (pp. 167–176). New York: ACM.

Schapire, R. E., & Singer, Y. (2000). BoosTexter: A boosting-based system for text categorization. *Machine learning, 39*(2–3), 135–168.

Tsoumakas, G., & Katakis, I. (2007). Multi-label classification: An overview. *International Journal of Data Warehousing (IJDWM), 3*(3), 1–13.

Tsoumakas, G., Katakis, I., & Vlahavas, I. (2008). Effective and efficient multilabel classification in domains with large number of labels. In *Proceedings of ECML/PKDD 2008 Workshop on Mining Multidimensional Data (MMD'08)* (pp. 30–44).

Tsoumakas, G., & Vlahavas, I. (2007). Random k-labelsets: An ensemble method for multilabel classification. In *Machine learning: ECML 2007* (pp. 406–417). Berlin/Heidelberg: Springer.

Ware, M., Frank, E., Holmes, G., Hall, M., & Witten, I. H. (2001). Interactive machine learning: Letting users build classifiers. *International Journal of Human-Computer Studies, 55*(3), 281–292.

Zhang, M. L., & Zhou, Z. H. (2007). ML-KNN: A lazy learning approach to multi-label learning. *Pattern Recognition, 40*(7), 2038–2048.

Visual Analysis of Topics in Twitter Based on Co-evolution of Terms

Lambert Pépin, Julien Blanchard, Fabrice Guillet, Pascale Kuntz, and Philippe Suignard

Abstract The analysis of Twitter short messages has become a key issue for companies seeking to understand consumer behaviour and expectations. However, automatic algorithms for topic tracking often extract general tendencies at a high granularity level and do not provide added value to experts who are looking for more subtle information. In this paper, we focus on the visualization of the co-evolution of terms in tweets in order to facilitate the analysis of the evolution of topics by a decision-maker. We take advantage of the perceptual quality of heatmaps to display our 3D data (term × time × score) in a 2D space. Furthermore, by computing an appropriate order to display the main terms on the heatmap, our methodology ensures an intuitive visualization of their co-evolution. An experiment was conducted on real-life datasets in collaboration with an expert in customer relationship management working at the French energy company EDF. The first results show three different kinds of co-evolution of terms: bursty features, reoccurring terms and long periods of activity.

1 Introduction

The sharp increase in the use of social web technology has led to an explosion of user-generated time-labelled texts such as news, on-line discussions and Twitter short messages. The analysis of this data has become a key issue for companies

L. Pépin (✉)
EDF R&D, Clamart, France

Equipe COD-LINA (UMR CNRS 6241), Université de Nantes, Nantes, France
e-mail: lambert.pepin@edf.fr; lambert.pepin@univ-nantes.fr

J. Blanchard • F. Guillet • P. Kuntz
Equipe COD-LINA (UMR CNRS 6241), Université de Nantes, Nantes, France
e-mail: julien.blanchard@univ-nantes.fr; fabrice.guillet@univ-nantes.fr;
pascale.kuntz@univ-nantes.fr

P. Suignard
EDF R&D, Clamart, France
e-mail: philippe.suignard@edf.fr

© Springer-Verlag Berlin Heidelberg 2015
B. Lausen et al. (eds.), *Data Science, Learning by Latent Structures, and Knowledge Discovery*, Studies in Classification, Data Analysis, and Knowledge Organization, DOI 10.1007/978-3-662-44983-7_15

seeking to understand consumer behaviour and expectations. In particular, Twitter, as a social networking and microblogging service, has become one of the most visited websites (http://www.alexa.com/topsites).

As a consequence, topic modelling (Blei and Lafferty 2006) and information tracking (Leskovec et al. 2009) in a time-labelled corpus are of renewed interest. Roughly speaking, most recent efforts have been focused on scalability challenges and the results of the automatic algorithms often provide general tendencies at a high granularity level. However, because of the volatile nature of trending topics on Twitter and the restricted format of the published messages, automatically modelling the evolution of their "semantic content" still remains an open problem. More specifically, observations of twitter messages show that: (1) most topics have a short life span "73% of the topics have a single active period and 31% of the periods are 1 day long" (Kwak et al. 2010) and (2) a high number of new words occur daily (see Fig. 1). The topic relevance is especially sensitive in practice where experts are well informed about the main trends of their customers' behaviour and are looking for more subtle information. An alternative to full-automation is to embed the user in the discovery process via an interactive visual support. Visual analytics which "is more than visualization and can rather be seen as an integrated approach combining visualization, human factors and data analysis" (Keim et al. 2010) has been proven efficient in guiding end-users to discover useful knowledge.

In this paper, we focus on the visualization of the co-evolution of Twitter terms in order to facilitate the analysis of the topic evolution by a decision maker. The visualization of terms with common behaviour might, indeed, favour the emergence of concepts. We propose a process in three main steps for co-evolution extraction: (1) scoring terms, (2) grouping the terms that evolve in the same way across a corpus and (3) visualizing the evolution of term scores on a heatmap. More precisely, we take advantage of the perceptual quality of heatmaps to display our 3D data (term × time × score) in a 2D space. Further, by computing an appropriate order to display the main terms on the heatmap, our methodology ensures an intuitive visualization

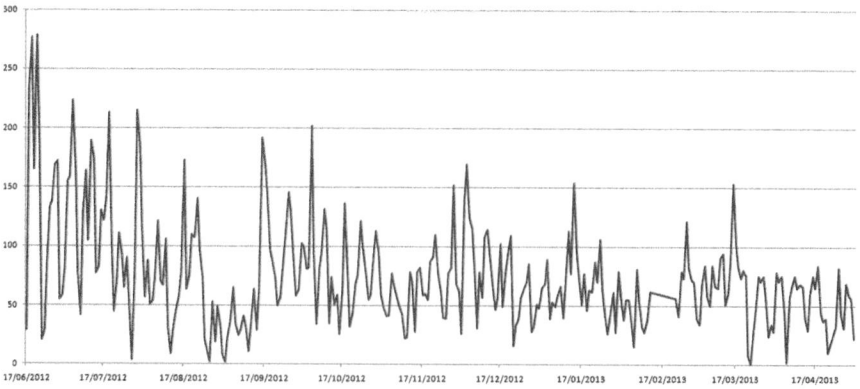

Fig. 1 Number of new words each day

of their co-evolution. Our approach was tested on a real-life dataset from the French energy company EDF.

The rest of the paper is organized as follows: Sect. 2 briefly reviews related work on topic detection and tracking (TDT) applied to Twitter and presents some visualizations adapted to time-oriented data. In Sect. 3, we put forward our methodology. In Sect. 4, we detail our experimental protocol and the results obtained in collaboration with an expert in customer relationship management.

2 Related Work

TDT (Allan et al. 1998) is concerned with three main problems: first, modelling the meanings of documents (the topics), then detecting events in a time-labelled corpus (giving boundaries to the addressed stories) and finally, tracking information over time.

Topic modelling which aims at understanding the semantic meanings of documents has been addressed in numerous papers (Kasiviswanathan et al. 2011; Caballero et al. 2012; Jo et al. 2011), from statistical approaches with Tf-Idf (Jones 1972) and Okapi-BM25 (Robertson et al. 1999) to probabilistic modelling with probabilistic-Latent Semantic Allocation (pLSA) (Hoffman 1999) and Latent Dirichlet Allocation (LDA) (Blei et al. 2003) through linear algebra with Latent semantic analysis (LSA) (Deerwester et al. 1990).

Event Detection and especially Peak Detection (Kleinberg 2002) also emphasizes the interest of time in understanding a corpus. Peak detection approaches take advantage of a time-labelled corpus to apply techniques from time series analysis and provide boundaries to the addressed stories (Marcus et al. 2011).

News Tracking addresses the problem of the evolution of topics over time. Some dynamic extensions of LDA (Hoffman et al. 2010) aim at tracking topic transitions. Graph clustering algorithms (Leskovec et al. 2009) may be used to construct groups of terms and see their evolution over time. Dynamic Clustering and Mapping algorithms (Gansner et al. 2012) may be used to construct a dynamic map which preserves the user's mental map.

Visualization. The increasing accessibility of time-oriented data has led to the development of various visual representations [see Aigner et al. (2011) for a recent overview]. The pioneering ones were dedicated to visualizing time series and data semantics were not taken into account. Over the last few years, different visualizations have been put forward to track topic evolution, but they do not explicitly take into account topic co-evolution. For instance, stacked areas charts (Leskovec et al. 2009) allow one to draw many distributions on the same chart (see Fig. 2), but are more efficient at displaying the evolution of one component with respect to the total flow than at highlighting the co-evolution of different components. Evolution graphs (Mei and Zhai 2005; Jo et al. 2011) display the evolution of topics over time, but the non-linearity of the evolution of topics (some may split in two, or two may merge into one) makes it difficult to visualize their

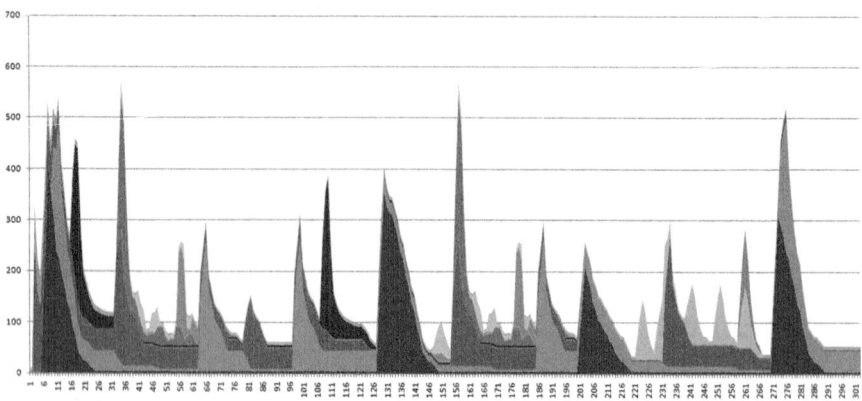

Fig. 2 Example of stacked areas chart

co-evolution. Dynamic clustering algorithms (Gansner et al. 2012) display data
on a map and continuously update the visualization to provide a succinct view
of evolving topics of interest. This kind of visualization is of great interest when
tracking the evolution of the whole graph, but because of the dynamic representation
of time, it may be difficult to visualize the co-evolution of clusters. In order to
overcome this limitation, we hereafter propose a representation based on heatmaps.
Clustered heatmaps have proven their efficiency in particular in bioinformatics: they
are one of the most popular means of visualizing genomic data (North et al. 2005).
However, as far as we know, their application to social networking analysis has not
yet been explored.

3   Methodology

Heatmaps represent the values contained in a matrix as colours, and allow one to
visualize 3D data in a 2D space. Because a lot of distributions can be displayed on
the same map with fixed size cells, heatmaps are well suited to detecting patterns in
the gradient of colours.

Central to ensuring the perceptual quality of heatmaps is the order in which data
are displayed. Different orders may, indeed, lead to different perceptions. To deal
with the above challenges, we propose a methodology for grouping terms that evolve
in the same way and visualize their evolution.

The full process is represented in Fig. 3 and may be summarized as follows :
First, we divide the total time span into fixed length time periods and select the terms
to be compared. For each term, we build a raw vector, called "evolution vector", with
a length equal to the number of periods and which contains the values of the scoring
function for this term and each period. Then, we compute the co-evolution matrix,
which indicates, for each pair of terms, how similar the temporal evolution vectors

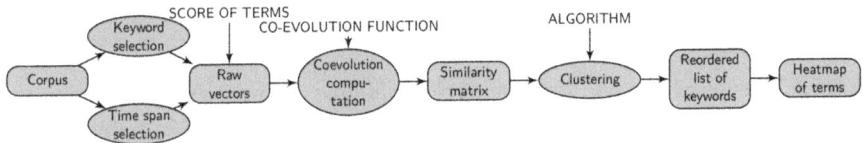

Fig. 3 Methodology summary

are. Thereafter, we apply a clustering algorithm to the co-evolution matrix in order to group together the terms that share common behaviour. Finally, we use the result of this clustering to order the terms and we visualize their evolution vectors with a heatmap.

3.1 Terms and Time Spans Selection

Given a set of tweets $\mathcal{T}$, recorded during a time interval t, we divide t into a set of fixed length time blocks $\{t_1, \ldots, t_n\}$. Accordingly, the set of all tweets $\mathcal{T}$ is split into n subsets $\{\mathcal{T}_1, \ldots, \mathcal{T}_n\}$ where $\mathcal{T}_i$ contains all the tweets of the period t_i. At the same time, the set of all tweets $\mathcal{T}$ is processed to extract the keywords to be compared. First, each tweet of $\mathcal{T}$ is tokenized into words, then any step of preprocessing may be applied to improve the quality of the clustering and hence the quality of the visualization. For instance, keywords may be unigrams or bigrams, and one may use lemme or stemme forms of words. The user may also feel it necessary to remove words from a predefined stop-list or with particularly high or low frequencies.

3.2 Scoring of Terms

The next step of our methodology is to define the measure used to represent the behaviour of each term across the corpus. Let $\mathcal{W}$ be the set of keywords to be tracked over the corpus. We call scoring-function, a function s such as

$$s : \mathcal{W} \times \{1, \ldots, n\} \to \mathbb{R} \tag{1}$$
$$w \times i \qquad \mapsto s(w, i)$$

i.e. for each word $w \in \mathcal{W}$ and each period t_i, $i \in \{1, \ldots, n\}$, s gives the value of the measure of interest of w for the period i. Depending on the application, s may be the number of documents containing w published during period i, or it may be a binary value corresponding to the fact that there is at least one document containing w in this period. The choice of the scoring-function is application driven and mainly depends on the interest of the user.

3.3 Co-evolution of Terms

For each keyword $w \in \mathcal{W}$, we define the evolution vector $\mathbf{ev}(w)$ of w by $\mathbf{ev_i}(w) = s(w, i)$, $\forall i \in \{1, \ldots, n\}$. For each pair of keywords (w_1, w_2), we compute the co-evolution of w_1 and w_2 as the co-evolution of their temporal evolution vectors $(\mathbf{ev}(w_1), \mathbf{ev}(w_2))$. The co-evolution function is symmetric. Intuitively, the more similar two vectors are, the higher their co-evolution is. Depending on the expert's objectives, the co-evolution function may be a similarity measure such as Jaccard, or a correlation measure.

By computing the co-evolution of each pair of keywords, we build a $|\mathcal{W}| \times |\mathcal{W}|$ co-evolution matrix, $\mathcal{M}_S$. Rows and columns of $\mathcal{M}_S$ are labelled with keywords ordered the same way, such as : $\forall i, j \in \{1, \ldots, |\mathcal{W}|\}$,

$$\mathcal{M}_S(i, j) = sim(\mathbf{ev}(w_i), \mathbf{ev}(w_j)) \tag{2}$$

3.4 Clustering

The key point here is that the order in which keywords are shown in $\mathcal{M}_S$ does not depict any correlation between terms. The goal of this step is to reorder terms of $\mathcal{W}$ with respect to their correlation. Hence, we apply a hierarchical clustering to $\mathcal{M}_S$ in order to build groups of terms and display co-occurring terms close to each other. At the beginning, each term is assigned to a distinct class and at each step the algorithm merges the two most similar classes, continuing until only a single class remains.

3.5 Heatmap

Once we get a meaningful order for the terms of our vocabulary $\mathcal{W}$, we build a heatmap of term scores, the columns of which correspond to the n periods, displayed in chronological order, and with $|\mathcal{W}|$ rows corresponding to the reordered terms of $\mathcal{W}$. The colour of each cell corresponds to the score provided by the scoring function for this term and this period.

4 Experimental Settings

Our experiments use a corpus of tweets exclusively written in French, but our methodology does not make any assumptions about the language and may be applied to any kinds of "keywords".

4.1 Dataset

The dataset used for this experiment is composed of all tweets containing the word "edf", in reference to the French energy company "Elecricité De France", published during the period June 17, 2012 to May 02, 2013. Because "edf" may also refer, in French, to "Equipe De France" (French team) a filter is applied to remove all the tweets that refer to sport. This filtering is not straightforward and will not be detailed here. This step leads to a corpus $\mathcal{T}$ of 73,023 tweets.

Each tweet is tokenized so that tokens are separated by a non letter-character. Mentions, URLs, "RTs" and accents are removed and the terms are converted to lowercase. The lemme of each term is obtained with TreeTagger[1] and we build our vocabulary $\mathcal{W}$ by retaining the 512 most frequent pairs of consecutive terms.

By computing the number of tweets published each hour, we observed very low activity at night, thus we divided our corpus into periods of a day spanning from 4 a.m. to 4 a.m. This leads to 307 blocks of time $\{t_1, \ldots, t_{307}\}$ and 307 subsets of our corpus $\{\mathcal{T}_1, \ldots, \mathcal{T}_{307}\}$, each $\mathcal{T}_i$ containing all tweets published during the period i.

4.2 Methods

In order to depict the evolution of the popularity of each *keyword* over time, we define the scoring function s such as

$$s : \mathcal{W} \times \{1, \ldots, n\} \rightarrow \mathbb{R} \tag{3}$$
$$w \times i \quad \mapsto N(w, i)$$

where $N(w, i)$ is the number of documents containing the *keyword* w published during the period t_i. At first glance, this function may seem too simplistic. But because we compute the correlation of each pair of *keywords* with a rank correlation coefficient, any monotonic transformation of the data would not have affected the results. Nonetheless, the choice of the scoring function is important and must be done with respect to the co-evolution function. This choice is mainly data driven and depends on the application. It could be a binary value depicting the presence or absence of the *keyword* or a normalization like TF-IDF.

This step leads to a temporary matrix of size 512×307 which contains the number of documents in which each term appears in each period. We use it to compute the correlation of each pair of terms with *Kendall*'s tau coefficient. This method, unlike, for example, *Pearson*'s and *Spearman*'s coefficients is not only sensitive to linear relationships between two terms, but also detects non-linear relationships.

---

[1] http://www.cis.uni-muenchen.de/~schmid/tools/TreeTagger/.

An agglomerative clustering algorithm is applied to the co-evolution matrix and we deduce from its result an order of the list of terms. At each step of the agglomerative clustering algorithm, the two most similar classes are merged using the *complete linkage criterion* chosen, here, for its simplicity and its tendency to favour compact clusters. For each class C_1, C_2,

$$\mathrm{sim}(C_1, C_2) = \min_{x \in C_1, y \in C_2} (\mathrm{sim}(x, y)).$$

This simple criterion has led to interpretable results but a comparison with other criteria is planned in the near future.

4.3 Results

This experiment was conducted in collaboration with an expert in natural language processing and customer relationship management working at the Research & Development department of EDF. Our expert was asked to visually detect groups of terms on the heatmap constructed by our methodology. Due to space constraints, we have only provided an extract of the heatmap, with a small number of *keywords* centred on their period of activity (see Fig. 4). This extract highlights three patterns of co-evolution visually detected by the expert.

In the first pattern (A) terms are associated with a high score for a few days (around August 17, 2012) and are not employed for the rest of the time. This pattern conforms with the short-life topics described by Kwak et al. (2010) and relies on

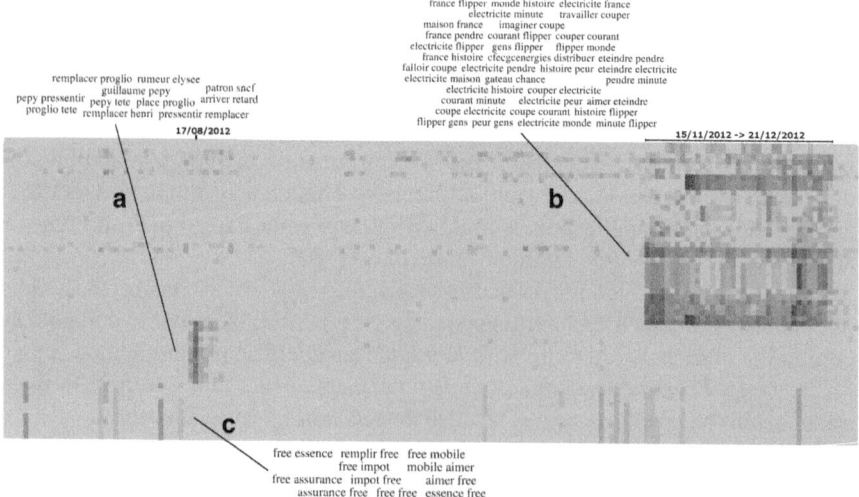

Fig. 4 Extract of the heatmap for a small number of terms centred on their period of activity

a rumour of change at the head of the executive board of EDF. This rumour was discussed in the traditional media and is not a novelty. However, detecting and quantifying the impact of this bursty feature in the social media remains valuable.

The second pattern (B) is a group of terms showing a long period of activity and may not be detected by a peak detection method. This pattern can be explained by a significant amount of documents about the end of the world announced in December 2012. In these messages Twitter users make jokes about the ancient Mayan prophecy and suggest switching off the power for 10 min to scare people. The joke stops short the day of the prophecy as it is no longer a topical issue.

Finally, in the third pattern (C), terms are employed together on many occasions and create what we call a recurring pattern. The volume of each appearance is low and may not trigger an alert, but put together these appearances provide a significant amount of information. These terms come from messages in which Twitter users humorously call for low prices in the energy sector following the lead of French telecommunications company Free. The interest of this pattern lies in the fact that, unlike the discussion about the Mayan prophecy, the joke is not circumstantial. This topic appears many times and is regularly discussed by Twitter users.

5 Conclusion

In this paper, we have presented a methodology for highlighting co-evolution patterns of Twitter terms on a visual support. Experiments on real-life data have led us to identify different classes of terms associated with characteristic behaviour. This first analysis plays the role of a proof-of-concept. We have presented here feedback from one expert only, and even if the first results are promising, we are faced with a problem of visualization evaluation. Generally speaking, this question is known to be highly problematic in visual analytics (Jankun-Kelly et al. 2007). However, here, we can rely on experience gained by using heatmaps in different domains. In addition to addressing the problem of evaluating the visualization, future works will include comparing co-evolution functions and clustering algorithms.

References

Aigner, W., Miksch, S., Schumann, H., & Tominski, C. (2011). *Visualization of time-oriented data*. London: Springer.

Allan, J., Carbonell, J. G., Doddington, G., Yamron, J., & Yang, Y. (1998). *Topic detection and tracking*. Pilot Study Final Report.

Blei, D. M., & Lafferty, J. D. (2006). Dynamic topic todels. In *Proceedings of the 23rd International Conference on Machine Learning* (pp. 113–120).

Blei, D. M., Ng, A. Y., & Jordan, M. I. (2003). Latent dirichlet allocation. *The Journal of Machine Learning Research, 3*, 993–1022.

Caballero, K. L., Barajas, J., & Akella, R. (2012). The generalized Dirichlet distribution in enhanced topic detection. In *Proceedings of the 21st ACM International Conference on Information and Knowledge Management (CIKM)* (pp. 773–782)

Deerwester, S. C., Dumais, S. T., Landauer, T. K., Furnas, G. W., & Harshman, R. A. (1990). Indexing by latent semantic analysis. *The Journal of the American Society for Information Science, 41*(6), 391–407.

Gansner, E., Hu, Y., & North, S. (2012). Visualizing streaming text data with dynamic maps. In *Proceedings of the 20th International Conference on Graph Drawing* (pp. 439–450).

Hoffman, M., Bach, F. R., & Blei, D. M. (2010). Online learning for latent Dirichlet allocation. In *Advances in Neural Information Processing Systems* (pp. 856–864).

Hoffman, T. (1999). Probabilistic latent semantic analysis. In *Proceedings of the 15th Conference on Uncertainty in Artificial Intelligence (UAI)* (pp. 289–296). San Francisco, CA: Morgan Kaufmann.

Jankun-Kelly, T. J., Ma, K.-L., & Gertz, M. (2007). A model and framework for visualization exploration. *IEEE Transactions on Visualization and Computer Graphics, 13*(2), 357–369.

Jo, Y., Hopcroft, J. E., & Lagoze, C. (2011). The web of topics: Discovering the topology of topic evolution in a corpus. In *Proceedings of the 20th International Conference on World Wide Web (W3C)* (pp. 257–266).

Jones, K. S. (1972). A statistical interpretation of term specificity and its application in retrieval. *The Journal of Documentation, 28*(1), 11–21.

Kasiviswanathan, S. P., Melville, P., Banerjee, A., & Sindhwani, V. (2011). Emerging topic detection using dictionary learning. In *Proceedings of the 20th ACM International Conference on Information and Knowledge Management (ICKM)* (pp. 745–754).

Keim, D. A., Mansmann, F., & Thomas, J. (2010). Visual analytics: How much visualization and how much analytics? *ACM SIGKDD Explorations Newsletter, 11*(2), 5–8.

Kleinberg, J. (2002). Bursty and hierarchical structure in streams. In *Proceedings of the 8th ACM SIGKDD International Conference on Knowledge Discovery and Data Mining (KDD)* (pp. 91–101).

Kwak, H., Lee, C., Park, H., & Moon, S. (2010). What is twitter, a social network or a news media? In *Proceedings of the 19th International Conference on World Wide Web (W3C)* (pp. 591–600).

Leskovec, J., Backstorm, L., & Kleinberg, J. (2009). Meme-tracking and the dynamics of the news cycle. In *Proceedings of the 15th ACM SIGKDD International Conference on Knowledge Discovery and Data Mining (KDD)* (pp. 497–506).

Marcus, A., Bernstein, M. S., Badar, O., Karger, D. R., Madden, S., & Miller, R. C. (2011). Twitinfo: Aggregating and visualizing microblogs for event exploration. In *Proceedings of the 2011 Annual Conference on Human Factors in Computing Systems (CHI)* (pp. 227–236).

Mei, Q., & Zhai, C. (2005). Discovering evolutionary theme patterns from text: An exploration of temporal text mining. In *Proceedings of the 11th ACM SIGKDD International Conference on Knowledge Discovery in Data Mining (KDD)* (pp. 198–207).

North, C., Rhyne, TM., & Duca, K. (2005). Bioinformatics visualization: Introduction to the special issue. *Information Visualization, 4*(3), 147–148.

Robertson, S. E., Walker, S., Beaulieu, M., & Willet, P. (1999). *Okapi at Trec-7: Automatic ad hoc, filtering, VLC and interactive track* (pp. 253–264). Gaithersburg: Nist Special Publication SP.

Incremental Weighted Naive Bays Classifiers for Data Stream

Christophe Salperwyck, Vincent Lemaire, and Carine Hue

Abstract A naive Bayes classifier is a simple probabilistic classifier based on applying Bayes' theorem with naive independence assumption. The explanatory variables (X_i) are assumed to be independent from the target variable (Y). Despite this strong assumption this classifier has proved to be very effective on many real applications and is often used on data stream for supervised classification. The naive Bayes classifier simply relies on the estimation of the univariate conditional probabilities $P(X_i|C)$. This estimation can be provided on a data stream using a "supervised quantiles summary." The literature shows that the naive Bayes classifier can be improved (1) using a variable selection method (2) weighting the explanatory variables. Most of these methods are related to batch (off-line) learning and need to store all the data in memory and/or require reading more than once each example. Therefore they cannot be used on data stream. This paper presents a new method based on a graphical model which computes the weights on the input variables using a stochastic estimation. The method is incremental and produces a Weighted Naive Bayes Classifier for data stream. This method will be compared to classical naive Bayes classifier on the Large Scale Learning challenge datasets.

1 Introduction

Since the 2000s, the data mining from streams became a standalone research topic. Many studies addressing this new problem have been proposed (Gama 2010). Among the solutions to the problems of learning on data streams, the incremental learning algorithms are one of the most used techniques. These algorithms are able to update their model using just the new examples.

C. Salperwyck (✉) • V. Lemaire • C. Hue
Orange Labs, 2 avenue Pierre Marzin, 22300 Lannion, France
e-mail: christophe.salperwyck@gmail.com

© Springer-Verlag Berlin Heidelberg 2015   179
B. Lausen et al. (eds.), *Data Science, Learning by Latent Structures,
and Knowledge Discovery*, Studies in Classification, Data Analysis,
and Knowledge Organization, DOI 10.1007/978-3-662-44983-7_16

In this article we focus on one of the most used classifiers in the literature: the naive Bayes classifier. We modify this classifier in order to make on-line supervised classification on data streams. This classifier only needs conditional probability $P(X_i|C)$, with X_i an explanatory variable and C a class of the classification problem. Its complexity to predict is very low which makes it suitable and widely used for stream mining prediction.

Nevertheless it has been proved for batch learning that selecting (Koller and Sahami 1996; Langley and Sage 1994) or weighting (Hoeting et al. 1999) variables can improve the classification results. Moreover Boullé in (Boullé 2006b) shows the close relation between weighting variables and averaging many naive Bayes classifier in the sense that at the end the two different processes produce a similar single model [see Eq. (1)]. In this paper we particularly focus on weighting variables for data streams for a naive Bayes classifier. This weighting produces a single model close to an "averaged naive Bayes classifier" a "weighted naive Bayes classifier." We propose a graphical model and a learning method to compute these weights.

The present work aims to study a new way to estimate incrementally the weights of a Weighted Naive Bayes classifier (WNB) using a graphical model close to a neural network. The paper is organized as follows: our graphical model and the way to compute its parameters (the weights) are presented in Sect. 2; Sect. 3 presents how the conditional density estimations, used as inputs of our model, are estimated; Sect. 4 realizes an experimental study of our Weighted Naive Bayes classifier trained incrementally on the large scale learning challenge datasets. Finally the last section concludes this paper.

2 Incremental Weighted Naive Bayes Classifiers

2.1 Introduction: Naive Bayes Classifier (NB) and Weighting Naive Bayes Classifiers (WNB)

The naive Bayes classifier (Langley et al. 1992) assumes that all the explanatory variables are independent knowing the target class. This assumption drastically reduces the necessary computations. Using the Bayes theorem, the expression to obtain the estimation of the conditional probability of a class C_k is: $P(C_k|X) = \frac{P(C_k)\prod_i P(X_i|C_k)}{\sum_{j=1}^{K}(P(C_j)\prod_i P(X_i|C_j))}$ where K is the number of classes, i the index of the explanatory variable.

The predicted class is the one which maximizes the conditional probabilities $P(C_k|X)$. The probabilities $P(X_i|C_k)$ are estimated using a conditional probability density estimation as, for example, using counts after discretization for numerical variables or grouping for categorical variables. The denominator of Eq. (1) normalizes the result so that $\sum_k P(C_k|X) = 1$. One of the advantages of this classifier in the context of data stream is its low complexity for deployment, which only depends on the number of explanatory variables. Its memory consumption is also low since it requires only one conditional probability density estimation per variable.

The literature shows that the naive Bayes classifier could be improved (1) using a variable selection method (Koller 1996; Langley and Sage 1994) (2) weighting the explanatory variables which amounts to a Bayesian Model Averaging (BMA) (Hoeting et al. 1999). These two processes can be mixed iteratively. The formulation of the conditional probabilities becomes:

$$P(C_k|X) = \frac{P(C_k) \prod_i P(X_i|C_k)^{w_i}}{\sum_{j=1}^{K} (P(C_j) \prod_i P(X_i|C_j)^{w_i})} \tag{1}$$

where each explanatory variable i is weighted by a weight w_i ($w_i \in [0-1]$).

2.2   The Proposed Approach

In off-line learning, the weights of the Weighted Naive Bayes Classifiers can be estimated using (1) an averaging of the Naive Bayes classifiers obtained (Hoeting et al. 1999) (2) an averaging of the Naive Bayes classifiers obtained using an MDL (Minimum Description Length) criterion (Boullé 2006b) (3) a direct estimation of the weights using a gradient descent (Guigourès and Boullé 2011). But all these methods require to have all the data in memory and to read them several times. The method proposed in this paper optimizes directly the weights of the classifier and is able to work on data stream.

The first step has been to elaborate a graphical model (see Fig. 1) dedicated to the optimization of the weights. This model allows the rewriting of Eq. (1) as a graphical model where the Weighted Naive Bayes classifier has a weight per class and per variable as presented in Eq. (2). The number of weights is therefore higher since the weights are no longer just associated with a variable, but with a variable conditionally to a class: w_{ik} is the weight associated with the variable i and the class k, b_k is the bias associated with the class k.

The first layer of our graphical model is a linear layer which realizes a weighted sum H_k for every class k, such as: $H_k = \sum_{i=1}^{d} w_{ik} \log(P(X_i|C_k)) + b_k$. The second layer is a *Softmax* such as: $P_k = \frac{e^{H_k}}{\sum_{j=1}^{K} e^{H_j}}$. Finally the graphical model, in the case where the inputs are based on the log conditional density estimation ($\log(p(X_i|C_k), \forall i, k)$), gives the values ($\forall k$) of the $P(C_k|X)$ such as:

$$P_k = \frac{e^{b_k + \sum_{i=1}^{d} w_{i,k} \log(p(X_i|C_k))}}{\sum_{j=1}^{K} e^{b_j + \sum_{i=1}^{d} w_{i,j} \log(p(X_i|C_j))}} \tag{2}$$

The input variables used as the inputs of this graphical model come from the on-line summaries described in Sect. 3.

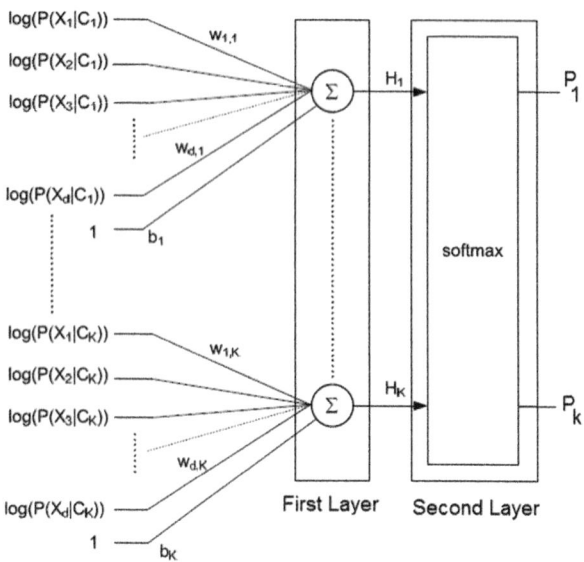

Fig. 1 Graphical model dedicated to the optimization of the weights

The optimization of the weights is done using a stochastic gradient descent for a given cost function. For a given example X the weights updates follow the formulae:

$$w_{ij}^{t+1} = w_{ij}^t - \eta \frac{\partial \text{Cost}^X}{\partial w_{ij}} \qquad (3)$$

where Cost^X is the cost function applied to the example X and $\frac{\partial \text{Cost}^X}{\partial w_{ij}}$ the derivative of the cost function w.r.t. the parameters of the model, here the weights w_{ij}.

The detail of this calculation is presented in the appendix of this paper, the result is simple:

$$\frac{\partial \text{Cost}}{\partial H_k} = P_k - T_k, \forall k \qquad (4)$$

where T_k is the desired probability and P_k the estimated probability. So the update of the weights has a very low complexity.

The method used to update the weights is the one used usually for a standard back-propagation and three main parameters have to be considered (Lecun et al. 1998) in the case of stochastic gradient descent: (1) the cost function; (2) the number of iterations; (3) the learning rate. For the cost function in the supervised classification context the best choice, since the output to learn takes only two values $\{0, 1\}$, is the log-likelihood (Bishop 1995) which optimizes $\log(P(C_k|X))$. The number of iterations in our case is set to 1 since each example of the stream is used only once for training. Finally the only parameter to set is the learning rate. A too small value will give a slow convergence but a high value may not allow reaching a global minimum. For off-line learning this value can be adjusted using a cross validation procedure but for on-line learning on data streams this procedure cannot be applied. In the experiments presented below the learning rate is fixed to $\eta = 10^{-4}$. However if we expect concept drift it could be interesting to have an adaptive learning rate as in Kuncheva and Plumpton (2008).

3   Conditional Density Estimation

This section presents how the conditional density probabilities used as inputs of our graphical model are estimated. This is not the main contribution of this paper, therefore the three methods used are briefly described. We use in the experimental part of this paper three methods (see Fig. 2) to compute the conditional density probability for each numerical explanatory variable and for each target class: (1) our two-layer incremental discretization method based on order statistics as described in Salperwyck and Lemaire (2013) (2) a two layer discretization method "cPiD" which is a modified version of the PiD method of Gama (Gama and Pinto 2006) (3) and a Gaussian approximation. The approach could be the same for categorical variable (not detailed in this paper) by putting in the first level the count-min sketch approach and in the second level the grouping MODL method.

The two-level approach is based on a first level which provides quantiles which are order statistics on the data stream for every explanatory variable i. Each quantile, q, is a tuple which contains: $< v_q^i, g_q^i, (g_{qj}^i)_{j=1,...,K} >$ where for each explanatory variable i of the data stream (1) v_q^i is a value (2) g_q^i corresponds to the number of values between v_{q-1}^i and v_q^i (3) (g_{qj}^i) is, for the K classes of the supervised classification problem, the number of elements in q belonging to the class j. The second level is a batch algorithm applied on the quantiles. In this paper the number of tuples is equal to 100 corresponding to the estimation of centiles. The tuning of the number of quantiles is discussed in Salperwyck (2012). cPid and GkClass, described below, are two different methods to obtain quantiles.

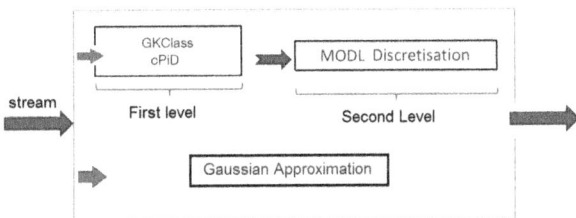

Fig. 2 On-line summaries for numerical variables: two-level approach for GKClass+MODL and cPiD+MODL and single-level approach for the Gaussian approximation

3.1 cPid

Gama and Pinto (2006), proposed a two-layer incremental discretization method. The first layer is a mix of a discretization based on the methods "Equal Width" and "Equal Frequency" (algorithm details: Gama and Pinto 2006, p. 663). This first layer is updated incrementally and needs to have much more bins than the second one. The second layer uses information of the first one to build a second discretization. Many methods can be used on the second layer such as: Equal Width, Equal Frequency, Entropy, Kmeans, etc. The advantage of this method is to have a fast first layer which can be used to build different discretizations on it (second layer).

In cPid we propose to change PiD to use a constant memory. In order to keep memory usage constant, after each split we merge the two consecutive intervals with the lowest counts sum. Thus the number of stored intervals remains constant. The second layer used is the MODL discretization (Boullé 2006a). For brevity PiD and cPiD are not compared in this paper but interested readers can find this comparison in Salperwyck (2012).

3.2 GKClass

This algorithm proposed by Greenwald and Khanna (2001) is an algorithm to compute quantiles using a memory of $O(\frac{1}{\epsilon} \log(\epsilon N))$ in the worst case. This method does not need to know the size of the data in advance and is insensitive to the arrival order of the examples. The algorithm can be configured either with the number of quantiles or with a bound on the error. We adapted the GK summary to store directly the class counts in tuples. The second layer used is the MODL discretization (Boullé 2006a).

The MODL discretization [5] and grouping are supervised and do not need any parameters. They are based on class counts and evaluate all possible intervals for the numerical variables and groups for the categorical variables. The quality evaluation of the model is based on a Bayesian approach. For numerical variables, its purpose is to find the best discretization parameters: number of intervals, frontiers of the intervals and class distribution in the intervals in a Bayesian sense. For categorical variables the method performs grouping in a similar way.

3.3 Gaussian Approximation

The main idea of this method relies on the hypothesis that the observed data distribution follows a Gaussian law. Only two parameters are needed to store a Gaussian law: its mean and its standard deviation. The incremental version requires one more parameter: the number of elements. This method has one of the lowest memory footprints and is constituted of a single layer but is dependent on the Gaussian assumption. It will be used as a reference method as the Large Scale Learning challenge datasets are generated using a Gaussian generator.

4 Experiments

4.1 Protocol

The Large Scale Learning challenge datasets (organized by the network of excellence PASCAL, http://jmlr.csail.mit.edu/papers/topic/large_scale_learning. html) have been used for the experiments. These datasets are constituted of 500,000 labeled examples which is large enough to evaluate an on-line classifier. The datasets alpha, beta, delta, and gamma contain 500 numerical explanatory variables and the dataset epsilon and zeta 2,000 numerical variables. The 100,000 first examples have been used as the test dataset and the remaining examples as the train dataset.

4.2 Results

For the first part of the experiments, a standard naive Bayes (without the weighting) is used on the top of the summaries described in Sect. 3. The results are presented in Table 1 and shows that the estimation of the conditional probabilities is accurate for all methods used since all the naive Bayes classifiers obtain good results. Despite the fact that the large scale learning datasets come from Gaussian generators the two others summaries obtain similar results without the Gaussian assumption. One additional advantage of the method based on GKClass summaries (level 1) with the MODL discretization (level 2) is that it guarantees a maximal error for a given memory on the quantile estimation. Therefore this two-level method is used for the second part of the experiments.

Knowing the results of Table 1, we know that the estimation of the conditional probabilities is accurate. Therefore the behavior of our Weighted Naive Bayes can be studied. The results are obtained using four classifiers: (1) a naive Bayes (NB) trained offline with the MODL discretization (Boullé 2006a) and all the data in memory; (2) an Averaged Naive Bayes (ANB) trained offline with the MODL discretization (Boullé 2006a), this algorithm is described in Boullé (2006b) to compute the weights and all the data in memory—this method is one of the best of the literature, see Guyon et al. (2009) (3) a Naive Bayes trained online with the two-level discretization method which uses GKClass (level 1) and the MODL discretization (level 2) (4) a Weighted Naive Bayes trained online with the two-level discretization method which uses GKClass (level 1) and the MODL discretization (level 2) and our method based on the graphical method to estimate the weights.

Table 2 shows that the results obtained by our On-line WNB are very interesting: it is better than the on-line NB except for the Gamma and Delta datasets. Its

Table 1 Accuracy of a Naive Bayes (without averaging) using the three methods to compute the conditional probabilities

#Train examples →	Alpha			Beta			Delta		
	40,000	100,000	380,000	40,000	100,000	380,000	40,000	100,000	380,000
GKClass	54.40	54.49	54.56	49.79	51.05	51.23	80.56	82.22	83.47
CPiD	54.36	54.52	54.57	49.79	51.09	51.12	80.66	82.39	83.77
Gaussien (niveau 1)	54.62	54.67	**54.67**	51.21	**51.50**	51.31	84.58	**85.10**	85.08

#Train examples →	Gamma			Epsilon			Zeta		
	40,000	100,000	380,000	40,000	100,000	380,000	40,000	100,000	380,000
GKClass	92.63	93.51	94.23	70.48	70.37	70.43	78.35	78.63	78.48
CPiD	92.93	93.83	94.41	**70.52**	70.38	70.36	78.50	78.48	78.42
Gaussian	95.09	95.10	**95.16**	70.66	70.57	70.43	**78.96**	78.77	78.52

The best result is in boldface.

Table 2 Comparison of the accuracy of Naive Bayes (NB), the averaged Naive Bayes (ANB) and the weighted Naive Bayes (WNB)

	Alpha			Beta			Delta		
#Train examples →	40,000	100,000	380,000	40,000	100,000	380,000	40,000	100,000	380,000
Offline NB (1)	54.60	54.61	54.61	49.79	51.36	51.19	80.78	82.44	83.91
Offline ANB (2)	66.13	67.30	*68.77*	49.79	51.39	*53.24*	80.80	82.46	*83.91*
Online NB (3)	54.40	54.49	54.56	49.79	51.05	51.23	80.56	82.22	**83.47**
Online ANB (4)	64.03	66.40	**67.61**	49.79	49.79	**52.20**	75.35	77.74	79.53

	Gamma			Epsilon			Zeta		
#Train examples →	40,000	100,000	380,000	40,000	100,000	380,000	40,000	100,000	380,000
Offline NB (1)	92.95	93.99	94.63	70.35	71.04	70.58	78.39	78.34	78.26
Offline ANB (2)	92.95	94.00	*94.64*	84.36	85.34	*86.01*	88.67	89.51	*90.43*
Online NB (3)	92.63	93.51	**94.23**	70.48	70.37	70.43	78.35	78.63	78.48
Online ANB (4)	90.25	91.05	91.76	69.25	74.76	**79.95**	73.82	79.91	**84.52**

The best "offline-result" is in italics and the best "online result" is in boldface.

performance is close to the Off-line ANB which is the "best results" which can be obtained on these datasets with Bayes type classifiers. But the classifier makes the assumption that all the data are stored in memory and readable several times whereas this is not possible on data streams. Our WNB approach uses a low amount of memory, thanks to the two-level approach to estimate the conditional densities, and is purely incremental, thanks to the graphical model and the stochastic gradient descent to estimate the weights. Therefore the results of our approach are very promising. The accuracy obtained versus the number of examples used to train the model seems to indicate that our on-line WNB (4) and the off-line ANB (2) would have the same accuracy if the number of examples was higher. The dimension could also be an explanation (Alpha, Beta, Delta, and Gama have 500 numerical variables, Epsilon and Delta have 2,000 variables) since the accuracy obtained by Online WNB increases significantly with the number of examples for Epsilon and Zeta. These points will be investigated in future works.

5   Conclusion

The results for our on-line weighted version of the Naive Bayes classifier are promising. It improves the performance compared to the non-weighted version and is close to the off-line averaged version of the naive Bayes classifier. However we think that its results could be improved in future works. The first idea would be to use the GK Class summaries as "mini-batch" (Cotter et al. 2011) and do several iterations to speed up the gradient descent. The second idea would be to use an adaptive learning rate: high at the beginning and low after, or to take into account the error rate as in Kuncheva and Plumpton (2008). Future works will be done in these directions.

Appendix: Derivative of the Cost Function

The graphical model is built to have directly the values of the $P(C_k|X)$ at the output. The goal is to maximize the likelihood and therefore to minimize the negative log likelihood. The first step in the calculation is to decompose the softmax considering that each output could be seen as the succession of two steps: an activation followed by a function of this activation.

Here the activation function could be seen as: $O_k = f(H_k) = \exp(H_k)$ and the output of the softmax part of our graphical model is: $P_k = \frac{O_k}{\sum_{j=1}^{K} O_j}$. The derivative of the activation function is:

$$\frac{\partial O_k}{\partial H_k} = f'(H_k) = \exp(H_k) = O_k \tag{5}$$

The cost function being the $-log\ likelihood$, we have to consider two cases: (1) the desired output is equal to 1 or (2) the desired output is equal to 0. For the following we note:

$$\frac{\partial \text{Cost}}{\partial H_k} = \frac{\partial C}{\partial P_k} \frac{\partial P_k}{\partial O_k} \frac{\partial O_k}{\partial H_k} \tag{6}$$

In the case where the desired output of the output k is equal to 1 by replacing (5) in (6):

$$\frac{\partial \text{Cost}}{\partial H_k} = \frac{\partial C}{\partial P_k} \frac{\partial P_k}{\partial O_k} \frac{\partial O_k}{\partial H_k} = \frac{-1}{P_k} \frac{\partial P_k}{\partial O_k} O_k \tag{7}$$

$$\frac{\partial \text{Cost}}{\partial H_k} = \frac{-1}{P_k} \left[\sum_{l=1,l\neq k}^{K} \left(\frac{O_l}{\left(\sum_{j=1}^{K} O_j\right)^2} \right) \right] O_k$$

$$= \frac{-1}{P_k} \left[\frac{\left(\sum_{j=1}^{K} O_j\right) - O_k}{\left(\sum_{j=1}^{K} O_j\right)^2} \right] O_k \tag{8}$$

$$\frac{\partial \text{Cost}}{\partial H_k} = \frac{-1}{P_k} \left[\frac{\left(\sum_{j=1}^{K} O_j\right) - O_k}{\left(\sum_{j=1}^{K} O_j\right)} \right] \frac{O_k}{\left(\sum_{j=1}^{K} O_j\right)}$$

$$= \frac{-1}{P_k} \left[1 - \frac{O_k}{\left(\sum_{j=1}^{K} O_j\right)} \right] \frac{O_k}{\left(\sum_{j=1}^{K} O_j\right)} \tag{9}$$

Therefore

$$\frac{\partial \text{Cost}}{\partial H_k} = \frac{-1}{P_k} [1 - P_k] P_k = P_k - 1 \tag{10}$$

In the case where the desired output of the output k is equal to 0 the error is only transmitted by the normalization part of the *softmax* function since the derivative for an output where the desired value is 0 is equal to 0. Therefore with similar steps we have: $\frac{\partial \text{Cost}}{\partial H_k} = P_k$

Finally we conclude: $\frac{\partial \text{Cost}}{\partial H_k} = P_k - T_k, \forall k$ where T_k is the desired probability and P_k the estimated probability. Then the rest of the calculation of $\frac{\partial \text{Cost}}{\partial w_{ik}}$ is straightforward.

References

Bishop, C. M. (1995). *Neural networks for pattern recognition*. New York: Oxford University Press.

Boullé, M. (2006a). MODL: A Bayes optimal discretization method for continuous attributes. *Machine Learning, 65*(1), 131–165.

Boullé, M. (2006b). Regularization and averaging of the selective naive bayes classifier. In *The 2006 IEEE International Joint Conference on Neural Network Proceedings* (pp. 1680–1688).

Cotter, A., Shamir, O., Srebro, N., & Sridharan, K. (2011). Better mini-batch algorithms via accelerated gradient methods. In J. Shawe-taylor, R.S. Zemel, P. Bartlett, F. C. N. Pereira, & K. Q. Weinberger (Eds.), *Advances in neural information processing systems* (Vol. 24, pp. 1647–1655). http://books.nips.cc/papers/files/nips24/NIPS2011_0942.pdf.

Gama, J. (2010). *Knowledge discovery from data streams*. Chapman and Hall/CRC Press

Gama, J., & Pinto, C. (2006). Discretization from data streams: applications to histograms and data mining. In *Proceedings of the 2006 ACM Symposium on Applied Computing* (pp. 662–667).

Greenwald, M., & Khanna, S. (2001). Space-efficient online computation of quantile summaries. *ACM SIGMOD Record, 30*(2), 58–66.

Guigourès, R., & Boullé, M. (2011). Optimisation directe des poids de modèles dans un prédicteur Bayésien naif moyenné. In *Extraction et gestion des connaissances EGC'2011* (pp. 77–82).

Guyon, I., Lemaire, V., Boullé, M., Dror, G., & Vogel, D. (2009). Analysis of the KDD cup 2009: Fast scoring on a large orange customer database. In *JMLR: Workshop and Conference Proceedings* (Vol. 7, pp. 1–22).

Hoeting, J., Madigan, D., & Raftery, A. (1999). Bayesian model averaging: a tutorial. *Statistical Science, 14*(4), 382–417.

Koller, D., & Sahami, M. (1996, May). Toward optimal feature selection. In *International Conference on Machine Learning* (pp. 284–292).

Kuncheva, L. I., & Plumpton, C. O. (2008). Adaptive learning rate for online linear discriminant classifiers. In *Proceedings of the 2008 Joint IAPR International Workshop on Structural, Syntactic, and Statistical Pattern Recognition* (pp. 510–519). Heidelberg: Springer.

Langley, P., Iba, W., & Thompson, K. (1992). An analysis of Bayesian classifiers. In *Proceedings of the National Conference on Artificial Intelligence* (pp. 223–228).

Langley, P., & Sage, S. (1994). Induction of selective Bayesian classifiers. In R. L. Mantaras & D. Poole (Eds.), *Proceedings of the Tenth Conference on Uncertainty in Artificial Intelligence* (pp. 399–406). Seattle, WA: Morgan Kaufmann.

Lecun, Y., Bottou, L., Orr, G. B., & Müller, K. R. (1998). Efficient BackProp. In G. Orr & K. Müller (Eds.), *Neural networks: Tricks of the trade. Lecture notes in computer science* (Vol. 1524, pp. 5–50). Heidelberg: Springer.

Salperwyck, C. (2012). *Apprentissage incrémental en ligne sur flux de données*. PhD thesis, University of Lille.

Salperwyck, C., & Lemaire, V. (2013). A two layers incremental discretization based on order statistics. In *Statistical models for data analysis* (pp. 315–323). Springer International Publishing. http://rd.springer.com/chapter/10.1007%2F978-3-319-00032-9_36

SVM Ensembles Are Better When Different Kernel Types Are Combined

Jörg Stork, Ricardo Ramos, Patrick Koch, and Wolfgang Konen

Abstract Support vector machines (SVM) are strong classifiers, but large datasets might lead to prohibitively long computation times and high memory requirements. SVM ensembles, where each single SVM sees only a fraction of the data, can be an approach to overcome this barrier. In continuation of related work in this field we construct SVM ensembles with Bagging and Boosting. As a new idea we analyze SVM ensembles with different kernel types (linear, polynomial, RBF) involved inside the ensemble. The goal is to train *one* strong SVM ensemble classifier for large datasets with less time and memory requirements than a single SVM on all data. From our experiments we find evidence for the following facts: Combining different kernel types can lead to an ensemble classifier stronger than each individual SVM on all training data and stronger than ensembles from a single kernel type alone. Boosting is only productive if we make each single SVM sufficiently weak, otherwise we observe overfitting. Even for very small training sample sizes—and thus greatly reduced time and memory requirements—the ensemble approach often delivers accuracies similar or close to a single SVM trained on all data.

1 Introduction

1.1 Related Work

Several researchers have studied support vector machines (SVM) ensembles during the last years (see, e.g., Wang et al. 2009 and Cortes et al. 2012 and the references therein). The goal of this research was mainly devoted to strengthen the overall accuracy, but not to handle large datasets with SVM. Kim et al. (2003), Lin and Li (2008), and Pavlov et al. (2000) propose SVM ensembles, but they do not use multiple kernels inside the ensemble. Yu et al. (2005) and Chang et al. (2008) present different approaches to tackle the large data aspect, like cluster-based data selection

J. Stork • R. Ramos • P. Koch (✉) • W. Konen
Cologne University of Applied Sciences, 51643 Gummersbach, Germany
e-mail: wolfgang.konen@fh-koeln.de

© Springer-Verlag Berlin Heidelberg 2015
B. Lausen et al. (eds.), *Data Science, Learning by Latent Structures, and Knowledge Discovery*, Studies in Classification, Data Analysis, and Knowledge Organization, DOI 10.1007/978-3-662-44983-7_17

or parallelization, but they do not use SVM ensembles. Recently, Meyer et al. (2013) analyzed two variants of SVM ensembles for large datasets, which were based on Bagging and Cascade SVM.

1.2 Research Questions

We analyze the following hypotheses in this article:

H-1 An ensemble classifier combining different kernel types performs similar than ensembles from a single kernel type alone.
H-2 Boosting is only productive if we make each single SVM sufficiently weak, otherwise we observe overfitting.
H-3 Even for very small training sample sizes—and thus greatly reduced time and memory requirements—the ensemble approach often delivers accuracies similar or close to a single SVM trained on all data.

2 Methods

We give a brief introduction to Support Vector Machines in Sect. 2.1 for classification tasks. Two well-known ensemble methods are incorporated in the experiments: the Bagging approach is introduced in Sect. 2.2, whereas the AdaBoost approach is delineated in Sect. 2.3.

2.1 Support Vector Machines

Support Vector Machines (Cortes and Vapnik 1995; Schölkopf and Smola 2002; Cristianini and Shawe-Taylor 2000) are state-of-the-art learning algorithms for classification and regression. In classification, data is usually written as a number of n observations

$$(\mathbf{x}_1, y_1), (\mathbf{x}_2, y_2), \ldots, (\mathbf{x}_n, y_n) \in \mathcal{X} \times \mathcal{Y} \tag{1}$$

where the set $\mathcal{X}$ defines the input values describing the patterns and the set $\mathcal{Y}$ comprises the corresponding class labels. In the simplest form, the output set only contains two elements, leading to binary classification, where the classes are often denoted by $\mathcal{Y} = \{-1, 1\}$

For linearly separable data, SVM fit a linear classifier, maximizing the margin between the classes in order to give the best generalization performance. But as data is often not linearly separable, it is the core of machine learning that two observations "being near in input space" should have a similar output value. Therefore, SVM incorporate kernel functions

$$k : \mathscr{X} \times \mathscr{X} \to \mathbb{R} \tag{2}$$

denoting the similarity of two observations. The kernel function k needs to suffice several conditions, e.g., symmetry and positive semi-definiteness. The function itself can be interpreted as a dot product in a high-dimensional space (Mercer 1909). It enhances the SVM learning algorithm by an implicit mapping of the input data into a higher-dimensional feature space, where a linear classifier is applicable.

In our experiments we incorporate a selection of the most commonly used kernel functions

linear: $k(\mathbf{x}, \mathbf{z}) = \mathbf{x}^T \cdot \mathbf{z}$
polynomial: $k(\mathbf{x}, \mathbf{z}) = (\mathbf{x}^T \cdot \mathbf{z} + c_0)^d$
radial: $k(\mathbf{x}, \mathbf{z}) = \exp(\gamma \cdot ||\mathbf{x} - \mathbf{z}||^2)$

where γ, c_0, and d are hyperparameters of the corresponding functions.

An optimal prediction model can now be determined by introducing the associated reproducing kernel Hilbert space H for the kernel function k and solving the optimization problem:

$$\hat{f} = \arg \inf_{f \in H, b \in \mathbb{R}} ||f||_H^2 + C \sum_{i=1}^{n} L(y_i, f(\mathbf{x}_i) + b) . \tag{3}$$

The first summand $||f||_H^2$ defines a penalty and in case of the 2-norm penalizes non-smooth functions. Because the function f maps into $\mathbb{R}$, the sign is calculated in the case of binary classification. Finally the second term measures the closeness of the predictions to the true outputs. The closeness is defined by a loss function, that is usually the Hinge loss $L(y,t) = L_h(y,t) = \max(0, 1 - yt)$ in case of classification. The Hinge loss is a convex, upper surrogate loss for the 0/1-loss (which is a desired loss function, but algorithmically intractable). A hyperparameter C controls the balance between the smoothness and the loss function.

SVMs are ideally suited for binary classification tasks, but can also handle more classes. Approaches for multi-class problems have been proposed by Weston and Watkins (1999), and Crammer and Singer (2002) gave an alternative formulation.

2.2 Bagging

Bagging (Breiman 1996), as a shorthand for bootstrap aggregation, is a well-known meta-algorithm to improve base classifiers in terms of stability and accuracy. The underlying idea is simple: Form several bootstrap samples by uniformly sampling T records with replacement from the full training set with N records ($T \le N$). Sub-classifiers are trained on each of these bootstrap samples, and the final classifier

---

Algorithm 1 Basic multi-class AdaBoost algorithm. See text for our modifications

---

1: Input: a training set $\Gamma = \{(\mathbf{x}_1, y_1), \ldots, (\mathbf{x}_N, y_N)\}$ with class labels y_i having K levels
2: Initialize: the weights $w_i^1 = 1/N$ for $i = 1, \ldots, N$
3: **for** $(t = 1, \ldots, T)$ **do**
4:     Draw a w_i^t-weighted training sample set S of size N with replacement from Γ.
5:     Train a weak learner h_t on S.
6:     Calculate training error $\epsilon_t = \sum_{i=1}^N w_i \, \Theta(h_t(\mathbf{x}_i) \neq y_i)$ on set Γ.
7:     Set quality of weak learner h_t as $\alpha_t = \frac{1}{2} \left(ln \left(\frac{1-\epsilon_t}{\epsilon_t} \right) + ln(K - 1) \right)$.
8:     Update weights: $w_i^{t+1} = w_i^t exp(\alpha_t \Theta(h_t(\mathbf{x}_i) \neq y_i))/Z$, where Z is a normalization constant such that $\sum_{i=1}^N w_i^{t+1} = 1$.
9: **end for**
10: Output: $f(\mathbf{x}) = \arg \max_c \left(\sum_{t=1}^T \alpha_t \Theta(h_t(\mathbf{x}) = c) \right)$. $\triangleright \Theta(P) = 1$ if P is true, 0 else.

---

makes its prediction by aggregating the predictions of all sub-classifiers. Typical aggregation methods are:

Majority voting   Predict the class most often predicted by the sub-classifiers (ties broken randomly).
Probability sum   If each sub-classifier delivers class probabilities, sum them up and predict the class with the highest probability sum.

Our Bagging approach has SVMs as sub-classifiers and investigates the benefits of combining different kernel types in one ensemble. Besides the pure types Lin (linear), Rad (radial, RBF), and Pol (polynomial) we form also mixed ensembles

LinPol   linear + polynomial
RadPol   radial + polynomial
LinRad   linear + radial
LinRadPol   linear + radial + polynomial

For example, given three ensembles Lin, Rad, Pol of ensemble size 10 each, a mixed ensemble LinRadPol of size 30 is built by joining these three.

2.3   AdaBoost

AdaBoost, as a shorthand for Adaptive Boosting, was formulated by Freund and Schapire (1995). The basic AdaBoost algorithm is shown in Algorithm 1. It works by repeatedly building and evaluating weak classifiers on the training set where each time a different sample from the training set distribution is drawn. Misclassified records in previous iterations get higher weights, leading to a stronger concentration for these records by the forthcoming classifiers. For each classifier h_t its quality $\alpha_t \in [0, \infty]$ on the original training set is evaluated. The final ensemble output is that class with the largest sum of α_t, where the sum is calculated for all classifiers voting for that class.

When applying AdaBoost to SVM as the base classifier, we propose three modifications:

1. As Wickramaratna et al. (2001) pointed out, it is essential for the classifiers to be *weak* in order to make AdaBoost productive. Since SVMs tend to be strong classifiers, it is necessary to weaken them. We sample in S for each classifier only a small fraction b of the set Γ in Algorithm 1, Step 4., e.g. $b = 0.1$ or $b = 0.01$. Note that the evaluation in Step 6. and weight update in Step 8. is done on the <u>full</u> set Γ: This gives a precise figure of merit for each classifier and keeps the weights in sync. Training on the set S with only bN records has the nice side effect that we can tackle large datasets with SVM, without being blocked by runtimes increasing approximately cubically with the number of training records.
2. To increase the diversity of the ensemble, we combine the results from different kernel types. We consider here the three well-known SVM kernel types radial (RBF), polynomial and linear. Two alternative ensemble-forming methods are proposed:

   Mixed   In each iteration, one of the three types is selected at random.
   Combined   First, three sub-ensembles of pure type (radial, polynomial, or linear) are formed, using the basic AdaBoost algorithm. The combined ensemble is formed by taking all classifiers h_t with their individual α_t from the three sub-ensembles. This ensemble predicts in the usual way the output on new cases.

3. As a further measure to increase diversity we choose for the radial SVMs in the ensemble the width γ randomly and uniformly from the .1 to .9 quantile range of $|\mathbf{x} - \mathbf{x'}|^2$, as suggested in Caputo et al. (2002), where x, x' are distinct data points. We found that this gives a better ensemble performance than using a tuned but fixed γ.

3   Experiment Setup

We tested our methods on several medium-sized machine learning datasets with 3,000–98,000 records from the UCI repository. Their characteristics are given in Table 1. In the cases where the original dataset provides a separation in training and test set, we used it in our experiments (column Train). Otherwise the dataset was randomly split in 2/3 training and 1/3 test data.

To compare accuracy and speedup we have run the following algorithms on the datasets: The basic SVM, modified AdaBoost-SVM with pure, mixed and combined ensembles (see Sect. 2.3) and a reduced training fraction (parameter b), Bagging-SVM with pure and mixed ensembles (Rad, RadPol, LinRadPol) and probability-based prediction, Bagging-SVM with the same ensembles and majority voting. For the underlying SVM we used the R package e1071 (Dimitriadou et al. 2008) which is based on the popular LIBSVM implementation (Chang et al. 2011).

Table 1 Datasets used in experiments: number of records, number of training records, number of features and number of classes

Name	Records	Train	Features	Classes	Remarks
Spam	4,601	3,036	57	2	
OptDigit	5,620	3,823	64	10	
Satellite	6,435	4,435	36	6	
Adult	45,222	30162	14	2	
Acoustic	98,528	78,823	50	3	
Acoustic2	98,528	78,823	50	2	Class 3 vs. rest

All algorithms were run repeatedly (10 times) with different training samples. The mean classification accuracy and its standard deviation is reported, as well as the training time on a single machine.

We performed a basic tuning of the SVM hyperparameters: The parameters c_0 and d were tuned by grid search and the parameter γ was chosen according to the recipe of Sect. 2.3, item 3.

4 Results

The effectiveness of Bagging is shown in Fig. 1. Ensembles of mixed type (LinRad, RadPol and LinRadPol) are significantly better in this case than ensembles of pure type (Lin, Rad and Pol). Even if we build from the best type (radial) an ensemble "Radialx3" with the same ensemble size as in LinRadPol, its accuracy is significantly below that of LinRadPol.

The complete results on all datasets are shown in Table 2. It is impressive to note that the ensembles—although each ensemble SVM is only trained on a much smaller training set—reach or even surpass the accuracy of the single SVM. For the ensembles we tested all three kernel types, but since radial and polynomial performed usually better than linear, we show only the former pure types. A result in column RadPol is underlined, if it was significantly ($\alpha = 5\%$) better in a t-test comparison to the "radial" or "polynom" results of the same method. It has to be noted that RadPol/LinRadPol ensembles have twice/three times as many SVMs as a pure ensemble, requiring much more computation time. If we compare LinRadPol in Table 1 with the column Radialx3, we find that it is sometimes better than even Radialx3 (dataset Spam), but sometimes not (datasets Satlog and Adult).

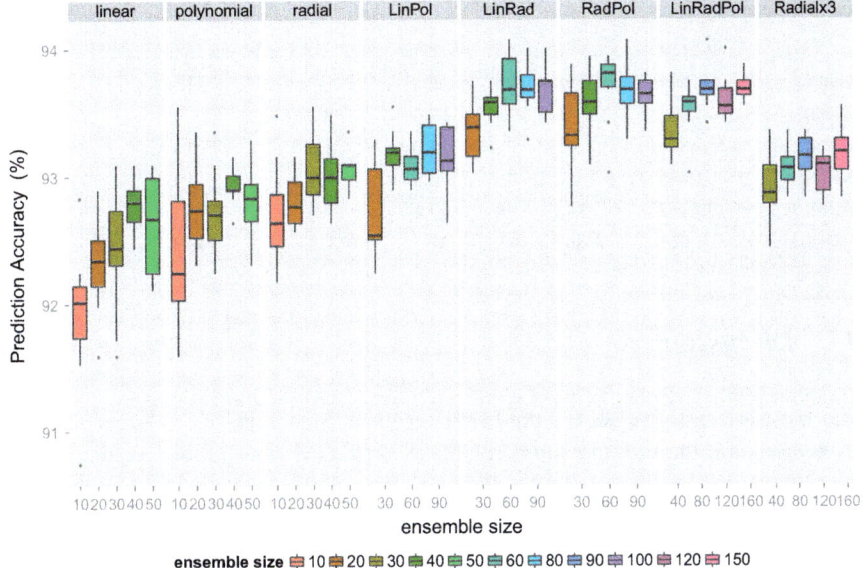

Fig. 1 Accuracy on task Spam with SVM-bagging. Each sub-classifier uses a training sample of only 300 records (roughly 10 % of the available training data). Different ensembles of pure and mixed kernel type are formed, as indicated in the figure head

Table 2 Accuracies of the different models in percent

Name	SVM	Bagging					Boosting			
		Radial	Polynom	RadPol	LinRadPol	Radialx3	Radial	polynom	Mixed	RadPol
ens.size	1	50	50	100	150	150	50	50	50	100
Spam	93.01	93.05	92.80	93.64	93.70	93.17	93.74	93.94	94.01	**94.10**
		0.14	0.26	0.13	0.10	0.21	0.33	0.24	0.19	0.32
OptDigit	**97.94**	96.36	96.19	96.33	96.15	96.38	97.69	97.38	97.11	97.69
		0.16	0.10	0.16	0.08	0.10	0.23	0.19	0.15	0.15
Satlog	**91.05**	87.51	86.49	87.21	86.16	87.76	89.96	88.73	89.69	90.31
		0.28	0.36	0.19	0.24	0.14	0.58	0.67	0.30	0.36
Adult	84.45	84.84	82.94	84.03	84.50	84.86	84.53	84.15	84.40	**84.92**
		0.07	0.13	0.10	0.05	0.05	0.30	0.22	0.42	0.26
Acoustic2	**90.70**	90.10	89.85	90.28	90.05	90.26	89.46	89.56	89.28	90.01
		0.13	0.10	0.11	0.05	0.08	0.37	0.18	0.05	0.15

All ensembles were trained on small stratified samples of size 300, while the single SVM was trained on all training data (3,000–79,000 records). The first line shows the mean test-set accuracy of ten runs with different training samples. The second line (in *italic numbers*) gives the standard deviations. The best result in each row is marked in **boldface**. A result in column *RadPol* is underlined if it is significantly better than the results in the preceding columns *radial* and *polynom*

Table 3 Training times in seconds (Intel Core i73632QM CPU, 64bit, 2.2 GHz, 8 GB RAM)

Name	# Train records	SVM on all records	Boosting on 300 records, ensemble size 50			
			Radial	Polynomial	Mixed	RadPol
Spam	3,067	1.2	11.0	8.0	9.5	19.1
OptDigit	3,823	2.2	13.0	11.5	12.4	24.5
Satlog	4,435	2.1	12.5	9.2	10.3	21.7
Adult	30,162	148.2	60.1	37.7	50.0	97.8
Acoustic2	78,823	3,866.5	231.3	167.2	199.4	398.6

4.1 Discussion

The most interesting effect is the lower resource consumption for larger datasets. If we compare the training times for the Acoustic2 dataset (Table 3), the RadPol ensemble needs less than $1/10$ and the radial ensemble needs only $1/16$ of the SVM (all data) training time. The price to pay is a somewhat longer prediction time of the ensemble, but SVM prediction is usually fast and the prediction time rises only linearly with ensemble size.

We analyzed the sample size settings used for the ensembles. Here, Bagging usually performs better with larger sample sizes, while Boosting tends to be more effective with smaller sample sizes. The reasons can be interpreted as follows: while for Bagging larger sample sizes lead to better predictions for the single SVM in the ensemble, consequently the ensemble performs better and is more robust. The opposite could be observed for boosting. In boosting, a sequential procedure is initiated, where the decisions in initial iterations directly influence the later behavior of the algorithm and the underlying prediction models. In our case the ensemble tends to focus on initial wrong predicted patterns, whereas the generalization of the remaining patterns is lost. Thus, the underlying problem can be seen as a certain kind of overfitting. A solution to this can now be given by reducing the sample sizes for the single SVM learners. By conducting a repeated resampling of the data, the overfitting for some individual patterns is avoided. The single learners trained with few patterns seem to be weak at first, but the whole ensemble achieves a better generalization performance by the right combination of several weak learners.

A nice side effect of ensemble learning with SVM is that the aggregation of multiple SVM learners seems to lead to a decreased impact of hyperparameter settings as C or γ. Usually these parameters are very sensitive and have a high influence on the prediction accuracy. The ensemble approach seems to weaken this influence by its subtle combination of the single learners. It is possible that due to the strength of the combined learning approach the influence of bad hyperparameter settings for single learners becomes less important. In our experiments the hyperparameters were determined by a simple internal estimation of the SVM algorithm. However, the role of hyperparameter settings in ensemble learning is also subject of future work.

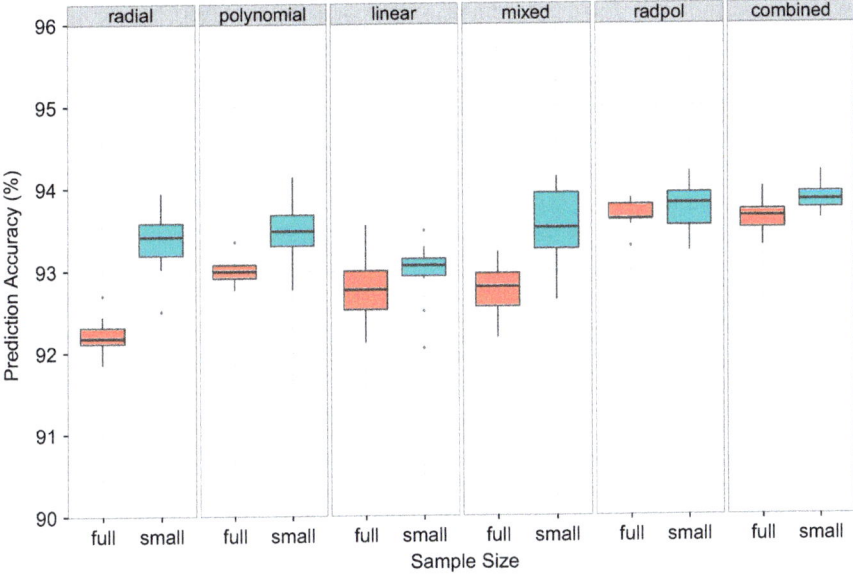

Fig. 2 Test-set accuracy on task Spam with SVM-Boosting. In case of training with the complete data (full: 3,036 records), the ensemble suffers from considerable overfitting, while at the same time the ensemble performs better with subsamples of the data (small: 300 records). Mixed ensembles like "radpol" and "combined" = (linear, polynomial, radial) are better than a pure "radial" ensemble

Meyer et al. (2013) did research on parallel approaches for SVMs with Bagging and cascade SVM. We can support their main result that Bagging is often effective in producing good accuracy within drastically reduced time. They had ensembles of size 9, having each 1/9 of the training data. We found that even lower sample sizes can be effective. In addition we found that AdaBoost gives **only** better results on SVM ensembles, if the training size is small enough.

Wang et al. (2009) considered a similar approach to ours in that they proposed a modified AdaBoost algorithm with reduced sample size for each classifier of the ensemble. However, they considered only a slight reduction (80% of all training data) which does not "weaken" the SVMs sufficiently, and they observe a strong degradation when the ensemble size is increased. We initially observed a similar behavior that the ensemble was less productive than a single SVM when the training sample was too big (see Fig. 2).

5   Conclusion

We performed an analysis of ensemble methods consisting of SVM learners. In our study we compared two ensemble approaches based on Bagging and Boosting. In an experimental study we observed comparable performances of the ensemble learners, with viable runtimes for the ensemble. This could be achieved by incorporating random resampling strategies, considering only parts of the training data for the ensemble learners.

In part of our experiments (Spam/Bagging and Adult/Boosting) the ensemble learners performed significantly better when different kernel types were combined. This is due to the fact that multiple kernel types inside the ensemble are beneficial for diversification, partially supporting our hypothesis H-1. In the other cases the combined ensembles did as well as the pure radial ensembles.

The ensemble performed well enough to compete with one single SVM learner trained on the full data, which corresponds with hypothesis H-3. Due to the bad scalability of SVM, such ensemble approaches are especially interesting for large datasets, where the runtime of a single SVM training becomes prohibitively long. Here, the approaches based on Bagging and Boosting are possible solutions. The most apparent advantage of Bagging is probably its nice potential for parallelization. Boosting could stress its good theoretical properties, but it was necessary to use small training sample sizes for making each SVM in the ensemble weak, as it was advocated by hypothesis H-2.

Acknowledgements This work has been partially supported by the Bundesministerium für Bildung und Forschung (BMBF) under the grant SOMA (AiF FKZ 17N1009) and by the Cologne University of Applied Sciences under the research focus grant COSA.

GEFÖRDERT VOM

Bundesministerium
für Bildung
und Forschung

References

Breiman, L. (1996): Bagging predictors. In: *Machine Learning, 24*(2), 123–140.

Caputo, B., SIM, K. Furesjo, F., & Smola, A. (2002). Appearance-based object recognition using SVMs: Which kernel should I use? In *Proceedings of Nips Workshop on Statistical Methods for Computational Experiments in Visual Processing and Computer Vision*, Whistler.

Chang, C., & Lin, C. (2011). LIBSVM: A library for support vector machines. *ACM Transactions on Intelligent Systems and Technology (TIST), 2*(3), 27.

Chang, E. Y., & Zhu, K., et al. (2008). PSVM: Parallelizing support vector machines on distributed computers. *Advances in Neural Information Processing Systems, 20*, 16.

Cortes, C., Mohri, M., & Rostamizadeh, A. (2012). Ensembles of kernel predictors. arXiv preprint:1202.3712, arxiv.org.

Cortes, C., & Vapnik, V. (1995). Support vector machine. *Machine Learning, 20*(3), 273–297.

Crammer, K., & Singer, Y. (2002). On the algorithmic implementation of multiclass kernel-based vector machines. *Journal of Machine Learning Research, 2*, 265–292.

Cristianini, N., & Shawe-Taylor, J. (2000). *Support vector machines*. Cambridge: Cambridge University Press.

Dimitriadou, E., Hornik, K., Leisch, F., Meyer, D., & Weingessel, A. (2008). Misc functions of the department of statistics (e1071), TU Wien. R package, version 1.5-18. http://CRAN.R-project. org/package=e1071.

Freund, Y., & Schapire, R. E. (1995). A decision-theoretic generalization of on-line learning and an application to boosting. In *Proceedings of the Second European Conference on Computational Learning Theory (EuroCOLT)* (pp. 23–37).

Kim, H.-C., Pang, S., Je, H.-M., Kim, D., & Yang Bang, S. (2003). Constructing support vector machine ensemble. *Pattern Recognition, 36*(12), 2757–2767.

Lin, H.-T., & Li, L. (2008). Support vector machinery for infinite ensemble learning. *The Journal of Machine Learning Research, 9*, 285–312.

Mercer, J. (1909). Functions of positive and negative type, and their connection with the theory of integral equations. *Philosophical Transactions of the Royal Society London, 209*, 415–446.

Meyer, O., Bischl, B., & Weihs, C. (2013). Support vector machines on large data sets: Simple parallel approaches. In M. Spiliopoulou, et al. (Eds.), *Data analysis, machine learning and knowledge discovery*. New York: Springer.

Pavlov, D., Mao, J., & Dom, B. (2000). Scaling-up Support Vector Machines using boosting algorithm. In *Proceedings of the 15th International Conference on Pattern Recognition* (Vol. 2, pp. 219–222). IEEE, Barcelona.

Schölkopf, B., & Smola, A. (2002). *Learning with kernels: Support vector machines, regularization, optimization and beyond*. Massachusetts: MIT Press.

Wang, S., Mathew, A., Chen, Y., Xi, L., Ma, L., & Lee, J. (2009). Empirical analysis of support vector machine ensemble classifiers. *Expert Systems with Applications, 36*(3), 6466–6476.

Weston, J., & Watkins, C. (1999). Support Vector Machines for multi-class pattern recognition. In *Proceedings of the 7th European Symposium on Artificial Neural Networks (ESANN)* (Vol. 99, pp. 61–72).

Wickramaratna, J., Holden, S. B., & Buxton, B. (2001). Performance degradation in boosting. In J. Kittler & F. Roli (Eds.), *Proceedings of the 2nd International Workshop on Multiple Classifier Systems* (pp. 11–21). Cambridge: Cambridge University Press.

Yu, H., Yang J., Han, J., & Li, X. (2005). Making SVMs scalable to large data sets using hierarchical cluster indexing. *Data Mining and Knowledge Discovery, 11*(3), 295–321.

Part IV
Data Analysis in Marketing

Ratings-/Rankings-Based Versus Choice-Based Conjoint Analysis for Predicting Choices

Daniel Baier, Marcin Pełka, Aneta Rybicka, and Stefanie Schreiber

Abstract Nowadays, for market simulation in consumer markets with multi-attributed products, choice-based conjoint analysis (CBC) is most popular. The popularity stems—on one side—from the possibility to use online-panels for affordable data collection and—on the other side—from the possibility to estimate part worths at the respondent level using only few observations. However, a still open question is, whether this money- and time-saving approach provides the same or even better results than ratings-/rankings-based alternatives. An experiment with 787 students from Poland and Germany is used to answer this question: Cola preferences are measured using CBC as well as ratings-/rankings-based alternatives. The results are compared using the Multitrait-Multimethod Matrix for the estimated part worths and first choice hit rates for holdout choice sets. The experiment shows a superiority of CBC, but also important differences between Polish and German cola consumers that outweigh methodological differences.

1 Introduction

Recently, some studies on the validity of commercial applications of conjoint analysis have posed the question whether the nowadays widespread choice-based conjoint analysis (CBC) is really a step forward. So, Selka et al. (2014) and Selka and Baier (2014) have analyzed 1,777 commercial applications of CBC and—disappointingly—found out that—in spite of many methodological improvements over time and more and more commercial applications by the market research institute under study—the predictive validity has deteriorated: In 2002 the average

D. Baier (✉) • S. Schreiber
Brandenburg University of Technology Cottbus-Senftenberg, Chair of Marketing and Innovation Management, Erich-Weinert-Straße 1, 03046 Cottbus, Germany
e-mail: daniel.baier@tu-cottbus.de; stefanie.schreiber@tu-cottbus.de

M. Pełka • A. Rybicka
Department of Econometrics and Computer Science, Wroclaw University of Economics, Nowowiejska 3, 58-500 Jelenia Góra, Poland
e-mail: marcin.pelka@ue.wroc.pl; aneta.rybicka@ue.wroc.pl

© Springer-Verlag Berlin Heidelberg 2015
B. Lausen et al. (eds.), *Data Science, Learning by Latent Structures, and Knowledge Discovery*, Studies in Classification, Data Analysis, and Knowledge Organization, DOI 10.1007/978-3-662-44983-7_18

first choice hit rates (FCHRs) for holdout choices was 0.626 ($n = 57$), in 2011 it reduced to 0.520 ($n = 457$). Various analyses showed that this deterioration over time is significant and independent from influences from, e.g., the product category, the problem type, the complexity of the application (number of attributes and levels), the usage of online-interviewing, -panels and multimedia, or the sample size (Selka and Baier 2014).

Does this mean that we have to reconsider the usage of alternatives like traditional conjoint analysis (TCA) and self-explicated models (SEM)? This paper tries to answer this question. In Sect. 2 CBC, SEM, and TCA are shortly discussed with known (dis)advantages and results from former comparisons and the research questions are derived. In Sect. 3 we discuss the experiments and results. Section 4 closes with a short conclusion and outlook.

2 Ratings-/Rankings-Based Versus Choice-Based Methods

2.1 Ratings-/Rankings-Based Methods: TCA and SEM

For many years, ratings-/ranking-based conjoint analysis and related methods have been widespread and often applied in marketing research (see, e.g., Wittink and Cattin 1989; Green et al. 2001). The methods provide a simple series of steps to measure consumers' preferences with respect to multi-attributed products: So, e.g., when using TCA, respondents are confronted with a set of alternative attribute-level-combinations (so-called stimuli) and asked to rate or rank them according to their buying preference. It should be mentioned that in this context, the question whether to collect ratings (of single stimuli) or rankings (across all stimuli, maybe in a two-step task where first "good" and "bad" stimuli are allocated on the left and right side of the table) lead to very similar results with respect to part worths, independent from the estimation procedure (ANOVA or MONANOVA, see Green and Srinivasan 1978). This is mostly ascribed to the typical misfit in TCA with few observations and many model parameters.

Figure 1 gives an example of such a data collection task, where a sample of respondents is asked to rank 29 alternative cola stimuli that differ with respect to the attributes brand, flavor, calorie content, caffeine content, price, and bottle size (To be concrete: They were first asked to sort them into desirability groups, then rank the stimuli within each group, then check for intergroup exchanges). After data collection, the responses can be used to estimate the (assumed additive) effects—the so-called part worths—of the attribute-levels using ANOVA, MONANOVA, or other estimation procedures. Then, finally, these derived part worths can be used to predict the future selection behavior of the sample of respondents or—after statistical projections—the complete market (market simulation). This prediction formalism is also used to evaluate the predictive accuracy of the application: Some of the stimuli (in Fig. 1 four of the 29 stimuli) were excluded from parameter

Please rank order the following alternatives:

Stimulus 24	Stimulus 27	Stimulus 19	Stimulus 18
Other Cola	Pepsi Cola	Pepsi Cola	Pepsi Cola
Cola with cherry	Cola	Cola with lemon	Cola
Zero	Zero	Zero	Light
Caffeine-free	Caffeinated	Caffeine-free	Caffeine-free
0.59 €/l	0.59 €/l	0.79 €/l	0.89 €/l
0.5 l	0.5 l	0.5 l	1 l

Stimulus 29	Stimulus 26	Stimulus 25	Stimulus 25	Stimulus 28
Other Cola	Other Cola	Coca Cola	Coca Cola	Coca Cola
Cola	Cola	Cola	Cola	Cola with lemon
Zero	Normal	Zero	Zero	Zero
Caffeinated	Caffeine-free	Caffeinated	Caffeinated	Caffeine-free
0.79 €/l	0.89 €/l	0.69 €/l	0.69 €/l	0.69 €/l
1.5 l	0.5 l	1.5 l	1.5 l	1 l

Fig. 1 Data collection in TCA: respondents are asked to rank or rate alternative attribute-level-combinations

Please select one of the following alternatives:

1	2	3	4	5
Other Cola	Other Cola	Coca Cola	Pepsi Cola	
Cola	Cola with cherry	Cola with orange	Cola with lemon	No choice
Normal	Zero	Normal	Normal	at the moment
Caffeinated	Caffeine-free	Caffeine-free	Caffeine-free	
0.79 €/l	0.59 €/l	0.79 €/l	0.69 €/l	
1 l	0.5 l	1.5 l	2 l	

Fig. 2 Collection of holdout choice data for evaluating the predictive validity: respondents are asked to select the most preferred stimulus in presented sets of alternative attribute-level-combinations (and a no-choice option)

estimation (as so-called holdout stimuli) to check the internal validity of the model. Additionally, after the data collection task described above, the respondents were shown some so-called holdout choice sets of attribute-level-combinations and asked to select the most preferred one. Figure 2 gives an example of such a choice task. Here, a so-called first choice hit (FCH) occurs, when the most preferred stimulus can be predicted by the estimated part worths (maximum utility rule). The FCHR gives the percentage of correctly predicted most preferred stimuli over all holdout choice sets and is a measure for the predictive validity of the application.

An alternative ratings-/rankings-based approach is SEM. Here, the respondents are asked to rate the different attribute-levels directly on like-dislike-scales, i.e. to give estimates of the part worth vectors by their own. Here, no estimation of part-worths is needed, since the respondents have explicated their preference structure by their own. However, often it is assumed that this SEM approach (belonging not directly to the conjoint analysis group of methods) leads to part worths of minor validity (e.g., over-estimated weights/importances for the price attribute and its levels) since the data collection doesn't reflect a real buying situation adequately where products are presented *conjointly* with all their attributes and levels.

2.2   Choice-Based Methods: CBC

However, also since some years, with CBC a strong methodological competitor to this ratings-/rankings-based methods has become popular. In this approach, the respondents are repeatedly confronted with sets of stimuli (the so-called choice sets, usual are 10–15 sets for one respondent, each with 3–4 stimuli and an additional no-choice option). They are asked to select in each choice situation the most preferred stimulus (see, e.g., Louviere and Woodworth 1983). This task is similar to the abovementioned holdout choice set task (see Fig. 2). Afterwards, the responses are analyzed at the aggregate level (using, e.g., multinomial logit models) or—more adapted to the usual conjoint analysis setting—at the respondent level by sharing information across respondents (using, e.g., the hierarchical Bayes versions of CBC). Again, during data collection, holdout choice sets are presented and FCH and FCHR can be used to evaluate the predictive validity of the application.

Surveys on the commercial application of conjoint analysis in marketing research show that the percentage of CBC applications has increased from 7 % (Wittink et al. 1994), 13 % (Baier 1999), and 47 % (Sattler and Hartmann 2008) to 94 % (Selka et al. 2014; Selka and Baier 2014) nowadays. This expansion seems to be triggered on one side by the availability of convenient software—e.g., the CBC series of systems provided by Sawtooth Software since 1996 (Sawtooth Software 2013), but also from some studies where ratings-/rankings-based and CBC were compared (see, e.g., Elrod et al. 1992; Oliphant et al. 1992; Vriens et al. 1998; Moore et al. 1998; Moore 2004; Karniouchina et al. 2009) and where—mostly—CBC was found to be superior to TCA and only slight differences between the resulting part worths were found.

2.3   Research Questions and Strategy to Answer Them

A closer look at the current literature shows ambivalent results: So, e.g., Elrod et al. (1992) and Moore (2004) found TCA superior to CBC, some other authors only found slight superiorities of CBC. Vriens et al. (1998) only used simulated data, and all other studies only used one sample of respondents (students) who were confronted with ratings-/rankings-based and choice-based data collection tasks consecutively, a situation that reflects data collection in real applications only in a limited way. Also, Karniouchina et al. (2009) found out that for some attributes and levels the resulting part worths from ratings-/rankings-based and choice-based models differed. So, e.g., they mention that *enriched* attributes (e.g., brand name) are more important in ratings-/rankings-based models, and *comparable* attributes (e.g., price) are more important in choice-based models.

This leads us to the following research questions:

- Is CBC superior to SEM and TCA with respect to predictive validity?
- Are the resulting part worths from CBC, SEM, TCA distinguishable?

To answer this question empirically we selected the cola market. This is an international market with many similar brands and product variants as well as target markets and consumer tastes across countries. So, we can test whether the differences in preference structures across countries outweigh the differences that arise from differences in data collection and analysis. We selected German and Polish students at universities as the population for our experiments and expect that they should show some differences in preference structures [see Ansari et al. (2012) for a discussion of food consumption and diet differences between students in Germany and Poland; here, e.g., the consumption of fast food was much higher in Germany than in Poland and much higher with male than with female students]: Polish students are assumed—on average—to be more price-sensitive than German students, also, one can assume that Polish students (especially the women) are more sensitive to control their outfit than German students (with implications, e.g., to the preference of calorie content). With respect to the differences stemming from the method applied, we assume—as above discussed—that in SEM the price is more important than in TCA and CBC (direct questioning) and in CBC more important than in TCA (comparable attribute). Also, we assume that the brand is more important in TCA than in CBC (enriched attribute).

To test whether the resulting part worths from the different methods and the populations are distinguishable and whether the differences stemming from the methods outweigh the differences stemming from the populations we use—besides *t*-tests—the Multitrait-Multimethod Matrix from Campbell and Fiske (1959) and Churchill (1979). In this matrix all correlations between the different measurements are collected and the ratio between *expected* and *unexpected* relations between these correlations can be used to answer this question. Also, we use FCH and FCHR as well as logistic regression to answer the question whether CBC is superior to SEM and TCA.

3   Empirical Studies: Experiments and Results

The product under investigation was—as already mentioned—cola to be bought in the supermarket with the attributes brand (with levels Coca Cola, Pepsi Cola, other brand), flavor (Cola, Cola with orange, Cola with lemon, Cola with cherry), calorie content (normal, light, zero), caffeine content (caffeinated, caffeine-free), price (0.59 €/l, 0.69 €/l, 0.79 €/l, 0.89 €/l), and bottle size (0.5 l, 1 l, 1.5 l, 2 l).

The attribute and levels were selected in the usual way, using focus groups with consumers and expert interviews to determine the most important ones for the consumers' choice decision in the supermarket. The data collection took place at two universities near the German-Polish border. In each country 200 students should be interviewed online (using the CBC format), 200 students offline (using the TCA and SEM format).

In order to keep the data collection comparable to commercial applications, in each experiment only one of the complicated conjoint experiments (TCA or CBC) was included. The first experiment in each country was an offline-experiment with an SEM and a TCA task. For TCA, 25 stimuli were generated using orthogonal plans as proposed by SPSS Conjoint. Moreover, four test stimuli were added to check the internal validity of the model (see Fig. 1). In Germany 199 respondents participated in the experiment, in Poland 194. The second experiment in each country was an online experiment with a CBC task. The respondents were confronted with 18 choice sets, each consisting of four attribute-level-combinations plus the no-choice option (see Fig. 2). The number of stimuli and choice sets is somewhat high in both experiments, but the students received an incentive and accepted the (complicated) tasks. In both cases the idea was to check later whether a reduced number of observations would lead to the same or even better results. In Germany 169 respondents participated in the experiment, in Poland 225. As an incentive for participating, all respondents received a voucher for a small bottle of cola in the cafeteria of the university. All experiments closed with the same holdout choice task (eight identical holdout choice sets were presented) to evaluate the predictive validity. All experiments were performed during 4 weeks in May and June 2013.

To summarize the data collection, one can state that data collection in all four experiments was possible without problems, which is somewhat in contrast to the nowadays usual assumption that conjoint experiments are getting less attractive to respondents. The respondents were interested and (mostly) followed the interviews with high interest and attention since the product was of interest for them. This supports the assumption of Selka et al. (2014) and others that we should perform only conjoint experiments with a very good fit between respondents and product. The CBC data collection was—as expected—much faster than the offline experiment which suffered from the long time of sorting 29 stimuli. TCA also proved very cumbersome when coding the questionnaire results, since the respondents recorded the numbers of the stimuli after sorting by themselves. Also—as expected—SEM proved to be the fastest and least cumbersome procedure for the respondents. However, in all experiments, there were cheating respondents when filling out the questionnaires: Some used constant ratings-/rankings-rules, some didn't sort the stimuli with great efforts. A rough estimate gives us 7–9 % of the respondents who were cheating when they filled out the questionnaires with a higher percentage within the Polish students. Even though it is possible to eliminate such respondents (by checking their predictive validity values), we decided to leave them all in the samples, since we wanted to evaluate the overall performance of the methods and not the performance with respect to a subsample of *good* answering respondents. Moreover, an elimination of respondents with low RLH-values (as a proxy for eliminating cheating people) didn't improve the predictive validity of the Polish sample dramatically.

For analyzing the responses, we used the usual software: SEM and TCA data were analyzed using SPSS, CBC data were analyzed using CBC/HB. For hierarchical Bayes estimation we used 10,000 draws as burn-ins and 10,000 draws for calculating the part worths. We also tested 100,000 burn-ins but found no differences in the results (Bayesian shrinkage). The internal validity of TCA was 0.982 (averaged Pearson correlation coefficient across respondents) for the German sample and 0.858 for the Polish sample, both are very good values. The internal validity of CBC was 0.647 (averaged root likelihood value across respondents) for the German sample and 0.598 for the Polish sample, again both are good values. Afterwards, for comparison reasons, the derived part worths at the respondent level were standardized in the usual way so that—for each respondent in each experiment—the maximum possible value for a stimulus is 1 and the minimum 0. Also the individual attribute importances in each experiment were calculated via the difference of the highest and the lowest part worth for levels of the corresponding attribute. Table 1 gives the averaged part worths for all six measurements (SEM, TCA, CBC in D and PL, D = Germany, PL = Poland) and also averaged importances in three two-group partitions (all SEM measurements vs. all TCA or CBC measurements, all TCA vs. all CBC measurements, all D vs. all PL measurements).

From Table 1 one can easily see that the assumed differences between the part worths are mainly supported: SEM overweighs the importance of price, also Polish respondents rate the price levels with higher part worths and—consequently—a higher importance. The expected differences between Polish and German students with respect to the calorie content are also significant. Not supported is the assumption that in CBC—compared to TCA—the price attribute has a higher importance (comparable attribute) and the brand a lower importance (enriched attribute).

To analyze the distinguishability of the different part worths in more detail, the correlations between them were calculated and organized in form of the Multitrait-Multimethod Matrix (see Table 2). The differences coming from the three methods SEM, TCA, and CBC should here be outweighed by the differences coming from the two populations/traits D and PL. Consequently, the correlation values in diagonals (with identical trait) should be higher than the corresponding rows and columns of the same submatrix (strong conditions) and the off-diagonals of the other submatrices (weak conditions). Comparisons lead to a fulfillment of 9 of 12 strong conditions and 9 of 18 weak conditions which can be overall summarized in that way that the differences stemming from the nationality overweighs the differences stemming from the methods. So, one can assume that all three methods lead to similar results (at least more similar across methods than across nationalities).

The last comparison deals with the predictive validity. Here the responses to the holdout choice tasks (identical in all experiments) have to be compared. There are two possibilities to calculate them: One could control all holdout choices (including the selections of the no-choice option, in total $n = 9,880$ selections) or one could

Table 1 Averaged standardized part worths, attribute importances, and standard deviations using different methods (*SEM* self-explicated model, *TCA* traditional conjoint analysis, *CBC* choice-based conjoint analysis) and nationalities (D = German, PL = Polish); only the importances differences were tested using *t*-tests

Attribute	Level	Averaged standardized part worths (standard deviation)					
		SEM, D	SEM, PL	TCA, D	TCA, PL	CBC, D	CBC, PL
Brand	Coca Cola	0.140(0.112)	0.077(0.131)	0.154(0.181)	0.115(0.142)	0.172(0.127)	0.144(0.172)
	Pepsi Cola	0.036(0.071)	0.072(0.113)	0.065(0.100)	0.096(0.121)	0.075(0.073)	0.121(0.141)
	Other	0.064(0.096)	0.038(0.091)	0.086(0.134)	0.104(0.131)	0.015(0.035)	0.088(0.164)
Flavor	Cola	0.168(0.096)	0.085(0.110)	0.214(0.181)	0.096(0.119)	0.241(0.140)	0.164(0.162)
	W. orange	0.067(0.098)	0.032(0.082)	0.113(0.145)	0.060(0.075)	0.094(0.111)	0.072(0.109)
	W. lemon	0.060(0.095)	0.070(0.118)	0.142(0.139)	0.105(0.106)	0.132(0.117)	0.105(0.111)
	W. cherry	0.046(0.095)	0.055(0.124)	0.089(0.121)	0.128(0.109)	0.060(0.100)	0.101(0.135)
Calorie	Normal	0.080(0.079)	0.055(0.057)	0.147(0.143)	0.080(0.083)	0.193(0.167)	0.064(0.080)
	Light	0.030(0.062)	0.022(0.051)	0.059(0.087)	0.078(0.072)	0.086(0.081)	0.037(0.056)
	Zero	0.029(0.067)	0.018(0.072)	0.049(0.081)	0.046(0.065)	0.046(0.096)	0.048(0.071)
Caf-feine	Caffeinated	0.136(0.091)	0.057(0.080)	0.107(0.124)	0.037(0.058)	0.083(0.092)	0.035(0.049)
	Caffeine-free	0.001(0.015)	0.029(0.076)	0.012(0.036)	0.035(0.049)	0.014(0.045)	0.026(0.060)
Price	0.59 €/l	0.236(0.118)	0.463(0.261)	0.066(0.083)	0.130(0.088)	0.087(0.095)	0.130(0.148)
	0.69 €/l	0.157(0.079)	0.309(0.174)	0.063(0.064)	0.079(0.086)	0.060(0.065)	0.100(0.074)
	0.79 €/l	0.079(0.039)	0.154(0.087)	0.057(0.053)	0.081(0.080)	0.038(0.044)	0.086(0.079)
	0.89 €/l	0.000(0.000)	0.000(0.000)	0.034(0.047)	0.072(0.077)	0.013(0.025)	0.044(0.094)
Bottle size	0.5l	0.098(0.091)	0.039(0.062)	0.084(0.093)	0.064(0.074)	0.027(0.049)	0.056(0.090)
	1l	0.086(0.071)	0.035(0.067)	0.075(0.086)	0.101(0.088)	0.061(0.052)	0.059(0.075)
	1.5l	0.052(0.063)	0.025(0.051)	0.055(0.066)	0.102(0.092)	0.049(0.050)	0.071(0.075)
	2l	0.021(0.055)	0.027(0.066)	0.036(0.051)	0.094(0.088)	0.048(0.056)	0.075(0.108)

Attribute	Averaged attribute importances (standard deviation)			
	SEM	TCA or CBC	TCA	CBC
Brand	0.157(0.130)	0.217***(0.166)	0.207(0.163)	0.227(0.167)
Flavor	0.177(0.134)	0.256***(0.152)	0.247(0.147)	0.265(0.155)
Calorie	0.093(0.084)	0.158***(0.122)	0.152(0.112)	0.163(0.131)
Caffeine	0.112***(0.095)	0.086(0.089)	0.095**(0.097)	0.076(0.078)
Price	0.348***(0.232)	0.148(0.108)	0.146(0.083)	0.151(0.128)
Bottle size	0.113(0.092)	0.135***(0.094)	0.153***(0.086)	0.118(0.102)

Attribute	Averaged attribute importances (standard deviation)	
	D	PL
Brand	0.188(0.140)	0.205(0.171)
Flavor	0.248***(0.142)	0.213(0.156)
Calorie	0.171***(0.133)	0.105(0.084)
Caffeine	0.118***(0.103)	0.072(0.073)
Price	0.152(0.115)	0.273***(0.217)
Bottle size	0.123(0.081)	0.132(0.107)

*Significant difference at $\alpha = 0.05$; significant difference ** at $\alpha = 0.01$; *** significant difference at $\alpha = 0.001$

Table 2 Multitrait-multimethod matrix: correlations are given for the three measurements (using the methods SEM, TCA, CBC) of the two traits (preference structure of the German students D and the Polish students PL)

Method	Trait	SEM		TCA		CBC	
		D	PL	D	PL	D	PL
SEM	D	1					
	PL	0.766**	1				
TCA	D	0.494*	−0.016	1			
	PL	0.327	0.363	0.267	1		
CBC	D	0.473*	0.089	0.895**	0.264	1	
	PL	0.588**	0.466*	0.596**	0.716**	0.588**	1

*Significant not-zero correlation at $\alpha = 0.05$; **Significant not-zero correlation at $\alpha = 0.01$

Table 3 Results of logistic regression with the first choice hits (FCHs) as dependent variable and the indicators German (1 = German yes, 0 = no), CBC (1 = data collection and analysis using CBC yes, 0 = no), and TCA (1 = using TCA, 0 = no) as independent variables; results are given with including the no-choices (with NC) and without them (without NC)

	Logistic regression coefficients	
	FCH with NC	FCH without NC
Const. effect	−0.948***	−0.846***
German	0.676***	1.284***
CBC	0.091	0.295***
TCA	0.196**	0.211***
Nagelkerkes R^2	0.037	0.126

*Significant effect at $\alpha = 0.05$; **significant effect at $\alpha = 0.01$; ***significant effect at $\alpha = 0.001$

control only the holdout choices where a holdout stimulus was selected (excluding the selection of the no-choice option, in total $n = 7,803$ selections). The fair comparison would be the second one (see, e.g., Karniouchina et al. 2009), since SEM and TCA don't collect data to predict the no-choice option, only CBC is able to give such predictions.

So, for the second one, in Germany, CBC outperforms TCA and SEM with FCHR values of 0.726 (for CBC), 0.626 (for TCA), and 0.573 (for SEM), whereas in Poland, TCA outperforms SEM and CBC with FCHR values of 0.357 (for TCA), 0.324 (for SEM), and 0.323 (for CBC). However, since the FCHR values for the Polish experiments are very low, one should nevertheless conclude that CBC is superior with respect to prediction. This assumption is supported by the logistic regression in Table 3 where the effects of nationality and CBC are significant in both cases (even though their is tested no direct superiority between CBC and TCA).

4 Conclusions and Outlook

The analyses have shown that—in answering the first research question—for choice predictions, CBC (concretely: CBC/HB) performed best. However, especially in markets with many cheating respondents, the ratings-/rankings-based approaches compete well, especially when the no-choice options are neglected. It should be mentioned that for the ratings-/rankings-based methods also hierarchical Bayes estimation procedures exist that weren't used in these analyses and could still improve the choice predictions of TCA.

Also—in answering the second research question—the cross validity of the various approaches is striking, especially when applied to (assumed) different preference structures: The distinguishability with respect to the different subsegments is higher than the distinguishability with respect to the methodological approaches.

Of course, the comparisons between ratings-/rankings-based and choice-based methods haven't come to an end, one needs more such comparisons. Also, from these comparisons, one can draw many ideas for methodological improvements with respect to the ratings-/rankings-based and choice-based experiments.

References

Ansari, W. E., Stock, C., & Mikolajczyk, T. (2012). Relationships between food consumption and living arrangements among university students in four European countries: A cross-sectional study. *Nutrition Journal, 28*, 1–7.

Baier, D. (1999). Methoden der Conjointanalyse in der Marktforschungs- und Marketingpraxis. In W. Gaul & M. Schader (Eds.), *Mathematische Methoden der Wirtschaftswissenschaften* (pp. 197–206). Heidelberg: Physica.

Campbell, D. R., & Fiske, D. W. (1959). Convergent and discriminant validation by the multitrait-multimethod matrix. *Psychological Bulletin, 56*, 81–105.

Churchill, G. A. (1979). A paradigm for developing better measures for marketing constructs. *Journal of Marketing Research, 16*(1), 64–73.

Elrod, T., Louviere, J., & Davey, K. (1992). An empirical comparison of ratings-based and choice-based conjoint models. *Journal of Marketing Research, 24*(3), 368–377.

Green, P. E., Krieger, A. M., & Wind, Y. (2001). Thirty years of conjoint analysis: Reflections and prospects. *Interfaces, 31*(3b), 56–73.

Green, P. E., & Srinivasan, V. (1978). Conjoint analysis in consumer research: Issues and outlook. *Journal of Consumer Research, 5*(2), 103–123.

Karniouchina, E. V., Moore, W. L., Van der Rhee, B., & Verma, R. (2009). Issues in the use of ratings-based versus choice-based conjoint analysis in operations management research. *European Journal of Operational Research, 197*(1), 340–348.

Louviere, J. J., & Woodworth, G. (1983). Design and analysis of simulated consumer choice or allocation experiments: An approach based on aggregate data. *Journal of Marketing Research, 20*(4), 350–367.

Moore, W. L. (2004). A cross-validity comparison of ratings-based and choice-based conjoint analysis models. *International Journal of Research in Marketing, 21*(3), 299–312.

Moore, W. L., Gray-Lee J., & Louviere, J. J. (1998). A cross-validity comparison of conjoint analysis and choice models at different levels of aggregation. *Marketing Letters, 9*(2), 195–208.

Oliphant, K., Eagle, T. G., Louviere, J. J., & Anderson, D. (1992). Cross-task comparison of ratings-based and choice-based conjoint. In M. Metegrano (Ed.), *Sawtooth Software Conference Proceedings* (pp. 383–404).

Sattler, H., & Hartmann, A. (2008). Commercial use of conjoint analysis. In K. I. Hoeck & M. Voigt (Eds.), *Operations management in theorie und praxis* (Vol. 1, pp. 103–119). Wiesbaden: Gabler.

Sawtooth Software. (2013). *The CBC system for choice-based conjoint analysis version 8*. Orem, UT: Sawtooth Software Inc.

Selka, S., & Baier, D. (2014). Kommerzielle Anwendung auswahlbasierter Verfahren der Conjointanalyse: Eine empirische Untersuchung zur Validitätsentwicklung. *Marketing ZFP, 36*(1), 54–64.

Selka, S., Baier, D., & Kurz, P. (2014). The validity development of conjoint analysis over time: An investigation of commercial studies. *Studies in Classification, Data Analysis, and Knowledge Organization, 48*, 227–234.

Vriens, M., Oppewal, H., & Wedel, M. (1998). Rating-based versus choice-based latent class conjoint models: An empirical comparison. *Journal of the Market Research Society, 40*(3), 237–248.

Wittink, D. R., & Cattin, P. (1989). Commercial use of conjoint analysis: An update. *Journal of Marketing, 53*(3), 91–96.

Wittink, D. R., Vriens, M., & Burhenne, W. (1994). Commercial use of conjoint analysis in Europe: Results and critical reflections. *International Journal of Research in Marketing, 11*(1), 41–52.

A Statistical Software Package for Image Data Analysis in Marketing

Thomas Böttcher, Daniel Baier, and Robert Naundorf

Abstract The strongly growing number of available images reveals a great opportunity for a new age in the field of statistical analysis. Today, several thousand digital images are taken and published every day but not used for marketing purposes. Common statistical tools like SPSS, SAS, R, MATLAB, or RapidMiner still provide none or insufficient image processing packages. In this paper we introduce IMADAC, a statistical software in expansion of Naundorf et al. (Computer science reports. Institute of Computer Science, Brandenburg University of Technology, Cottbus, 2012) and Zellhöfer et al. (Proceedings of the 2nd ACM international conference on multimedia retrieval, ICMR '12, pp. 59–60, 2012). IMADAC, designed for experts as well as users without image processing background, combines statistical analysis on both, common statistical data (e.g., age or gender) and image processing methods. This paper demonstrates the usage of low level image features for statistical purposes (e.g., clustering or multi-dimensional scaling). To improve marketing analysis results, we further show how to combine image features with other statistical data and how it can be done in a graphical user interface (GUI).

1 Introduction

Nowadays, the usage of digital images is becoming more and more popular. Today, several thousand digital images are taken and published every day. Those images reveal a great opportunity for modern statistical analysis, but the usage for marketing purposes is almost non-existent. Working on a large set of images, the question for an appropriate method for analyzing and evaluating photos using a statistical software has to be answered. Considering huge amounts of images or image databases, several approaches can be used to get more information from images itself. To structure these images or to group them into classes a clustering approach can be applied. Additionally the usage of user information (e.g., which

T. Böttcher (✉) • D. Baier • R. Naundorf
Institute of Business Administration and Economics, Brandenburg University of Technology
Cottbus-Senftenberg, Postbox 101344, 03013 Cottbus, Germany
e-mail: tboettcher@tu-cottbus.de; daniel.baier@tu-cottbus.de; robert.naundorf@tu-cottbus.de

© Springer-Verlag Berlin Heidelberg 2015
B. Lausen et al. (eds.), *Data Science, Learning by Latent Structures,
and Knowledge Discovery*, Studies in Classification, Data Analysis,
and Knowledge Organization, DOI 10.1007/978-3-662-44983-7_19

user provides which images) allows an advanced analysis on the image data (see Sect. 4). As mentioned in a previous work (Naundorf et al. 2012) a market segmentation based on photos provided by the customers could be another use case (Baier and Daniel 2010). Here photos replace or supplement the common survey data. Asking customers in an online survey to upload images (e.g., typical holiday photos) provides the opportunity to distinguish different holiday interests. Different photos most likely indicate different interests (Naundorf et al. 2012). Extending these approach, in Baier et al. (2012) and Daniel and Baier (2013) a new method for lifestyle segmentation was introduced. Uploaded images from respondents should describe interests, activities, or opinions in different categories.

Such image analysis needs content-based methods (respectively the pixel information of the image). Here the so-called low-level information of the images is used (Lew et al. 2006). In contrast high-level features like metadata such as keywords, tags, or descriptions are very powerful, but hard to extract and if manually set, very subjective. For the scope of this paper high-level features are not considered anyway but they can be easily extracted and applied in IMADAC too. Using content-based methods to analyze sets of images is a large research topic over several fields. In the last years a various number of visual image features are developed. Especially in the field of image retrieval lots of developments are done (Datta et al. 2008), but in the case of marketing purposes a great lack exists.

As a part of our research work, we want to explore the possibilities of visual image features in combination with traditional marketing purposes. Our main goal is to use visual image features and demonstrate the effectiveness of classical methods like clustering, starting with images only. Furthermore, we try to fuse traditional marketing purposes and image processing methods and create a more powerful mechanism to analyze survey data and images.

The remainder of this paper is structured as follows: Sect. 2 gives an overview of a comparison on well-known statistical software packages and their possibilities for processing and analyzing images. Several requirements for marketing purposes are defined and compared with strength of IMADAC. Section 3 gives a short introduction to the IMADAC system and a basic description of image processing. The next section introduces the combination of traditional marketing purposes and image analyses and shows how a sample application can be done. Finally, we give a conclusion and give some ideas for future work.

2 Statistical Software Packages for Data Analysis in Marketing

As a result of a recently taken analysis in Naundorf et al. (2012), we showed that current available software packages did not provide sufficient methods for image processing in marketing. Because of the importance of these results, we give a short summary of the results.

For sure, common statistical software comes up with several clustering algorithms and numerous distance measures. But analyzing images by their content, traditional approaches cannot be used directly. Often, extracted low-level feature data contain different structures in contrast to traditional data, i.e., the length of the feature data can differ from image to image using the same visual feature. For those features special distance measures are needed and traditional approaches fail.

The three main preferred statistical software packages (Rexer 2010) which were analyzed are SPSS Statistics (IBM Corporation 2013), RapidMiner (Rapid-I 2010), and R (Venables et al. 2011). Additionally MATLAB (Mathworks Inc. 2013) and the also well-known SAS (SAS Institute Inc. 2013) software package should be considered, when comparing current statistical software. According to the Rexer Analytics Annual Data Miner Survey (Rexer 2010) all of them belong to the preferred tools for Data Mining and Predictive Analytics. Various different analysis and comparisons on all mentioned software packages are done (Rexer 2010; KDNuggets 2011; MaCcallum 1983; Acock 2005; Wahbeh et al. 2011). Table 1 gives a short summary of image processing capabilities for three common statistical software packages.

SPSS did neither provide methods or packages for multimedia processing nor algorithms for image feature extraction and therefore it seems to be inapplicable for the usage of multimedia data. In contrast RapidMiner's and R's image feature extraction algorithms are provided by additional packages or extensions. Such packages like *biOps* and *EBImage* provide various algorithms to extract common color histograms, color moments, or image features including textures or edges.

Analyzing the two other software packages, Matlab and SAS, we get similar results. Like SPSS SAS did not provide any multimedia processing or image feature

Table 1 Excerpt of a comparison of SPSS statistics, RapidMiner and R according to multimedia processing and image feature extraction capabilities (Naundorf et al. 2012)

	SPSS	RapidMiner	R
Packages/ extensions	Three different proprietary editions containing up to 15 modules, e.g., advanced statistics, bootstrapping, conjoint	Ten official extensions plus a few additional extensions from the community, e.g., R-, Weka-, reporting- or text-extension	> 4, 300 contributed packages ($\approx$ 3, 200 at CRAN), e.g., boot, class, cluster, fields, multicore, stats, survey, survival, XML
Multimedia processing capabilities	None	Grayscale and color image, e.g., via image processing extension (Burget et al. 2010), image mining extension	Image data: e.g., adimpro, biOps, biOpsGUI, dcemriS4, EBImage, ImageMetrics, RImageJ, ReadImages
Image feature extraction	None	Global features (e.g., color moments), segment features (e.g., dominant color) and local features	via biOps package: several edge features and via EBImage package: hull features, edge features, image moments

extraction algorithms. Matlab's capabilities are comparable with these of R. Matlab comes up with an image processing toolbox (Mathworks Inc. 2013), which provides various image feature extraction algorithms, e.g., color histograms, edge detection, or image segmentation algorithms.

As a conclusion all presented software packages are not easy to use. Provided methods need special knowledge about the algorithms behind, if they are available at all. Furthermore, the data management makes it hard for the user to analyze lots of data in several ways and multiple times, because there is no way to store extracted image features in any software package. Even if these restrictions can be overcome by the user, several other barriers exist, e.g., the distance calculation between two images or the combination of multiple (visual) image features. Finally, an easy usable end-user interface is missing.

To overcome all these problems and barriers, IMADAC was developed as an extension of (Naundorf et al. 2012; Zellhöfer et al. 2012). The goal is to provide persistent feature extraction for large data sets combined with various image processing capabilities embedded in an easy to use software application.

3   A New Package: IMADAC

Based on the limits of the mentioned software packages, we tried to create a new software application which was inspired by SPSS user interaction design, but extended with several methods and algorithms for multimedia and image processing capabilities in marketing. IMADAC allows analyzing image databases statistically and includes packages for image feature extraction as well as clustering, classification, or numerous other statistical evaluations.

In this section, we give an overview of the so far developed application and demonstrate the usage of feature extraction and handling, image analyzing (e.g., clustering) and finally the handling of traditional survey data is described.

3.1   Overview

IMADAC was implemented in a modular design to provide maximum flexibility for developments of new packages. The visualization part is as much as possible separated from the underlying architecture. This allows to develop extensions in the back-end system (e.g., new feature extraction algorithms) without knowing details of visualization and to develop visualization parts without explicitly knowing the background structure of the System.

Basically IMADAC consists of four main components. The input and data management component provides intuitive methods for handling image data. For image include all typical formats (JPEG, PNG, etc.) are supported. To define your working set of images the user can select single images or complete directories. All

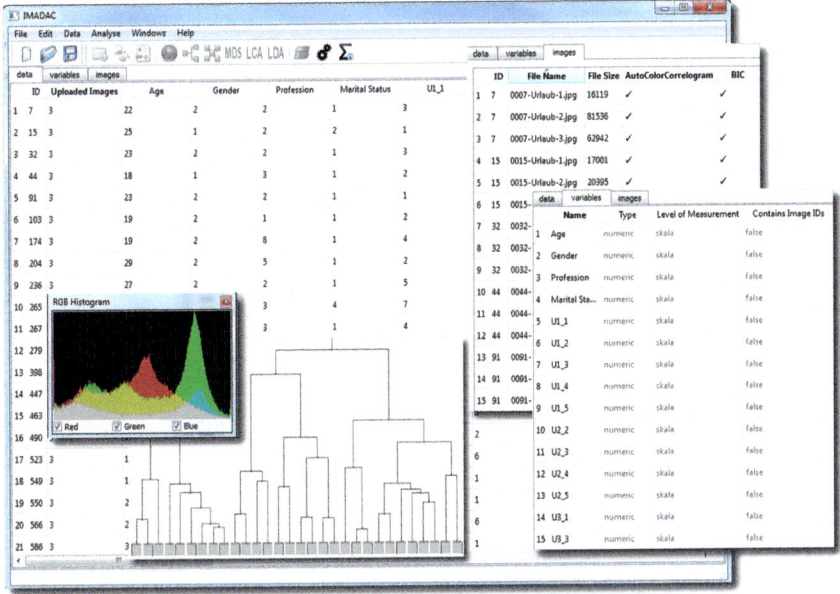

Fig. 1 Screenshot of IMADAC showing different working sheets and a dendrogram result view

imported images are then listed in a spreadsheet view (see top right subwindow on Fig. 1). The second main component provides various image processing methods, including a feature extraction and algorithms like clustering, multi-dimensional scaling, or latent class analysis (see Sects. 3.2 and 3.3). The third component allows the user to work on common survey data. Therefore IMADAC provides an SPSS like spreadsheet view (see Sect. 3.4). Finally, the visualization component provides several methods to display analyzing results, e.g., to visualize clustering results IMADAC provides the common used dendrogram view or an easy to read table view.

3.2   Image Feature Extraction and Handling

In the field of content based image processing, computational complexity of feature extraction and distance calculation is a big problem, especially on a large set of several hundreds or thousands of images. Executing numerous researching, which demands a repeatedly running of analyzing steps with different combinations of image features and distance measurements, makes it necessary to reduce computational time. To manage this problem IMADAC will eliminate the need for rerunning feature extraction. Further, calculated distances can be stored for a later reuse. A typical workflow for an image clustering analysis is shown in Fig. 2.

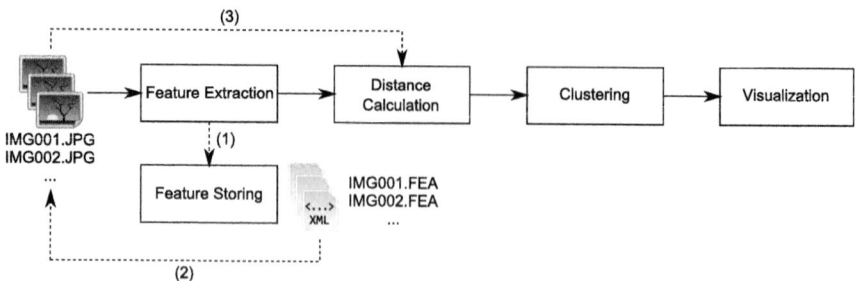

Fig. 2 Storing already extracted image features and referring to them in subsequent clustering tasks reduces computational effort considerably (Naundorf et al. 2012)

Table 2 Excerpt of available features, type and origin (Böttcher et al. 2013)

Name	Type	Origin
Auto color correlogram (ACC)	Color-related, global	Huang et al. (1997)
BIC	Color-related, global	Stehling et al. (2002)
CEDD	Texture/color-related, global	Chatzichristofis et al. (2008a)
Color histogram (region based)	Color-related, pseudo-local	Balko and Schmitt (2012)
Color histogram	Global	512 bin RGB histogram (own impl.)
Color layout[a]	Color-related, global	Cieplinski et al. (2001)
Color structure[a]	Color-related, global	Cieplinski et al. (2001)
Dominant color[a]	Color-related, global	Cieplinski et al. (2001)
Edge histogram[a]	Edge-related, global	Cieplinski et al. (2001)
FCTH	Texture/color-related, global	Chatzichristofis and Boutalis (2008b)
Scalable color[a]	Color-related, global	Cieplinski et al. (2001)
Tamura	Texture-related, global	Tamura et al. (1978)
Region-based shape[a]	Global	Cieplinski et al. (2001)

[a]Features in the scope of MPEG-7 (Manjunath et al. (2002), Ohm et al. (2002))

The general image clustering workflow, consisting of feature extraction, distance calculation clustering, and visualization of the results, is depicted along solid arrows. The auxiliary steps are displayed along dashed arrows. Whenever the software extracts a feature from an image, it is saved to a separate file in the file system next to the image file (see (1) in Fig. 2). Thus after finishing the feature extraction step, every image file has its corresponding feature file (2). When later running another image analyzing task, the software can use already existing feature files. If all required features are already contained in the image's feature file, the extraction can be skipped (3) (Naundorf et al. 2012).

For image processing IMADAC provides 37 different image features (17 global and 20 local). A small excerpt of existing global features is shown in Table 2.

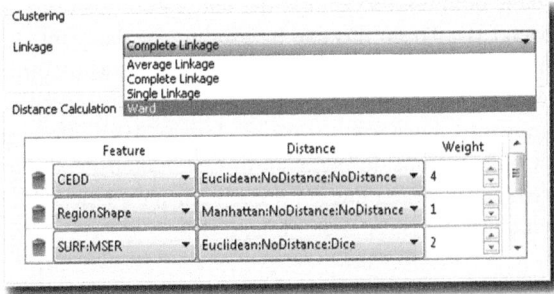

Fig. 3 Hierarchical clustering with a combination of multiple features and weighting

3.3   Calculation of Distances Between Sets of Images

As already mentioned IMADAC provides several algorithms to analyze sets of images. While analyzing images, one of the most important steps is to calculate distances between images. IMADAC provide numerous distance measures, starting with simple L_p-norms and going up to complex feature specific distance measures.

For statistical analysis IMADAC provides algorithms for clustering (hierarchical and partitioning clustering), multi-dimensional scaling, latent class analysis, or linear discriminant analysis. For all of these methods any combination of visual image features can be used. Furthermore, each feature can get a special weight to define features as more or less important. Additionally for each feature a set of usable distance measurements is choosable. Figure 3 demonstrates the feature selection and weighting for an hierarchical clustering approach. Here three features are selected with different weights, which defines CEDD (Chatzichristofis et al. 2008a) as the most important feature. In this hierarchical clustering approach a distance matrix (based on the default selected measurements) for each feature of the given set of images is calculated. Finally, an aggregation function, which uses the defined weights, merges all matrices. The final distance matrix is then used to calculate the cluster by the Ward approach (Ward 1963). For approaches which do not need distance measurements IMADAC can also provide the raw feature data.

3.4   Data Analysis

The analysis of sets of images reveals a great opportunity to generate additional information. But typically images are not the complete set of data the user will rely on. Normally at least demographic data are also recorded. To handle these common survey data, IMDAC provides SPSS like spreadsheet views. These tables can be easily used to handle your survey data and to define the attribute types (such as numeric, float, etc.) and level of measurement (nominal, ordinal or scale).

To include image data references into the data table, IMADAC references each image with a special ID. These IDs can be used in the data table to define a set of images belonging to one respondent. In the variables table a column can be set with a flag that indicates that these columns contain image IDs. Furthermore, hierarchical-based, partitioning-based clustering or latent class analysis allows an export of the results to the data table. Considering a set of respondents, each uploaded a set of images. When clustering these images for each cluster a new column is inserted in the data table, which contains the users' image IDs.

4 Sample Application

So far, we discussed the usage of image data for statistical analysis. As mentioned in the last section images regarding a set of users are of great interest for marketing purposes. Therefore IMADAC provides special algorithms to calculate the similarity (or the distance) between users by using their provided (or uploaded) images. It plays no role if the user provides only one or a multiple images. To demonstrate an excerpt for an possible use case we consider an online survey on holiday habits. The respondents provide demographic data as well as answers about their holiday interests. Furthermore they upload images of their favorite holidays.

In a first analysis we only use the uploaded images of the respondents to create different user groups. The distance between two users is calculated in two steps. First, distances of the provided images are calculated. This can be done as described in Sect. 3. So any combination of features connected with weightings can be used. The second step needs an aggregation method to calculate the overall distance between two users. IMADAC supports various approaches to solve this challenge. Finally, a distance matrix is generated, which contains distances between all users. This allows further analysis like clustering on the users. A possible result can be that some users prefer holidays in the mountains, others uploaded images are very heterogeneous and therefore a great range of holiday destinations are qualified.

In a second use case the fusion of common survey data with image features is presented. Our software package allows any combination of all attributes, whether their type is nominal, ordinal, scale or image data. For each type of attributes the user can choose a preferred distance measurement. For image data additionally the features, weights and the distance measure can be chosen. IMADAC allows different feature combination for each image data set (e.g., a different combination for holiday images and for lifestyle images). The result of the whole calculation is a relationship between all users regarding all chosen attributes. Figure 4 demonstrate the usage for an extended hierarchical clustering based on a fusion of traditional survey data and image features. Here we search for a correlation of users profession and their uploaded holiday images.

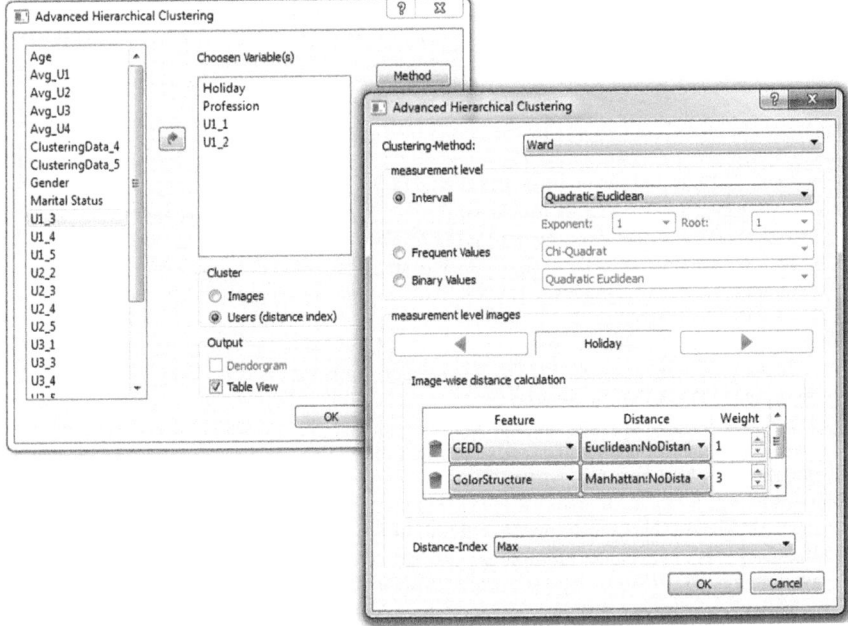

Fig. 4 Extended hierarchical clustering using a combination of survey data and image features

5 Conclusion and Future Work

In this paper we showed that images are of great interest for marketing purposes but no actual available statistical software package can fulfill the requirements of professional image processing. To close these gap IMADAC was developed. Our prototype comes up with various packages to extend traditional marketing researches. Analysis on images itself or in a combination of survey data with image features. The implemented GUI was designed for experts as well as users without image processing background.

Although we discussed several use cases for marketing, they were just a few out of many possible applications. In further research we plan an evaluation of real-world data in cooperation with marketing researchers, i.e., to extract customer segments using image features. To satisfy special demands (e.g., recognize company logos in images) additionally improvements and extension on the existing software will be done.

References

Acock, A. (2005). SAS, Stata, SPSS: A comparison. *Journal of Marriage and Family, 67*(4), 1093–1095.

Baier, D., & Daniel, I. (2010). Image clustering for marketing purposes. In *Proceedings of the 34th Annual Conference of the Gesellschaft für Klassifikation* (pp. 487–494), Karlsruhe. Heidelberg/Berlin: Springer. ISBN 978-3-642-24465-0.

Baier, D., Daniel, I., Frost, S., & Naundorf, R. (2012). Image data analysis and classification in marketing. *Advances in Data Analysis and Classification, 6*(4), 253–276.

Balko, S., & Schmitt, I. (2012). Signature indexing and self-refinement in metric spaces. In *Computer science reports* (Vol. 6). Cottbus: Institute of Computer Science, Brandenburg University of Technology.

Böttcher, T., Zellhöfer, D., & Schmitt, I. (2013). BTU DBIS' personal photo retrieval runs at ImageCLEF 2013. In *CLEF 2013 Labs and Workshop, Notebook Papers* (pp. 23–26), September 2013, Valencia, Spain.

Burget, R., Karásek, J., Smékal, Z., Uher, V., & Dostál, O. (2010). Rapidminer image processing extension: A platform for collaborative research. In *International Conference on Communications and Signal Processing*, Baden, Vienna.

Chatzichristofis, S. A., & Boutalis, Y. S. (2008b). FCTH: Fuzzy color and texture histogram - A low level feature for accurate image retrieval. In *Proceedings of the 2008 Ninth International Workshop on Image Analysis for Multimedia Interactive Services* (pp. 191–196), WIAMIS '08. Washington, DC: IEEE Computer Society. ISBN 978-0-7695-3130-4.

Chatzichristofis, S. A., Boutalis, Y. S., Gasteratos, A., Vincze, M., & Tsotsos, J. K. (2008a). CEDD: Color and edge directivity descriptor - A compact descriptor for image indexing and retrieval. In *Proceedings of the 6th International Conference on Computer Vision Systems* (pp. 312–322). ISBN 978-3-540-79546-9.

Cieplinski, L., Jeannin, S., Ohm, J.R., Kim, M., Pickering, M., Yamada, A., (2001). MPEG-7 Visual XM version 8.1. Pisa, Italy

Daniel, I., & Baier, D. (2013). Lifestyle segmentation based on contents of preferred images versus ratings of items. In *Studies in Classification, Data Analysis, and Knowledge Organization, 46*, (vol. 4, pp. 439–448).

Datta, R., Joshi, D., Li, J., & Wang, J. Z. (2008). Image retrieval: Ideas, influences and trends of the new age. *ACM Computational Survey, 40*(2). Article 5, 1–60.

Huang, J., Kumar, R. S., Mitra, M., Zhu, W. J., & Zabih, R. (1997). Image indexing using color correlograms. In *Proceedings of the 1997 Conference on Computer Vision and Pattern Recognition (CVPR '97)* (pp. 762–768). Washington, DC: IEEE Computer Society. ISBN 0-8186-7822-4.

IBM Corporation. (2013). *IBM SPSS statistics 20 core system user's guide*. http://www.library.uvm.edu/services/statistics/SPSS20Manuals/IBM_SPSS_Statistics_Core_System_User_Guide.pdf. Accessed 24 Sept 2013.

KDNuggets. (2011). *KDNuggets polls*. http://www.kdnuggets.com/polls/index.html. Accessed 24 Sept 2013.

Lew, M. S., Sebe, N., Djeraba, C. & Jain, R. (2006). Content-based multimedia information retrieval: State of the art and challenges. *ACM Transactions on Multimedia Computing, Communications, and Applications, 2*(1), 1–19.

Maccallum, R. (1983). A comparison of factor analysis programs in SPSS, BMDP, and SAS. *Psychometrika, 48*(2), 223–231.

Manjunath, B., Salembier, P., & Sikora, T. (2002). *Introduction to MPEG-7: Multimedia content description interface*. New York, NY, USA: Wiley. ISBN 0-4714-8678-7.

MathWorks Inc. (2013). *Image processing toolbox - User's guide 2013b*. http://www.mathworks.com/help/pdf_doc/images/images_tb.pdf. Accessed 24 Sept 2013.

Naundorf, R., Baier, D., & Schmitt, I. (2012). Statistical software for clustering images. In *Computer science reports*. Cottbus: Institute of Computer Science, Brandenburg University of Technology. ISSN:1473–7969.

Ohm, J. R., Cieplinski, L., Kim, H. J., Krishnamachari, S., Manjunath, B. S., Messing, D. S., et al. (2002). *Color descriptors: Introduction to MPEG-7 – Multimedia content description interface* (pp. 187–212). New York: Wiley.

Rapid-I. (2010). *RapidMiner 5.0 user manual*. http://www.sourceforge.net/projects/rapidminer/files/1.%20RapidMiner/5.0/rapidminer-5.0-manual-english_v1.0.pdf/download. Accessed 24 Sept 2013.

Rexer, K. (2010). Fourth annual rexer analytics data miner survey. In *Predictive Analytics World*, Washington, DC, Oct 2010.

SAS Institute Inc. (2013). *SAS 9.3 language reference*. http://www.support.sas.com/documentation/cdl/en/lrcon/65287/PDF/default/lrcon.pdf. Accessed 24 Sept 2013.

Stehling, O. R., Nascimento, A. M., & Falcao, X. A. (2002). A compact and effcient image retrieval approach based on border/interior pixel classification. In *Proceedings of the 11th International Conference on Information and Knowledge Management* (pp. 102–109), CIKM '02.

Tamura, H., Mori, S., & Yamawaki, T. (1978). Texture features corresponding to visual perception. *IEEE Transactions on System, Man and Cybernatic, 8*(6), 460–472.

Venables, W. N., Smith, D. M., & R Development Core Team. (2011). *An introduction to R: Notes on R, A programming environment for data analysis and graphics*. Vienna: Department of Statistics and Mathematics, Vienna University of Economics and Business.

Wahbeh, A. H., Al-Radaideh, Q. A., Al-Kabi, M. N., & Al-Shawakfa, E. M. (2011). A comparison study between data mining tools over some classification methods. *International Journal of Advanced Computer Science and Applications - IJACSA, 2*(8), 18–26.

Ward, J. H. (1963). Hierarchical grouping to optimize an objective function. *Journal of the American Statistical Association, 58*, 236–244.

Zellhöfer, D., Bertram, M., Böttcher, T., Schmidt, C., Tillmann, C., & Schmitt, I. (2012). PythiaSearch – A multiple search strategy-supportive multimedia retrieval system. In *Proceedings of the 2nd ACM International Conference on Multimedia Retrieval, ICMR '12* (pp. 59–60). ISBN 978-1-4503-1329-2.

The Bass Model as Integrative Diffusion Model: A Comparison of Parameter Influences

Michael Brusch, Sebastian Fischer, and Stephan Szuppa

Abstract New technologies are permanently developed and introduced into markets. Although their adoption process is extremely volatile and varies from case to case, it is of extreme interest to companies to somehow plan and especially to estimate the development. For these estimations so-called diffusion models are utilized. A well-known and often used one is the Bass model, which incorporates different parameters and their specific influences.

Our paper analyzes what kind of parameters (e.g., coefficient of innovation, underlying distribution) have what kind of influence (e.g., number of adoptions, standard deviation from adoption time) on the diffusion estimations. For the analysis the market of electric vehicles with its politically motivated objectives and current sales quantities serves as an application example. For the analysis itself, a factorial design with synthetically generated and disturbed data is applied.

1 Introduction

The development of new technologies and their subsequent introduction into markets is a continuous process. The corresponding adoption process is neither permanent nor stable. It is rather an extremely volatile process that varies from technology to technology. However, it is of extreme interest to companies to somehow plan and especially to estimate this adoption process. To enable such estimations so-called diffusion models have been developed. A well-known and quite powerful one is the Bass model (Bass 1969; Peres et al. 2010), which incorporates different parameters and their specific influences.

M. Brusch (✉)
Chair of Business Administration, Marketing and Corporate Planning, Anhalt University
of Applied Sciences, Bernburger Str. 57, 06366 Köthen, Germany
e-mail: m.brusch@emw.hs-anhalt.de

S. Fischer • S. Szuppa
Siemens Professional Education, Siemens Technik Akademie Berlin, Siemens AG,
Nonnendammallee 104, 13629 Berlin, Germany
e-mail: sebastian.fischer@education-siemens.com; stephan.szuppa@siemens.com

© Springer-Verlag Berlin Heidelberg 2015
B. Lausen et al. (eds.), *Data Science, Learning by Latent Structures,
and Knowledge Discovery*, Studies in Classification, Data Analysis,
and Knowledge Organization, DOI 10.1007/978-3-662-44983-7_20

Integrated in this framework, our paper analyzes what kind of parameters (e.g., coefficient of innovation, underlying distribution) have what kind of influence (e.g., number of adoptions in a specific period, standard deviation from adoption time) on the diffusion estimations. The variation of the analyzed parameters describes the need of necessary market research and/or diffusion experience. Representing a new technology, for the final analysis the market of electric vehicles (e.g., Hebes et al. 2011) with its politically motivated objectives and current sales quantities serves as an application example. For the analysis itself, a factorial design with synthetically generated and disturbed data is applied.

Therefore, this paper is structured as follows. In Sect. 2 an overview about diffusion models in marketing including the favored Bass model will be given. In a first analysis in Sect. 3 the model parameter will be estimated within a Monte Carlo comparison with a factorial design using synthetic data. For the second analysis in Sect. 4 data from the German electric vehicles market is used. The discussion in Sect. 5 closes this contribution.

2 Diffusion Models in Marketing

For a successful market launch of new products adoption and diffusion processes are relevant. The adoption, i.e. the decision of an individual to take on an innovation, with its related process can be divided into different phases: starting with (1) awareness, over (2) interest, (3) evaluation and (4) trial to (5) adoption. The diffusion, i.e. the distribution of an innovation from its sources to its final users, with its related process describes the acceptance of a new product by innovative buyers and its penetration of the overall market (see, e.g., Homburg et al. 2013; Rogers 2003).

Consequently, diffusion models summarize individual purchases and can be an alternative for predicting the diffusion rate of a new product. Here, three basic types of models can be differentiated (see, e.g., Mahajan et al. 1990) and are currently well known (see, e.g., Homburg et al. 2013):

- models for innovative purchase behavior (e.g., Fourt and Woodlock 1960),
- models for imitative purchase behavior (e.g., Mansfield 1961), and
- models for innovative and imitative purchase behavior (e.g., Bass 1969).

The model of Fourt and Woodlock (1960) considers only purchases of innovators. With α as the penetration rate (or innovation coefficient), $\bar{X}$ as the number of potential buyers (or market potential), t as an index for the time period, and X_{t-1} as the number of buyers who adopted the innovation till $t-1$ the number of first-time buyers x in period t can be calculated as follows:

$$x_t = \alpha(\bar{X} - X_{t-1}).$$
(1)

In contrast, the model of Mansfield (1961) considers only purchases of imitators (or followers, i.e. of customers who buy the new product if others have already purchased this product). With $\bar{X}$, t, X_{t-1} and x_t being similar to the model of Fourt and Woodlock (1960) and β as intensity of imitative behavior (or imitation coefficient) the number of first-time buyers results from

$$x_t = \beta \frac{X_{t-1}}{\bar{X}} (\bar{X} - X_{t-1}). \tag{2}$$

The model of Bass (1969) considers both innovative and imitative purchase behavior. This very prominent alternative of an integrative diffusion models is an additive combination of the two previously described models. With all variables in accordance with these other models the number of first-time buyers x in period t can be calculated as follows:

$$x_t = \alpha(\bar{X} - X_{t-1}) + \beta \frac{X_{t-1}}{\bar{X}} (\bar{X} - X_{t-1}). \tag{3}$$

For estimation reasons formula (3) can be rewritten as

$$x_t = b_0 + b_1 X_{t-1} + b_2 X_{t-1}^2 \quad \text{with} \quad b_0 = \alpha \bar{X}, \ b_1 = \beta - \alpha, \ b_2 = \frac{-\beta}{\bar{X}}. \tag{4}$$

This quadratic function of x_t can be applied—if a sufficiently long time series of sales in the past is given—for the estimation of the unknown parameters b_0, b_1, b_2 using regression analysis and—in the following—for an approximation of the parameters α and β.

The classical Bass model was subject of some discussions and further developments for specific types of products or situations, e.g., for continuous repeat purchasing (Norton and Bass 1987) or for products with multiple technological generations (Chanda and Bardhan 2008). However, the Bass model in general could show a very good fit to many diffusion patterns (Lilien et al. 2000). In combination with the simple utilization of the Bass model (only a limited amount of empirical data of the past is necessary) this method seems to be very interesting for diffusion estimation purposes. Other known methods require, for example, the knowledge about the influence (i.e., direction and weight) of other (also endogenous) variables (as, e.g., System Dynamics) or need additional (but in a certain way biased, i.e. subjective) data (as, e.g., integration of expert teams).

However, the main problem is to set the model parameters to a specific value to be able to predict the diffusion of innovations. Therefore, the parameter values of a product that is as similar as possible can be taken, e.g., using (see, for a broader overview, Albers 2004):

- published parameter values of specific innovations (for an overview see, e.g., Mahajan et al. 2000).
- existing meta-analyses, e.g., with the main results: $\alpha = 0.03$, $\beta = 0.38$ (Sultan et al. 1990).

- derived parameter values from more than one time series of diffusion data as well as from a larger cross-section of related products (e.g., Gatignon et al. 1989).
- pooled data across several entities to explain differences in the parameter values caused by the presence of specific product attributes (e.g., Bähr-Sepplfricke 1999).
- combined forecast data of several analogous products (due to the fact that a combination of forecasts exceed single method based forecasts, see Meade and Islam 1998).

In the following the influences of different parameters of the Bass model will be analyzed within a Monte Carlo comparison. On the one hand, these are the model parameters α and β (representing the general shape of the curve). On the other hand, the existence of different kinds of data will be analyzed through varying the existing time periods and the quality of the—affected by errors—data collection (representing the need for time of experience and/or effort of market research). This is done in a first step using synthetic (and disturbed) data (Sect. 3) and in a second step using data from the German market of electric vehicles (Sect. 4).

3 Estimations Using Synthetic Data

In a first analysis synthetically generated data will be used. This enables the analysis of the (statistical) relevance of different factors and factor levels independent from specific research objects (i.e., specific innovations).

It will be assumed that the sales data of first-time buys of eight periods exist $(x_1 = 93,100, x_2 = 122,795, x_3 = 159,272, x_4 = 202,029, x_5 = 248,860, x_6 = 295,165, x_7 = 333,799, x_8 = 356,196)$. Based on this data and using regression analysis as in formula (4) the parameters of the Bass model can be estimated as follows: $\alpha = 0.0266, \beta = 0.3549998, \bar{X} \approx 3,500,000$ (with $x_t \approx 0$ in $t = 40$). Then, the resulting curves for the number of first-time buys x_t and the total number of sales in $t-1$ X_{t-1} can be approximately determined. The analysis of the found regression function shows a very good model fit measured as coefficient of determination R^2 of 0.9999974 and a very good predictability measured as total sum of distances (TSD) with a quantity of 13.82.

For the following Monte Carlo comparison (with the results in Tables 1 and 2) a factorial design with seven factors—each with three levels—was used. The first two factors consider the existence and the (because of the synthetic generation necessary) disturbance of the "true" data and vary the

- existing knowledge (with very few, few or moderate percentages of collected data periods of the expected diffusion time)—factor 1,
- measurement error with respect to (w.r.t.) the sales x_t (with small, medium, or large standard deviations for generating additive error)—factor 2.

Table 1 Monte Carlo comparison concerning the impact of the influence factors using synthetic data w.r.t. correlations ($n = 21, 870$ datasets)

| Factor | Level | Mean correlation of data using observed parameters and | | |
		Estimated parameters	Suggested parameters	Overall
Existing knowledge	10 %	0.5855	0.8420***	0.7138
	15 %	0.5893	0.8398***	0.7145^{ns}
	20 %	0.5878	0.8399***	0.7138
Measurement error	$\sigma = 0.01$	0.6460	0.8408***	0.7434***
w.r.t. x_t	$\sigma = 0.05$	0.5880	0.8413***	0.7147
	$\sigma = 0.10$	0.5286	0.8396***	0.6841
Measurement error	$\sigma = 0.1$	0.5733	0.8413***	0.7073
w.r.t. α	$\sigma = 0.2$	0.5869	0.8414***	0.7141
	$\sigma = 0.3$	0.6024	0.8390***	0.7207***
Measurement error	$\sigma = 0.1$	0.5966	0.8389***	0.7177^{ns}
w.r.t. β	$\sigma = 0.2$	0.5829	0.8421***	0.7125
	$\sigma = 0.3$	0.5831	0.8407***	0.7119
Measurement error	$\sigma = 0.1$	0.7131	0.8405***	0.7768***
w.r.t. the market	$\sigma = 0.2$	0.5761	0.8432***	0.7097
potential	$\sigma = 0.3$	0.4734	0.8380***	0.6557
Systematical error	0 %	0.6822	0.8281***	0.7551***
w.r.t. the market	5 %	0.5910	0.8409***	0.7160
potential	10 %	0.4894	0.8526***	0.6710
Underlying	Normal	0.7621	0.8311***	0.7966***
distribution	Uniform	0.6458	0.8538***	0.7498
	Lognormal	0.3547	0.8367***	0.5957
Overall		0.5875	0.8406***	

***Significant differences within rows (t-test) and columns (F-test) at the $p < 0.001$ level; ns not significant

The factors 3–6 are related to the disturbance of the "estimated" data and vary the

- measurement error w.r.t. the innovation coefficient α (with small, medium, or large standard deviations for generating additive error)—factor 3,
- measurement error w.r.t. the imitation coefficient β (with small, medium, or large standard deviations for generating additive error)—factor 4,
- measurement error w.r.t. the market potential $\bar{X}$ (with small, medium, or large standard deviations for generating additive error)—factor 5,
- systematical error w.r.t. the market potential $\bar{X}$ (with none, small, and moderate displace of market potential estimations)—factor 6.

Both types of data (i.e., "true" and "estimated" data) were disturbed while varying the underlying distribution (with normal, uniform, or lognormal distribution)—factor 7.

Table 2 Monte Carlo comparison concerning the impact of the influence factors using synthetic data w.r.t. TSD ($n = 21,870$ datasets)

Factor	Level	TSD of data using observed parameters and		
		Estimated parameters	Suggested parameters	Overall
Existing knowledge	10%	9,998,476***	22,329,553	16,164,015
	15%	10,094,024***	22,458,814	16,276,419
	20%	9,953,911***	22,161,241	16,057,576[ns]
Measurement error	$\sigma = 0.01$	9,400,298***	22,248,296	15,824,297**
w.r.t. x_t	$\sigma = 0.05$	10,020,898***	22,534,344	16,277,621
	$\sigma = 0.10$	10,625,216***	22,166,968	16,396,092
Measurement error	$\sigma = 0.1$	9,886,820***	22,187,543	16,037,182[ns]
w.r.t. α	$\sigma = 0.2$	10,069,120***	22,290,158	16,179,639
	$\sigma = 0.3$	10,090,472***	22,471,907	16,281,189
Measurement error	$\sigma = 0.1$	9,215,425***	21,380,980	15,298,203***
w.r.t. β	$\sigma = 0.2$	9,885,503***	22,010,590	15,948,046
	$\sigma = 0.3$	10,945,483***	23,558,038	17,251,761
Measurement error	$\sigma = 0.1$	7,009,489***	23,322,233	15,165,861***
w.r.t. the market	$\sigma = 0.2$	9,839,110***	21,600,235	15,719,672
potential	$\sigma = 0.3$	13,197,813***	22,027,140	17,612,476
Systematical error	0%	9,082,254***	25,322,579	17,202,416
w.r.t. the market	5%	10,063,579***	22,378,792	16,221,186
potential	10%	10,900,579***	19,248,237	15,074,408***
Underlying	Normal	7,756,872***	27,710,198	17,733,535
distribution	Uniform	6,725,560***	20,476,006	13,600,783***
	Lognormal	15,563,979***	18,763,403	17,163,691
Overall		10,015,471***	22,316,536	

***Significant differences within rows (t-test) and columns (F-test) at the $p < 0.001$ level; *at the $p < 0.1$ level; ns not significant

During the Monte Carlo comparison different transformation and data analysis steps were carried out in SAS/IML:

- Step 1: Forming of the independent data matrices with (a) all data without restrictions, and (b) data which is varied w.r.t. factors 1, 2, and 7.
- Step 2: Estimating the parameter of the regression model using (a) all data and (b) varied data.
- Step 3: Calculating the parameter of the Bass model α, β, and $\bar{X}$ using (a) observed, (b) predicted, (c) suggested values; as parameters have been used: for (a) the estimated values of step 2(a), for (b) the estimated values of step 2(b), for (c) suggested values with $\alpha = 0.03$, $\beta = 0.38$, $\bar{X} = 2,100,000$ (with $\bar{X} = 3,500,000 \times 60\% =$ number of all similar products expected to be sold $\times$ expected possible market share, as suggested by the meta-analyses, see, e.g., Sultan et al. 1990; Albers 2004 and Sect. 2).
- Step 4: Disturbing the found parameter w.r.t. to factors 3–7.

- Step 5: Calculating x_t using (a), (b), and (c) for the whole diffusion time.
- Step 6: Comparing (a) with (b), and (a) with (c) w.r.t. thecorrelation r and the TSD.

The results of this comparisons with tenfold replication are shown in Tables 1 and 2. It can be seen that the usage of "suggested" parameters outperforms the usage of estimated parameters w.r.t. correlation (always significant). However, a look at the difference in predictability measured as TSD in absolute numbers shows the ability to predict the correct sales is much worse in case of suggested parameters. Overall it can be identified that the existing knowledge (factor 1) and the measurement error w.r.t. α are not significant. These general findings, based on synthetic data, should be investigated more in detail for a real-world problem using a real dataset.

4 Estimations Using German EV Market Data

The electric vehicle (EV) is still in the initial stages of its development, both from a technology and a market perspective. The modelling of possible adoption scenarios and the analysis of underlying parameter influences can be of great value to the various stakeholders of the automotive market. Accordingly, the German government identified the EV as a potential accelerator for the integration of renewable and smart energy in the country. This potential is formulated in the ambitious plan to increase the number of EVs in Germany to 1 million by 2020 and to 6 million by 2030 (Federal Government 2009). The EV is the next step in automotive development. Its innovative potential is based on the increasing electrification of the vehicle's driving technology.

In this second analysis the data of this innovative product will be used. Up to now sales data of first-time buys of EV of eight periods since 2005 exist (47, 19, 8, 36, 162, 541, 2,154, 2,956; see Federal Motor Transport Authority 2013). To improve the possibility to estimate the parameters of the Bass model the data of a product that is as similar as possible can be considered (see Sect. 2).

Although no data of a real similar product is available for the EV, the hybrid electric vehicle (HEV) is the intermediate vehicle form in the development from the conventional combustion engine to the fully electrified engine and therefore is an analogous technology to the EV. Following this, the corresponding sales data of the years since 2005 are considered (3,589, 5,278, 7,581, 6,464, 8,374, 10,661, 12,622, 21,438; see Federal Motor Transport Authority 2013).

For the following Monte Carlo comparison a factorial design with five factors—each with three levels—was used (the first two factors of Sect. 3 "existing knowledge" and "measurement error w.r.t. the sales x_t" will not be applied). Only the factors related to the disturbance of the "estimated" data will be used and varied w.r.t. the

- measurement error w.r.t. the innovation coefficient α (with small, medium, or large standard deviations for generating additive error)—factor 1,

- measurement error w.r.t. the imitation coefficient β (with small, medium, or large standard deviations for generating additive error)—factor 2,
- measurement error w.r.t. the market potential $\bar{X}$ (with small, medium, or large standard deviations for generating additive error)—factor 3,
- systematical error w.r.t. the market potential $\bar{X}$ (with none, small, and moderate displace of market potential estimations)—factor 4,
- underlying distribution (with normal, uniform, or lognormal distribution)—factor 5.

During the Monte Carlo comparison different transformation and data analysis steps were carried out again in SAS/IML:

- Step 1: Forming of the independent data matrices with data of (a) the sold EV, and (b) the sold HEV between 2005 and 2012.
- Step 2: Estimating the parameter of the regression model using all data of (a) and (b).
- Step 3: Calculating the parameter of the Bass model α, β, and $\bar{X}$ using (a) observed, (b) predicted, (c) suggested values; as parameters have been used: for (a) the estimated values of step 2(a), for (b) the estimated values of step 2(b), for (c) suggested values with $\alpha = 0.03$, $\beta = 0.38$, $\bar{X} = 26,054,406$ (with $\bar{X} = (43,431,124 - 7,114) \times 60\% = $ (number of all cars − number of electric vehicles in June 2013) × expected possible market share, as suggested by the meta-analyses, see, e.g., Sultan et al. 1990; Albers 2004 and Sect. 2).
- Step 4: Disturbing the found parameter w.r.t. to factors 1–5.
- Step 5: Calculating x_t using (a), (b), and (c) for the whole diffusion time.
- Step 6: Comparing (a) with (b), and (a) with (c) w.r.t. the correlation r and the TSD.

The following analyses of these comparisons w.r.t. the correlation values (next paragraph) and the TSD (Table 3) are based on a tenfold replication.

The analysis of the correlation values shows that they are on a high level, but they are in the average mainly the same with 0.8319 for the first correlation (i.e., between "observed" and "estimated") and with 0.8324 for the second correlation (i.e., between "observed" and "suggested"). In a similar manner, the overall minimum correlation ($r_{min} = 0.8295050$) and the overall maximum correlation ($r_{max} = 0.8325992$) show no relevant differences. The analysis of the effect of different factor levels shows that the measurement errors w.r.t. α and the market potential, and the systematical error w.r.t. the market potential are not significant, whereas the measurement error w.r.t. β is slightly significant. Only the underlying distribution has a strong influence.

In contrast to the correlation values, the values of TSD in Table 3 show important differences between the used data basis (using estimated or suggested parameters): The data basis using estimated parameters is much closer to the real (observed) data, i.e., the values of TSD are much lower. However, these values are also quite bad keeping in mind that the absolute sales numbers are approximately 3,000 cars a year up to now.

Table 3 Monte Carlo comparison concerning the impact of the influence factors using German EV market data w.r.t. TSD ($n = 2,430$ datasets)

Factor	Level	TSD of data using observed parameters and		
		Estimated parameters	Suggested parameters	Overall
Measurement error w.r.t. α	$\sigma = 0.1$	40,017***	8,096,355	4,068,186[ns]
	$\sigma = 0.15$	41,945***	8,436,397	4,239,171
	$\sigma = 0.2$	43,236***	8,661,693	4,352,464
Measurement error w.r.t. β	$\sigma = 0.1$	41,779***	8,406,720	4,224,250
	$\sigma = 0.2$	42,050***	8,454,675	4,248,363
	$\sigma = 0.3$	41,368***	8,333,050	4,187,209[ns]
Measurement error w.r.t. the market potential	$\sigma = 0.1$	37,880***	7,708,925	3,873,403***
	$\sigma = 0.2$	42,444***	8,526,632	4,284,538
	$\sigma = 0.3$	44,873***	8,958,888	4,501,881
Systematical error w.r.t. the market potential	0%	39,961***	8,079,583	4,059,772[ns]
	5%	41,324***	8,323,827	4,182,575
	10%	43,913***	8,791,035	4,417,474
Underlying distribution	Normal	32,293***	6,712,629	3,372,461***
	Uniform	37,825***	7,703,087	3,870,456
	Lognormal	55,080***	10,778,729	5,416,904
Overall		41,733***	8,398,148	

*** Significant differences within rows (t-test) and columns (F-test) at the $p < 0.001$ level; ** at the $p < 0.01$ level; * at the $p < 0.1$ level; ns not significant

Here, the overall best ($\text{TSD}_{\text{min}} = 6,703$) and the overall worst ($\text{TSD}_{\text{max}} = 101,205,704$) predictions show that in at least some cases it was possible to estimate sales which are close to the real sales numbers a year.

5 Conclusion and Outlook

In this contribution the influencing parameters of a diffusion model could be identified (especially for the Bass model) and verified (especially the underlying distribution). The resulting data quality, for using an analogous technology, could be tested with acceptably high correlation values in general, but dramatic errors, especially w.r.t. the market potential.

The implications for practical applications are that the Bass model is able to predict the general shape of the curve (because of the high correlation values)—but it is not able to estimate the exact sales quantities for real innovations (because of the large difference between expected and real sales, measured as TSD). Furthermore, when estimating the sales (i.e., diffusion) curve of an innovation the researcher should especially keep the structure of the data (i.e., the underlying distribution of sales) in mind. This is the most important factor (compared to, e.g., measurement or systematical errors of data collection).

In further research new empirical comparisons are necessary to identify the most relevant factor levels for testing the found results in practice. In addition, further Monte Carlo comparisons with consideration of more aspects could be helpful, e.g., with different number of diffusion periods and levels of newness.

References

Albers, S. (2004). Forecasting the diffusion on an innovation prior to launch. In S. Albers (Ed.), *Cross-functional innovation management. Perspectives from different disciplines* (pp. 243–258). Wiesbaden: Gabler.

Bähr-Sepplfricke, U. (1999). *Diffusion neuer Produkte: Der Einfluss von Produkteigenschaften.* Wiesbaden: DUV.

Bass, F. M. (1969). A new product growth for model consumer durables. *Management Science, 15*(5), 215–227.

Chanda, U., & Bardhan, A. K. (2008). Modelling innovation and imitation sales of products with multiple technological generations. *The Journal of High Technology Management Research, 18*(2), 173–190.

Federal Government. (2009). Nationaler Entwicklungsplan Elektromobilität der Bundesregierung, Germany. http://www.bmbf.de/pubRD/nationaler_entwicklungsplan_elektromobilitaet.pdf. Accessed 1 June 2013.

Federal Motor Transport Authority. (2013). Central Vehicle Register, Flensburg, Germany. http://www.kba.de/cln_031/nn_269000/DE/Statistik/Fahrzeuge/Bestand/Umwelt/b__umwelt__z__teil__1.html. Accessed 1 June 2013

Fourt, L. A., & Woodlock, J. W. (1960). Early prediction of market success for new grocery products. *Journal of Marketing, 25*(2), 31–38.

Gatignon, H., Eliashberg, J., & Robertson, T. S. (1989). Modeling multinational diffusion patterns: An efficient methodology. *Marketing Science, 8*, 231–247.

Hebes, P., Kihm, A., Mehlin, M., & Trommer, S. (2011). *Policy driven demand for sales of plug-in hybrid electric vehicles and battery-electric vehicles in Germany.* Berlin: German Aerospace Center, Institute of Transport Research.

Homburg, C., Kuester, S., & Krohmer, H. (2013). *Marketing management. A contemporary perspective.* London: McGraw-Hill.

Lilien, G. L., Rangaswamy, A., & Van den Bulte, C. (2000). Diffusion models: Managerial applications and software. In V. Mahajan, E. Muller, & Y. Wind (Eds.), *New product diffusion models* (pp. 295–311). Boston: Kluwer.

Mahajan, V., Muller, E., & Bass, F. M. (1990). New product diffusion models in marketing: A review and directions for research. *Journal of Marketing, 54*(1), 1–26.

Mahajan, V., Muller, E., & Wind, Y. (Eds.) (2000). *New-product diffusion models.* Boston: Kluwer.

Mansfield, E. (1961). Technical change and the rate of imitation. *Econometrica, 29*(4), 741–766.

Meade, N., & Islam, T. (1998). Technological forecasting – Model selection, model stability, and combining models. *Management Science, 44*(8), 1115–1130.

Norton, J. A., & Bass, F. M. (1987). A diffusion theory model of adoption and substitution for successive generations of high-technology products. *Management Science, 33*(9), 1068–1086.

Peres, R., Muller, E., & Mahajan, V. (2010). Innovation diffusion and new product growth models: A critical review and research directions. *International Journal of Research in Marketing, 27*, 91–106.

Rogers, E. M. (2003). *Diffusion of innovations.* New York: Free Press.

Sultan, F., Farley, J. U., & Lehmann, D. R. (1990). A meta-analysis of applications of diffusion models. *Journal of Marketing Research, 27*, 70–77.

Preference Measurement in Complex Product Development: A Comparison of Two-Staged SEM Approaches

Jörgen Eimecke and Daniel Baier

Abstract Since many years, preference measurement has been used to understand the importance that customers ascribe to alternative possible product attribute-levels. Available for this purpose are, e.g., compositional approaches based on the self-explicated-model (SEM) as well as decompositional ones based on conjoint analysis (CA). Typically, in SEM approaches, customers evaluate the importance of product attributes one by one whereas in decompositional approaches, they evaluate possible alternative products (attribute-level combinations) followed by a derivation of the importances. The SEM approaches seem to be superior when products are complex and the number of attributes is high. However, there are still improvement possibilities. In this paper two innovative two-staged SEM approaches are proposed and tested. The complex products under study are small remotely piloted aircraft systems (small RPAS) for German search and rescue (SAR) forces.

1 Introduction

One of the major challenges for a successful product development is the early design of a specification that meets customers' requirements (Sattler 2006). Here, a popular strategy is to collect preferential evaluations of alternatives by a sample of potential customers and to use these evaluations as a basis for specification design (Eckert and Schaaf 2009). Mainly, two groups of approaches exist: Decompositional (e.g., CA) and compositional ones (e.g., SEM). Eckert and Schaaf (2009) indicated in a survey that only seven of 40 discussed two-staged SEM approaches have been empirically tested up to the year 2009. This paper tries to close this gap by proposing and

J. Eimecke (✉) • D. Baier
Institute of Business Administration and Economics, Brandenburg University of Technology Cottbus, Cottbus, Germany
e-mail: joergen.eimecke@tu-cottbus.de; daniel.baier@tu-cottbus.de

© Springer-Verlag Berlin Heidelberg 2015   239
B. Lausen et al. (eds.), *Data Science, Learning by Latent Structures, and Knowledge Discovery*, Studies in Classification, Data Analysis, and Knowledge Organization, DOI 10.1007/978-3-662-44983-7_21

testing two additional two-staged SEM approaches. The predictive validity of these approaches is judged by using Spearman's rank correlation coefficient with bindings on an individual level (calculated per respondant and then averaged, in the following just "Spearman").

Section 2 introduces the used methods. An overview of SEM and the advantages over alternatives, the preference modelling, and an overview of the known SEM approaches and their advantages are given. The complex product under study and the resulting research questions are explained in Sect. 3. Section 4 discusses the realized surveys. The first survey, an expert survey, uses a two-staged SEM with constant sum scale on both stages and returns no satisfying predictive validity. The second survey (the "online survey") solved this problem by using a rank order at the first stage. Results are shown in Sect. 5. Conclusions and a outlook of further possibilities for improving the predictive validity are discussed in Sect. 6.

2 SEM for Preference Measurement

2.1 Overview and Advantages Over Alternatives

CA and SEM both show advantages with respect to validity: SEM with its direct evaluations of single attributes and levels is assumed to consist of easy to understand tasks for the respondents, CA with its evaluations of complete product alternatives is assumed to consist of difficult tasks for the respondents but is more close to real buying situations (Green et al. 1981). So, not surprisingly, there are a mixed results with respect to advantages of SEM over CA and vice versa (e.g., Green et al. 1993). Nevertheless, the SEM approaches are under-represented in theory and practice (Eckert and Schaaf 2009) in spite of its "surprising robustness" (Srinivasan and Park 1997).

This paper tries to fill this gap, especially since the product under study is a complex product with many describing attributes and levels for which SEM is often said to have advantages (Green and Srinivasan 1990). Here, CA is assumed to lead to an information overload for the respondents (Höpfl and Huber 1970) and to be inapplicable due to time and cost constraints (Srinivasan and Park 1997).

2.2 Preference Modelling

For SEM three stages could be used. Stage zero consists of the elimination of unacceptable attribute-levels (Srinivasan 1988) but some studies, e.g., by Dorsch and Teas (1992), stated that respondents eliminate levels too fast. So, stage zero is not used in this paper's surveys. Stage one consists of the assessment of the importance of attribute-levels and stage two of the assessment of the importance of attributes. The part-worths for the attribute-levels are summed for conveying the overall utility as shown in the following model [see Eckert and Schaaf (2009) for a similar notation]:

The part-worth $y_{m_j,h}$ of the mth attribute-level of the jth attribute for the hth consumer is defined as:

$$y_{m_j,h} = w_{j,h} \cdot b_{m_j,h} \qquad (j \in J, m_j \in M_j, h \in H) \qquad (1)$$

with

$w_{j,h}$:   importance of the jth attribute for the hth consumer,
$b_{m_j,h}$:   evaluation of the mth attribute-level of the jth attribute by the hth consumer,
J: index set for attributes,
M_j: index set for attribute-levels of the jth attribute,
H: index set for consumers.

The overall utility $u_{p,h}$ of the pth product for the hth consumer is defined as:

$$u_{p,h} = \sum_{j \in 1} \sum_{m_j \in M_j} y_{m_j,h} \cdot x_{p,m_j} \qquad (h \in H, p \in P) \qquad (2)$$

with

x_{p,m_j}:   1   in case mth attribute-level of the jth attribute of the product p,
        0   otherwise,
P: index set for the products under study.

2.3 Data Collection Alternatives

SEM data collection alternatives are manifold (Green and Srinivasan 1990).

A methodical overview by Eckert and Schaaf (2009) stated that only seven of 40 possible combinations of evaluation methods in stage one and two were empirically implemented (see Table 1). The methods are divided in non-comparative (individual assessment) and comparative (assessment of one in relation to another) with assessment of the differences and without assessment of the differences. The assessment of differences results more information but the cognitive challenge is higher (Eckert and Schaaf 2009). Based on Table 1 it is possible to figure out which combination of evaluation methods in stage one and two would be innovative. The overview is supplemented with the adaptive SEM (ASE) by Netzer and Srinivasan (2011) which combines on its stage two rank order and constant sum scale over pairs and the realized combination of stage one with rank order and stage two with constant sum scale in the online survey. Also combinations which are not shown in the overview like the constant sum scale on both stages (realized in the expert

Table 1 Overview of existing and potential two-staged SEM-approaches

(underlined are one-staged approaches)

		Stage one				
			Non comparative evaluation methods	Comparative evaluation methods with assessment of differences		
			Rating scales		Rating scales	
Stage two		No stage one	No fixed point	Rank order	One fixed point	Two fixed points
Non comparative evaluation methods — No assessment of differences — Rating scales	No fixed point	√	√√	√√	√√	–
	Unlimited scaling	√				
	Dollar metric	√				
	Rank order	√				
	Q-sort	√				
	Pair comparison	√				
	Maximum difference method	√				
	Constant sum scale	√	√	+√		√
	Constant sum scale over pairs	–	√	+√		√
Assessment of differences — Rating scales	One fixed point	√				
Comparative evaluation methods						

(*Source*: extending Eckert and Schaaf 2009)

√ = used in empirical studies investigated by Eckert and Schaaf (2009)

+√ = supplement by this study and by Netzer and Srinivasan (2011)

survey) or the pairwise comparison-based preference measurement (PCPM) by Scholz et al. (2010) with pairwise comparison on both stages seem to be innovative and it is to be mentioned that methods like the rank order were rarely used in a two-staged SEM.

Because the predictive validity of the used methods of stage one came off well by some empirical findings and no significant differences could be stated between them (Pullman et al. 1999). Eckert and Schaaf (2009) stated that there is no need for new methods.

Eckert and Schaaf (2009) also had chosen different evaluation criteria for the methods in stage two (see Table 2). They gave an overview for stage two methods by the quality related criteria validity and discrimination and application orientated criteria expenditure of time and applicability for the market researcher. Table 2 shows that the constant sum scale has the highest score over all criteria (score 13) and rank order, Q-sort, maximum difference scale, constant sum over pairs and rating scale with one fixed point rated second with score 11. For constant sum scale a good validity, discrimination, applicability, and expenditure of time are identified and it is also to prefer for application-orientated fields of application (Eckert and Schaaf 2009).

3   Empirical Test: Small RPAS as an Example of a Complex Product

For the realized surveys the complex product under study is small RPAS ("small drones" under 150 kg maximum take-off-weight) for German search and rescue (SAR) forces as an arrangement for the comprehensive German research project "MIFU-SAR" ("Mobile Information Systems for Leading-Assistance in SAR"). This field of application has to be described as very application-orientated because the complex product is already not in use and it has big influence on the leading structure and tasks of planning and enforcement of SAR-missions.

Therefore we searched to figure out which attributes and attribute-levels of a small RPAS as a complex, new and innovative product for operations in SAR as a complex and very application-orientated field of application are preferred by, e.g., disaster relief forces. The resulting research question was: Which attributes, attribute-levels and preference measurement approach with which evaluation methods for a complex, new and innovative product in a complex and very application-orientated field of application have to be chosen.

Table 2 Assessment of existing evaluation methods for stage two

		Stage two evaluation methods	Quality related criteria		Application orientated criteria		Sum
			Validity	Discrimination degree	Expenditure of time	Applicability	
Noncomparative evaluation methods		Rating scales-no fixed point	+ ✓	+ ✓	++++ ✓	++++	10
		Unlimited scaling	+ ✓	++ ✓	+++ ✓	++	8
		Dollar metric	+	+	+++	+++	8
Comparative evaluation methods	No assessment of differences	Rank order	++ ✓	+++	++ ✓	++++	11
		Q-sort	+++ ✓	+++ ✓	+++ ✓	++	11
		Pair comparison	++++ ✓	++++ ✓	+ ✓	+	10
	Assessment of differences	Maximum difference scale	++++ ✓	++++ ✓	++ ✓	+	11
		Constant sum scale	+++ ✓	+++ ✓	+++ ✓	++++	13
		Constant sum scale over pairs	++++ ✓	++++ ✓	++	+	11
		Rating scales-one fixed point	++ ✓	++ ✓	+++ ✓	++++	11

(*Source*: according to Eckert and Schaaf 2009)
✓ = evaluation based on empirical findings

4 Empirical Test: Data Collection

It was the first step to figure out which attributes and attribute-levels of a small RPAS had to be chosen. Evaluations were gathered by expert-interviews and published long-term investigations.

On the first survey a two-staged SEM with constant sum scale on both stages returned no satisfying predictive validity when correlating observed and predicted utilities for the holdout products. The second survey solved this problem by using a rank order in stage one.

Both surveys were constructed in five parts. They start with some introductional questions (technical affinity, acceptance and the knowledge of the high-water problems 2013 in Germany), then the respondents were asked to evaluate the importance of three attribute-levels (SEM stage one) for each attribute and at stage two they were asked for the assessment of the importance of the seven attributes. For examination of the predictive validity holdout stimuli were constructed in the fourth part (assessment by constant sum scale) and the surveys came with some sociodemographic questions to an end.

In this paper the term "predictive validity" is used in the meaning of Akaah and Korgorgaonkar (1983) described as the correlation between the model's predictions and the actual preference scores for a holdout sample. The comparison of the predicted order (based on the overall utilities) with observed order (based on the holdout stimuli) is used for validating the prediction using Spearman's rank correlation coefficient.

4.1 The Expert Survey

The first survey (the expert survey) was implemented on the Conference of the German Fire Protection Association 2013 where—for this field of application—some of the best informed respondents (experts) could be found. At the end 25 experts in small RPAS and SAR have been interviewed (fire workers, scientists, political decision makers, small RPAS suppliers).

Based on Tables 1 and 2 for an innovative two-staged SEM approach an empirically rarely implemented evaluation method, constant sum scale on both stages, was chosen. It is feasible for application-orientated fields of application (Eckert and Schaaf 2009) and assesses differences (more information) while having strong validity, discrimination, low time exposure, and good applicability (see Table 2). Respondents have to allocate a given number of scores (here 100) for the attributes or attribute-levels (Torgerson 1958). In this case the respondents could assess different objectives equally. Three holdout stimuli named and arranged like real possible products were constructed (see Fig. 1) for examination.

Attributes	Attribute-levels			
1. Handling	Easy handling	Updated for emergency application	Augmented reality	Updated for emergency application
2. System integration	No integration	Ground control integrated	Ground control and aircraft integrated	Ground control integrated
3. Autonomy	No autonomy	Partial autonomy	See & avoid	See & avoid
4. Robustness	Flight in rain	Flight in smoke and dust	CBRN-capability	Flight in rain
5. Air safety	Redundant rotors	Automatic emergency landing	See & avoid / additional safetysystems	See & avoid / additional safetysystems
6. Sensors	Video camera	Thermal imaging camera	CBRN-sensors	Thermal imaging camera
7. Flight peformance	Long flight duration / small payload (60 min / 0,5 kg)	Average flight duration / average payload (45 min / 0,75 kg)	Short flight duration / heavy payload (30 min / 1,0 kg)	Average flight duration / average payload (45 min / 0,75 kg)
Holdout stimuli expert-survey	P1 - Simple areal reconnaissance	P2 - Leadership system	P3 - CBRN-system	-
Holdout stimuli online-survey	P1	P2	P3	P4

Fig. 1 Holdout stimuli of the expert and the online survey

4.2 The Online Survey

The online survey—as the name suggests—was implemented online using the LimeSurvey tool. There were about 148 respondents who started to answer the questions but only 48 of them completed the questionary. This could be a sign that the survey was still too long and/or the questionary was too difficult. All respondents were related to SAR (40 are working on SAR) or small RPAS as well as in the expert survey.

With regard to the worse predictive validity the evaluation method for stage one was changed to the rank order. The rank order method forces the respondents to bring the attribute-levels in preference order with no choice for assessing two attribute-levels equally. There is no information about how much the respondents differentiate between two ranks. The same difference between ranks has to be assumed (which disagrees Table 1 in stage one for rank order and the scores in Table 2). So the mirrored ranks could be used as numerical values for the part-worths (Green and Krieger 1993). Advantages of the rank order are a smaller cognitive challenge (stated by respondents who completed both surveys), a low expenditure of time, an excellent applicability, and a moderate validity (see Table 2). Also the combination rank order and constant sum was not empirically surveyed (see Table 1).

The online survey was constructed with much less text (introduction, questions, explanations) in order to decrease the response time and the cognitive challenge. The holdout stimuli were shown as a picture (not in textual way like in the expert survey) and without names, so that there were no references because of the names.

Additional a fourth holdout stimuli was constructed out of the results of the expert survey (best assessed attribute-levels for product P4, see Fig. 1). In stage two also the constant sum scale was used.

5 Empirical Test: Results

For the predictive validity only a low averaged Spearman value across individuals could be calculated (mean $r_s = 0.2093$). Because the respondents were directly interviewed some reasons were found to explain the bad predictive validity. First of all respondents had problems by the method of asking for holdout stimuli itself (e.g., "Why do I have to answer about the same questions again?"), the construction of the holdout stimuli (e.g., "I would take attribute-levels of P2 and P3 but because of the given names I afford different requirements.") and the information overload (e.g., "It is hard to evaluate 21 attribute-levels of seven attributes combined in three products which arranged and named for different applications."). Thus the predictive validity was strongly influenced by the complexity of the assessment of the holdout stimuli and also the preferences because of the given names. These problems with the expert survey were tried to solve by the online survey.

Because of the improvements of the expert survey the online survey shows a much better predictive validity with respect to the mean Spearman value ($r_s = 0.5184$ on average). In Fig. 2 a histogram of Spearman values is shown and only three respondents have negative values. Without these three negative values the averaged Spearman value would raise to 0.5746.

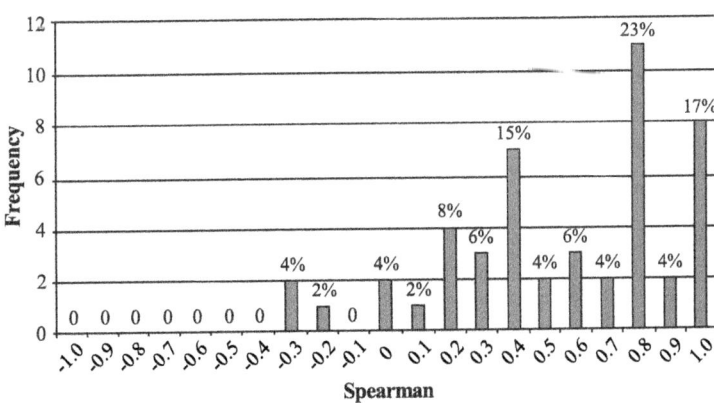

Fig. 2 Spearman frequency scale

6   Conclusion and Outlook

As a result of these studies it is to be stated that two new two-staged SEM approaches were implemented. The first survey with constant sum scale on both stages returned no acceptable predictive validity. This problem was solved by the second two-staged SEM, a combination of rank order on stage one and constant sum scale on stage two.

Furthermore it has to be stated that the scoring of Eckert and Schaaf (2009) for validity of rank order and constant sum scale (see Table 2) could not be confirmed by the implemented studies. Respondents judged a smaller cognitive challenge for the rank order than the constant sum scale. The increasing of the predictive validity is constituted by the smaller expenditure of time (less text and rank order on stage one) and the smaller cognitive challenge [rank order on stage one, no names for holdout stimuli and a fourth holdout stimuli (from the expert survey as a pretest)]. The results of the two realized surveys show that the constant sum scale has a too high cognitive challenge particularly if the number of attributes is high (especially at the assessment of the holdout stimuli). Some literature also stated that with a higher number of objectives probands will be mentally overstrained by rank order and constant sum scale (e.g., Myers 1999). Myers (1999) stated a higher number as ten objectives as mentally overstrained for constant sum scale. So the holdout task itself has a too high cognitive challenge with seven attributes and each attribute three attribute-levels. The smaller the cognitive challenge of the questionary (especially of the holdout) the better the predictive validity. Assessment of holdout stimuli could be very confusing for people who are not involved in marketing. Even experts on small RPAS and SAR could not recognize which holdout stimulus they "have to" prefer. A feasible solution could be a more flexible method. For example, a method like a simple product configurator in which the probands only have to decide for the attribute-levels they prefer. Another way could be that one of the holdout stimuli is constructed dynamically by the given answers before. This would be much simpler for respondents and the predictive validity would increase. So the way of assessing holdout stimuli is probably counterproductive for a good predictive validity. Furthermore to give the holdout stimuli names with a real background has to be avoided because this forces the respondents to ignore the chosen attribute levels.

The main point of criticism for this study surely is the small number of respondents. It is to explain by the complex and specific product and field of application and the influence on leading, structure and organisation in SAR that could only be estimated by experts. This could also be put in another way: preferences for a product which is not in use are very difficult to measure because respondents have no real preferences even if they are experts.

The aggregation of the knowledge of already implemented studies leads to the hypothesis that the implemented two-staged SEM approaches would be feasible for complex products. Especially for the second survey, the combination of rank order and constant sum scale, this can be acknowledged.

There has to be some further research in preference measurement of new, complex, and application-orientated products especially to decrease the cognitive challenge. For new products whose implementation in a field of application creates new applications and procedures like small RPAS in SAR another alternative of holdout tasks or assessment of them has to be identified to increase predictive validity. Developments of the last years like the ASE by Netzer and Srinivasan (2011) or the PCPM by Scholz et al. (2010) could be more feasible than the traditional two-staged SEM. Like Meissner et al. (2011) proposed an empirical comparison of the ASE by Netzer and Srinivasan (2011) with measuring of preferences by constant sum scale could be a further research project and the two realized surveys of this study deliver first empirical data for this.

References

Akaah, I. P., & Korgorgaonkar, P. K. (1983). An empirical comparison of the predictive validity of self-explicated, huber-hybrid, traditional conjoint, and hybrid conjoint models. *Journal of Marketing Research, 20*(2), 187–197.

Dorsch, M. J., & Teas, R. K. (1992). A test of the convergent validity of self-explicated and decompositional conjoint measurement. *Journal of the Academy of Marketing Science, 20*(1), 37–48.

Eckert, J., & Schaaf, R. (2009). Verfahren zur Präferenzmessung – Eine Übersicht und Beurteilung existierender und möglicher neuer Self-Explicated-Verfahren. *Journal für Betriebswirtschaft, 59*(1), 31–56.

Green, P. E., Goldberg, S. M., & Montemayor, M. (1981). A hybrid utility estimation model for conjoint analysis. *Journal of Marketing, 45*(1), 33–41.

Green, P. E., & Krieger, A. M. (1993). Conjoint analysis with product positioning applications. In Eliashberg, J. & Lilien, G. (Eds.), *Handbook in operations research and management science* (Vol. 5, pp. 467–515). Amsterdam: Elsevier.

Green, P. E., & Srinivasan, V. (1990). Conjoint analysis in marketing: New developments with implications for research and practise. *Journal of Marketing, 54*(4), 3–19.

Green, R. E., Krieger, A. M., & Agarwal, M. K. (1993). A cross validation test of our models for quantifying multiattribute preferences. *Marketing Letters, 4*(4), 369–380.

Höpfl, R. T., & Huber, P. H. (1970). A study of self-explicated utility models. *Behavioral Sciences, 15*(5), 408–414.

Meissner, M., Decker, R., & Adam, N. (2011). Ein empirischer Validitätsvergleich zwischen adaptive self-explicated approach (ASE), pairwise comparison-based preference measurement (PCPM) und adaptive conjoint analysis (ACA). *Zeitschrift für Betriebswirtschaft, 81*, 423–446.

Myers, J. H. (1999). *Measuring customer satisfaction: Hot buttons and other measurement issues.* Chicago: American Marketing Association.

Netzer, O., & Srinivasan, V. (2011). Adaptive self-explication of multi-attributed preferences. *Journal of Marketing Research, 48*(1), 140–156.

Pullman, M. E., Dodson, K. J., & Moore, W. L. (1999). A comparison of conjoint methods when there are many attributes. *Marketing Letters, 10*(2), 125–138.

Sattler, H. (2006). Methoden zur Messung von Präferenzen für Innovationen. *Zeitschrift für betriebswirtschaftliche Forschung, 54*(6), 154–176.

Scholz, S., Meissner, M. & Decker, R. (2010). Measuring consumer preferences for complex products: A compositional approach based on paired comparisons. *Journal of Marketing Research, 47*(4), 685–698.

Srinivasan, V. (1988). A conjunctive-compensatory approach to the self-explication of multiattributed preferences. *Decision Sciences, 19*(2), 295–305.

Srinivasan, V., & Park, C. S. (1997). Surprising robustness of the self-explicated approach to customer preference structure measurement. *Journal of Marketing Research, 34*(2), 286–291.

Torgerson, W. S. (1958). *Theory and method of scaling*. New York: Wiley.

Combination of Distances and Image Features for Clustering Image Data Bases

Sarah Frost and Daniel Baier

Abstract Daily, millions of pictures are released online but it is hard to analyze them automatically for marketing purposes. This paper tries to show how methods from the content-based image retrieval could be used to classify image data and make them usable for marketing applications. There are a number of different image features which can be extracted from the images to calculate dissimilarities between them afterwards with different kinds of distance measures (Manjunath et al. 2001). We focus especially on mass-transportation-problems, like the Earth Mover's Distance (EMD) (Rubner et al., Int J Comput Vis 40(2):99–121, 2000), because they fit the human perception on dissimilarities. Furthermore there are already some studies that show that they are robust to disturbances like changes in resolution, contrast, or noise (Frost and Baier, Algorithms from and for nature and life. Studies in classification, data analysis, and knowledge organization, vol 45. Springer, Heidelberg, 2013). We compare some approximations of the EMD (e.g., Pele and Werman 2009) with an approximation algorithm developed by ourselves. The aim is to find a combination of features and distances which allows to cluster large image data bases in a way that fits the human perception.

1 Introduction

One can use image clustering, for example, for customer segmentation. Instead of asking for activities, interests, and opinions one can use uploaded pictures from these topics (e.g., holiday pictures), because customers with similar images may also have similar interests or activities. Or one can combine questionnaires and images to find out which type of people takes which types of photos. Another application of the calculated similarity of images is to position brands via their poster advertisings. This method could help to erase confusion of brands or to reposition brands with new campaigns.

S. Frost (✉) • D. Baier
Institute of Business Administration and Economics, Brandenburg University of Technology
Cottbus, Postbox 101344, 03013 Cottbus, Germany
e-mail: sarah.frost@tu-cottbus.de; daniel.baier@tu-cottbus.de

© Springer-Verlag Berlin Heidelberg 2015
B. Lausen et al. (eds.), *Data Science, Learning by Latent Structures, and Knowledge Discovery*, Studies in Classification, Data Analysis, and Knowledge Organization, DOI 10.1007/978-3-662-44983-7_22

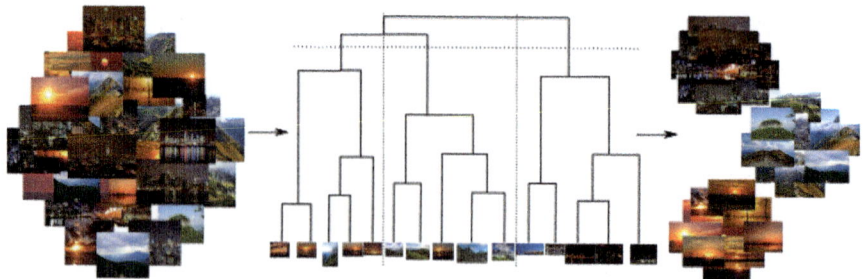

Fig. 1 Example for classifying an image data base using hierarchical clustering

The aim of our research is to classify image data bases to find, e.g. customer- or market segments. Therefore we extract various image feature vectors from each image and calculate the distances between them. Afterwards we use a hierarchical clustering method to find the clusters. Figure 1 shows an example for organizing a huge image data base in three clusters.

The paper is organized as follows: in Sect. 2 the image features we use in our experiments are presented. Section 3 lists the distance measures which were used to calculate the dissimilarity of the images. In Sect. 4 the empirical study will be presented which was used to evaluate the image features and the distance measures. Section 5 concludes our findings and gives a short outlook for future work.

2 Image Features

The most common way to calculate the dissimilarity of images is to extract feature vectors first. After the extraction these feature vectors are handled like points in the metric space, where a distance between two points can be calculated. In this section there will be a presentation of some common feature descriptors. Some of them were developed by the MPEG-7 consortium (Manjunath and Sikora 2002).

Color histograms are discrete frequency distributions of image colors. They are robust to translation and rotation (Swain and Ballard 1991) and yield robust clustering results when resolution is reduced or noise is added (Frost and Baier 2013). For the experiments in this paper the L*a*b* color space (Wyszecki and Stiles 2000) is used and the color distributions are stored in 128 bin vectors.

The *color and edge directivity descriptor* (CEDD) was developed by Chatzichristofis and Boutalis (2008a). It combines a 24 bin color histogram in the HSV (Hue, Saturation, Value) color space and a 6 bin edge directivity histogram with the six directions: horizontal, vertical, diagonal (45°), diagonal (135°), non-directional, and no edge. To retrieve this histogram the image is divided into 1,600 image blocks where the number of image blocks with the same color and the same edge directivity is stored in the bins.

The *fuzzy edge and texture histogram* (FCTH) is similar to the CEDD but instead of edge directivities there are eight types of textures used to combine color and structure of an image (Chatzichristofis and Boutalis 2008b). The assignment to a bin of the FCTH follows a fuzzy logic. That means that an image block can be 40 % black and 60 % gray. Because of very good image data base clustering results in some pretests we developed a ground distance for CEDD and FCTH, which is the distance between two bins of the histogram (see Sect. 3).

The *color structure descriptor* (CSD) represents the color distribution of an image as well as the spatial structure of color (Buturovic 2005). Therefore a structuring element visits every position of the image and counts if a color exists within this element. The size of the structuring element depends on the image size. The result is that widely spread colors are counted more often than compact color areas. Yet we did not develop a ground distance for this feature, because the spatial information cannot be read out of the histogram.

The *scalable color descriptor* is a compact Haar transform-based encoding scheme basing on a HSV color histogram (Manjunath et al. 2001).

Tamura feature is a texture feature, measuring coarseness, contrast, directionality, line-likeness, regularity, and roughness (Tamura et al. 1978).

The *Gabor texture feature* uses wavelet transformations to analyze the texture (Zhang et al. 2000). It is also used for some MPEG-7 texture descriptors (Homogeneous Texture Descriptor and Texture Browsing Descriptor).

The *edge histogram* divides the image into 16 sub-images and counts the number of image blocks with the same edge directionality (Park et al. 2000).

The next section gives an introduction to the distance measures we used to compare images.

3 Distance Measures

Because the features we use are all in histogram form we distinguish our distance measures in bin-by-bin and cross-bin distances. Bin-by bin distances compare only corresponding bins whereas cross-bin distances compare all bins from one histogram with all bins from another. The main advantage of bin-by-bin distances is the linear computation complexity. Figure 2 shows the difference between bin-by-bin and cross-bin distances. In (a) a bin-by-bin distance compares only the corresponding bins. That is why $d(h_1, k_1) > d(h_1, k_2)$ even if the human perception is contrary. In (b) all bins are compared with each other and in (c) the histograms are transformed into each other like EMD does. In (c) $d(h_1, k_1) < d(h_1, k_3)$, which fits the human perception.

Table 1 lists the distance measures we used. All distances are calculated between the histograms $\mathbf{h} = (h_1, \ldots, h_N)$ and $\mathbf{k} = (k_1, \ldots, k_N)$ where $m_i = (h_i + k_i)/2$ and $\hat{h}_i = \sum_{j=1}^{i} h_j$ for all $i \in \{1, \ldots, N\}$. Especially the bin-by-bin distances are quite easy to understand but other distances need a more detailed explanation.

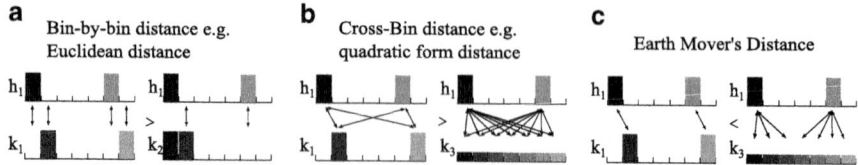

a Bin-by-bin distance e.g.
Euclidean distance

b Cross-Bin distance e.g.
quadratic form distance

c Earth Mover's Distance

Fig. 2 Example for calculating the increasing radius algorithm in a two-dimensional histogram

Table 1 Used distance measures

Distance	Formula	Source
Bin-by-bin distances		
Euclidean distance	$d(\mathbf{h}, \mathbf{k}) = \sqrt{\left(\sum_i |h_i - k_i|^2\right)}$	Minkowski (1910)
Cosine distance	$d(\mathbf{h}, \mathbf{k}) = 1 - \frac{\mathbf{h} \cdot \mathbf{k}}{\|\mathbf{h}\| \cdot \|\mathbf{k}\|}$	Deza and Deza (2009)
Chebyshev distance	$d(\mathbf{h}, \mathbf{k}) = \max_i |h_i - k_i|$	Minkowski (1910)
Histogram intersection	$d(\mathbf{h}, \mathbf{k}) = 1 - \frac{\sum_i \min_i (h_i, k_i)}{\sum_i k_i}$	Swain and Ballard (1991)
Jeffrey divergence	$d(\mathbf{h}, \mathbf{k}) = \sum_i \left(h_i \log \frac{h_i}{m_i} + k_i \log \frac{k_i}{m_i}\right)$	Ojala et al. (1996)
χ^2 statistic	$d(\mathbf{h}, \mathbf{k}) = \sum_i \frac{(h_i - m_i)^2}{m_i}$	Zhang and Lu (2003)
Cross-bin distances		
Quadratic form	$d(\mathbf{h}, \mathbf{k}) = \sqrt{(\mathbf{h} - \mathbf{k})^\top \mathbf{A}(\mathbf{h} - \mathbf{k})}$	Niblack et al. (1993) Hafner et al. (1995)
Match distance	$d(\mathbf{h}, \mathbf{k}) = \sum_i |\hat{h}_i - \hat{k}_i|$	Werman et al. (1985)
Kolmogorov–Smirnov	$d(\mathbf{h}, \mathbf{k}) = \max_i \left(|\hat{h}_i - \hat{k}_i|\right)$	Geman et al. (1990)
EMD	$d(\mathbf{h}, \mathbf{k}) = \min \sum_i \sum_j g_{ij} f_{ij}$	Rubner et al. (1997)
IR-algorithm	$d(\mathbf{h}, \mathbf{k}) \approx \min \sum_i \sum_j g_{ij} f_{ij}$	Frost and Baier (2013)
$\widehat{EMD}$	$d(\mathbf{h}, \mathbf{k}) \approx \min \sum_i \sum_j g_{ij} f_{ij}$	Pele and Werman (2009)

The quadratic form distance uses a similarity matrix $\mathbf{A} = [a_{ij}]$ to regard the spatial distribution of bins in the underlying feature space. The distance between bins in a joint feature space is called ground distance $\mathbf{G} = [g_{ij}]$ where g_{ij} is the distance between bins i and j. The similarity matrix $\mathbf{A}$ is calculated as $a_{ij} = 1 - g_{ij}/g_{max}$ where g_{max} is the maximal ground distance between any two bins (Rubner et al. 1997).

A special focus of this paper lays on the so-called Earth Mover's Distance (EMD), it bases on the Kantorovich–Rubinstein transshipment problem. The problem of transforming one histogram into the other is solved with the transportation simplex program. Therefore one histogram can be considered as piles of earth and the other as holes. The aim is to transport the whole earth from the piles into the holes with the minimal amount of work (Rubner et al. 1997). The amount of earth transported from bin i to bin j is called the *flow* f_{ij}. Like for the quadratic form we call the distance between i and j ground distance g_{ij}. There are some constraints to be fulfilled [Eqs. (1)–(4)]. Note that for the EMD the histograms $\mathbf{h}$ and $\mathbf{k}$ can have different sizes, so we use m to be the length of $\mathbf{h}$ and n for $\mathbf{k}$. Equation (1) means that only transportations from piles to holes are allowed. Equations (2) and (3) ensure that it is impossible to transport more earth from a pile than it contains or more earth into a hole than it can take.

The main disadvantage of the EMD is the long computation complexity of $\mathcal{O}(n^3 \log n^3)$ where n is the number of nonempty bins (in at least one histogram). Therefore a lot of approximations were developed in recent time, but some of them cannot handle histograms with different sizes or different ground distances. That is why we developed an approximation algorithm that meets these requirements.

$$f_{ij} \geq 0 \quad \forall i \in \{1,\ldots,m\}, \; j\{1,\ldots,n\} \tag{1}$$

$$\sum_{j=1}^{n} f_{ij} \leq h_i \quad \forall i \in \{1,\ldots,m\} \tag{2}$$

$$\sum_{i=1}^{m} f_{ij} \leq k_j \quad \forall j \in \{1,\ldots,n\} \tag{3}$$

$$\sum_{i=1}^{m}\sum_{j=1}^{n} f_{ij} = \min\left\{\sum_{i=1}^{m} h_i, \sum_{j=1}^{n} k_j\right\} \tag{4}$$

The increasing radius algorithm (IR-alg.) is a self-developed approximation of the EMD where no optimization problem has to be solved (Frost and Baier 2013). The first step of the algorithm is to build the difference histogram, so we get all piles and holes in one histogram (see Fig. 3). The starting radius is equal to 1 and the algorithm iterates through the histogram and fills all holes that are in the circuit of a pile with a radius of 1. Then the radius will be increased. At the end the distance is the summation of all flows times the radius. In worst case the algorithm has to iterate $\lceil r_{\max} \rceil$ times through a histogram of length n, where $\lceil r_{\max} \rceil$ is the rounded up maximum possible radius. It depends on the type of the ground distance (e.g., Euclidean distance or others) and on the size and dimensionality of the histogram.

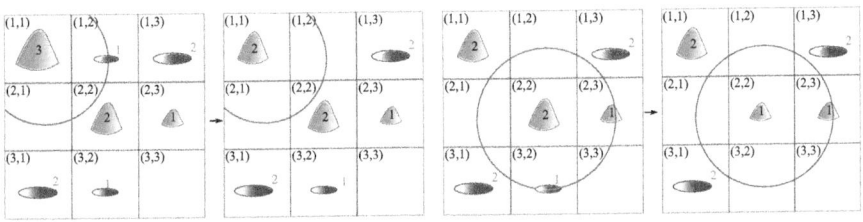

Fig. 3 Example for calculating the increasing radius algorithm in a two-dimensional histogram

Thus the computation complexity is $\mathcal{O}(\lceil r_{max} \rceil \cdot n)$. In a previous study the IR-alg. showed to be robust against changes of resolution, contrast, color, brightness, and noise of an image (Frost and Baier 2013).

The last presented distance measure is the $\widehat{EMD}$ developed by Pele and Werman (2009). Instead of allocating piles to all possible holes in an optimal way, piles are allocated firstly to the corresponding hole because this transportation does not have costs and after that to neighboring holes. The ground distance for transportations has to be smaller than a given threshold t (mostly $t = 2$). The rest of the piles are allocated via a so-called transportation vertex. All of these transportations have transportation costs of t. Therefore the $\widehat{EMD}$ between two histograms can be less than the EMD.

In the following we call EMD, IR-alg., and $\widehat{EMD}$ transportation distances.

4 Experiments

4.1 Experimental Design

In a previous study we classified only 15 holiday pictures from three categories because we wanted to get an overview which image features are most likely to be appropriate. In the first step we only took distances that do not need a ground distance and measured the adjusted Rand Index (Hubert and Arabie 1985) to evaluate the clustering results. A 1 means perfect clustering, when all images from one category are in one cluster. In this pretest we found that CEDD and FCTH yielded the best results, that is why we developed a ground distance for them to prove if the EMD and its approximations work even better than the other distances.

The ground distances for CEDD and FCTH are combinations of the ground distances in the HSV color space and the dissimilarity of edges or textures, respectively. We used the Euclidean distance as ground distance in the color space in both cases. The dissimilarity of edge types is ranked from 0 to 4 and of textures from 0 to 6. The highest dissimilarity is between "no edge" and "any edge" or "no texture" and "any texture" and the lowest between edges with only 45° angle between them or between textures with two out of three identical directionalities, respectively. The two types of ground distances were linked by summing up both ground distances. Therefore different weights were tried for both parts to find out which weight yields the best results. For the CEDD ground distance we set the weight of the color ground distance to be 1 and the edge ground distance to be 0.1. In case of FCTH the color weight also was set to be 1 and the texture weight has to be 0.25. Now we were able to use CEDD and FCTH in combination with the transportation distances.

For our experiment we downloaded 2,700 images with a Flickr download tool using the three keywords "skyline at night," "mountains," and "sunset." The first 900 images for each category were used. Now we extracted the eight image features (see Sect. 2) with the software tool IMADAC developed at the chair of Marketing at the Brandenburg University of Technology (Naundorf et al. 2012).

4.2 Results

The results of our experiments are shown in Table 2. The best result over all features and distances was yielded with the CSD and the Jeffrey divergence (bold printed) where 2,309 out of 2,700 images are clustered correctly. The image features with the best average results are color histogram, FCTH, and CEDD. Our self-developed IR-algorithm was the most appropriate distance measure for the color histogram and the other EMD approximation ($\widehat{EMD}$) achieved the best adjusted Rand index when CEDD feature was used. There are some features like Gabor, SCD, and Tamura feature, which showed that they do not achieve good results with any of the used distance measures.

The two last columns in Table 2 show the average adjusted Rand indices over all features or over all features that have a ground distance, respectively. Here one can see that the χ^2 statistic, the Jeffrey divergence, and the histogram intersection (which is equivalent to the city block distance in our case) performed best but also the $\widehat{EMD}$ got better average indices than other distances.

To compare the usability of the feature distance combinations we also timed how long the computations took. Table 3 shows the computation time to calculate all distances between the 2,700 images (3,643,650 calculations) in minutes. All calculation were made on a common desktop PC (Intel i7 processor with 8 cores and 12 GB RAM).

Table 2 Adjusted Rand indices for clustering 2,700 images basing on different features and distances using Ward's method

	CEDD	CoHi	CSD	Edge	FCTH	Gabor	SCD	Tamura	Mean[1]	Mean[2]
Chebyshev	0.28	0.14	0.17	1.18	0.38	0.04	0.05	0.03	0.14	0.27
Cosine	0.40	0.26	0.25	0.24	0.28	0.04	0.04	0.06	0.18	0.31
Euclidean	0.38	0.20	0.26	0.12	0.40	0.06	0.07	0.04	0.18	0.33
HI	0.41	0.37	0.33	0.24	0.51	0.06	0.06	0.05	0.23	0.43
Jeffrey	0.42	0.43	**0.62**	0.13	0.36	0.04	0.02	0.05	0.23	0.40
KS	0.08	0.33	0.09	0.05	0.16	0.02	0.05	0.02	0.09	0.19
Match	0.09	0.30	0.09	0.05	0.09	0.04	0.06	0.04	0.09	0.16
χ^2	0.41	0.37	0.42	0.23	0.48	0.04	0.02	0.04	0.23	0.42
QF	0.13	0.30	0.11	0.06	0.14	0.03	0.05	0.04	0.10	0.19
IR-alg.	0.38	0.46	–	–	0.26	–	–	–	–	0.37
$\widehat{EMD}$	0.47	0.35	–	–	0.42	–	–	–	–	0.41
EMD	0.23	0.45	–	–	0.25	–	–	–	–	0.31
Mean	0.31	0.33	0.26	0.15	0.31	0.04	0.05	0.04		

Key: *CoHi* color histogram, *Mean*[1] average adj. Rand index over all features, *Mean*[2] average adj. Rand index over features with a ground distance, *HI* histogram intersection, *KS* Kolmogorov–Smirnov distance, *QF* quadratic form, –: no ground distance available

Table 3 Time (in minutes) to calculate 3,643,650 distances with the particular feature distance combination

	CEDD	CoHi	CSD	Edge	FCTH	Gabor	SCD	Tamura
Chebyshev	1.57	1.65	2.71	2.18	2.78	2.70	0.66	2.30
Cosine	1.88	2.09	2.90	2.31	2.97	2.84	0.71	2.89
Euclidean	1.54	1.63	2.64	2.15	2.81	2.68	0.63	2.27
HI	1.82	1.95	2.85	2.32	3.12	2.83	0.76	2.67
Jeffrey	2.10	2.09	3.47	2.89	3.84	3.68	0.90	2.96
KS	1.63	1.72	2.68	2.23	2.92	2.75	0.71	2.37
Match	1.60	1.69	2.69	2.23	2.89	2.69	0.69	2.34
χ^2	1.94	2.47	3.24	2.62	3.30	3.39	0.85	2.67
QF	124.91	216.40	99.19	24.42	35.93	36.58	3.33	379.79
IR-alg.	22.04	5.97	–	–	38.78	–	–	–
$\widehat{EMD}$	21.05	5.29	–	–	37.46	–	–	–
EMD	107.54	16.29	–	–	40.29	–	–	–

Key: *CoHi* color histogram, *HI* histogram intersection, *KS* Kolmogorov–Smirnov distance, *QF* quadratic form, –: no ground distance available

The upper eight distances in Table 3 are more or less similar to each other within one column (one feature), the calculation time varies just a little. The transportation distances are much slower. In these cases the calculation time depends not only on the length of the histogram but also on the number of nonempty bins [see Rubner et al. (2000) for definition of the so-called signatures]. That is the reason why the quadratic form distance is even slower than the transportation distances.

5 Conclusion and Outlook

The aim of this study was to find out which feature distance combination is most suitable to classify image data bases. Even if none of the used combinations achieved perfect clustering (adj. Rand index that is 1.00) we found some combinations that yielded satisfying results. For example, the combination of Jeffrey divergence and CSD found a classification where 2,309 out of 2,700 images were classified correctly. For every tested image feature we found a distance measure that works best and now we know that using only texture information like Gabor or Tamura features we do not get satisfactory results.

The next step of testing feature distance combinations is to combine many features with each other. Every feature with its most fitting distance measure. We also plan to learn weights for each feature to make the classification more independent on the image data base (this time we only used colored images of landscapes).

References

Buturovic, A. (2005). *MPEG-7 color structure descriptor: For visual information retrieval project VizIR*. Institute for Software Technology and Interactive Systems Technical University Vienna, Technical Report.

Chatzichristofis, S., & Boutalis, Y. (2008a). CEDD: Color and edge directivity descriotor - A compact description for image indexing and retrieval. In A. Gasteratos, M. Vincze, & J. Tsotsos (Eds.), *Computer vision systems* (pp. 312–322). Heidelberg: Springer.

Chatzichristofis, S., & Boutalis, Y. (2008b). FCTH: Fuzzy color and texture histogram - A low level feature for accurate image retrieval. In *9th International Workshop on Image Analysis for Multimedia Interactive Services* (pp. 191–196).

Deza, M., & Deza, E. (2009). *Encyclopedia of distances*. Berlin: Springer.

Frost, S., & Baier, D. (2013). Comparing earth mover's distance and its approximations for clustering images. In B. Lausen, D. van den Poel, & A. Ultsch (Eds.), *Algorithms from and for nature and life. Studies in classification, data analysis, and knowledge organization* (Vol. 45). Heidelberg: Springer

Geman, D., Geman, S., Graffigne, C., & Dong, P. (1990). Boundary detection by constrained optimization. *IEEE Transactions on Pattern Analysis and Machine Intelligence, 12*(7), 609–628.

Hafner, J., Sawhney, H., Equitz, W., Flickner, M., & Niblack, W. (1995). Efficient color histogram indexing for quadratic form distance functions. *IEEE Transactions on Pattern Analysis and Machine Intelligence, 17*(7), 729–736.

Hubert, L., & Arabie, P. (1985). Comparing partitions. *Journal of Classification, 2*, 193–218.

Manjunath, B., Ohm, J.-R., Vasudevan, V., & Yamada, A. (2001). Color and texture descriptors. *IEEE Transactions on Circuits and Systems for Video Technology, 11*(6), 703–715.

Manjunath, B., & Sikora, T. (2002). Overview over visual descriptors. In B. S. Manjunath, P. Salembier, & T. Sikora (Eds.), *Introduction to MPEG-7* (pp. 179–185). Chichester: Wiley.

Minkowski, H. (1910). *Geometrie der Zahlen* (2nd ed.). Leipzig: Teubner.

Naundorf, R., Baier, D., & Schmitt, I. (2012). *Statistical software for clustering images*. Institute of Computer Science, Brandenburg University of Technology Cottbus, Technical Report.

Niblack, W., Barber, R., Equitz, W., Flickner, M., Glasman, E., Petkovic, D., et al. (1993). The QBIC project: Querying images by content using color, texture, and shape. In *SPIE Storage and Retrieval for Image and Video Databases, 1908* (pp. 173–187).

Ojala, J. R., Pietikänien, M., & Harwood, D. (1996). A comparative study of texture measures with classification-based on feature distributions. *Pattern Recognition, 29*(1), 51–59.

Park, D., Jeon, Y., & Won, C. (2000). Efficient use of local edge histogram descriptor. In *MULTIMEDIA '00 Proceedings of the 2000 ACM Workshops on Multimedia* (pp. 51–54).

Pele, O., & Werman, M. (2009). Fast and robust earth mover's distances. In *IEEE 12th International Conference on Computervision, 12*(1), 460–467.

Rubner, Y., Guibas, L., & Tomasi, C. (1997). The earth mover's distance, multi- dimensional scaling, and color-based image retrieval. In *Proceedings of the ARPA Image Understanding Workshop* (pp. 661–668).

Rubner, Y., Tomasi, C., & Guibas, L. J. (2000). The earth mover's distance as a metric for image retrieval. *International Journal of Computer Vision, 40*(2), 99–121.

Swain, M., & Ballard, D. (1991). Color indexing. *International Journal of Computer Vision, 7*(1), 11–32.

Tamura, H., Mori, S., & Yamawaki, T. (1978). Textural features corresponding to visual perception. *IEEE Transactions on Systems, Man, and Cybernetics, 8*(6), 460–473.

Werman, M., Peleg, S., & Rosenfeld, A. (1985). A distance metric for multidimensional histograms. *Computer Vision, Graphics, and Image Processing, 32*, 328–336.

Wyszecki, G., & Stiles, W. (2000). *Color science: Concepts and methods, quantitative data and formulae* (2nd ed.). New York: Wiley.

Zhang, D., & Lu, G. (2003). Evaluation of similarity measurements for image retrieval. In *IEEE International Conference on Neural Networks and Signal Processing* (pp. 928–931).

Zhang, D., Wong, A., Indrawan, M., & Lu, G. (2000). Content-based image retrieval using gabor texture features. In *First IEEE Pacific-Rim Conference on Multimedia* (pp. 392–395).

A Game Theoretic Product Design Approach Considering Stochastic Partworth Functions

Daniel Krausche and Daniel Baier

Abstract Developing new products is a necessary but costly and risky adventure. Therefore, the customers' point of view and the prospective competitive environment have to be considered. Here, conjoint analysis has proven to be helpful since this preference modeling approach can be used to predict market shares [see, e.g., Baier and Gaul (J Econ 89(1–2):365–392, 1999), Baier and Gaul (Conjoint measurement: methods and applications. Springer, Berlin, pp. 47–66, 2007)]. When, additionally, competitive reactions must be considered, game theoretic approaches are a helpful extension [see, e.g., Choi and Desarbo (Market Lett 4(4):337–348, 1993), Steiner and Hruschka (OR Spektrum 22:71–95, 2000), Steiner (OR Spectrum 32:21–48, 2010)]. However, recently, new Bayesian procedures have been developed for conjoint analysis that allow to model customers' partworth functions in a stochastic fashion. The idea is that customers have different preferences over time. In this paper we propose a new game theoretic approach that considers this new aspect. The new approach is applied to a (fictive) product design setting. A comparison to a traditional approach is presented.

1 Introduction

For companies, developing new products is necessary to stay competitive in the current marketplace. Therefore new product development has been and continues to be essential to remain successful (Urban et al. 1996). The innovative process in developing new products has effects on a company's value but also its risk (Sorescu and Spanjol 2008). To ensure success and minimize the risk of those products failing market researchers have increased the number of studies to find the most important components of a new product's success. One major component are the product characteristics and within those customer needs or preferences of the product (Henard and Szymanski 2001). One of the most prominently featured

D. Krausche (✉) • D. Baier
Institute of Business Administration and Economics, Brandenburg University of Technology
Cottbus, Postbox 101344, 03013 Cottbus, Germany
e-mail: daniel.krausche@tu-cottbus.de; daniel.baier@tu-cottbus.de

© Springer-Verlag Berlin Heidelberg 2015
B. Lausen et al. (eds.), *Data Science, Learning by Latent Structures,
and Knowledge Discovery*, Studies in Classification, Data Analysis,
and Knowledge Organization, DOI 10.1007/978-3-662-44983-7_23

methods of modeling customer preferences in relation to designing new products has been the conjoint analysis, first introduced into marketing by Green and Rao (1971). Since its introduction a lot of research has been devoted to conjoint analysis, Green and Kieger (1987), Gaul et al. (1995), Steiner and Hruschka (2000), Baier and Gaul (2007) Albritton and Mcmullen (2007), and Steiner (2010) are just a few examples. Therefore it is still a very relevant topic in research and practice (Teichert and Shehu 2010). Because companies operate in a competitive market it is also important to anticipate how rival companies will react if a new or changed product is established. The competitive market can be seen as a game with each company seen as a player, therefore game theoretic approaches could be used to model the competitive market.

With the help of game theory all kinds of different market scenarios can be modeled. From the number of players (companies), if communication is allowed between the players to the possibility of binding contracts and so on. There have been a number of previous conjoint approaches that tried to account for competitive reactions. An overview is given in Table 1 and a summary can be found in Steiner and Baumgartner (2009). Like in previous work we focus on a non-cooperative game where the players make rational decisions based on their objective function (here profit maximization) and each player offers one product. The goal is to find a Nash equilibrium in the given context as solution to the game. Unlike the previous approaches where the partworths are assumed to stay constant, here, we propose that the partworths are modeled as a stochastic variable which follows a certain distribution function because of changing customer preferences as these changes affect customer behavior [see Balk (1989), Hoch and Loewenstein (1991) among others].

In this paper, we are extending the work of Choi and DeSarbo (1993) and Steiner and Hruschka (2000) by accounting for changes in customer preferences. We then compare our results with those from Steiner and Hruschka (2000). The next section will provide a brief overview how game theory can be implemented while modeling product competition and how stochastic partworth functions can enhance this model. In Sect. 3 a Nash model for new product design considering stochastic partworth functions is given, followed in Sect. 4 by our empirical findings where we compare our results with the previous work done by Steiner and Hruschka (2000). Section 5 will provide our conclusions and an outlook as well.

Table 1 Overview of previous work combining conjoint analysis and game theory

Number of products	Type of equilibrium	Reference
Single product	Nash equilibrium	Choi and DeSarbo (1993)
Single product	Nash-equilibrium	Gutsche (1995)
Single product	Nash equilibrium	Green and Kieger (1997)
Single product/product lines	Nash equilibrium	Steiner and Hruschka (2000)
Single product	Stackelberg-Nash equilibrium	Steiner (2010)

2 Game Theory and Stochastic Partworth Functions

2.1 Game Theory and Product Competition

Product price and product positioning therefore product design are very important tools while modeling product competition [see Baier and Gaul (1999), Wittink et al. (1994) among others]. With the help of conjoint analysis the best product designs can be found based on customer preferences. To incorporate competitive reactions strategies employed by the competing companies are based on those product designs. Because of the discrete nature of conjoint data competitive strategies in a Nash equilibrium cannot be found analytically, therefore numeric or heuristic methods have to be applied (see, e.g., Steiner and Hruschka 2000). One method of finding a Nash equilibrium requires a process to repeatedly find the best response from each competitor to the current market situation called tatonnement process. There are two types of tatonnement processes, simultaneous and sequential. The simultaneous tatonnement process assumes that all competitors react at the same time based on the previous strategies and an equilibrium is found when all competitors stay with their previous strategies while in the sequential tatonnement process the competitors react in a predetermined order and an equilibrium is found when no competitor changes his strategy in one round.

However, there are some issues with finding a discrete Nash equilibrium with either tatonnement process as it is hard to verify the existence of such an equilibrium (Choi and DeSarbo 1993). We will focus on the sequential tatonnement process as it closely represents real market scenarios and has a better chance of finding an equilibrium if one exists (see, e.g., Choi and DeSarbo 1993; Steiner 2010). Also, if an equilibrium is found, there is no certainty that it is unique as different starting combinations can result in different equilibria. Therefore, to find all equilibria all possible starting combinations have to be considered (Steiner 2010).

The previously proposed models show various similarities in their basic assumptions. Except Steiner (2010) they all represent Nash models which use simultaneous decisions on product design (i.e., price and non-price attribute levels), a probabilistic choice rule (either Bradley-Terry-Luce share-of-utility rule or the logit choice rule), profit maximizing companies are assumed and a linear-additive partworth-model is applied.

2.2 Stochastic Partworth Functions

Stochastic partworth functions are designed to accommodate possible preference changes of customers. For example, when a scholar graduates and gets his first job his preferences may shift to more expensive products of higher quality or a worker gets laid off which could result in a shift towards cheaper products. Like the previous models we use a linear-additive partworth-model where the individual partworths are estimated via conjoint analysis. The segment utility function can be described as the following equation:

$$u_{ij} = \sum_{k=1}^{K} \sum_{l=1}^{L} \beta_{ikl} x_{jkl} \quad i \in I, j \in J \tag{1}$$

where $i = 1, \ldots, I$ segments, $j = 1, \ldots, J$ products (competitors), $k = 1, \ldots, K$ attributes, $l = 1, \ldots, L_k$ levels of attribute k, u_{ij} is the utility of a consumer in segment i of product j, β_{ikl} is the partworth of attribute k with level l and x_{jkl} is either 0 or one, depending on whether the product j possessess level l of attribute k. The β_{ikl} values are calculated via conjoint analysis. As previously mentioned, the goal now is to account for changes in customer behavior. In order to do that we do not view the β_{ikl} as fixed values but instead as stochastic variables β_{ikl}^* that follow certain distributions with expected values β_{ikl} and standard deviations σ_{ikl}. The distribution functions including the standard deviations are determined by the expected shift of customer preferences for the levels of attributes in each segment while the expected values are determined by conjoint analysis. Depending on the expected shift in customer preferences different distributions might be considered, for example a normal distribution could be used where preferences are expected to stay the same as no change or very little change is most likely or a uniform distribution could be used where preferences are rather unpredictable as the probability of any value of a predetermined interval is the same.

3   A Nash Model for New Product Design Considering Stochastic Partworth Functions

In order to be able to compare our results with the results of Steiner and Hruschka (2000) we use the same assumptions. We assume single-brand companies, competing companies decide simultaneously on their product design, the logit choice rule is applied as probabilistic choice rule and we also assume profit maximizing companies. Additionally, Steiner and Hruschka (2000) used a linear-additive partworth-model and although we also use a linear-additive partworth-model, we now use the segment utility function with β_{ikl}^* following a certain distribution, depending on the expected shift in customer preferences for the levels of attributes.

$$u_{ij} = \sum_{k=1}^{K} \sum_{l=1}^{L} \beta_{ikl}^* x_{jkl} \quad i \in I, j \in J \tag{2}$$

The product design problem presuming profit-maximizing companies for a competitor who offers product j becomes the following (cf. Choi and DeSarbo 1993; Steiner and Hruschka 2000; Steiner 2010):

$$\Pi_j = (p_j - c_j^{(var)} \mathbf{x}_j) \sum_{i=1}^{I} (Q_i \cdot prob_{ij}) \rightarrow max \tag{3}$$

subject to

$$prob_{ij} = \frac{\exp(\mu \cdot u_{ij})}{\sum\limits_{m=1}^{J} \exp(\mu \cdot u_{im})} \qquad i = 1, \ldots, I \tag{4}$$

$$\sum_{l=1}^{L_k} x_{jkl} = 1 \qquad k = 1, \ldots, K \tag{5}$$

$$x \in \{0, 1\} \qquad j \in J, \ k = 1, \ldots, K, \ l = 1, \ldots, L_k \tag{6}$$

where p_j is the price of product j, $c_j^{(var)}$ is the variable cost of product j, $\mathbf{x}_j$ is a feasible design-price vector, Q_i is the sales volume of segment i, $prob_{ij}$ is the probability that a customer in segment i chooses product j and μ is a scaling parameter of the logit model with $\mu > 0$. Equation (3) relates to the predicted profits Π_j of competitor j by offering his product design. The profit is maximized with regard to product price and product design and is calculated by unit contribution times the sum of units sold across segments. The probability $prob_{ij}$ that a customer in segment i purchases the product of competitor j is given in Eq. (4) where all competing products are represented by their respective conjoint utilities u_{im} calculated via Eq. (2). Equation (5) ensures that only feasible products are offered and Eq. (6) represents the binary restrictions to the attribute levels as an attribute level can either be chosen or not be chosen.

The unit cost function $c_j^{(var)} \mathbf{x}_j$ is assumed to be a linear-additive model of individual attribute level costs $c_{jkl}^{(var)}$ shown in Eq. (7) following Gaul et al. (1995), Choi and DeSarbo (1993), and Steiner and Hruschka (2000) among others.

$$c_j^{(var)} \mathbf{x}_j = \sum_{k=1}^{K} \sum_{l=1}^{L_k} c_{jkl}^{(var)} x_{jkl} \tag{7}$$

To model the product competition endogenous product prices and product designs of the competitors are assumed and therefore simultaneous position-price equilibria can be found. A simultaneous position-price equilibrium can now be represented as a strategy vector (p^*, x^*). A Nash equilibrium is found when Eq. (8) applies for all strategies (p_j, x_j).

$$\Pi_j((p_j^*, x_j^*), (p_{-j}^*, x_{-j}^*)) \geq \Pi_j((p_j, x_j), (p_{-j}^*, x_{-j}^*)) \quad j \in J \quad (8)$$

Here (p_j^*, x_j^*) is the current strategy of competitor j, (p_{-j}^*, x_{-j}^*) are the strategies of all competitors except competitor j.

Because of the discrete nature of the variables a sequential tatonnement process is applied to find the Nash equilibrium of one exists. Therefore Eqs. (3)–(6) are solved for each competitor in a predetermined order. Every competitor decides on the best strategy based on the current position and price decisions of the other competitors. This process will be repeated until either a Nash equilibrium is found or a maximal number of iterations is reached (cf. Steiner and Hruschka 2000).

4 Empirical Application and Comparisons

In this sections we want to present our findings comparing the previous game theoretic approach of Steiner and Hruschka (2000) with the game theoretic approach considering stochastic partworth functions. To be able to compare the results we used the same data as Steiner and Hruschka (2000) with the product category being sneakers with two attributes (price and cushioning), as well as partworth utilities for three customer segments. Each attribute has five levels with price having feasible levels 90, 110, 130, 150, and 170 in DM and cushioning having feasible levels of STA = Base System, AIR = Air-cushioning, HEX = Hexalite-cushioning, GEL = Gel-cushioning, and SEL = Self-adjustment. Furthermore the individual attribute level costs for the non-price attribute, i.e. cushioning are assumed to be identical for all competitors. The data is shown in Table 2. Note that we updated the monetary unit from DM to Euro (€). We used this data as basis to calculate the stochastic parthworths. For our simulation we assumed the same distribution for both attributes price and cushioning, in this case a normal distribution with expected value β_{ikl} and standard deviation 0.7. The normal distribution was chosen to illustrate that even though little change is most likely it can have effects on the calculated equilibria

Table 2 Segment-level partworth utilities and individual attribute level costs

Attributes	Price (in €)					Cushioning				
Level	90	110	130	150	170	STA	AIR	HEX	GEL	SEL
Segment A	0.000	−1.186	−1.927	−2.611	−3.119	1.082	0.896	0.634	0.377	0.000
Segment B	0.000	−0.237	−0.441	−0.675	−0.878	0.000	0.495	0.792	1.045	1.760
Segment C	0.000	−0.316	−0.643	−2.651	−3.319	0.000	0.893	1.344	0.779	0.137
Individual attribute level costs (in €)										
	–	–	–	–	–	52	86	89	92	114

Table 3 Stochastic segment-level partworth utilities

Attributes	Price (in €)					Cushioning				
Level	90	110	130	150	170	STA	AIR	HEX	GEL	SEL
Segment A1	0.344	−2.093	−2.673	−2.950	−1.036	0.900	1.158	0.191	0.159	−0.248
Segment A2	0.605	−1.326	−1.232	−2.423	−4.151	1.206	0.135	0.712	0.294	−0.030
⋮	⋮	⋮	⋮	⋮	⋮	⋮	⋮	⋮	⋮	⋮
Segment B1	−0.291	−1.306	−1.336	−0.413	−0.536	−0.234	1.219	0.242	2.105	0.918
Segment B2	0.137	0.085	−0.508	−0.633	−0.445	−0.500	1.052	−0.253	0.772	2.358
⋮	⋮	⋮	⋮	⋮	⋮	⋮	⋮	⋮	⋮	⋮
Segment C1	0.612	0.865	−1.008	−2.626	−1.880	−0.410	0.644	1.116	1.441	−0.026
Segment C2	−1.810	−0.293	−0.698	−2.001	−4.032	0.589	−0.531	0.854	1.295	0.157
⋮	⋮	⋮	⋮	⋮	⋮	⋮	⋮	⋮	⋮	⋮

and a standard deviation of 0.7 was chosen so that there is a reasonable chance of changes happening. Note that now the price utility is no longer monotonically decreasing which could possibly be explained by enhancing customers' perception of the product's quality relative to the selling price (e.g., Dodds et al. 1991). To calculate estimations of β_{ikl}^{*} of Eq. (2) the segments were divided into 20 part segments and random numbers from our given normal distribution were generated to fill the segments. The numbers were generated in a matter where the estimator for β_{ikl}^{*} for the corresponding segments equals β_{ikl}. All segment values Q_i were assumed to be equal.

We generated 25 datasets out of the original data and calculated the equilibria for the same settings as Steiner and Hruschka (2000). Table 3 showcases parts of one dataset of the partworth utilities generated where β_{ikl}^{*} is normally distributed. In the following we will illustrate the results for our example.

In Table 4 the results for our example from Table 3 are shown for the setting of two competing companies. The left column shows which segments were considered for the calculations. New equilibria were found on two separate occasions, considering only segment A or considering segment B and C. In the other cases either the same equilibria were found (segment B) or former inefficient equilibria were eliminated (segment C, segments AB, and segments BC). Table 5 shows the results for the setting of three competing companies.

In the setting of three competing companies we can see that for segments B and C as well as for segments AB and AC different equilibria were found. In the segment BC the inefficient equilibrium was eliminated and only in segment A both approaches find the same equilibrium. Now also different product offerings (segments B, AB, and AC) can be found in between the competing companies.

Table 4 Comparison of Steiner and Hruschka and stochastic partworth approach

	Steiner and Hruschka (2000)		Stochastic partworth approach	
Setting	Company 1	Company 2	Company 1	Company 2
A	90 STA	90 STA	130 STA	130 STA
			110 STA	110 STA
			90 STA	90 STA
B	170 STA	170 STA	170 STA	170 STA
	170 GEL	170 GEL	170 GEL	170 GEL
	170 SEL	170 SEL	170 SEL	170 SEL
C	130 STA	130 STA	130 STA	130 STA
	130 HEX	130 HEX		
AB	170 STA	170 STA	170 STA	170 STA
	150 STA	150 STA		
	130 STA	130 STA		
AC	130 STA	130 STA	130 STA	130 STA
	110 STA	110 STA		
BC	130 STA	130 STA	170 STA	170 STA
	130 HEX	130 HEX		

Table 5 Comparison of Steiner and Hruschka and stochastic partworth approach

	Steiner and Hruschka (2000)			Stochastic partworth approach		
Setting	Company 1	Company 2	Company 3	Company 1	Company 2	Company 3
A	90 STA	90 STA	90 STA	90 STA	90 STA	90 STA
B	170 SEL	170 SEL	170 SEL	170 SEL	170 SEL	170 GEL
C	130 HEX	130 HEX	130 HEX	130 STA	130 STA	130 STA
				130 HEX	130 HEX	130 HEX
AB	130 STA	130 STA	130 STA	130 STA	130 STA	170 STA
	110 STA	110 STA	110 STA			
	90 STA	90 STA	170 SEL			
AC	90 STA	90 STA	90 STA	110 STA	110 STA	130 STA
BC	130 STA	130 STA	130 STA	130 STA	130 STA	130 STA
	130 HEX	130 HEX	130 HEX			

In the previous work only the segments AB, AC and BC were considered for four competing companies. The results can be seen in Table 6. For all segments different equilibria were found and the variety of products offered increased as three different products are offered in all but one equilibrium.

Table 6 Comparison of Steiner and Hruschka and stochastic partworth approach

Setting	Steiner and Hruschka (2000)				Stochastic partworth approach			
	Co. 1	Co. 2	Co. 3	Co. 4	Co. 1	Co. 2	Co. 3	Co. 4
AB	90 STA	90 STA	170 SEL	170 SEL	110 STA	110 STA	130 STA	170 AIR
					90 STA	90 STA	130 STA	170 SEL
AC	90 STA	90 STA	90 STA	90 STA	90 STA	90 STA	130 STA	130 HEX
BC	130 HEX	130 HEX	130 HEX	170 SEL	130 STA	130 STA	130 STA	170 HEX
					130 HEX	130 HEX	130 STA	170 GEL

5   Conclusions and Outlook

Our example showed that by considering the possibility of changing customer preferences different equilibrium solutions were found despite the fact that only small changes in preferences were most likely. The other 24 generated stochastic segment-level partworth utilities showed similar results as different equilibrium solutions were found or inefficient ones were eliminated. Therefore companies thinking long-term need to consider changing customer preferences while designing new products otherwise they could lose profits. Also, in the equilibria found more distinguished products are now possible because of the different scenarios considered.

Future challenges are to accurately predict changes in customer preferences and also to consider product lines and other possible competitive market situations, like leader and followers leading to a Stackelberg–Nash equilibrium. Also, in three of our experiments no equilibriums were found for certain segments. This needs to be investigated further as it is assumed that sequential tatonnement processes always converge (Steiner 2010).

References

Albritton, M. D., & Mcmullen, P. R. (2007). Optimal product design using a colony of virtual ants. *European Journal of Operational Research, 176*, 498–520.

Baier, D., & Gaul, W. (1999). Optimal product positioning based on paired comparison data. *Journal of Econometrics, 89*(1–2), 365–392.

Baier, D., & Gaul, W. (2007). Market simulation using a probabilistic ideal vector model for conjoint data. In A. Gustafsson, A. Herrmann, & F. Huber (Eds.), *Conjoint measurement: Methods and applications* (pp. 47–66). Berlin: Springer.

Balk, B. M. (1989). Changing consumer preferences and the cost-of-living index: Theory and nonparametric expressions. *Journal of Economics, 50*, 157–169.

Choi, S. C., & Desarbo, W. S. (1993). Garne theoretic derivations of competitive strategies in conjoint analysis. *Marketing Letters, 4*(4), 337–348.

Dodds, W. B., Monroe, K. B., & Grewal, D. (1991). The effects of price, brand, and store information on Buyers' product evaluations. *Journal of Marketing Research, 28*, 307–19.

Gaul, W., Aust, E., & Baier, D. (1995). Gewinnorientierte Produktliniengestaltung unter Berück-sichtigung des Kundennutzens. *Zeitschrift für Betriebswirtschaft, 65(8)*, 835–855.

Green, P. E., & Kieger, A. M. (1987). A simple heuristic for selecting 'good' products in conjoint analysis. *Advances in Management Science, 5*, 131–153.

Green, P. E., & Kieger, A. M. (1997). Using conjoint analysis to view competitive interaction through the customer's eyes. In G. S. Day & D. J. Reibstein (Eds.), *Wharton on dynamic competitive strategy* (pp. 343–368). New York: Wiley.

Green, P. E., & Rao, V. R. (1971). Conjoint measurement for quantifying judgmental data. *Journal of Marketing Research, 8*, 355–363.

Gutsche, J. (1995). Produktpräferenzanalyse. *Ein modelltheoretisches und methodisches Konzept zur Marktsimulation mittels Präferenzerfassungsmodellen.* Berlin: Duncker & Humblot

Henard, D. H., & Szymanski, D. M. (2001). Why some new products are more successful than others. *Journal of Marketing Research, 38(3)*, 362–375.

Hoch, S. J., & Loewenstein, G. F. (1991). Time-inconsistent preferences and consumer self-control. *Journal of Consumer Research, 17*, 492–507.

Sorescu, A. B., & Spanjol, J. (2008). Innovation's effect on firm value and risk: Insights from consumer packaged goods. *Journal of Marketing, 72*, 114–132.

Steiner, W. J. (2010). A Stackelberg-Nash model for new product design. *OR Spectrum, 32*, 21–48.

Steiner, W. J., & Baumgartner, B. (2009). Spieltheoretische ansätze in der conjointanalyse. In D. Baier & M. Brusch (Eds.), *Conjointanalyse: Methoden-Anwendungen-Praxisbeispiele, 2009* (pp. 183–198). Heidelberg: Springer.

Steiner, W. J., & Hruschka, H. (2000). Conjointanalyse-basierte Produkt(linien)gestaltung unter Berücksichtigung von Konkurrenzreaktionen. *OR Spektrum, 22*, 71–95.

Teichert, T., & Shehu, E. (2010). Investigating research streams of conjoint analysis: A bibliometric study. *BuR - Business Research, 3*, 49–68.

Urban, G. L., Weinberg, B. D., & Hauser, J. R. (1996). Premarket forecasting of really-new products. *Journal of Marketing, 60*, 47–60.

Wittink, D. R., Vriens, M., & Burhenne, W. (1994). Commercial use of conjoint analysis in Europe: Results and critical reflections. *International Journal of Research in Marketing, 11*, 41–52.

Key Success-Determinants of Crowdfunded Projects: An Exploratory Analysis

Thomas Müllerleile and Dieter William Joenssen

Abstract Crowdfunding, a process with which enterprises or individuals seek to secure project funding, has received much attention recently, not only from the media. The boon in visibility provided to crowdfunding by Internet platforms has made securing project funding, by soliciting pledges from potential donors, simpler than ever. A popular way of allocating funding, and thus bypassing traditional venture capital providers, is by setting a reserve pledge-sum. If this pledge-sum is achieved, the promised pledges are collected from the project supporters. Upon project completion, these pledgers receive a compensation, which is usually non-monetary and based on the magnitude of their contribution. Projects funded in this way include a wide topic variety, ranging from hardware manufacturing to fine arts and even disaster relief. This study investigates possible key success factors for attaining the reserve pledge-sum. To this end, data on 45,400 crowdfunding campaigns was collected and key success factors were analyzed using the results of a logistic-regression. The results indicate that communications and professionalism have a high impact on funding success, and that such communication measures as having a unique website set a minimum standard. Further conclusions allow practitioners to positively influence the campaign outcome and researchers to build upon the results of this study.

1 Introduction

Innovators struggle to secure adequate funding for their projects. Traditionally, this funding is provided by venture capitalists, banks, share-holders, or philanthropists. However, securing these funds remains difficult, because the aforementioned groups are constituted of few people. A possible alternative, for innovators, is to directly ask prospective buyers for project funding. Crowdfunding (CF), a financing scheme utilizing this decentralized approach, has received much attention in recent years due to the important benefits it offers. These benefits include independence from

T. Müllerleile (✉) • D.W. Joenssen
Ilmenau University of Technology, Helmholtzplatz 3, 98693 Ilmenau, Germany
e-mail: Thomas.Muellerleile@TU-Ilmenau.de; Dieter-William.Joenssen@TU-Ilmenau.de

© Springer-Verlag Berlin Heidelberg 2015
B. Lausen et al. (eds.), *Data Science, Learning by Latent Structures, and Knowledge Discovery*, Studies in Classification, Data Analysis, and Knowledge Organization, DOI 10.1007/978-3-662-44983-7_24

said venture capitalists, early tests for market demand, and the possibility to build a close relationship with prospective clients.

CF, which is used in different commercial and noncommercial domains, enables innovators to reduce the risks linked to the development and market introduction of an innovation. Successful project financing no longer hinges on engaging a few, powerful intermediaries, but on engaging many people who can directly support projects with small amounts of money. Furthermore, project initiators will receive direct feedback from the crowd, and may hence better estimate their idea's market potential. CF dynamics, as well as geographic crowd dispersion, enables project initiators to overcome financing barriers and utilize globalization for successful financing. For instance, since its inception in 2009, more than 988 million dollars have been pledged, for more than 130,000 projects in 13 different categories from more than 5.5 million people, on the current market leader of CF platforms, kickstarter.com (Kickstarter 2013).

Even though CF has existed for several years, little attention has been paid in literature to quantitative key success factors. The investigation of these factors has been neglected in favor of studies utilizing qualitative methods (cf. Ordanini et al. 2011) or formulating conceptual models (cf. Belleflamme et al. 2011). To ameliorate this neglect, this study investigates the impact of certain factors on CF project success. A common definition of CF and an overview of existing studies, which focus on CF, is presented in Sect. 2. Section 3 details the exploratory analysis performed on a collected sample of 45,400 projects, while Sect. 4 presents not only insights and actionable recommendations, but also an outlook for further investigations that could be preformed on similar data sets.

2 Crowdfunding Projects

The following subsections give an overview of research on CF to date. In Sect. 2.1 a comprehensive definition for CF is derived from literature, while Sect. 2.2 details potential success factors and the guiding research question developed from further literature and the definition.

2.1 Definition of Crowdfunding

Identifying factors influential on CF project success, not only requires processing available data, but also theoretical considerations, which are greatly facilitated by the availability of a common definition for CF. A comprehensive CF definition will result in a well-defined research object.

The two definitions of CF available in literature are given by Ley and Weaven (2011) and Belleflamme et al. (2011). Ley and Weaven (2011, p. 86) consider CF from a venture capital perspective and define it as a "(…) source of start-up equity capital pooled via small contributions from supporting individuals collaborating through social media." Viewing CF from this perspective necessarily constrains the spectrum of the definition. Nonetheless, the crowd in any CF context is neither limited to individuals, nor are these individuals limited to collaboration through social media. While word-of-mouth does play an important role in advertising CF projects, collaboration between supporters is not in any way a prerequisite for funding a project. These shortcomings are ameliorated by the more general definition of Belleflamme et al. (2011, pp. 5–6). They consider CF "(…) an open call, essentially through the internet, for the provision of financial resources either in form of a donation or in exchange for some form of reward and/or voting rights." However, this definition also falls short in some of CFs key aspects.

Quite correctly, Belleflamme et al. (2011) state that pledgers either donate their support or receive material or non-material rewards (e.g., voting rights) in return for their support. However, this support must not be of financial nature. Especially projects requiring community involvement seek supporters to pledge their time or other non-monetary resources, such as access to land or machinery (cf. Cellan-Jones 2013).

Just as the support offered by pledgers may be non-monetary in nature, the primary motivation of project initiators may not be to secure funding. While CF does suggest that funding acquisition is of primary interest, contextual objectives of project initiators may be different. CF may be utilized to assess an idea's market potential and to build customer relationships. The former may be determined through the amount and size of pledges, even if they are not sufficient to fund the project, and the latter may be achieved through the communication forum offered by the project. This possibility for feedback may be used to establish customer relationships and reputation, not only at a project, but also at product level.

A further shortcoming of the Belleflamme et al. (2011) definition is that it constrains the concept of CF to an online context. Admittedly, considering the whole internet, not only social media, broadens Ley's and Weaven's (2011) view of CF, but also a CF campaign may also be conducted offline without substantially changing the nature of the project. The Internet simply facilitates communication and thus should not constrain the definition of CF.

Beyond these shortcomings, the current definitions also fail to broach two essential subjects. First, while the call for CF is open, it is not open-ended. The time frame for promoting a project and raising resources may be set freely by the project initiators, but must be constrained. Second, the chosen payout scheme for the pledged resources is an important aspect of any CF project. Whether a "threshold pledge model" (Hemer 2011) is chosen, pledges are always payed-out or stretch goals are defined, the chosen payout scheme influences project initiator and pledger behavior. ·

Considering all aforementioned elements of CF, including those of definitions by Ley and Weaven (2011) and Belleflamme et al. (2011), the following definition of CF may be formulated:

> Crowdfunding is a process where commercial or non-commercial projects are initiated in a public announcement by organizations or individuals to receive funding, assess the market potential, and build customer relationships. Pledgers may then contribute individual amounts of monetary or non-monetary resources, during a specified time-frame, using offline or online campaign platforms that utilize different payout schemes, in exchange for a product specific or unspecific, material or immaterial reward.

2.2 Potential Success Factors

Despite its growing importance for innovators, consumers and researchers alike, the topic of CF remains relatively unexplored in literature to date. Rather than examining CF based on empirical data, the vast majority of past work has focused on conceptual models describing CF from a qualitative perspective, to provide a theoretical background. These literature streams concentrate either on the financial aspects, donation features or innovation economic facets of CF.

Pope (2011) offers insight into the legal problems accompanying the financial funding process. The legal problems stem from the fact that small offerings, below $1,000, are "over-regulated" by the SEC to prevent fraud. Ley and Weaven (2011) discuss CF from a venture capitalist perspective. They address agency dynamics at work and its associated problems in CF. Further, mechanisms such as project screening, to adequately manage prospective CF projects, are identified. Wojciechowski (2009) investigates how online platforms could support calls for donations. He considers the "threshold pledge model" (Hemer 2011, p. 15) payout scheme, which only pays out if a certain threshold is exceeded, a key benefit for social, donation driven projects. Belleflamme et al. (2011) shed light on the CF phenomenon from a micro-economic standpoint. Based on a theoretical model and assumptions, they deduce managerial implications and recommend "equity-share" style CF for larger projects and "pre-order" style CF for smaller projects.

The few empirical investigations, available in literature, focus on heterogeneous topics. For example, Agrawal et al. (2011) show that the local and distant crowd differ in terms of when they decide to fund the project. Ordanini et al. (2011) reveal that behavior patterns of the crowd differ, depending on the project category. Kuppuswamy and Bayus (2013) also investigate pledgers and project initiator behavior. They show that potential pledgers feel responsible to contribute to a project that has not received much support and stipulate that update-frequency, from the project initiators, increases towards the end of a funding period. Factors influencing funding success have, so far, been neglected in the context of CF, and may thus be of interest to researchers and practitioners alike.

On the basis of the CF definition and the preceding discussion, research questions may be readily generated by considering aspects of the research object. As is clear, from the definition, an archetype CF project's success is not only driven by the requested funding. The ability to build customer relationships and how much competition is on the campaign platform are also deciding factors. Further, the rewards offered for pledging, how long pledging may be performed, and how active the campaign platform is during this time differentiate one CF project from another. Thus, the research question driving the analysis is as follows:

> How do the aforementioned factors, of an archetype CF project, influence the funding success probability?

3   Empirical Study

The following subsections detail the exploratory analysis performed on the collected data to answer the previously defined research question. Sections 3.1 and 3.2, respectively, describe the sample and the analysis performed. The final Sect. 3.3 discusses the results of the logistic-regression-model.

3.1   Sample Description

The current market leader in online CF campaign platforms, kickstarter.com, was selected as a data source to answer the research question. Data on a total of 45,400 projects were collected between May 16th and 19th 2013. This publicly available data on kickstarter.com was extracted by a custom web crawler. Variables collected included the requested and pledged funding amount, the funding period length in days, the associated project website, the number of updates performed by the project initiators, the amount of comments made by the contributors, the levels at which contributors could pledge, and the funding period start date. Upon completion, necessary data transformation was performed to yield variables suitable for statistical analyses. For example, by comparing requested and pledged funding amounts, it was inferred whether or not the project is funded successfully. Other transformations included inferring whether or not a given project website is unique among all websites in the data and how many other projects were initiated on the same day and category.

After data collection and transformation, it was deemed necessary to constrain the analysis to projects initiated in the past 14 months (cf. Fig. 1). This constraint was implemented due to the increased attention CF received at that time, which resulted in a structural change in the market. Based on this constrained data

Fig. 1 Funding periods
initiated

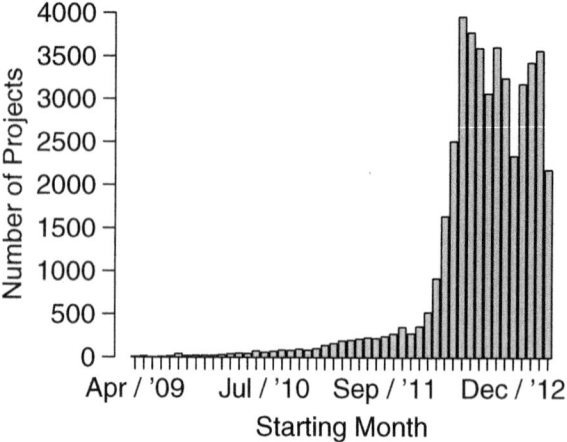

Fig. 2 Funding in % of goal

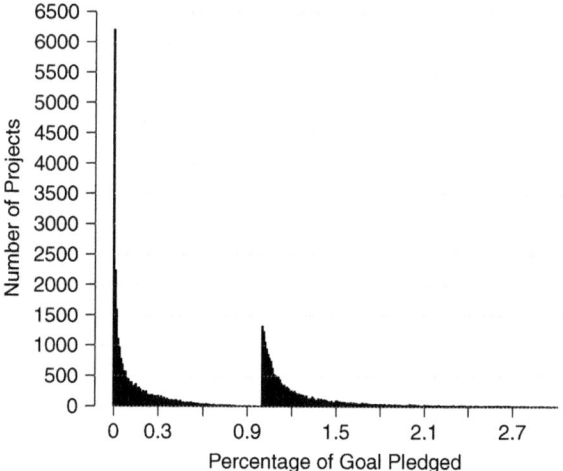

set, further data cleaning was performed. First, four projects with missing data, namely missing funding period length, were eliminated from the sample. Further, projects deemed outliers, i.e., those not representative of normal projects, were detected and eliminated. Projects eliminated included those with excessive funding wishes and excessive amounts of comments. Additionally, projects having no updates, comments, and pledged amounts are considered orphaned projects, and thus removed. Application of these criteria yielded 37,726 projects for analysis, of which 48.3 % are successfully funded (cf. Fig. 2).

3.2 Analysis

Answering the chosen research question requires determining how the independent variables influence the probability of successful project funding. This was performed via logistic-regression-analysis.

Regressors are chosen as follows: the number of comments made by pledgers (x_1) represents how active the campaign is; the number of different levels at which contributors may pledge (x_2) are the rewards offered for pledging; the number of updates made by the project initiators (x_3) represents the efforts made in customer relationship building; the amount of funds requested by the project initiators (in thousand USD, x_4) indicate the funding request; how many other projects were initiated on the same day in the same category (x_5) is used as a measure of competition on the platform; whether the named project website is unique (x_6) also represents efforts in building customer relationships; how long the funding period was set (x_7) is indicative of the influence that the chosen time frame has. The rational for the regressor selection is threefold. First and foremost, these variables are in line with the arguments set forth in Sect. 2.2. Second, all variates may readily be influenced by project initiators, thus, any recommendations made upon the results will be actionable by practitioners. Finally, further regressors were not retrievable given the technical constraints imposed by kickstarter.com.

As with any regression analysis, multicollinearity may pose a serious challenge for coefficient estimation in logistic-regression. Literature on the subject suggests calculating the tolerance for each independent variable and investigating the Pearson correlation between the independent variables (Menard 1995). For the tolerance, 0.2 is the threshold below which multicollinearity is a cause for concern; the Pearson correlations must not be too large. Since tolerances for all independent variables are well above the threshold of 0.2 and the Pearson correlations are all "small," as shown in Table 1, problems associated with multicollinearity should not be an issue for coefficient estimation.

Table 1 Pearson correlations and tolerances for the independent variables

	x_1	x_3	x_3	x_4	x_5	x_6	x_7	Tol
x_1 ~comments	1	0.211	0.466	0.112	−0.118	0.011	0.038	0.763
x_2 ~pledge levels		1	0.346	0.131	0.000	0.088	0.071	0.855
x_3 ~updates			1	0.011	−0.077	0.045	0.039	0.713
x_4 ~goal				1	−0.020	0.046	0.115	0.956
x_5 ~uniqueness on start day					1	−0.062	−0.002	0.980
x_6 ~unique website						1	0.010	0.987
x_7 ~funding period length							1	0.983

3.3 Results

Interpreting logistic-regression results involves a three-step approach. After the assessment of coefficient significance and model fit, coefficients must further be interpreted. The relationship between the independent and dependent variables is nonlinear in the logistic-regression-model.

The parameter estimation results, given in Table 2, indicate that the effects of all chosen variables on the expected success probability are highly significant ($p \approx 0$). These results hold when using either the Wald z-statistic or testing the deviance from the null-model for each parameter, using a chi-squared test.

Model fit, in the context of logistic-regression, is considered based on deviance to the null-model. Various measures of model fit exist, prominent among them are McFadden's (1973) and Nagelkerke's (1991) R^2 statistics. McFadden (1978, p. 307) states that a pseudo-R^2 of more than 0.2 indicates a good fit and over 0.4 indicates an excellent fit. Nagelkerke's R^2 is standardized to between zero and one and thus rules of thumb for regular regression analysis may be applied. McFadden's and Nagelkerke's R^2 for the model are 0.336 and 0.496, respectively, indicating a very good model fit. These measures are complimented by a predictive accuracy of about 79.7 %, indicating that further explanatory variables with significant influence may exist.

Coefficient interpretation for the current case is not only hampered by the non-linear link function between the independent variables and the dependent variable but also by the mixture of metric and dichotomous variables. Thus, a comparison in change of success probability due to a marginal change in an independent variable offers valuable insight (Long 1997). To this end, changes in success probability due to a variation of factors are shown in Table 3. Median project values, changed by one marginal unit, ceteris paribus, are one comment, eight pledge levels, two updates, a goal of five thousand dollars, 16 projects initiated on the same day in the same category, the project website is unique, and a funding period length of 30 days having a predicted success probability of 46.54 %.

Table 2 Logistic-regression results

	Coefficient	Wald z-statistic	Deviance statistic
(Intercept)	−0.336	−6.525***	−
x_1 ~comments	0.060	28.185***	2807.9***
x_2 ~pledge levels	0.018	5.836***	138.5***
x_3 ~updates	0.294	66.062***	7467.0***
x_4 ~goal	−0.092	−51.981***	6238.7***
x_5 ~uniqueness on start day	0.013	14.698***	177.4***
x_6 ~unique website	0.451	16.845***	285.2***
x_7 ~funding period length	−0.027	−20.763***	450.5***

Significance codes: *** $p < 0.001$; ** $p < 0.01$; * $p < 0.05$; $p < 1$

Table 3 Discrete change in the success probability for the logit-model (in ppts)

	Median case	-1	$+1$
$x_1 \sim$ comments	1	-1.50	1.51
$x_2 \sim$ pledge levels	8	-0.44	0.44
$x_3 \sim$ updates	2	-7.20	7.34
$x_4 \sim$ goal (in thousand USD)	5	2.30	-2.28
$x_5 \sim$ uniqueness on start day	16	-0.33	0.33
$x_6 \sim$ unique website	1	-10.86	–
$x_7 \sim$ funding period length	30	0.66	-0.66

Changes are computed with other values held constant

As the values indicate, changes in success probability are nearly linear around the median project. Changes in the success probability due to a change in the independent variables are substantially different. While adding one meaningful update increases success probability by about 7.3 percentage points (ppts) to 53.88 %, not providing a unique website for the project causes a reduction by about 11 ppts to 35.68 %. Other influential factors are the amount of comments elicited from pledgers and the set goal. Here, increasing the set goal by one thousand dollars reduces the success probability by 2.3 ppts, while one more comment will increase the probability of success by 1.5 ppts. The influence of the remaining factors is too small to substantially change the success probability.

4 Conclusions

The goal of this paper is to not only contribute theory to CF literature, which is scant to date, but also to provide empirical evidence of which factors are critical to CF success. These contributions are twofold. First, and most notably, this study offers a comprehensive definition of CF, including various dimensions neglected by previous definitions, from which it is built upon. Second, success factors are identified using logistic-regression-analysis.

The analysis shows that the most important success factors for the model fit are the number of performed updates, the set financing goal, and the number of comments. Less important, but nonetheless influential, are the chosen funding period length and the availability of a unique project website. The amount of competition on the launch date and the number of pledge levels offered are of least importance for the model fit. This ranking of success factors sheds light on which factors are of primary importance for project initiators.

First, it is of utmost importance for project creators to communicate with pledgers and potential pledgers. If sufficient content for the creation of additional updates is available, these should be performed. This could be planned prior to project initiation, in form of a communication strategy. Nonetheless, it cannot be advised to simply perform an update as an end to itself. Even splitting an update cannot be guaranteed to increase the success probability.

Second, funding goals should be set realistically. Surely, reducing the required funds makes achieving financing easier, and thus financing goals should be set aggressively. Nevertheless, this recommendation must also be considered critically. Setting a project goal to low will hurt project plausibility. Thus, a point of inflection may exist for each project, where further reducing the project goal will reduce the success probability.

Third, the importance that the amount of generated comments holds indicates that a promising communication strategy must be in place for the project. The project must appear active and facilitate communication between backers in a visible forum, to be more successful. Fundamentally important is a certain degree of professionalism. This is especially apparent in that most projects have a unique project website. Projects not featuring their own website or one on a social media platform have a substantially lower success probability.

Interesting, in a different sense, is the minor influence that the amount of pledge levels have. In a sense, pledge levels offer means for price-discrimination or - differentiation. Economic theory dictates that price-differentiation has a positive influence on sales, and thus contributes positively to a project success. The role and strength of this factor may thus require further research.

However, some limitations are worth noting. First, the study did, due to space constraints, not consider whether project success factors behave differently between the project categories. Second, only one CF platform was regarded. Results again may differ for different online or offline forums. Third, only directly quantifiable variables were used. Other, not directly measurable, success factors, such as technical maturity, feasibility, and uniqueness of the idea itself and the initiators reputation, could be considered in further studies. Fourth, only funding success probability is investigated. Additionally, other measures of success, such as stretch goal achievement or over-funding, could be considered. Finally, in future work, other aspects of CF, such as whether a project is completed, on time or at all, could be investigated.

References

Agrawal, A., Catalini, C., & Goldfarb, A. (2011). *The geography of crowdfunding*. Working Paper No. 10-08, NET Institute.

Belleflamme, P., Lambert, T., & Schwienbacher, A. (2011). *Crowdfunding: Tapping the right crowd*. CORE Discussion Paper, 2011/32.

Cellan-Jones, R. (2013). *Fast fibre: A community shows the way*. www.bbc.co.uk/news/ technology-21442348. Accessed 28 July 2013.

Hemer, J. (2011). *A snapshot on crowdfunding*. Working Papers Firms and Region No. R2/2011, Frauenhofer Institute for Systems and Innovation Research.

Kickstarter (2013). *Kickstarter stats*. www.kickstarter/help/stats. Accessed 21 February 2014.

Kuppuswamy, V., & Bayus, B. L. (2013). *Crowdfunding creative ideas: The dynamics of project backers in kickstarter.* Working Paper. dx.doi.org/10.2139/ssrn.2234765. Accessed 28 July 2013.

Ley, A., & Weaven, S. (2011). Exploring agency dynamics of crowdfunding in start-up capital financing. *Academy of Entrepreneurship Journal, 17*, 85–110.

Long, J. S. (1997). *Regression models for categorical and limited dependent variables.* Thousand Oaks: Sage.

Mcfadden, D. (1973). Conditional logit analysis of qualitative choice behavior. In P. Zarembka (Ed.), *Frontiers in econometrics* (pp. 105–142). New York: Academic Press.

Mcfadden, D. (1978). Quantitative methods for analyzing travel behaviour of individuals: Some recent developments. In D. Hensher & P. Stopher (Eds.), *Behavioural travel modelling* (pp. 279–318). London: Croom Helm London.

Menard, S. W. (1995). *Applied logistic regression analysis.* Thousand Oaks: Sage.

Nagelkerke, N. J. D. (1991). A note on a general definition of the coefficient of determination. *Biometrika, 78*, 691–693.

Ordanini, A., Miceli, L., Pizzetti, M., & Parasuraman, A. (2011). Crowdfunding: Transforming customers into investors through innovative service platforms. *Journal of Service Management, 22*, 443–470.

Pope, N. (2011). Crowdfunding microstartups: It's time for the securities and exchange commission to approve a small offering exemption. *Journal of Business Law, 13*, 101–129.

Wojciechowski, A. (2009). Models of charity donations and project funding in social networks. *Computer Science, 5872*, 454–463.

Preferences Interdependence Among Family Members: Case III/APIM Approach

Adam Sagan

Abstract The purpose of the paper is to identify the preference structures in the framework of actor-partner interdependence (APIM) model based on paired-comparison or ranking data (Thurstone Case III/V model). The households preferences of the income allocation between consumption, savings and investments are considered. Then, the preference structures among the families are identified on the basis of Thurstonian Case III preference model. Latent preferences are used to modeling the actor-partner interdependencies between the household members.

1 Thurstonian Models of Income Allocation Preferences

1.1 Strategies of Income Allocation Among Household Members

In contemporary marketing, and behavioral economics, the households behavior and preference models are developed taking into account two broad perspectives.

The first one represents an "atomic view" on the household behavior where families or households represent single units of action. Preference formation can be analyzed on the household level only. In most microeconomic studies, the household is treated unitarily as an individual. Therefore, the family decision-making strategy is embedded with altruistic, unified or "benevolent dictator" model of consumer behavior where the "head of the family" represents the common, collective, cooperation-based preferences of such a single unit of decision maker (Chiappori 1988; Apps and Rees 1993; Browning and Chiappori 1998).

The second and much modern perspective stresses the "molecular view" in which the interplay between family member's preferences takes place. Decision-making strategy is rooted in two broad classes of models: (1) cooperative-bargaining (McElroy and Horney 1981; Manser and Brown 1980) and (2) non-cooperative

---

A. Sagan (✉)
Cracow University of Economics, Rakowicka 27, 31-510 Cracow, Poland
e-mail: sagana@uek.krakow.pl

© Springer-Verlag Berlin Heidelberg 2015
B. Lausen et al. (eds.), *Data Science, Learning by Latent Structures, and Knowledge Discovery*, Studies in Classification, Data Analysis, and Knowledge Organization, DOI 10.1007/978-3-662-44983-7_25

bargaining (Lundberg and Pollak 1993). In these competitive-rivalry models the individuals can influence the preference formation within the household (Doss 1996). The understanding of households' preferences of income (or in general—scarce resources) allocation strategies is therefore one of the key problems in marketing, consumer behavior, economics of households, and family decision making. In the literature, several research approaches models of intrahousehold preference formation have been proposed (Lazear and Michael 1988; Lise and Seitz 2011; Commuri and Gentry 2005). However, the typologies and classification schemes of income allocation strategies are rarely explicated.

In order to fill this gap, the research on preference and values with respect to consumption, savings and investments of Polish households supported by National Scientific Center (research grant UNO-2011/01B/HS4/04812) was undertaken. In the process of operationalization of savings and investment decisions two criteria were selected and provided to the respondents: (a) security (savings instruments like bank accounts are regarded as more secure in comparison with investment instruments), (b) goals (savings for the future consumption and investments for the future wealth). The data was based on the representative sample of 410 Polish households and 1,100 individual respondents (in each household up to three interviews with parents and the oldest child were conducted). The households were chosen on the basis of multistage area sample. On the basis of literature review and preliminary qualitative studies, four basic strategies of income allocation have been identified:

1. sequential,
2. branched,
3. forked,
4. parallel.

These four types of the income allocation strategies are depicted in Fig. 1.

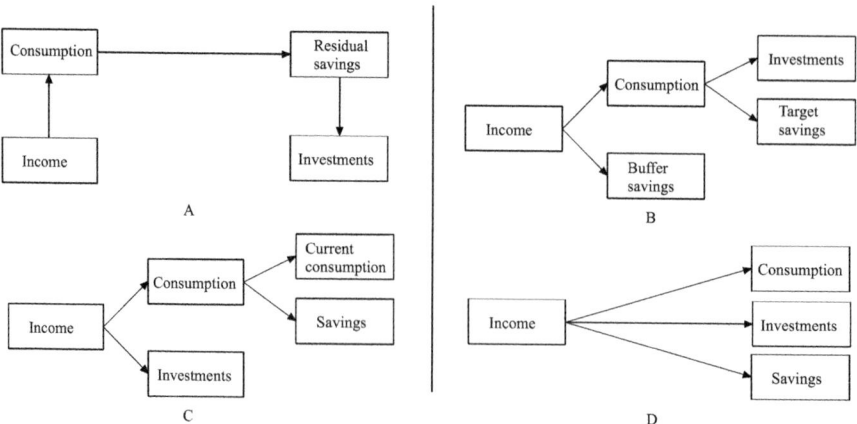

Fig. 1 The strategies of income allocation

In the sequential strategy (A), income is divided a priori into the current consumption and residual unintended savings that are regarded as a buffer savings for the unexpected expenditures. The residual existing remains of income and savings are finally devoted to the investments. Sequential strategy is used in low income and low education families as well as among household members with high hedonistic values.

In branched strategy (B) the scarce resources are divided primarily into consumption and savings, then the latter can be partly invested or devoted to deferred consumption or target savings. Branched strategy is popular in the families with moderate income and low propensity to consume. Also, the households members with high risk aversion tend to use this kind of income allocation.

Forked strategy (C) is based on splitting income first into consumption and investments. In the next step, the consumption is divided into current and future, so the savings are regarded just as means for increasing the future consumption. The forked strategy is characteristic for high income families with high propensity to consume with relatively low risk aversion.

Parallel strategy (D) means a division of income into three predefined sets, where savings are treated both as means of future consumption and as ends of buffer savings.This kind of strategy is common among high income families and consumers with the balanced and integrated values.

Strategy selection depends not only on individual psychographic, social and demographic characteristics of the family members, but also on economic factors on the households level (i.e., stage of the family life cycle, wealth, and well being).

1.2 Thurstonian Preference Models

The measurement of preferences usually involves choice-based and conjoint-like approaches where the outcome variables are situated on nominal or ordinal level of measurement. Well-known latent variable models that are developed for solving that problem are: (1) integrated choice and latent variable models (ICLV) (Temme et al. 2008), and (2) Thurstonian models for preference measurement using ranking and paired-comparisons data.

Preference measurement is based on Thurstonian family (Case V, Case III, and Thurstone–Takane model) of discrete latent utility model of paired comparisons of A, B, C, and D strategies (Maydeu-Olivares 2003; Maydeu-Olivares and Böckenholt 2005).

In the first step, the selected preference models were estimated and examined:

1. Unrestricted model[free parameters (except identification constraints)],
2. Thurstone Case III model (independent preference model with no correlations between latent preferences),
3. Thurstone Case V model (with equality restrictions of preference variances),

4. Thurstone–Takane model (paired comparisons with intransitive preference model with random errors).

In unrestricted preference model, one of the item means is fixed to 0 (as a reference item), all covariances involving the last utility are fixed to 0 and the variances of the first and the last variable are fixed to 1.

In Case III Thurstonian model one of the item means is fixed to 0, all covariances are fixed to 0 and the last latent variable variance is set as 1.

The most restrictive case V model has one of the item means set as 0, all covariances between factors are fixed to 0 and all variances of the latent variables are set to 1. All of the three models are estimated and compared with respect to total sample, and wives and husbands sub-samples.

The goodness of fit of these models is depicted in Table 1.

The case III model as a measurement model of preferences was used for subsequent analysis. Population-based index (Root Mean Square Error of Approximation—RMSEA) is acceptable and depicts the close fit to population covariance matrix. Comparative fit index (CFI) shows that actual model is much better than independent null model. There is no significant difference between unrestricted and Case III model. Both of them have relatively good fit. χ^2 difference test is insignificant between these two models. However, the more restricted Case V model has unacceptable fit and has been rejected. The Case III model parameters are shown in Table 2.

Table 1 Cross-sectional analysis of preferences

Goodness of fit	General	Wives	Husbands
Unrestricted model			
Chi-square, df., *p*-level	11.19, 13, 0.59	24.10,13, 0.03	14.81,13, 0.31
RMSEA, CFI	0.000, 1.00	0.046, 0.976	0.019, 0.996
Case III model			
Chi-square, df., *p*-level	19.05, 15, 0.21	26.36, 15, 0.03	17.98, 15, 0.26
RMSEA, CFI	0.016, 0.99	0.043, 0.98	0.023, 0.99
Case V model			
Chi-square, df., *p*-level	293.78, 17, 0.00	141.22, 18, 0.00	123.70, 18, 0.00
RMSEA, CFI	0.118, 0.77	0.129, 0.738	0.124, 0.755

Case III model is selected

Table 2 Case III model parameters

Parameters	Strategy	General	Wives	Husbands
Means	Sequential (A)	0.32	0.23	0.28
	Branched (B)	0.49	0.44	0.41
	Forked (C)	0.48	0.45	0.41
Variance	Sequential (A)	0.43	0.50	0.36
	Branched (B)	0.13	0.12	0.10
	Forked (C)	0.18	0.16	0.17

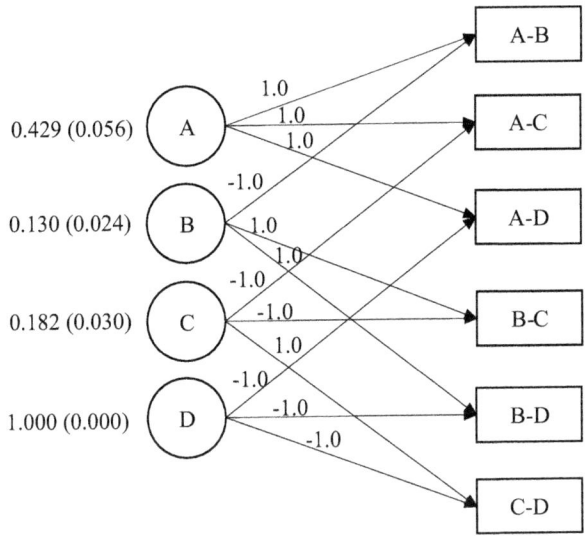

Fig. 2 CASE III Thurstonian model

The parameters in Table 2 represent the means and variances of latent preferences estimated in Case III model. Four latent variables represent the unobserved preferences and six indicators correspond to combination of pairs in paired-comparison of allocation strategies (four indicators give six combinations of pairs).

Factor loadings are fixed to 1 or −1 depending on which item dominates in each pair. Figure 2 displays the structure of Case III model. The model parameters indicate relatively similar preferences among family members. The most preferred allocation strategy is branched (B) and forked (A) with a slight differences across family members. The least preferred strategies are sequential and parallel (as a reference strategy of income allocation).

2 Actor-Partner Models of Households' Preferences

2.1 Structure of API Models

Family decision making and preference formation causes the problem of preferences' interdependencies among family/household members (Simpson et al. 2012). Modeling interdependent consumer preferences may involve the network approach [i.e., spatial autoregressive discrete-choice model (Yang and Allenby 2003; Rossi et al. 2005)]. However, in order to explain the mutual relationships of the households members we adopt the Actor-Partner Interdependence Model (APIM) based on paired-comparison/ranking data.

The APIM on the individual level involves dyadic relations between distinguished pairs (mother–father), that are explained by individual level predictors (i.e., buying role of household's member). The objective of the model is to diagnose the actor and partner effects (Kenny et al. 2006; Cook and Kenny 2005; Olsen and Kenny 2006). On the family level, the individual preferences are explained by the group level variable (i.e., stage of family life cycle).

The structure of APIM preference model consists of five types of variables:

1. Y_{ij}—propensity to spend money (of wives and husbands) that is the dependent (outcome) variable,
2. WA–WC—wives allocation strategy preferences (preference distance of strategy A, B, and C from the strategy D),
3. HA–HC—husbands allocation strategy preferences (preference distance from strategy D),
4. FLC3–FLC4—stages of family life cycle (as group-level variable),
5. Gender (effect-coded qualitative variable).

The identification of the effects in APIM models is based on the two approaches: (1) interaction APIM and (2) Two-intercept model (Kashy and Donnellan 2012).

In the first approach (interaction APIM), mixed regression is used to test the main effect of gender (G) and the interaction with actor (XA × G) and partner (XP × G) effects. The significant interaction terms check the hypothesis that whether the dyad members are truly distinguishable in their actor and partner effects. Level I model (fixed effect) equations are given below:

$$Y_{ij} = b_{0j} + b_{1j}\mathrm{XA}_{ij} + b_{2j}\mathrm{XP}_{ij} + b_{3j}G_{ij} + b_{4j}\mathrm{XA}_{ij}G_{ij} + b_{5j}\mathrm{XP}_{ij}G_{ij} + e_{ij}, \quad (1)$$

where:

Y_{ij}—outcome variable
b_{0j}—intercepts for dyad j
b_{1j}—coefficient for actor's predictor (actor effect)
b_{2j}—coefficient for partner's predictor (partner effect)
b_{3j}—coefficient for gender effect
b_{4j}—coefficient for actor and gender interaction
b_{5j}—coefficient for partner and gender interaction
e_{ij}—error term

Level II equation (random effects) shows the variability of intercepts given in equations on upper level. This type of mixed regression model is also known as multilevel random intercepts-as-outcome model. Models with random slopes are rarely used because of very small number of units in each cluster (family).

$$b_{0j} = a_0 + d_j, b_{1j} = c_0, b_{2j} = b_0, b_{3j} = k_0, b_{4j} = m_0, b_{5j} = p_0 \quad (2)$$

where:

b_{0j}—intercept
c_0—actor effect
b_0—partner effect
k_0—average gender difference with respect to outcome variables
m_0—actor effect difference as a function of gender
p_0—partner effect difference as a function of gender

In the second approach (two-intercept APIM), two dummy variables for wives (W) and husbands (H) are introduced, and model has two intercepts—both for wives and husbands. The two-intercept model is given by following equation (there is no error term in the model):

$$Y_{ij} = b_{1j}W_{ij} + b_{2j}H_{ij} + b_{3j}W_{ij}XA_{ij} + b_{4j}H_{ij}XA_{ij} + b_{5j}W_{ij}XP_{ij} + b_{6j}H_{ij}XP_{ij}, \quad (3)$$

where:

Y_{ij}—outcome variable
b_{1j}—intercepts for wives in dyad j
b_{2j}—intercepts for husbands in dyad j
b_{3j}—coefficient for actor effects for wives
b_{4j}—coefficient for actor effects for husbands
b_{5j}—coefficient for partner effects for wives
b_{6j}—coefficient for partner effects for husbands

2.2 Actor-Partner Model of the Preferences Among Polish Households

In the first step of modeling the interplay between preferences, the χ^2 difference test of empirical distinguishability was used to check whether the members of dyads are empirically distinguishable. The data has form of a pairwise structure, where both long and wide format of data arrangement were used for APIM, multilevel modeling. In the pairwise data structure the cases are doubled and actor's scores are repeated for partner's respective variables.

The result shows that $\chi^2 = 25.118(6)$, with p-level $= 0.003$. Intraclass correlation coefficients for wives and husbands are 0.06 (wives) and 0.07 (husbands), respectively. This means that there is not enough variability on level I that is explained by level II variation (ICC expresses the ratio of between cluster variance to total variance). The proportion of the total variance attributed to household level (clustering of partners) is relatively small.

In the first step, the mixed regression model was used for an identification of actor and partner effects (without covariates on the second level).

Fixed effects equations for husbands and wives are as follows [Eqs. (4) and (5)]:
(1) Wifes propensity to spend money (Y_{wj})

$$Y_{wj} = -109.4 + 0.26\text{AA} + 0.12\text{AP} - 201.4\text{G}^* + 0.65\text{AA} \times \text{G}^*$$
$$+0.05\text{AP} \times \text{G} + 1.02\text{BA} + 0.47\text{BP} + 2.55\text{BA} \times \text{G}^* + 0.19\text{BPG}$$
$$+0.53\text{CA} + 0.25\text{CP} + 1.30\text{CA} \times \text{G}^* + 0.09\text{CP} \times \text{G} \qquad (4)$$

The fixed-effect model shows that only gender (G) and actor effects (AA × G, BA × G and CA × G) have significant regression coefficients (*). It means that only wives allocation strategy preferences have positive influence on her own propensity to spend. The branched strategy of wives has the strongest influence on their own propensity to spend money.
(2) Husbands propensity to spend money (Y_{hj})

$$Y_{hj} = -118.4 + 1.09\text{AA} + 0.14\text{AP} + 192.16\text{G}^* - 0.63\text{AA} \times \text{G}^*$$
$$-0.03\text{AP} \times \text{G} + 1.09\text{BA} + 0.53\text{BP} - 2.40\text{BA} \times \text{G}^* - 0.14\text{BP} \times \text{G}$$
$$+0.57\text{CA} + 0.27\text{CP} - 1.17\text{CA} \times \text{G}^* + 0.07\text{CP} \times \text{G} \qquad (5)$$

The description of the remaining variables in the models is given below:

1. AA—actor effect with respect to strategy A,
2. AP—partner effect with respect to strategy A,
3. BA—actor effect with respect to strategy B,
4. BP—partner effect with respect to strategy B,
5. CA—actor effect with respect to strategy C,
6. CP—partner effect with respect to strategy C,
7. G—gender,

The results of husbands' allocation strategies are similar to wives. Also, only the actor effects are significant (*) and branched strategy has positive influence for husbands' own propensity to spend. In both cases no partner effects are significant, so wives/husbands allocation strategy does not influence partners' (husbands/wives) propensity to spend money. Because of the effect coding of gender, the signs of parameters coefficients have an opposite directions.

Random effects equation for husbands and wives is given in Eq. (6):

$$\tau_{00\text{Wdec}} = 0.53^*, \, \tau_{00\text{Hdec}} = 0.54^* \qquad (6)$$

The variability of intercepts for wives and husbands is significant, so it can be explained by level II factors. For group-level analysis, the Bayesian multilevel structural equation model was adopted (Mplus 7.1 package was used for the estimation). Bayesian (and also robust maximum likelihood) estimation was applied because of a relatively small sample size and informative priors concerning normal distribution of the Thurstonian preferences.

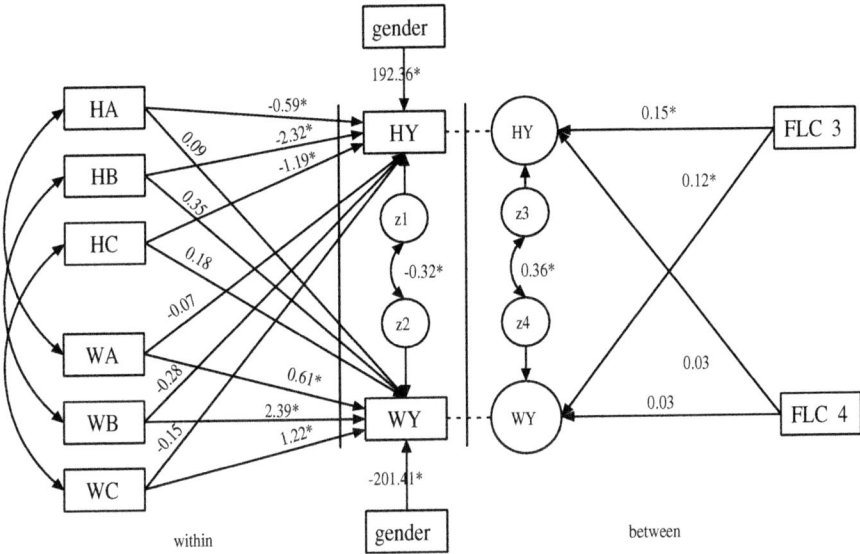

Fig. 3 The households strategies of income allocation

The Thurstonian case III models served as a measurement model of the preferences in the APIM analysis. The scale values (factor scores) of the latent preferences for total sample, and wives and husbands sub-samples were the input data during the model estimation.

Figure 3 presents the multilevel APIM model with the dummy-coded stages of family life cycle as a predictors on "between" level. The model is saturated. Significant positive correlation (0.66) between wives/females and husbands' sequential strategy of income allocation and strong negative correlation (−0.95) between parallel strategies of partners were observed.

The results show the significant and positive actor (wives/females decision making role) effects on propensity to spend. Additionally, group-level predictor (3rd stage of the family life cycle—the families with mature children) has significant influence on spending propensity in the families.

3 Summary

The results indicate that the splitting income strategies (branched and forked) are the most preferred strategy of income allocation among Polish households. It may be caused by relatively high expectation of economic insecurity (both for low- and high-risk avoiders).

No partner effects are significant, so it may suggest that there is no adaptive expectation in preference strategy formation. Integrated Case III/APIM models allow for explanation of the interdependencies between consumer/household preferences with the latent variable framework and extension of classic ICLV models (choice based data) to general preference model with latent variables assuming joint actions of household members in the institutional (higher level) context of the analysis.

References

Apps, P. F., & Rees, R. (1993). *Labor supply, household production and intra-family welfare distribution*. Working Papers in Economics and Econometrics, Canberra, Australia: Australian National University, 248.

Browning, M., & Chiappori, P. A. (1998). Efficient intrahousehold allocation: A characterisation and tests. *Econometrica, 66*(6), 1241–1278.

Chiappori, P. A. (1988). Rational household labor supply. *Econometrica, 56*(1), 63–90.

Commuri, S., & Gentry, J. W. (2005). Resource allocation in households with women as chief wage earners. *Journal of Consumer Research, 32*(2), 185–195.

Cook, W. L., & Kenny, D. A.(2005). The actor-partner interdependence model: A model of bidirectional effects in developmental studies. *International Journal of Behavioral Development, 29*(2), 1001–1094.

Doss, Ch. (1996). Testing among models of intrahousehold resource allocation. *World Development, 24*(10), 1597–1609.

Kashy, D. A., & Donnellan, M. B. (2012). Conceptual and methodological issues in the analysis of data from dyads and groups. In K. Deaux & M. Snyder (Eds.), *The Oxford handbook of personality and social psychology* (pp. 209–238). Oxford: Oxford University Press.

Kenny, D. A., Kashy, D. A., & Cook, W. L. (2006). *Dyadic data analysis*. New York: Guilford Press.

Lazear, E. P., & Michael, R. T., (1988). *Allocation of income within the household*. Chicago: University of Chicago Press.

Lise, J., & Seitz, S. (2011). Consumption inequality and intra-household allocations. *Review of Economic Studies, 78*, 328–355.

Lundberg, S., & Pollak, R. (1993). Separate spheres bargaining and the marriage market. *Journal of Political Economy, 101*(6), 988–1010.

Manser, M., & Brown, M. (1980). Marriage and household decision-making: A bargaining approach. *International Economic Review, 21*(1), 31–44.

Maydeu-Olivares, A. (2003). On Thurstone's model for paired comparisons and ranking data. In H. Yanai, A. Okada, K. Shigematu, Y. Kano & J.J. Meulman (Eds.), *New Developments in Psychometrics* (pp. 519–526). Tokyo: Springer.

Maydeu-Olivares, A., & Böckenholt, U. (2005). Structural equation modeling of paired-comparison and ranking data. *Psychological Methods, 10*(3), 285–304.

Mcelroy, M., & Horney, M. J. (1981). Nash-bargained household decisions: Toward a generalization of the theory of demand. *International Economic Review, 22*(2), 333–349.

Olsen, J., & Kenny, D. A. (2006). Structural equation modeling with interchangeable dyads. *Psychological Methods, 11*(2), 127–141.

Rossi, P. E., Allenby, G., & McCulloch, R. (2005). *Bayesian statistics and marketing*. West Sussex: Wiley.

Simpson, J. A., Griskevicius, V., & Rothman, A. J. (2012). Bringing relationships into consumer decision-making. *Journal of Consumer Psychology, 22*, 329–331.

Temme, D., Paulssen, M. & Dannewald, T. (2008). Incorporating latent variables into discrete choice models. *A Simultaneous Estimation Approach Using SEM Software, Business Research, 1*(2), 220–237.

Yang, S., & Allenby, G. M. (2003). Modeling interdependent consumer preferences. *Journal of Marketing Research, 40*(3), 282–294.

Part V
Data Analysis in Biostatistics and Bioinformatics

Evaluation of Cell Line Suitability for Disease Specific Perturbation Experiments

**Maria Biryukov, Paul Antony, Abhimanyu Krishna, Patrick May,
and Christophe Trefois**

Abstract Cell lines are widely used in translational biomedical research to study
the genetic basis of diseases. A major approach for experimental disease modeling
are genetic perturbation experiments that aim to trigger selected cellular disease
states. In this type of experiments it is crucial to ensure that the targeted disease-
related genes and pathways are intact in the used cell line. In this work we are
developing a framework which integrates genetic sequence information and disease-
specific network analysis for evaluating disease-specific cell line suitability.

1 Introduction

Cell lines are widely used for perturbation experiments that aim to understand disease mechanisms at a cellular scale. Cells used in such experimental setups are only
rarely a *sine qua non* biological model for the disease of interest. Most commonly,
genetic or environmental perturbations are required to create a cell fate which can
serve as experimental disease model. Prior to the era of whole genome sequencing
it was often acceptable to mimic a disease state by specifically perturbing a single
gene of interest. The understanding of diseases as network perturbations (Zhu et al.
2007; del Sol et al. 2010) and especially as genetic network perturbations, however,
requires considering not only single genes of interest but also whole disease
networks. It is known that many cell lines are carrying major genetic variations
which would be lethal for humans already at the level of prenatal development. The
main advantage of highly proliferative cell lines in comparison with primary cells
and induced pluripotent stem cells is a drastically better capacity for experiments
that require big amounts of cells with identical genetic background, such as in
high throughput experiments, proteomics, and metabolomics. The availability of
big amounts of cells with identical genetic background also opens the possibility

M. Biryukov (✉) • P. Antony • A. Krishna • P. May • C. Trefois
Luxembourg Centre for Systems Biomedicine, University of Luxembourg, Esch-sur-Alzette,
Luxembourg
e-mail: maria.biryukov@uni.lu; paul.antony@uni.lu; abhimanyu.krishna@uni.lu;
patrick.may@uni.lu; christophe.trefois@uni.lu

© Springer-Verlag Berlin Heidelberg 2015
B. Lausen et al. (eds.), *Data Science, Learning by Latent Structures,
and Knowledge Discovery*, Studies in Classification, Data Analysis,
and Knowledge Organization, DOI 10.1007/978-3-662-44983-7_26

for comparative perturbation experiments aiming to compare phenotypic outputs derived from a set of single node perturbations (Gonçalves and Warnick 2009). The interpretation of such experiments can strongly profit from the assumption that the compared systems differ from each other by only a single genetic variation.

The main caveat for an educated design of perturbation experiments in highly proliferative cell lines is the need to identify intact modules of genetic disease networks. In this study, the module is considered to be intact when all its genetic elements are free of damage in the cell line. Perturbations such as knock-down applied to genes within defective network modules can be expected to produce different phenotypic outputs than identical perturbations applied to intact network modules. While a module integrity check can serve as a guide for knock-down experiments, it can also serve as a guide for rescue experiments. In the rare scenario where a genetic variation of the cell line mimics a disease specific genetic defect, a repair of the genetic defect with tools such as Zinc finger nucleases or transcription activator-like effector nucleases (TALENs) (Joung and Sander 2013) could be used to restore a healthy control state via generating a new cell line.

Here we introduce a network analysis methodology for the evaluation of cell line suitability in the context of perturbation experiments targeting diseases and disease specific processes. For illustrating the methodology, we evaluate SH-SY5Y in the context of neurodegenerative diseases and provide a specific example where the biological question is whether the cell line SH-SY5Y is an appropriate experimental model for RNA interference screening targeting mitochondrial dysfunction in Parkinson's disease. Indeed, this neuroblastoma cell line is very commonly used in neurodegenerative disease research and differentiation protocols allow performing high throughput experiments in neuron-type cells (Encinas et al. 2000; Korecka et al. 2013). However, to our knowledge the integrity of neurodegeneration related genetic networks has not yet been evaluated in a consistent manner in this cell line. The methodology that we used to evaluate the integrity of genetic networks underlying the selected diseases and processes builds upon four components:

- *Whole genome sequencing of the cell line* which provides the list of genes carrying mutations or associated with chromosome regions with non-diploid copy number.
- *Text mining* of full text articles and abstracts discussing application of SH-SY5Y, from which a list of diseases studied with the cell line is inferred.
- *OMIM*, a catalog of human genes and genetic disorders, from which we get the list of genes, associated with the respective diseases.
- *Protein–protein interaction repository*, i.e. STRING (Franceschini et al. 2013), which is used to build a high level network model of the disease or process in question.

We combine information from the above components in order to quantitatively assess the impact of genes damaged in the cell line in the context of a selected disease network. This assessment will help the experimental biologist to estimate how appropriate the cell line is for the questions under study. Our methodology is developed on the SH-SY5Y cell line, but can easily be adapted to other cell lines, given that more and more cell lines will be sequenced in the future. We believe that this methodology can help researchers worldwide to make an educated choice of experimental cell culture models for translational research. In the following section we describe each of the components in more detail.

1.1 *Whole Genome Sequencing of SH-SY5Y*

Whole-genome sequencing (WGS) of the undifferentiated SH-SY5Y cell line was performed by Complete Genomics (CG) (Mountain View, CA, USA) using a proprietary paired-end nanoarray-based sequencing-by-ligation technology (Drmanac et al. 2010). The DNA sample sequencing, sequencing data quality control, mapping and variant calling were performed by CG Sequencing reads were mapped against the NCBI build 37. For gene annotations NCBI build 37.2 (RefSeq) was used. Variants were then annotated using the Ingenuity Variant AnalysisTM tool (version 2.1.20130621) and further filtered for deleterious variants. We kept variant calls that are experimentally observed to be associated with one of the following phenotypes: *pathogenic, possibly pathogenic, unknown significance*. We further kept variants with published gain of function, gene fusions, activating mutations inferred by Ingenuity, predicted gain of function by BSIFT, and variants in a microRNA binding site. Additionally, we filtered for exonic variants representing frameshifts, in-frame indels, stop codon changes, missense variants, or variants disrupting splice sites up to 2.0 bases into introns. For each gene we counted the number of deleterious variants per category and total variants. In this study we consider the genes with mutations in gene coding region and the genes associated with the chromosome regions that have abnormal ($\neq 2$) number of copies.

1.2 *Text Mining of SH-SY5Y-Related Literature*

The role of text mining in this study is to identify the diseases that have frequently been studied using the SH-SY5Y cell line. In order to avoid publications that mention but do not study the cell line or describe third-party cell line-based experiments, we aimed at finding articles in which SH-SY5Y occurred in either title/abstract, list of keywords, section headings, or table/figure captions. These portions of research articles are likely to contain principal aspects of its content (McDonald and Hsinchun 2002; Cohen et al. 2010). With this criteria in mind we searched PubMed and PMC archives of the MEDLINE database (NML) and downloaded relevant abstracts. Furthermore, we downloaded full text articles that mentioned "SH-SY5Y" or synonyms from the Elsevier repository, for which we have an authorized access. In order to identify articles that pertain to the cell line in question we implemented a full text parser and retained only those articles that satisfied the relevancy requirements. The resulting collection of 5,353 items was

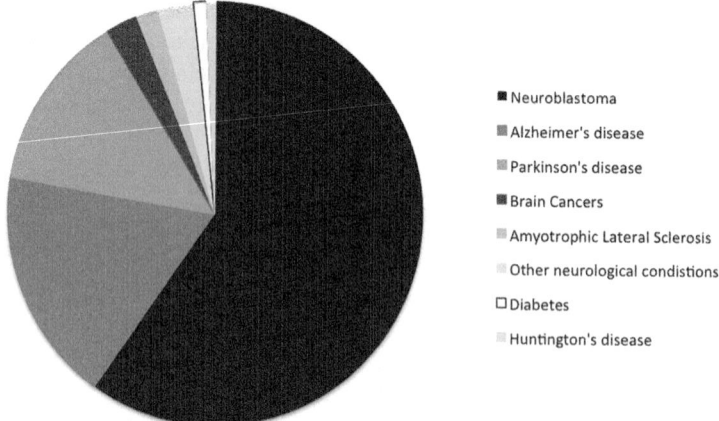

Fig. 1 Diseases, frequently studied with SH-SY5Y cell line

further annotated for disease names with Reflect (Pafilis et al. 2009). Applying the same heuristics as above, we labeled and grouped articles by the corresponding disease name, and calculated disease frequency distributions in the entire collection of SH-SY5Y-related articles. The results are summarized in Fig. 1. Notice that after Neuroblastoma, from which the cell line was obtained, neurodegenerative diseases such as Alzheimer's, Parkinson's, Amyotrophic Lateral Sclerosis and more constitute the next largest class of pathologies studied with the cell line. Although the search of articles was not restricted to neurodegeneration, such publications prevail. Indeed, the biochemical properties of the cell line, which were described in the 1970s, motivated their widespread in neuroscience (Biedler et al. 1978; Cheung et al. 2009; Yusuf et al. 2013).

1.3   Extraction of the Disease Core Genes

In addition to the identification of a cell line specific disease list inferred from text mining, we used Morbid Map of the Online Mendelian Inheritance in Man (OMIM) to associate core genes with each of those diseases. Importantly, the source distinguishes between correlation and causality in terms of genotype/phenotype associations. In order to avoid unconfirmed gene-disease associations, we select only the genes in which mutations have been found and shown to have causative effects on the disorder of interest.

1.4 Expansion of the Core Genes into a Disease-Related Network

Information about the protein–protein interactions is derived from STRING—an up-to-date state-of-the-art database of known and predicted protein interactions, that covers 5,214,234 proteins and 1,133 organisms (Franceschini et al. 2013). Each interaction in STRING is associated with a probabilistic confidence score, derived from a range of evidence sources (i.e., associations in completely sequenced genomes, high-throughput experimental data, curated databases of pathways and post-protein modifications, PubMed abstracts, and freely available PMC articles).

We downloaded STRING in February 2013 and extracted a human-related undirected network with 18,600 nodes and 1,640,707 edges. In order to minimize the chance of inclusion of non-existing or low probable protein–protein interactions, we retained edges with confidence score $C \geq 0.7$ in this study, classified as high level (ibid.), mostly coming from curated databases. Once the edges have been selected, we mapped disease-related genes onto the network and formed a disease-related protein–protein interaction network by expanding the core gene set with their neighbors at distance one. Due to the high node degree (the number of edges incident to the node) of such networks we limited the expansion to the directly connected proteins. Note that instead of representing the entire disease as a network, one can zoom into a specific disease-related cellular process or pathway by extracting a network based on a subset of related genes and their expansion.

2 Cell Line Suitability Scoring

According to our initial assumption about the cell line suitability as a study model, ideally no nodes in the network that represents the process under study should correspond to the mutated genes in the cell line. Therefore we evaluate the adequacy of the cell line by estimating the impact of the mutated genes on the network that represents the process under study. Among the wide range of network analysis metrics we choose the one which is called "Betweenness centrality." Betweenness centrality of a node expresses how much information flows through that node. Betweenness centrality g of a node v in the network G is given by Eq. (1),

$$g(v) = \sum_{s \neq v \neq t} \frac{\sigma_{st}(v)}{\sigma_{st}}, \tag{1}$$

where σ_{st} is the *total* number of shortest paths (also called "geodesics") between the nodes $s, t \in G$, and $\sigma_{st}(v)$ is the number of shortest paths that go through v.

In order to quantitatively assess the cell line suitability we extend the metric from individual node characteristic to the characteristic of node types. First of all we map the genes mutated in the cell line onto the network under study and label all the nodes as either "Damaged" or "Intact." Next we try to assess the impact of the damaged nodes on the network. Intuitively this can be achieved by comparing the overall betweenness centrality scores of the damaged and intact nodes. We call this metric *BC-ratio* μ, which is formally defined as follows:

$$\mu(N_{\text{Disease}}|\text{cell line}) = \frac{\sum_{v \in \text{Damaged}} g(v)}{\sum_{v \in \text{Intact}} g(v)}, \tag{2}$$

where the nominator represents the sum of betweenness centrality scores of all damaged nodes, and the denominator stands for the sum of the betweenness centrality scores of all intact nodes. The higher is the BC-ratio the more influence damaged nodes have on the network.

3   Results and Discussion

Our methodology tries to help biologist in two ways. One is the visual support which allows to display a selected disease or process network along with the involved cell line genes. Figure 2 shows such a network for Neuroblastoma in SH-SY5Y. Black nodes correspond to the genes damaged in the cell line, and light gray nodes correspond to intact genes. The role of the visualization is to give an idea how damaged genes are distributed in the network, what are the genes they are connected to, and how they might alter processes of interest. Another idea consists in comparative assessment of cell lines. The BC-ratio can be used to rank various diseases according to their appropriateness for being studied with the given cell line. At the same time, if more than one (sequenced) cell line is available, one can also use the scores to choose the one which seems to suit best for the study. Here we show SH-SY5Y cell line scoring with regard to five different diseases for which it has been widely used: *Neuroblastoma (NB), Alzheimer's Disease (AD), Parkinson's Disease (PD), Amyotrophic Lateral Sclerosis (ALS), and Huntington Disease (HD)*. Table 1 shows network statistics along with the BC-ratio for each disease.

Diseases are shown in order of increasing BC-ratio (column 2), from Parkinson's to Neuroblastoma. It can be seen that the networks vary in size and involvement of the genes mutated in the cell line (columns 3–6). The BC-ratio shows the impact of the damaged nodes on the network: the higher it is, the more the disease network is affected. The disease with the highest BC-ratio is Neuroblastoma. This is not surprising given that the SH-SY5Y cell line is derived from an NB patient. Here it can be assumed also that the genetic variations of the cell line mimic the genetic defect of the disease. The main question is, however, whether the selected disease specific networks correspond to the healthy state in these cells. The aim is to avoid network states which are affected by the genes damaged in the cell line

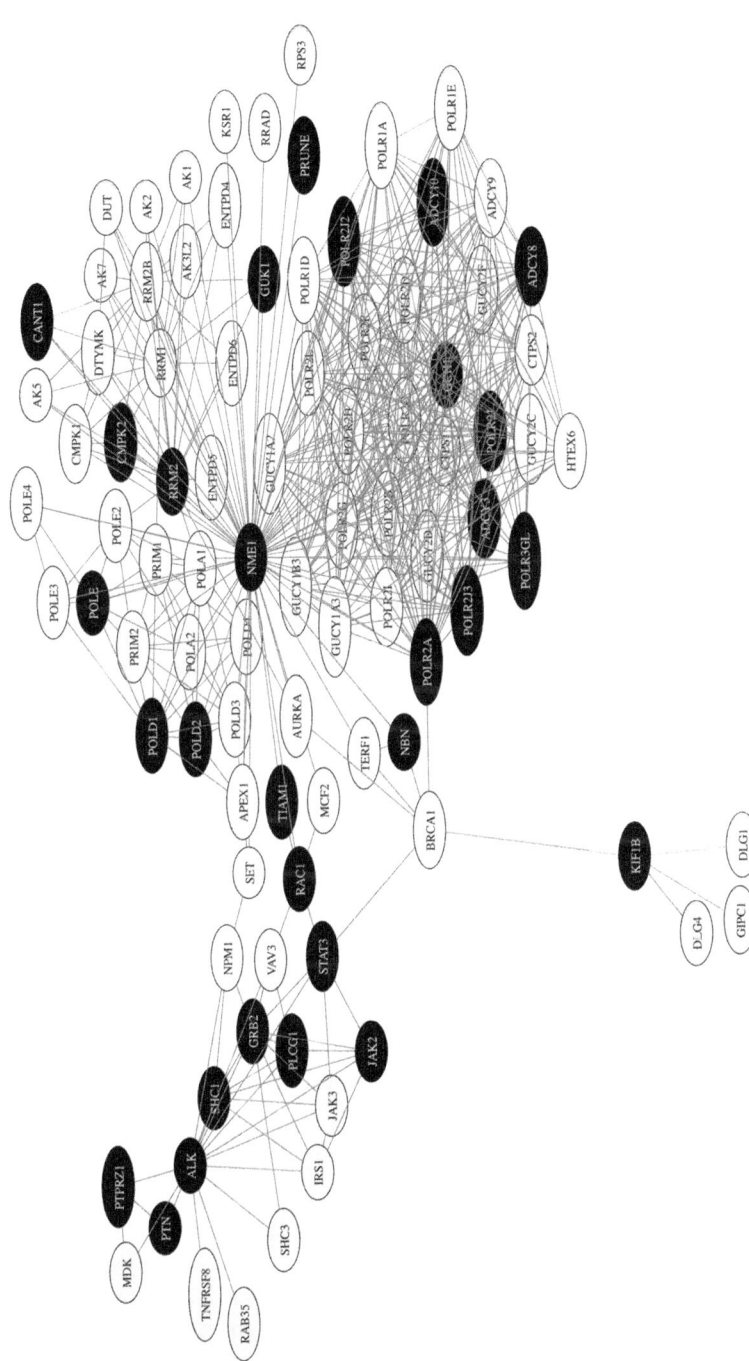

Fig. 2 Neuroblastoma-related network. *Black nodes* correspond to the genes damaged in the cell line. *Light gray nodes* correspond to intact genes.
BC-ratio = 3.28

Table 1 Network statistics and SH-SY5Y scoring with regard to various diseases

Disease name	BC-ratio	Overall nodes	Overall edges	Damaged genes, $g(v) > 0$	Intact genes, $g(v) > 0$	D_σ(Network)
PD	0.22	526	7050	72	329	0.15
AD	0.24	544	15591	58	262	0.04
ALS	0.30	397	7977	60	238	0.60
HD	0.78	63	153	8	22	1.40
NB	3.28	123	1821	34	75	5.40

and thus do not correspond to the healthy state. Unknown mutations could cause a gain of function, a loss of function, or a modified function of the gene product. These uncontrolled parameters could significantly interfere with the interpretation of experiments. In consequence, perturbations applied to the defective network modules may yield different phenotypic outputs than the same perturbations applied to the intact network modules. The lower the BC-ratio, the better is the suitability of the cell line for disease or process-specific genetic perturbation experiments. In contrast, high BC-ratios are not sufficient to conclude that a cell line can be used as disease model as such, without the need for experimental perturbations in order to create a cellular disease state. Indeed, any mutation in the genes of interest would raise the BC-ratio, independent of the functional relationships between sequenced mutations and the disease or process of interest. In the case of SH-SY5Y, we have additional evidence concerning the origin of the cell line (Biedler et al. 1978), and hence, it can be assumed that the high BC-ratio reflects an intrinsic Neuroblastoma cell state.

3.1 Metric Evaluation

We evaluate our BC-ratio metric by producing 1,000 randomized networks N_R for each disease, for which we keep the topology and counts of *Damaged* and *Intact* nodes, but randomize label assignment, and calculate the BC-ratio for each randomization. Then we compute the normalized deviation $D_\sigma(N)$ as:

$$D_\sigma(N) = \frac{E(\mu(N_R)) - \mu(N_{\text{Disease}}|\text{cell line})}{\sigma_{N_R}}, \tag{3}$$

where $E(\mu(N_R))$ and σ_{N_R} are the mean and standard deviation of N_R, and $\mu(N_{\text{Disease}}|\text{cell line})$ is the BC-ratio of the real data as defined by Eq. (2). D_σ(Network) values are shown in the last column of Table 1. They indicate to which extent mutated genes in the cell line correlate with the nodes that have high betweenness centrality in the disease network. We observe that in case of

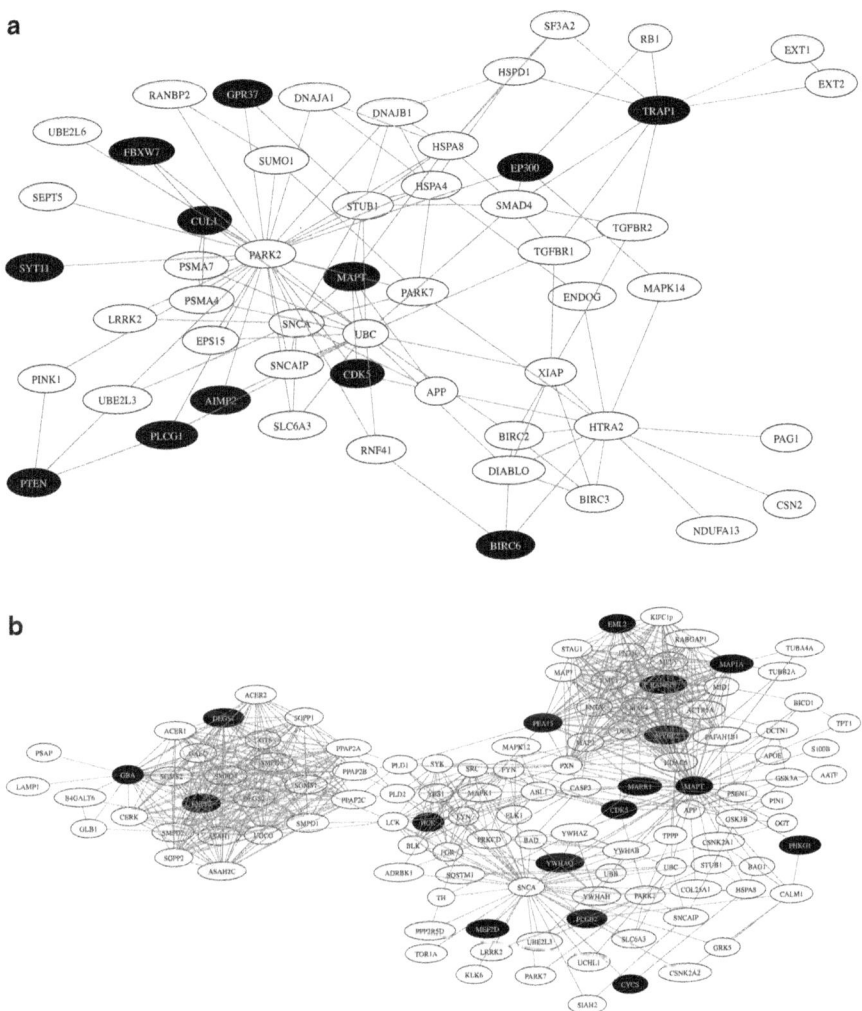

Fig. 3 Process-related networks of Parkinson's disease. Cell line mutated genes are shown in *black*; intact nodes are shown in *light gray* (**a**) Mitochondria function-related network. BC-ratio $= 0.15$. $D_\sigma(N) = 0.65$. (**b**) α-Synuclein misfolding-related network. BC-ratio $= 0.34$. $D_\sigma(N) = 0.9$

Neuroblastoma and Huntington disease D_σ(Network) is > 1 which indicates that nodes with high betweenness-centrality in the disease network significantly correlate with the genes mutated in the cell line. This is not the case for other diseases and PD-related processes (please see Fig. 3a, b for the latter). In such cases priority will be given to the expert's interpretation of how important the damage is with regard to the process to be studied.

4 Case Study

So far we discussed the cell line as model for studying diseases as a whole. However it seems more realistic that one would study a particular process within a disease. Our methodology applies easily for such scenarios. We illustrate them on the example of Parkinson's disease. One process is related to the mitochondrial function (MF) and involves *PINK1, LRRK2, HTRA2,* and *TRAP1.* Another one is based on α-Synuclein misfolding (SM), and involves *MAPT, GBA,* and *SNCA.* The networks are shown in Fig. 3. Note that in the MF network (Fig. 3a) the damaged genes are distributed rather on the periphery, while in the SM network (Fig. 3b), *MAPT, PLD2* and *GBA*—all are damaged in the cell line—are among the nodes with the highest betweenness-centrality. This explains why the BC-ratio for α-Synuclein misfolding (0.34) is higher than BC-ratio of the mitochondrial function network (0.15). In conclusion, our method predicts that in Parkinson's disease, the cell line SH-SY5Y is better suited for perturbation experiments targeting mitochondrial dysfunction than for the same type of study on α-Synuclein misfolding.

5 Conclusions and Future Work

In this study we introduce a novel methodology for cell line evaluation in terms of suitability for experimental modeling of selected diseases or disease-related processes via perturbation experiments. Our approach accounts for the network-based interpretation of the pathogenetic processes and allows one to assess the impact of the cell line genome on the integrity of the network under study. In the example of SH-SY5Y and a set of neurodegenerative diseases, we have shown that the methodology applies to disease-related networks, as well as their subparts which correspond to particular processes. As input our approach requires a genomic characterization of the cell line, a core set of disease-related genes, and a protein–protein interaction network. Therefore it is not restricted to either a specific cell line or disease, and can be applied every time when a gene perturbation experiment is planned.

In the future we plan to extend our approach as to quantitatively express severeness of damages which may be caused by various mutation types and to include directionality information on the network edges. As sequencing data for various cell lines is becoming increasingly available, we plan to systematically apply our method to additional cell-lines and to identify the most appropriate ones for our future experiments.

References

Biedler, J.L., Roffler-Tarlov, S., Schachner, M., & Freedman, L.S. (1978). Multiple neurotransmitter synthesis by human neuroblastoma cell lines and clones. *Cancer Research, 38*, 3751–3757.

Cheung, Y. T., Lau, W.K., Yu, M.S., Lai, C.S., Yeung, S.C., So, K.F., et al. (2009). Effects of all-trans-retinoic acid on human SH-SY5Y neuroblastoma as in vitro model in neurotoxicity research. *NeuroToxicology, 30*, 127–135.

Cohen, K.B., Johnson, H.L., Verspoor, K., Roeder, C., & Hunter, L.E. (2010). The structural and content aspects of abstracts versus bodies of full text journal articles are different. *BMC Bioinformatics, 11*, 492.

del Sol, A., Balling, R., Hood, L., & Galas, D. (2010). Diseases as network perturbations. *Current Opinion in Biotechnology, 21*(4), 566–571.

Drmanac, R., Sparks, A.B., Callow, M.J., Halpern, A.L., Burns, N.L., Kermani, B.G., et al. (2010). Human genome sequencing using unchained base reads on self-assembling DNA nanoarrays. *Science, 327*, 78–81.

Encinas, M., Iglesias, M., Liu, Y., Wang, H., Muhaisen, A., Ceña, V., et al. (2000). Sequential treatment of SH-SY5Y cells with retinoic acid and brain-derived neurotrophic factor gives rise to fully differentiated, neurotrophic factor-dependent, human neuron-like cells. *Journal of Neurochemistry, 75*(3), 991–1003.

Franceschini, A., Szklarczyk, D., Frankild, S., Kuhn, M., Simonovic, M., Roth, A., et al. (2013). STRING v9.1: Protein-protein interaction networks, with increased coverage and integration. *Nucleic Acids Research, 41*, D-80815.

Gonçalves, J., & Warnick, S. (2009). System theoretic approaches to network reconstruction. In *Control theory and systems biology* (1st ed., pp. 262–292). Cambridge: MIT Press.

Joung, K., & Sander, J. D. (2013). TALENs: A widely applicable technology for targeted genome editing. *Nature Reviews Molecular Cell Biology, 14*, 49–55.

Korecka, J.A., van Kesteren, R.E., Blaas, E., Spitzer, S.O., Kamstra, J.H., Smit, A.B., et al. (2013). Phenotypic characterization of retinoic acid differentiated SH-SY5Y cells by transcriptional profiling. *PLoS One, 8*(5), e63862.

Mcdonald, D., & Hsinchun, C. (2002). Using sentence-selection heuristics to rank textsegments in TXTRACTOR. In *Proceedings of the 2nd ACM/IEEE-CS Joint Conference on Digital libraries* (pp. 28–35). New-York: ACM.

NML. (2014). MEDLINE, PubMed and PMC: How they are different? http://www.nlm.nih.gov/pubs/factsheets/dif_med_pub.html

OMIM: Online Mendelian inheritance in man. (2015). McKusick-Nathans Institute of Genetic Medicine, Johns Hopkins University (Baltimore, MD). http://omim.org/

Pafilis, E., O'Donoghue, S.I., Jensen, L.J., Horn, H., Kuhn, M., et al. (2009). Reflect: Augmented browsing for the life scientist. *Nature Biotechnology, 27*, 308–310.

Yusuf, M., Leung, K., Morris, K.J., & Volpi, E.V. (2013). Comprehensive cytogenomic profile of the in vitro neuronal model SH-SY5Y. *Neurogenetics, 14*(1), 63–70.

Zhu, X., Gerstein, M., & Snyder, M. (2007). Getting connected: Analysis and principles of biological networks. *Genes and Development, 21*(9), 1010–1024.

Effect of Hundreds Sequenced Genomes on the Classification of Human Papillomaviruses

Bruno Daigle, Vladimir Makarenkov, and Abdoulaye Baniré Diallo

Abstract The classification of the hundreds of papillomaviruses (PVs) still constitutes a major issue in virology, disease diagnosis, and therapy. Since 2003, PVs are classified within three levels of hierarchical clusters according to their similarity and their position in the phylogenetic tree, using the DNA sequence of the L1 gene. With the increased number of sequenced genomes, the boundaries of the different clusters within the different levels might overlap and the topology of the associated tree could change, thus avoiding a unique and coherent classification. Here, we studied the classification of 560 currently available human PVs (HPV) with respect to the criteria established 10 years ago as well as novel ones. The results highlight that current taxonomic identification does fit with the monophyletic criteria for the L1 gene, but the sequence similarity criteria violates the established boundaries to classify PVs. Finally, we argue that the substitution of L1 gene similarity by the whole genome similarity would allow to have less overlap between the different clusters and provide a better classification.

1 Introduction

The *Papillomaviridae* (PV) virus family is constituted of hundreds of variants infecting vertebrates (Antonsson et al. 2000). About 150 types of PVs infecting humans have been identified as human papillomaviruses (HPV). Among those, 15 types are considered as high risk (Muñoz et al. 2003). For example, the HPV16 and HPV18 groups are associated with 70 % of known cervical cancers. Accurate identification of the HPV type is thus of primary importance for disease diagnosis and therapy (Tota et al. 2011). The virus genome is a double stranded DNA of about 8,000 nucleotides long. It typically consists of the genes coding the capsid proteins, L1 and L2, the transcription and replication genes, E1 and E2, and the transformation process genes, E5, E6, and E7 (Zheng and Baker 2006).

B. Daigle • V. Makarenkov • A.B. Diallo (✉)
Department of Computer Science, Université du Québec à Montréal, P.O. Box 8888, Downtown Station, Montreal, QC, Canada H3C 3P8
e-mail: diallo.abdoulaye@uqam.ca

© Springer-Verlag Berlin Heidelberg 2015
B. Lausen et al. (eds.), *Data Science, Learning by Latent Structures, and Knowledge Discovery*, Studies in Classification, Data Analysis, and Knowledge Organization, DOI 10.1007/978-3-662-44983-7_27

Table 1 L1 gene similarity threshold classification

Cluster	Intra-cluster (%)	Inter-cluster (%)
Genera	>= 60	< 60
Species	>= 70	60–70
Types	>= 90	71–89

The current HPV classification is built solely from the L1 gene, based on DNA sequence similarity, the topological position of the taxon within the phylogenetic tree, and three taxonomic levels: genera, species, and types. Viruses within a genera share over 60 % similarity of their L1 gene DNA sequence; species are below genera and share over 70 % similarity; types are below species and share over 90 % similarity. Table 1 summarizes the similarity criteria.

Several criticisms on the current classification are rising with the sequencing of several hundreds of new strands of PVs that violate the set of previous knowledge on the monophyletic aspect of the PVs according to several kinds of attributes derived from their consequences. Recently, Bernard et al. (2010) also suggested that PV classification should take into account the genomic organization, the biology and the pathogenicity. In fact, the relative similarity measures are sensible to the addition of more data. Hence, it may change the overall classes as well as the phylogenetic relationships between viruses. Rebuilding such a classification necessitates more insight in the class property of the different genomic regions (each gene and the whole genome).

In this paper, we assessed the quality of such a classification included in the HPV taxonomic tree if we incorporate newly sequenced PV taxa. To this end, we extracted 560 HPV genomic sequences from GenBank (Benson et al. 2013) as well as their taxonomic classification based on de Villiers et al. (2004) criteria. The current set of sequences were assessed based on their conformance to the defined similarity thresholds and the topological position in the phylogenetic tree. To analyze the similarity criteria, the HPV classification can be regarded as a hierarchical clustering and one could consider using some of the tens of existing validation measures to estimate its quality, mainly the degree of compactness and separation (Handl et al. 2005; Liu et al. 2010). Unfortunately, external validation measures like the *F-measure* (Rijsbergen 1979) and the *Rand Index* (Rand 1971) or others are not directly helpful because they require a *gold standard* for the comparison. In a sense, the actual classification constitutes itself the gold standard. On the other hand, internal validation measures such as the *Silhouette Index* (Rousseeuw 1987) could be exploited.

To better assess and interpret the incongruences of the classification, we modeled a discrete measure that is similar to the Silhouette index (Rousseeuw 1987). While Silhouette index includes the distance measure for object comparison, the new index will consider discrete comparison (boolean) making it easier to interpret and insensible to sequence distance bias. The new internal validation measure so-called *cohesion index* permits to unambiguously compare the compactness and separation of the clusters constituting the actual HPV classification and to assess the similarity criterion for each genomic region.

2 Methods

2.1 Genome Acquisition, Pairwise Similarities and Phylogenies

To perform the study, we extracted HPV complete genomes from GenBank database (Benson et al. 2013) and PaVE database (Van Doorslaer et al. 2013). We conserved only unique genomes with length between 6 and 10 kb, leaving a total number of 560 sequences containing 150 types, 38 species, and 5 genera (Table 2). The taxonomic tree of the selected genomes was also extracted from GenBank. For each genomic region, a global alignment was produced using the popular MUSCLE program (Edgar 2004). The multiple alignments were corrected with the Gblock program (Castresana 2000). Similarity measures were computed as the ratio of the *number of identical characters* over the *length of the alignment*, for each pair of genomic region globally aligned. We reconstructed the phylogenetic tree of each genomic region with BPV1 (bovine papillomavirus 1) as outgroup using the PhyML program (Guindon and Gascuel 2003) with 128 bootstrap replicates and HKY evolutionary model (Hasegawa et al. 1985).

Two methods of tree comparison were exploited. The first is the well-known Robinson and Foulds (RF) topological distance (Robinson and Foulds 1981) to compute the pair distance between the reconstructed HPV gene trees and their comparison against the whole genome phylogenetic tree. However, due to the multifurcation of the taxonomic tree, its comparison against the inferred phylogenetic trees cannot be done using RF. Hence we designed an algorithm to indicate if a given taxonomic tree agrees with a given phylogenetic tree based on their clustering. This is done by repetitive removal of common subtrees—internal nodes (and their children nodes) that correspond to genera, species, and types according to their height in the tree—based on their taxonomic identification. The algorithm traverses in post-order the internal nodes of the taxonomic tree. If the exact same set of leaves under the node could be found under a node of the phylogenetic tree, then both internal nodes (and all the nodes under) would be removed from the trees. The algorithm returns TRUE when the trees become empty (meaning the trees agree). Otherwise the remaining trees contain the discrepancies indicating that the phylogenetic position of some HPV violates the taxonomic classification (the pseudo code is available in supplementary material S1).

Table 2 Summary of the number of HPV genomic sequences retrieved from GenBank by genera, with number of species and types

	Genera	Species	Types	Sequences
	alpha	13	63	464
	beta	5	45	52
	gamma	17	39	39
	mu	2	2	4
	nu	1	1	1
Total	5	38	150	560

2.2   Cohesion Index

Let T—a tree; $\ell(T)$—the leaves of T; T_i—a subtree of T; subtrees(T)—the set of subtrees of T; $\text{sim}(x, y)$—similarity between leaves x and y; $\sigma(x, y, T_i)$— a discrete value according to whether similarity of x and y is greater than the similarity between x and every other leaves outside T_i:

$$\sigma(x, y, T_i) = \begin{cases} 1 & \text{if } \text{sim}(x, y) > \max_{z \in \ell(T) \backslash \ell(T_i)} \text{sim}(x, z) \\ 0 & \text{otherwise,} \end{cases}$$

factor(x, T_i)—a belonging factor of x to T_i computed as the mean of the σ value of x with respect to every other leaves of T_i:

$$\text{factor}(x, T_i) = \frac{1}{|\ell(T_i)| - 1} \sum_{y \in T_i \backslash x} \sigma(x, y, T_i),$$

then the index of T_i is defined as the mean of the factor of T_i for each of its leaves:

$$\text{index}(T_i) = \frac{1}{|T_i|} \sum_{x \in T_i} \text{factor}(x, T_i).$$

The global index (whole tree) is defined as the weighted mean of the indexes of the subtrees :

$$\text{Global Index}(T) = \frac{\displaystyle\sum_{T_i \in \text{subtrees}(T)} \text{index}(T_i) |\ell(T_i)|}{\displaystyle\sum_{T_i \in \text{subtrees}(T)} |\ell(T_i)|}.$$

Nota: The interval of the index is between 0 and 1. An index of 1 indicates that the similarity between any two elements within a cluster is always greater than the similarity with an element outside.

3   Results

The comparison of the L1 gene pairwise similarity distribution of 189 PVs from Bernard et al. (2010) and those based on the extracted 560 HPVs reveals major discrepancies, as can be seen in Fig. 1. The local minimum expected to split genera around 60 % in the 189 PVs is not present for the 560 GenBank HPVs. Furthermore, we computed the intra-cluster and inter-cluster similarities for each taxa, respectively, within the three levels of clustering (genera, species, and type)

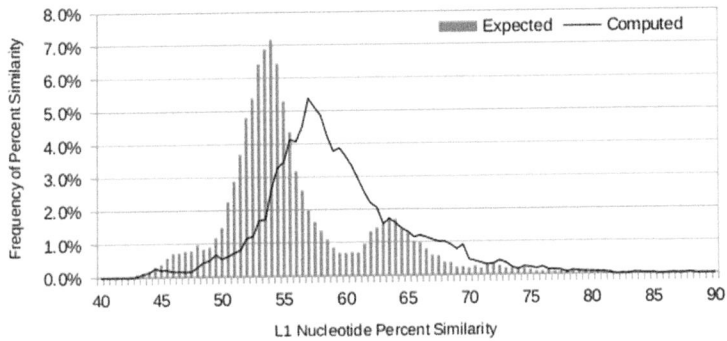

Fig. 1 Distribution of L1 gene similarity between 189 PVs from Bernard et al. (2010) and 560 HPVs

Table 3 Intra-cluster and inter-cluster pairwise similarities exceeding thresholds

Region	Cluster	Measure	Threshold	Min	NbMin	Max	NbMax	NbComp	%
L1	Genera	Intra	60–100	48	5909	–	–	105,254	5.61
		Inter	0–60	43	0	64	7,520	89,248	8.43
	Species	Intra	70–100	57	2,459	–	–	34,452	7.14
		Inter	60–70	48	11,804	72	2,272	141,604	9.94
	Type	Intra	90–100	88	36	–	–	8,716	0.41
		Inter	71–89	57	7,414	88	0	51,472	14.49
Genome	Genera	Intra	53.5–100	44	178	–	–	105,254	0.17
		Inter	0–53.5	43	0	55	218	89,248	0.24
	Species	Intra	62–100	56	2	–	–	34,452	0.00
		Inter	53.5–63	44	356	66	462	141,604	0.58
	Type	Intra	90–100	90	1	–	–	8,716	0.01
		Inter	63–90	56	6	90	2	51,472	0.02

Top half of the table shows results for the L1 gene and bottom half for the whole genome. For each cluster included in the three clustering levels (genera, species, and type), the number of sequences violating the boundaries of a cluster according to its similarities is shown. *Min*—represents the minimum value observed within the comparison, *NbMin*—the number of pairwise comparisons below the lower boundary, *Max*—the maximum value observed, *NbMax*—the number of pairwise comparisons above the higher boundary, *NbComp*—the total number of pairwise comparisons made and the %—the fraction of comparisons violating the boundaries (*NbMin* + *NbMax*)/*NbComp*

and among all other sister clusters (same level in the taxonomic tree). The results are shown in Table 3. These results show that for the L1 gene, more than 5 % of the comparisons violate the intra-cluster or inter-cluster boundaries (except the *Intra-cluster of the Type* with value of 0.41 %, due to almost identical similarities). The minimum observed pairwise values could be 10 % lower than the actual known boundaries. However, the whole genome pairwise similarities exhibit less than 0.6 % for all classes of comparison. It allows the intra-cluster similarities of species and type to violate the class boundaries with less than 0.02 %. These results suggest that the addition of more taxa induces a wide violation of the class boundaries when

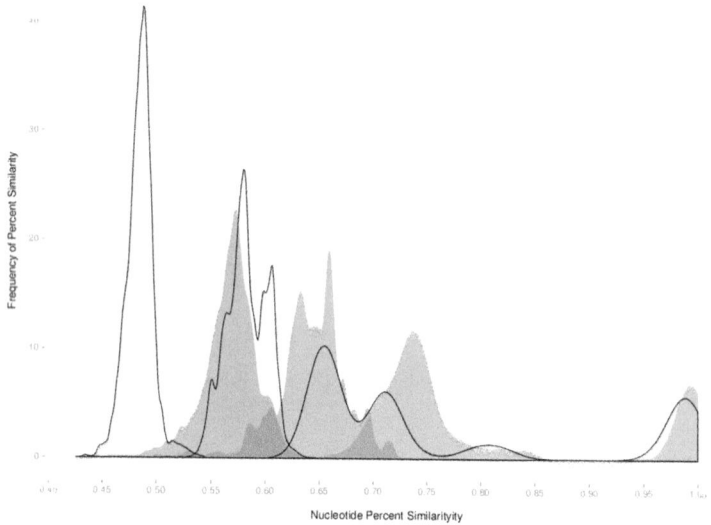

Fig. 2 Distribution of intergeneric, interspecies and intraspecies similarity of the L1 gene (*filled curves*) and the whole genome (*black lines*) of HPV from GenBank

the actual measure of similarity is based on the L1 gene. But considering the whole genome could drastically reduce the class boundary violation and would allow a better class separation. It appears clearly that the overlaps of the boundaries between the genera, species, and type classes are lower in the whole genome compared to the L1 gene (see Fig. 2). Furthermore, the intra-cluster similarities distribution within the genera and species classes are shifted about 10 % off the L1 gene distribution allowing a large interval to the type class that could improve the classification within this level of clustering where genomes are more similar and more difficult to classify.

We also explored the capacity of separation of the gene similarities. We computed the *cohesion index* for each cluster and sub-cluster within the three levels of clustering. This index represents the fraction of viruses that have their minimum pairwise similarities inside their belonging cluster. A class with one member (e.g., nu) and the top level class (e.g., all HPV) will always have an index of 1, by default. Since the index is a fraction, it will vary with the size of the cluster and the number of minimum pairwise similarities outside the cluster. If 1 minimum pairwise similarity is outside the cluster, the index will be 1/2 for a 2-member cluster, while it will be 99/100 for a 100-member cluster. The results shown in Table 4 present values of the cohesion index for each gene as well as the global index for the whole taxonomic tree. All the genomic regions used have several clusters with an index different of 1 (total agreement of viruses to their cluster). Furthermore, no genomic region has the best index for all the classes. However, the global index of the whole genome is the highest with a value of 0.994. Hence using the whole genome similarity enhances the cluster separation for more than 5 % with 99.4 % of the viruses confidently

Table 4 The *Cohesion index* for the clusters as well as the global index of the whole taxonomic tree, for each genomic region

Taxon	L1	L2	E1	E2	E4	E5	E6	E7	HPV
Alpha	0.965	0.998	0.992	0.988	0.756	0.987	0.936	0.918	0.998
Alpha1							0.333		
Alpha2	0.973				0.982				
Alpha3	0.818	0.833			0.538		0.977	0.712	0.833
Alpha5		0.833			0.333	0.833		0.917	0.833
Alpha6	0.987		0.954	0.986					
Alpha7	0.769	0.998			0.919	0.995			0.998
Alpha8	0.750						0.667		
Alpha9	0.866	0.999	0.838	0.941	0.717	0.537	0.869		0.999
Alpha10					0.926				
Alpha14	0.905	0.905			0.810				0.905
Beta	0.993			0.918	0.750		0.739	0.807	
Beta1	0.965	0.985		0.985	0.553		0.706	0.768	0.995
Beta2	0.875	0.775	0.976	0.942	0.867		0.801	0.620	0.975
Beta3							0.833	0.667	
Beta5							0		
Gamma	0.490	0.752	0.806	0.928	0.518		0.502	0.535	0.846
Gamma6	0.833				0				
Gamma7					0.738		0.905	0.929	
Gamma8					0.500				
Gamma9				0.500	0		0.500		
Gamma11					0.250			0.917	
Gamma12		0.500		0.500			0		0.500
HPV1a									0.500
HPV35				0.929					
HPV42							0		
HPV53			0.933						
HPV56				0.857					
HPV94		0.667							0.667
HPV120							0.917		
Global index	0.941	0.985	0.963	0.977	0.834	0.870	0.924	0.944	0.994

Taxon with indexes less than 1 (blank case) for at least one region are presented

classified. Similar trends have been observed using the Silhouette index (see results in supplementary material S2). Due to the distance effects in the later index, it is difficult to scale and compare the different genomic regions. But the Pearson correlation between the two indexes is 0.45 with P-value of 2.2e−16 indicating their proximity.

Table 5 Robinson and Foulds distance between trees expressed normalized as the distance/(2n–4) where $n = 507$ is the number of taxa common to all genomic regions

	E1	E2	E4	E6	E7	L1	L2	HPV
E1	0.00	0.33	0.34	0.33	0.31	0.30	0.40	0.37
E2	0.33	0.00	0.32	0.31	0.32	0.31	0.40	0.42
E4	0.34	0.32	0.00	0.29	0.29	0.33	0.38	0.46
E6	0.33	0.31	0.29	0.00	0.28	0.29	0.36	0.43
E7	0.31	0.32	0.29	0.28	0.00	0.26	0.34	0.40
L1	0.30	0.31	0.33	0.29	0.26	0.00	0.35	0.35
L2	0.40	0.40	0.38	0.36	0.34	0.35	0.00	0.46
HPV	0.37	0.42	0.46	0.43	0.40	0.35	0.46	0.00
Mean	0.34	0.35	0.35	0.33	0.31	0.31	0.38	0.41

The last line is the mean of the RF distance for each column

3.1 Phylogenetic Tree Comparison

Using the previous algorithm, we then compared the agreement between the actual taxonomic tree and the reconstructed gene trees as well as the whole genome tree. Although the L1, L2, and the whole genome (HPV) trees are different according to the Robinson and Foulds Topological distance (see Table 5), these inferred trees agree with the actual multifurcated taxonomic tree. However, bootstrap supports are higher in the whole genome tree suggesting a more robust tree (all trees are available in Newick format in supplementary material). The lower bootstrap values in the gene tree could be explained by the high number of sequences for the same type (e.g., HPV16 and HPV18) or the same species (e.g., alpha9) having more than 99 % similarity, thus reducing the phylogenetic signal. The whole genome tree carries more phylogenetic signal thus better bootstrap support for the chosen bifurcations. Furthermore, the differences of topology observed among the gene trees and the whole genome tree have already been highlighted for smaller datasets (e.g., Burk et al. 2009; Narechania et al. 2005; Diallo et al. 2009), suggesting distinct evolutionary histories. Hence adding more taxa emphasizes the discrepancies among the gene trees, while the whole genome tree provides a more robust tree.

4   Conclusion

In this study, we have shown that the massive sequencing of new HPVs could lead to a misleading classification when the actual criteria of classification are exploited. Neither the phylogenetic tree robustness of the L1 gene, nor its pairwise similarities provide an explicit criteria in favor of its use as main criteria. The compactness and the separability of groups are violated in each level of the classification (genera, species, type). Previous studies have highlighted that the similarity threshold percentages are not strictly respected, but here, we also noticed an increase of the overlap between clusters. However, we highlighted that exploiting the whole genome pairwise similarity measure as criteria instead of the actual one based on the L1 gene pairwise similarity provides enhancement on the cluster compactness and separability. The whole genome reduces the overlap of the boundaries, leading to better separability. This trend is confirmed by the compactness and separability analysis done with the cohesion and Silhouette indexes. But still, these results showed that all clusters do not have an optimal compactness and separation. Even though the L1 gene tree agrees with the taxonomic tree, the weakness of its bootstrap support makes it difficult to design an algorithm based solely on the current similarity given the whole set of sequence at once (proof not shown). But the phylogenetic inference method as described here, using the whole genome, is indeed able to reproduce exactly the taxonomic tree automatically with a computer algorithm without further manual curation, due to its phylogenetic signal. Once the phylogenetic tree is built, the classification process can be completed by automatically assigning taxa to internal nodes of the tree, according to criteria such as the whole genome similarity thresholds. It would be important to automate this process in the future to help improving PVs classification. Finally, some idiosyncrasies exist in the current taxonomic tree, by closely relating PVs that have different biological or pathogenic properties (Bernard et al. 2010). It would be interesting in the future work to investigate new similarity measures taking into account other properties of the genomes like the infected species, the type of disease, the carcinogenicity (Diallo et al. 2009) and/or exploring the different gene tree histories as to propose an improved classification. Supplementary material is available at: ancestors.bioinfo.uqam.ca/hpvECDA2013.zip.

References

Antonsson, A., Forslund, O., Ekberg, H., Sterner, G., & Hansson, B. G. (2000). The ubiquity and impressive genomic diversity of human skin papillomaviruses suggest a commensalic nature of these viruses. *Journal of Virology, 74*, 11636–11641.

Benson, D. A., Cavanaugh, M., Clark, K., Karsch-Mizrachi, I., Lipman, D. J., Ostell, J., et al. (2013). GenBank. *Nucleic Acids Research, 41*, D36–D42.

Bernard, H. U., Burk, R. D., Chen, Z., Van Doorslaer, K., Zur Hausen, H., & De Villiers, E. M. (2010). Classification of papillomaviruses (PVs) based on 189 PV types and proposal of taxonomic amendments. *Virology, 401*, 70–79.

Burk, R. D., Chen, Z., & Van Doorslaer, K. (2009). Human papillomaviruses: Genetic basis of carcinogenicity. *Public Health Genomics, 12*, 281–290.

Castresana, J. (2000). Selection of conserved blocks from multiple alignments for their use in phylogenetic analysis. *Molecular Biology and Evolution, 17*, 540–552.

De Villiers, E. M., Fauquet, C., Broker, T. R., Bernard, H. U., & Zur Hausen, H. (2004). Classification of papillomaviruses. *Virology, 324*, 17–27.

Diallo, A. B., Badescu, D., Blanchette, M., & Makarenkov, V. (2009). A whole genome study and identification of specific carcinogenic regions of the human papilloma viruses. *Journal of Computational Biology, 16*, 1461–1473.

Edgar, R. C. (2004). MUSCLE: Multiple sequence alignment with high accuracy and high throughput. *Nucleic Acids Research, 32*, 1792–1797.

Guindon, S., & Gascuel, O. (2003). A simple, fast, and accurate algorithm to estimate large phylogenies by maximum likelihood. *Systematic Biology, 52*(5), 696–704.

Handl, J., Knowles, J., & Kell, D. B. (2005). Computational cluster validation in post-genomic data analysis. *Bioinformatics, 21*, 3201–3212.

Hasegawa, M., Kishino, H., & Yano, T. (1985). Dating of the human-ape splitting by a molecular clock of mitochondrial DNA. *Journal of Molecular Evolution, 22*(2), 160–174.

Liu, Y., Li, Z., Xiong, H., Gao, X., & Wu, J. (2010). Understanding of internal clustering validation measures. In *2010 IEEE 10th International Conference on Data Mining (ICDM)* (pp. 911–916).

Muñoz, N., Bosch, F. X., De Sanjosé, S., Herrero, R., Castellsagué, X., Shah, K. V., et al. (2003). Epidemiologic classification of human papillomavirus types associated with cervical cancer. *New England Journal of Medicine, 348*, 518–527.

Narechania, A., Chen, Z., Desalle, R., & Burk, R. D. (2005). Phylogenetic incongruence among oncogenic genital alpha human papillomaviruses. *Journal of Virology, 79*, 15503–15510.

Rand, W. M. (1971). Objective criteria for the evaluation of clustering methods. *Journal of the American Statistical Association, 66*, 846–850.

Rijsbergen, C. J. V. (1979). *Information retrieval* (2nd ed.). Newton: Butterworth-Heinemann.

Robinson, D. F., & Foulds, L. R. (1981). Comparison of phylogenetic trees. *Mathematical Biosciences, 53*, 131–147.

Rousseeuw, P. (1987). Silhouettes: A graphical aid to the interpretation and validation of cluster analysis. *Journal of Computational and Applied Mathematics, 20*, 53–65.

Tota, J. E., Chevarie-Davis, M., Richardson, L. A., Devries, M., & Franco, E. L. (2011). Epidemiology and burden of HPV infection and related diseases: Implications for prevention strategies. *Preventive Medicine, 53*(1), S12–S21.

Van Doorslaer, K., Tan, Q., Xirasagar, S., Bandaru, S., Gopalan, V., Mohamoud, Y., et al. (2013). The papillomavirus episteme: A central resource for papillomavirus sequence data and analysis. *Nucleic Acids Research, 41*, D571–D578.

Zheng, Z. M., & Baker, C. C. (2006). Papillomavirus genome structure, expression, and post-transcriptional regulation. *Frontiers in Bioscience: A Journal and Virtual Library, 11*, 2286–2302.

Donor Limited Hot Deck Imputation: A Constrained Optimization Problem

Dieter William Joenssen

Abstract Hot deck methods impute missing data by matching records that are complete to those that are missing values. Observations absent within the recipient are then replaced by replicating the values from the matched donor. Some hot deck procedures constrain the frequency with which any donor may be matched to increase the precision of post-imputation parameter-estimates. This constraint, called a donor limit, also mitigates risks of exclusively using one donor for all imputations or using one donor with an extreme value or values "too often." Despite these desirable properties, imputation results of a donor limited hot deck are dependent on the recipients' order of imputation, an undesirable property. For nearest neighbor type hot deck procedures, the implementation of a constraint on donor usage causes the stepwise matching between each recipient and its closest donor to no longer minimize the sum of all donor–recipient distances. Thus, imputation results may further be improved by procedures that minimize the total donor–recipient distance-sum. The discrete optimization problem is formulated and a simulation detailing possible improvements when solving this integer program is presented.

1 Introduction

Missing data are a ubiquitous problem in the social sciences and present challenges wherever information is collected. Compensating missing values properly is pivotal to any analysis not only because typical statistical methods are designed for complete data, but also because any biases introduced upstream from the analysis will propagate downstream.

While the absence of missing observations within a data set is desirable, economic and technical constraints prohibit that all missing values are always resolved through logical inference or a two-phase sampling scheme. Thus, explicit provisions must be made for the remaining missing values. Provisions include

D.W. Joenssen (✉)
Ilmenau University of Technology, Helmholtzplatz 3, 98693 Ilmenau, Germany
e-mail: Dieter-William.Joenssen@TU-Ilmenau.de

© Springer-Verlag Berlin Heidelberg 2015
B. Lausen et al. (eds.), *Data Science, Learning by Latent Structures, and Knowledge Discovery*, Studies in Classification, Data Analysis, and Knowledge Organization, DOI 10.1007/978-3-662-44983-7_28

319

elimination, imputation, and parameter estimation methods. While elimination procedures simply remove missing observations or variables, and parameter estimation methods seek to directly estimate parameters of interest under distributional assumptions, imputation methods replace missing values with predictions. These missing data methods need to be applied in light of the stochastic process governing the appearance of unobserved values. Understanding of this process is critical in mitigating any possible non-response bias and precision reduction. To this end, Rubin (1976) classified the stochastic processes, determining the presence of missing values, into three groups, called missingness mechanisms. The appropriate missing data method is determined by which of these missingness mechanisms is present in the data.

The first missingness mechanism is missing-completely-at-random (MCAR). If the data are MCAR, then the available data represent a sub-sample of the intended sample, and no prediction is necessary, although possible. Under this most stringent assumption, observations or variables containing missing values may be eliminated, and primarily the consequences of a reduced sample size remain. The second class of missingness mechanisms is subsumed under missing-at-random (MAR). Here a connection between the presence of missing values in one variable and the values of another, fully observed, variable exists. Thus, unavailable values can and must be predicted from available covariates to eliminate the non-response bias. The final mechanism, not explicitly named by Rubin (1976), has subsequently been dubbed either not-missing-at-random (NMAR) (Little and Rubin 2002) or missing-not-at-random (MNAR) (Collins et al. 2001; Enders 2010) in literature. With this mechanism present, absent data may not be predicted from the data on hand, because, for example, the probability that a value is unobserved is dependent on what would be observed. To eliminate bias due to missingness under the NMAR mechanism, data and mechanism must be extensively modeled prior to imputation. Alternatively, highly correlated, complete covariates may be added to the analysis to make a MAR mechanism plausible (cf. Collins et al. 2001; Schafer and Graham 2002; Enders 2010).

One group of imputation algorithms are hot deck methods, the definition of which goes back to Ford (1983) and Sande (1983). Both define hot deck methods as imputation procedures where available values from one or more records (donors) are used to complete another, incomplete record (recipient) from the same data set. Selecting an arbitrary donor can be expected to lead to valid inferences when data is MCAR. However, matching donors and recipients based on some measure of similarity improves not only the imputation results for MCAR data, but also allows for valid inferences under the MAR mechanism. Similarity, in this context, is usually defined via a distance function, membership in a previously determined imputation class, or a combination of both. Andridge and Little (2010) demonstrate that membership in a certain imputation class may be expressed as a distance function. Thus, all three methods of defining similarity may be reduced to donor–recipient allocation based on a matching-cost matrix.

Hot deck methods have a number of desirable properties. Not only are they computationally simple, but they also make minimal distributional assumptions and are thus robust against model misspecification (Andridge and Little 2010). Further, the replication property guarantees not only that unique distributional features are conserved, but also that replacement values will be from the imputed variable's domain. Nonetheless, this replication property may lead to undesirable consequences because, fundamentally, a donor may be matched to multiple recipients. This poses the inherent risks, that "too many" or even all recipients are imputed by a single donor or that an outlier is chosen as a donor "too often." The probability of these risks manifesting is aggravated by the way imputation classes are defined. While more sophisticated methods of determining imputation classes exist, these classes are usually defined by cross-classifying a number of categorical covariates (Andridge and Little 2010). This cross-classification, also known as the adjustment cell method, often leads to many imputation classes, which do not seldom suffer from an acute lack of donors.

To mitigate the aforementioned risks, some hot deck methods limit how many times any donor may be chosen to donate its values. Although this donor limit is in line with the computationally simplistic way practitioners solve practical problems, and donor selection without replacement, the most stringent donor limit, is known to reduce imputation variance (Kalton and Kish 1984), it introduces a new problem. Generally speaking, a constraint on donor-usage makes results dependent on the order in which the recipients are imputed (Bankhofer and Joenssen 2014; Kovar and Whitridge 1995). This presents two challenges. First, if recipients are imputed strictly by order of appearance in the data, reordering the data may lead to different imputation results. This is particularly undesirable for deterministic hot decks, as reproducibility is one of the key arguments for deterministic imputation methods. The second challenge is that an optimal donor–recipient allocation for each recipient in sequence is not guaranteed to provide a globally optimal solution, regardless of how the recipient order is chosen.

This paper's objectives are twofold. The first objective is to formulate the integer-program that, when solved, yields not only a sequence independent, but also a globally optimal donor–recipient mapping for donor limited hot deck imputation methods. The second objective is to gauge improvements that may be achieved by solving the previously formulated integer-program using a sequence independent heuristic. To this end, the optimization problem is formulated in Sect. 2. Section 3 details the simulation study and its results to complete the second objective. Conclusions and directions for further research are discussed in the final section.

2   Optimization Problem Formulation

Without loss of generality, assume that the donors and recipients form two disjoint
sets of sizes d and r, respectively. Further, let c_{ij} denote the $d \times r$ distance, and thus
matching-cost, matrix for the $n \times m$ data matrix. Then, if the chosen constant usage
limit for all donors is denoted as dl, the following integer program can be defined:

$$g\left(x_{ij}\right) = \sum_{i=1}^{d} \sum_{j=1}^{r} c_{ij} x_{ij} \rightarrow \min$$

$$\sum_{j=1}^{r} x_{ij} \leq dl, \quad \forall i = 1, \ldots, d$$

$$\sum_{i=1}^{d} x_{ij} = 1, \quad \forall j = 1, \ldots, r$$

$$x_{ij} \in \{0; 1\}, \quad dl \in \{\lceil r/d \rceil, \ldots, r\}$$

(1)

where x_{ij} denotes the allocation of donor i to recipient j, and $\lceil \bullet \rceil$ denotes the
ceiling function. The second line of Eq. (1) indicates the constraints introduced by
implementing a donor limit. It is this set of constraints that make the conventional,
step-wise matching procedure sub-optimal. The third line of Formula (1) represents
the fundamental requirement that all missing data be resolved through matching. It
thus becomes apparent, through combination of the requirements and constraints,
that the range of possible donor limits is constrained to integer values between
certain bounds. The upper bounds are given by the fact that a donor may at most
be matched to each recipient, in which case the optimization problem becomes
unconstrained. The lower bounds are defined by the necessity to solve the recipient
requirements.

Optimizing the total donor–recipient distance-sum has several appealing properties, in addition to resolving the sequence dependency described above. First, it
is the logical consequence of constraining the supply of donors, for if no donor
limit were implemented, the step-wise allocation of donors to recipients would
also be sequence independent, and lead to a minimal distance-sum. Second, while
fewer recipients receive the best possible match, the matches are on average
better, which should have positive effects for small sub-populations, as extreme
deviations between the true, unobserved values and the imputed values will occur
less frequently. Third, when donors become scarce or a more stringent donor limit
is chosen, the selection of outliers for value donation becomes unavoidable. Here, in
an almost paradoxical case, the selection of outliers, against which the donor limit
should guard, is enforced. In this situation, the optimization guarantees that outliers
are distributed among the recipients so that the least possible bias is introduced.

The basic structure of the integer program, given in Eq. (1), is that of the classical capacitated transportation problem. While some special structure exists, the problem may readily be solved by those algorithms conceived to solve the standard problem. One notable heuristic for solving this problem is the column-minimum method. If the distance matrix is formulated, as is the case here, with rows representing donors and columns representing recipients, this heuristic is equivalent to the naïve approach currently practiced, selecting the minimum-distance donor, among the remaining donors, for each recipient in turn. Another notable heuristic for solving allocation problems, as defined above, is Vogel's approximation method (Reinfeld and Vogel 1958). Vogel's approximation method has the distinction of being sequence independent, as the selection criterion is calculated iteratively over all distance matrix rows and columns for each matching step (cf. Domschke 1995). Algorithms for finding guaranteed optimal solutions include the Network Simplex and graph based methods, for which Domschke (1995) offers exhaustive explanations and detailed discussions.

3 A Simulation Study

To gauge possible improvements that may be achieved by solving the integer-program formulated in Sect. 2, a limited simulation study is presented in the following subsections. Section 3.1 presents the design parameters, including the selected factors and quality criterion. The results are divulged in Sect. 3.2.

3.1 Study Design

For an assessment of possible improvements a more sophisticated, order independent nearest-neighbor hot deck offers, imputation quality is compared for simulated data. The simulation is performed using a framework implemented in R version 2.15.2 (R Core Team 2013). Imputation routines are available through the function *impute.NN_HD* from the contributed R-Package **HotDeckImputation** (Joenssen 2013). The factors varied, with some simplification, are akin to those chosen by Bankhofer and Joenssen (2014).

The basic data structure consists of two, equally sized, bivariate-normal clusters, centered around $(-1, -1)$ and $(1, 1)$, generated using the function *rmvnorm* from the R-package **mvtnorm** (Genz et al. 2013). The size of the data matrix is chosen to be either 50 or 100 total objects with three different within-class-correlations (0.00, 0.35, or 0.70). These data matrices, which are generated 1,000 times each, may be understood as either the full sample, or as a single imputation class of a larger sample.

The resulting 6,000 complete data sets are then subjugated to three different missing data mechanisms, each with four different proportions of missing data. This is repeated 1,000 times, yielding 90,000,000 matrices with missing data. Proportions of missing data, as calculated over the single afflicted variable, are 10 to 50 % in steps of 10 %. The first missingness mechanism, applied to the complete data, randomly deletes the defined proportion of values, creating MCAR data. The second mechanism, of type MAR, is defined so that the amount of non-response in one cluster is always twice that of the other. The last mechanism, also of type MAR, is akin to the second except here the between cluster non-response ratio is set to 1:4.

Euclidean distances used for matching are calculated over the binary, cluster-indicator variable and the second, complete covariate. Donor–recipient matching is then performed for all 90,000,000 matrices with missing data using the current standard, naïve approach and Vogels approximation method, the simple, but sequence independent heuristic. The selection of matching algorithms is, on the one hand, due to constraints in computation time and algorithm availability, and on the other hand to gauge a minimum of possible improvement that may be achieved.

Since sample surveys, the largest use-case of hot deck imputation, are usually conducted with the intent of estimating distribution parameters (cf. Little and Rubin 2002), imputation quality is judged based on the estimation of seven parameters. These parameters are calculated for all 6,000 complete and 180,000,000 imputed data matrices, equivalent to more than 308.4 gigabytes of data. The set of parameters includes four univariate parameters, calculated on the imputed variable, and three multivariate parameters, calculated between the imputed variable and the complete, continuous covariate. The parameter set contains estimations of the sample mean ($\bar{x}$), standard deviation (sd_x), skewness (s_x), kurtosis (k_x), Mardia's skewness ($b_{1,xy}$) and kurtosis ($b_{2,xy}$), and the Pearson correlation (r_{xy}). Considering every parameter in this set, the root-mean-square-error (RMSE), between the true parameter p_T, based on the data before deletion, and the estimated parameter value p_I, based on the imputed data matrix, is calculated as follows:

$$\text{RMSE} = \sqrt{\frac{1}{1,000,000} \sum_{i=1}^{1,000,000} (p_T - p_I)^2} \tag{2}$$

RMSEs are then averaged as needed to yield main effect estimates for the results.

3.2 Results

The following section reports the results of the previously described Monte Carlo simulation. Values shown in Tables 1, 2 and 3 are differences in RMSE between the two donor–recipient matching methods. Differences are chosen to be between the naïve and Vogel's approximation matching method. Thus, negative values indicate that the column-minimum method is better, and positive values indicate that Vogel's

Table 1 RMSE differences for the *MCAR* cases, main effects

Factor	Factor levels	Univariate				Multivariate		
		$\bar{x}$	sd_x	s_x	k_x	r_{xy}	$b_{1,xy}$	$b_{2,xy}$
Object count	50	–	–	–	–	0.004	0.086	0.121
	100	–	–	–	–	0.002	0.067	0.122
Within-class-correlation	0.00	–	–	–	–	0.001	0.018	0.034
	0.35	–	–	–	–	0.003	0.051	0.092
	0.70	–	–	–	–	0.005	0.160	0.238
Proportion missingness	10 %	–	–	–	–	–	–	0.001
	20 %	–	–	–	–	–	0.001	0.003
	30 %	–	–	–	–	–	0.004	0.013
	40 %	–	–	–	–	0.001	0.039	0.080
	50 %	–	–	–	–	0.013	0.338	0.510

Table 2 RMSE differences for the *MAR 1:2* cases, main effects

Factor	Factor levels	Univariate				Multivariate		
		$\bar{x}$	sd_x	s_x	k_x	r_{xy}	$b_{1,xy}$	$b_{2,xy}$
Object count	50	–	–	–	–	0.009	0.144	0.160
	100	–	–	–	–	0.007	0.132	0.150
Within-class-correlation	0.00	–	–	–	–	0.003	0.038	0.035
	0.35	–	–	–	–	0.008	0.096	0.105
	0.70	–	–	–	–	0.012	0.279	0.324
Proportion missingness	10 %	–	–	–	–	–	–	0.001
	20 %	–	–	–	–	–	0.001	0.004
	30 %	–	–	–	–	–	0.005	0.015
	40 %	–	–	–	–	0.004	0.166	0.342
	50 %	–	–	–	–	0.034	0.516	0.414

Table 3 RMSE differences for the *MAR 1:4* cases, main effects

Factor	Factor levels	Univariate				Multivariate		
		$\bar{x}$	sd_x	s_x	k_x	r_{xy}	$b_{1,xy}$	$b_{2,xy}$
Object count	50	–	–	–	–	0.018	0.149	0.133
	100	–	–	–	–	0.015	0.150	0.141
Within-class-correlation	0.00	–	–	–	–	0.008	0.040	0.030
	0.35	–	–	–	–	0.016	0.107	0.095
	0.70	–	–	–	–	0.025	0.302	0.286
Proportion missingness	10 %	–	–	–	–	–	–	0.001
	20 %	–	–	–	–	–	0.002	0.005
	30 %	–	–	–	–	0.001	0.028	0.069
	40 %	–	–	–	–	0.013	0.357	0.120
	50 %	–	–	–	–	0.069	0.360	0.492

approximation method leads to superior imputation results. The reported values are truncated to the first three significant figures. Values with the first three significant digits equal to zero are indicated as hyphens to improve readability.

3.2.1 Simulation Results for *MCAR* Missing Data

The results in Table 1 show that optimizing the donor–recipient sum of distances is never an inferior strategy, when the missingness mechanism is *MCAR*. Beyond this, the most noticeable result is that the estimation of univariate parameters is not, or rather only marginally influenced by varying the matching methods. Neither mean, standard deviation, skewness, nor kurtosis of the imputed variable can be improved substantially. The converse holds true for the post-imputation-estimation of the multivariate parameters. Estimates of the Pearson correlation and the Mardia skewness and kurtosis measures are improved for any amount of *MCAR* missing data. Attainable improvements increase monotonically with the within-class-correlation and the proportion of missing data. There is also some indication that smaller data sets profit from a global optimization of donor–recipient matches.

3.2.2 Simulation Results for *MAR 1:2* Missing Data

Table 2 displays the results for the *MAR 1:2* missingness mechanism, the case that one cluster has twice the missingness of the other. Again, these results indicate that optimizing the total donor–recipient distance-sum is never inferior, and that only the estimation of the multivariate parameters is improved. None of the univariate parameters are substantially affected by a choice in donor selection strategy. Further, the same effects, only stronger, are apparent when contrasting the values in Tables 1 and 2. The optimization becomes more important as the amount of missing values and the within-class-correlation increases, and the amount of objects considered decreases.

3.2.3 Simulation Results for *MAR 1:4* Missing Data

Upon considering Table 3, the homogeneous picture set forth by the first missing data mechanisms is completed. For the third time, Vogel's approximation method is never inferior to the naïve method of matching. The *MAR 1:4* mechanism is more severe than the *MAR 1:2* mechanism, because one cluster contains only a fifth of all missing values, while the other cluster contains the remaining four-fifths. This difference in severity leads to even more pronounced improvements through optimizing the donor–recipient distance-sum. Again, only multivariate-parameter estimation is improved and improvement tendencies are the same as with the *MCAR* and *MAR 1:2* cases.

4   Conclusions

This paper discusses an aspect of hot deck imputation known as the donor limit and how the implementation thereof leads to the failure of the conventional matching method to achieve a globally optimal solution. In addition to the formulation of the optimization problem, results of a simulation study are presented in which the conventional method and another heuristic are compared.

The results of the Monte Carlo simulation show that the optimized method is never inferior to the conventional matching method, when a donor limit is in place. Tendencies of these advantages to increase are consistent over all three missingness mechanisms considered. Advantages always increase with increasing donor-sparsity and variate association within the clusters. Advantages decrease as the sample size decreases and as the missingness mechanism becomes less severe.

Perhaps the most interesting result is that only the estimation of multivariate parameters is affected by changing the matching procedure. This indicates that the primary value in minimizing the total donor–recipient distance-sum, rather than only minimizing the donor–recipient distance in each step, lies in preserving the multivariate structure of the data. This is especially important if the data should be subject to multivariate analysis, e.g., multiple regression, subsequent to imputation.

Results of this paper are expected to generalize well for a wide range of situations and applications, not only because the most important factors are considered, but also because the assumptions made are few and weak. Nonetheless, this study does have some important limitations. First, while the simulated missingness mechanisms, where two groups have different response propensities, cover a broad range of situations, many more MAR mechanism types are conceivable. Second, in practice, due to the way imputation classes are constructed or if multiple variables exhibit missing values, donor-sparsity may be considerably higher than in the situations simulated, possibly making a donor limit of $dl = 1$ unfeasible. Third, many more data structures are possible, especially for real data. More complex data structures may be more sensitive to choosing the optimized approach. Further, the original approach to matching is only compared to a heuristic that promises order independent, near optimal results rather than optimal results. Future work may compare optimal matching results to the two heuristics, not only based on improvements offered in parameter estimation, but also based on computational costs. Finally, there may exist some special structure within the optimization problem that would allow specialized, efficient algorithms to be used for solving the integer program. However, the investigation of these factors is well beyond the scope afforded by this paper.

The conclusions and results of this paper should not only be of interest to an academic audience, but also to practitioners using multivariate analysis on hot deck imputed data.

References

Andridge, R. R., & Little, R. J. A. (2010). A review of hot deck imputation for survey nonresponse. *International Statistical Review, 78,* 40–64.

Bankhofer, U., & Joenssen, D. W. (2014). On limiting donor usage for imputation of missing data via hot deck methods. In M. Spiliopoulou, L. Schmidt–Thieme, & R. Jannings (Eds.), *Data analysis, machine learning and knowledge discovery* (pp. 3–11). Berlin: Springer.

Collins, L., Schafer, J., & Kam, C. (2001). A comparison of inclusive and restrictive strategies in modern missing data procedures. *Psychological Methods, 6,* 330–351.

Domschke, W. (1995). *Logistik: Transport.* München: Oldenbourg.

Enders, C. K. (2010). *Applied missing data analysis.* New York: Guilford.

Ford B. (1983). An overview of hot-deck procedures. In W. Madow, H. Nisselson, & I. Olkin (Eds.), *Incomplete data in sample surveys* (pp. 185–207). New York: Academic Press.

Genz, A., Bretz, F., Miwa, T., Mi, X., Leisch, F., Scheipl, F., et al. (2013). mvtnorm: Multivariate normal and distributions. R package version 0.9-9995. http://CRAN.R-project.org/package=mvtnorm.

Joenssen, D. W. (2013). HotDeckImputation: Hot deck imputation methods for missing data. R package version 0.1.0. http://CRAN.R-project.org/package=HotDeckImputation.

Kalton, G., & Kish, L. (1984). Some efficient random imputation methods. *Communications in Statistics Theory and Methods, 13,* 1919–1939.

Kovar, J. G., & Whitridge, J. (1995). Imputation of business survey data. In B. G. Cox, D. A. Binder, B. N. Chinnappa, A. Christianson, M. J. Colledge, & P. S. Kott (Eds.), *Business survey methods* (pp. 403–423). New York: Wiley.

Little, R. J. A., & Rubin, D. B. (2002). *Statistical analysis with missing data.* Hoboken: Wiley.

R Core Team. (2013). *R: A language and environment for statistical computing.* R Vienna: Foundation for Statistical Computing. http://www.R-project.org/

Reinfeld, N. V., & Vogel, W. R. (1958). *Mathematical programming.* New Jersey: Prentice-Hall.

Rubin, D. B. (1976). Inference and missing data (with discussion). *Biometrika, 63,* 581–592.

Sande I. (1983). Hot-deck imputation procedures. In W. Madow, H. Nisselson, & I. Olkin (Eds.), *Incomplete data in sample surveys* (pp. 339–349). New York: Academic Press.

Schafer, J., & Graham, J. (2002). Missing data: Our view of the state of the art. *Psychological Methods, 7,* 147–177.

Ensembles of Representative Prototype Sets for Classification and Data Set Analysis

Christoph Müssel, Ludwig Lausser, and Hans A. Kestler

Abstract The drawback of many state-of-the-art classifiers is that their models are not easily interpretable. We recently introduced Representative Prototype Sets (RPS), which are simple base classifiers that allow for a systematic description of data sets by exhaustive enumeration of all possible classifiers.

The major focus of the previous work was on a descriptive characterization of low-cardinality data sets. In the context of prediction, a lack of accuracy of the simple RPS model can be compensated by accumulating the decisions of several classifiers. Here, we now investigate ensembles of RPS base classifiers in a predictive setting on data sets of high dimensionality and low cardinality. The performance of several selection and fusion strategies is evaluated. We visualize the decisions of the ensembles in an exemplary scenario and illustrate links between visual data set inspection and prediction.

1 Introduction

Supervised classification methods provide valuable insights into the structure of data by recognizing patterns that are far too complex for human cognition. This particularly applies to high-dimensional measurements from biomolecular high-throughput experiments. For example, microarrays usually measure the expression levels of

Christoph Müssel and Ludwig Lausser have contributed equally to this work.

C. Müssel
Medical Systems Biology and Institute of Neural Information Processing, Ulm University, 89069 Ulm, Germany

L. Lausser • H.A. Kestler (✉)
Medical Systems Biology and Institute of Neural Information Processing, Ulm University, 89069 Ulm, Germany

Leibniz Institute for Age Research – Fritz-Lipman Institute, 07745 Jena, Germany
e-mail: hans.kestler@uni-ulm.de; hkestler@fli-leibniz.de

© Springer-Verlag Berlin Heidelberg 2015
B. Lausen et al. (eds.), *Data Science, Learning by Latent Structures, and Knowledge Discovery*, Studies in Classification, Data Analysis, and Knowledge Organization, DOI 10.1007/978-3-662-44983-7_29

several thousands of genes simultaneously. At the same time, the complexity of the identified structures can also be seen as one of the major drawbacks of most state-of-the-art classification methods: While these classifiers may predict unseen samples with a high accuracy, it often remains unclear on which information the decisions are based. In many respects, the interpretability of decision criteria can be as important as the decision itself. For example, interpretable classification models of biomolecular data can provide decision support to life scientists when judging new probes.

We recently proposed an approach that aims at integrating both characterization and prediction of data sets (Lausser et al. 2012). This method is based on simple data-dependent prototype set classifiers (Representative Prototype Sets, RPS) that are comprised of exactly one sample from each class. Our approach is related to other classifiers that directly select prototypes from a data set. The k-Nearest Neighbour algorithm (k-NN, Fix and Hodges 1951) utilizes all given training samples as prototypes for prediction. Other data-dependent strategies are based on the prediction scheme of k-NN, but edit the training set by removing dispensable samples. An overview of such approaches is given by Brighton and Mellish (2002) or by Dasarathy (1991). A first editing strategy, the Condensed k-NN, was presented by Hart (1968). More recently, Lausser et al. (2013) presented the predictive hubs k-NN, which lays a particular focus on borderline samples. This strategy has also been observed for support vector machines (Schwenker and Kestler 2002). Other well-known classification algorithms use prototypes that do not directly correspond to training samples. Examples for such methods are learning vector quantization (Kohonen 1988), the nearest centroid classifier (e.g., Webb 2002), or the nearest shrunken centroid classifier (Tibshirani et al. 2002).

Among the above methods, the RPS approach utilizes the simplest classification model. A major advantage of this simplicity is the fact that all possible classifiers can be enumerated for low-cardinality data sets, such as microarray data. In this way, the full set of classifiers can be employed to analyze characteristics of a data set. For example, they can aid in the identification of samples that are typical representatives of their respective classes. Often, single RPS classifiers are too simple to achieve a competitive classification accuracy. In Lausser et al. (2012), we evaluate basic ensembles of RPS classifiers. In this paper, we investigate more sophisticated methods of constructing ensembles. The ensemble methods are evaluated on several well-known microarray data sets. In an exemplary case, we show how observations from visualizations can be related to the final predictions made by the ensembles.

2   Representative Prototype Set Classification

2.1   *Representative Prototype Sets*

A RPS classifier comprises exactly one prototype per class, which is chosen among the training samples of the class. The labels of unseen data points are predicted according to the label of the nearest prototype in the set.

We define a training set with instances of k classes as $\mathcal{T} = \bigcup_{i=1,\ldots,k} \mathcal{T}_i$, where each $\mathcal{T}_i$ comprises samples $(\mathbf{x}, i)$ consisting of a feature vector $\mathbf{x} \in \mathbf{R}^n$ and an associated class label i. A Representative Prototype Classifier consists of a set of prototypes

$$\mathcal{P} = \{p_1, \ldots, p_k \mid p_i \in \mathcal{T}_i\}$$

with $p_i = (\mathbf{x}_i, i)$. The classifier predicts an unseen sample $\mathbf{v}$ by choosing the prototype p_i with the smallest (Euclidean) distance $d(\mathbf{v}, \mathbf{x}_i)$, i.e.

$$\text{RPS}_{\mathcal{P}}(\mathbf{v}) = \text{argmin}_{i=1,\ldots,k} \, d(\mathbf{v}, \mathbf{x}_i).$$

Although RPS classifiers are a general concept and are described here as such, we lay a particular focus on the two-class case ($k = 2$). This is the most typical scenario in microarray analysis (e.g., disease vs. healthy). The two-class setting also enables particular visualization techniques that are not applicable to multi-class data sets (see Lausser et al. 2012).

The complete set of RPS for a training set $\mathcal{T}$ is given by:

$$\mathcal{C} = \bigcup_{p_1 \in \mathcal{T}_1} \bigcup_{p_2 \in \mathcal{T}_2} \cdots \bigcup_{p_k \in \mathcal{T}_k} \{\{p_1, \ldots, p_k\}\}.$$

Hence, the total number of possible prototype sets for a training set $\mathcal{T}$ is $\prod_{i=1,\ldots,k} |\mathcal{T}_i|$. This means that it is often feasible to analyze the complete set of RPS classifiers in an exhaustive manner for typical microarray data sets that often comprise fewer than 100 samples. By considering all possible classifier models, a data set can be characterized (see Lausser et al. 2012). For example, the empirical accuracies of all RPS classifiers on the training samples can be measured and analyzed. If a sample achieves high accuracies in many combinations, it is probably a good representative for its class. Furthermore, it is possible to utilize the empirical distribution of accuracies for a statistical test that assesses the suitability of RPS for a specific data set: For a specific accuracy a, the probability that a value drawn from this distribution is less than a can be interpreted as the p-value of achieving an accuracy of at least a, given a randomly chosen RPS classifier.

2.2 Ensembles of Representative Prototype Sets

Single RPS classifiers are simple base learners and may not achieve satisfactory accuracies in some settings. In particular, they are unsuitable for scenarios where a single class consists of more than one cluster, as there is only a single prototype per class. A fusion of multiple RPS classifiers can yield a more flexible and robust ensemble classifier. Some preliminary results for a simple ensemble type have already been shown in Lausser et al. (2012). Here, we investigate further ensemble methods that use more advanced selection techniques, but still yield interpretable ensembles and decisions.

In general, ensembles are established by choosing a subset of m prototype sets

$$\mathcal{E}_m = \{\mathcal{P}_1, \ldots, \mathcal{P}_m\} \subseteq \mathcal{C},$$

from the complete set of RPS for a training set according to a selection strategy. Here, m is the prespecified ensemble size.

To classify an unseen sample $\mathbf{v}$, a *fusion rule* $F_{\mathcal{E}_m}$ merges the decisions of the ensemble members. We evaluate the following selection strategies and fusion rules.

2.2.1 Rank by Accuracy

We apply a simple ensemble method that chooses ensemble members according to their classification accuracies on the training set. In detail, we calculate the accuracies

$$A(\mathcal{P}) = \frac{1}{|\mathcal{T}|} \sum_{(\mathbf{x},i) \in \mathcal{T}} \mathbb{I}_{[\text{RPS}_{\mathcal{P}}(\mathbf{x})=i]}$$

for all $\mathcal{P} \in \mathcal{C}$. To form an ensemble with m members, we then choose the m prototype sets with the highest accuracies A.

As a fusion rule, we use unweighted majority voting. That is,

$$F_{\mathcal{E}_m}(\mathbf{v}) = \text{argmax}_{i=1,\ldots,k} \, |\{\mathcal{P} \in \mathcal{E}_m \mid \text{RPS}_{\mathcal{P}}(\mathbf{v}) = i\}|$$

As a second fusion rule, we additionally apply the Behavior Knowledge Space (BKS) approach (Huang and Suen 1995). BKS is a trainable fusion strategy that is based on a set of m trained base classifiers. It groups the training samples in $\mathcal{T}$ according to the m-dimensional prediction vectors of the ensemble members. This can be interpreted as a segmentation of the feature space. Each segment is labeled with the class label that is most frequent among the training samples in the segment. To predict an unseen sample $\mathbf{v}$, its prediction vector is calculated, and the label of the segment that corresponds to the prediction vector is returned. If no training samples are assigned to the segment, the sample is predicted according to a nearest neighbour approach.

2.2.2 Rank by Average Class-Wise Accuracies

Basing the selection of ensemble members on the overall training set accuracies can yield unsuitable ensembles in case of unbalanced data sets. In this case, the accuracies of the ensemble members on the smaller classes tend to be neglected in favour of the larger classes. To compensate for this effect, we apply a balanced selection criterion that calculates the accuracies separately for the classes, i.e.

$$A_{\text{bal}}(\mathcal{P}) = \frac{1}{k} \sum_{i=1}^{k} \frac{1}{|\mathcal{T}_i|} \sum_{(\mathbf{x},i) \in \mathcal{T}_i} \mathbb{I}_{[\text{RPS}_\mathcal{P}(\mathbf{x})=i]}$$

As above, the decisions of the ensemble members are fused according to a majority voting and BKS.

2.2.3 AdaBoost

The boosting algorithm AdaBoost is an iterative training scheme that increases the ensemble size in a stepwise manner (Freund and Schapire 1995). The selection procedure for additional base classifiers focuses on samples that are misclassified by the current ensemble members. The current importance of a training sample $(\mathbf{x}_j, i_j) \in \mathcal{T}$ is given by the size of a weight $w_j \in \mathbb{R}_+$. The base classifier that maximizes the weighted accuracy is chosen as the next ensemble member.

$$A_{\text{boo}}(\mathcal{P}) = \frac{1}{k} \sum_{i=1}^{k} \frac{1}{|\mathcal{T}_i|} \sum_{(\mathbf{x}_j, i_j) \in \mathcal{T}_i} w_j \mathbb{I}_{[\text{RPS}_\mathcal{P}(\mathbf{x}_j)=i_j]}$$

2.2.4 Bagging

Bagging (bootstrap aggregating) aims at generating base classifiers that misclassify different (independent) parts of the data (Breiman 1996). The ensemble members are combined via unweighted majority voting. It is assumed that the majority of the ensemble members makes correct predictions for an unseen sample. Each base classifier of a bagging ensemble is trained on a bootstrap replicate $\mathcal{T}'$ of the original data set $\mathcal{T}$, with $|\mathcal{T}| = |\mathcal{T}'|$. The prototype set that maximizes the accuracy on the bootstrap replicate is selected for inclusion in the ensemble.

$$A_{\text{bag}}(\mathcal{P}) = \frac{1}{|\mathcal{T}nn|} \sum_{(\mathbf{x},k) \in \mathcal{T}'} \mathbb{I}_{[\text{RPS}_\mathcal{P}(\mathbf{x})=k]}$$

3 Evaluation

We evaluated the ensemble methods on three well-known microarray data sets:

The *Bittner data set* (Bittner et al. 2000) measures 8,067 gene expression values for 31 melanoma profiles and 7 control profiles. The initial analysis of this data identified a stable cluster of 19 of the melanomas using hierarchical clustering. Here, we denote this cluster as class ML1, while the 19 remaining samples (melanomas and controls) are denoted as class ML2.

The *Wang data set* (Wang et al. 2012, GEO Accession no. GSE19826) comprises expression profiles of 12 gastric cancer tissues (GC) and 15 healthy tissues (N) with 54,613 features each. The data set consists of 12 pairs samples taken from adjacent normal and tumor tissues in the same individual and three additional normal tissue samples.

The *Shipp data set* (Shipp et al. 2002) consists of 77 samples two types of B-cell lymphoma (58 diffuse large B-cell lymphoma (DLBCL) and 19 follicular lymphoma (FL)). For each profile, the expression levels of 7,129 genes were measured.

The *Notterman data set* (Notterman et al. 2001) contains 18 paired samples of colon adenocarcinomas (CA) and normal tissues (N) taken from the same individuals. Each profile comprises 7,457 features.

To estimate the classification performance of the ensemble classifiers, we applied a cross-validation with $f = 10$ partitions and $r = 10$ independent repetitions. The error E_c was averaged over the r repetitions. In general, we evaluated ensemble sizes $m \in \{1, 3, 5, 7, 9, 11, 13, 15, 25, 35, 35, 55\}$. We chose odd ensemble sizes to eliminate the possibility of ties in the majority voting. For the BKS ensembles, the size of the behaviour knowledge space grows exponentially with the ensemble size. As the number of non-empty cells is limited by the training set size, this means that large ensemble sizes result in large numbers of empty cells. Therefore, we restricted the BKS ensemble sizes to $m \in \{3, 5, 7, 9\}$. All experiments were performed using the TunePareto software (Müssel et al. 2012).

Figure 1 shows the average cross-validation errors of the different RPS ensemble types for different ensemble sizes m on the three microarray data sets. For comparison, we also applied linear support vector machines (Vapnik 1998) with different cost parameters $C \in \{0.001, 0.01, 0.1, 1, 10.50.100\}$, L1-norm regularized SVMs with the same parameters, support vector machines with an RBF kernel and all combinations of cost parameters $C \in \{0.001, 0.01, 0.1, 1, 10.50.100\}$ and γ parameters $\gamma \in \{10^{-9}, 10^{-8}, \ldots, 1\}$, and the k-NN classifier (Fix and Hodges 1951) with $k \in \{1, 3, 5, 7, 9, 11\}$. Among the tested methods, the regularized SVM is the only approach that incorporates a feature reduction by introducing a penalty term. For each of the four reference classifiers, the configuration with the smallest cross-validation error is depicted as a grey dashed line.

On the Bittner data set, the overall best result is achieved by a regularized SVM with an error of 0.018. Apart from this feature-reducing classifier, the RPS ensembles clearly outperform the other reference classifiers. The best RPS methods is a bagging ensemble with 45 members achieving an error of $E_c = 0.076$, followed by a AdaBoost ensemble with 55 members and $E_c = 0.078$. The best single RPS base classifier achieves an error of $E_c = 0.089$, which is beaten only by a few large ensembles. This is related to our previous observation (Lausser et al. 2012) that there are only few highly accurate prototype sets on this, and that it is hard to find suitable ensembles due to the lack of diversity among the RPS base learners.

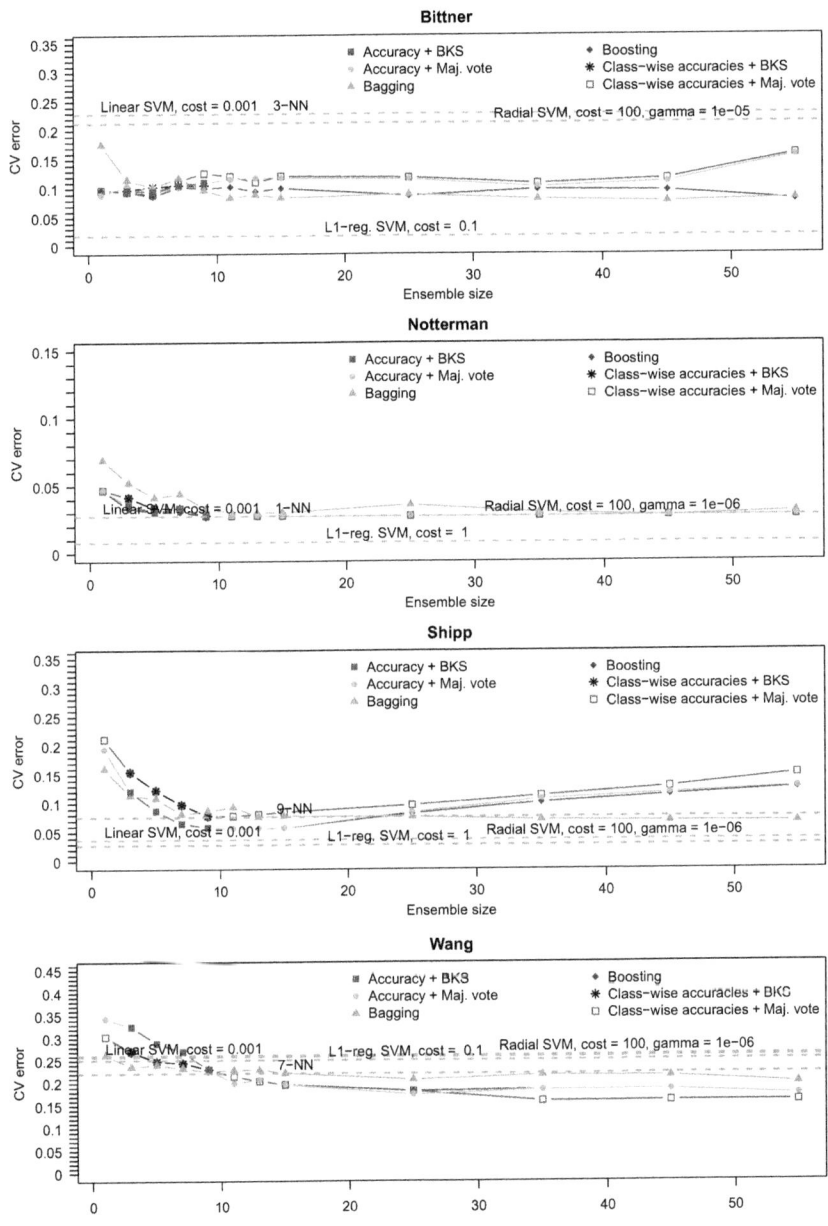

Fig. 1 Cross-validation errors of different RPS ensemble types and reference classifiers on the Bittner data set, the Wang data set, the Shipp data set and the Notterman data set. On the x axis, the sizes of the RPS ensembles are varied from 1 to 55. The y axis corresponds to the mean cross-validation error. For the reference classifiers (SVMs with linear and RBF kernels, L1-norm regularized SVMs, k-NN), the accuracies of the best tested configurations are shown as *grey dashed lines*

On the Wang data set, RPS ensembles outperform all comparison classifiers. This data set is rather hard to classify, with the best error rate of $E_c = 0.159$ achieved by ensembles selected according to class-wise accuracies with 35–55 ensemble members. Possibly, employing the class-wise accuracies as a selection criterion enforces the correct prediction of the normal tissue class, which is slightly larger than the other class. AdaBoost and rankings by accuracy select mostly the same ensemble members on this data set and also achieve good accuracies ($E_c = 0.174$ for 55 ensemble members).

On the Shipp data set, prototype set learners can clearly benefit from ensemble classification. Among the evaluated ensemble types, the smallest cross-validation error of $E_c = 0.057$ is achieved by ensembles that were chosen according to a ranking by accuracy and by AdaBoost, both for ensemble sizes of 11, 13 and 15. Here, these two ensemble types choose the same members for smaller ensemble sizes. Furthermore, bagging ensembles achieve good classification accuracies particularly for large ensemble sizes ($E_c = 0.076$ for 45 ensemble members. While most RPS ensembles outperform the Nearest Neighbour classifiers, the support vector machines perform slightly better than the ensembles on this data set.

The Notterman data set is comparatively easy to classify, possibly because it consists of paired samples from the same individuals. On this data set, nearly all classifiers achieve the same performance of $E_c = 0.028$, which corresponds to exactly one misclassified sample in each cross-validation run. This applies to most configurations of RPS ensembles as well as the best reference classifiers. Only a single configuration of the L1-regularized SVM (cost $= 1$) is able to attain a lower error of $E_c = 0.008$.

To investigate the behaviour of the ensembles on the Notterman data set further, Fig. 2 gives a visualization of the predictions made by ensembles whose members were chosen according to a ranking by accuracy in a reclassification scenario. Here, each row corresponds to one sample in the data set, and each column corresponds to a single ensemble member (i.e., and RPS classifier). The leftmost classifier is chosen first, and the ensemble size increases to the right. Cells are shown in light grey (class N) or black (class CA) respectively if they are predicted correctly by the corresponding ensemble member. They are shown in white if the ensemble member misclassifies the sample. For each ensemble member (column), the chosen prototypes for the two classes are highlighted as shaded cells. The figure clearly shows that there is a distinct sample (N_6) that is misclassified by almost all ensemble members. An investigation of the cross-validation results showed that the same sample is misclassified in many of the cross-validation experiments and is responsible for the typical error rate of $E_c = 0.028$ that is seen for most of the ensemble types (not only the accuracy rankings) and for most reference classifiers. The fact that the only classifier that achieves a lower error is the regularized SVM indicates that feature reduction may enable the correct prediction of this sample by eliminating noisy features.

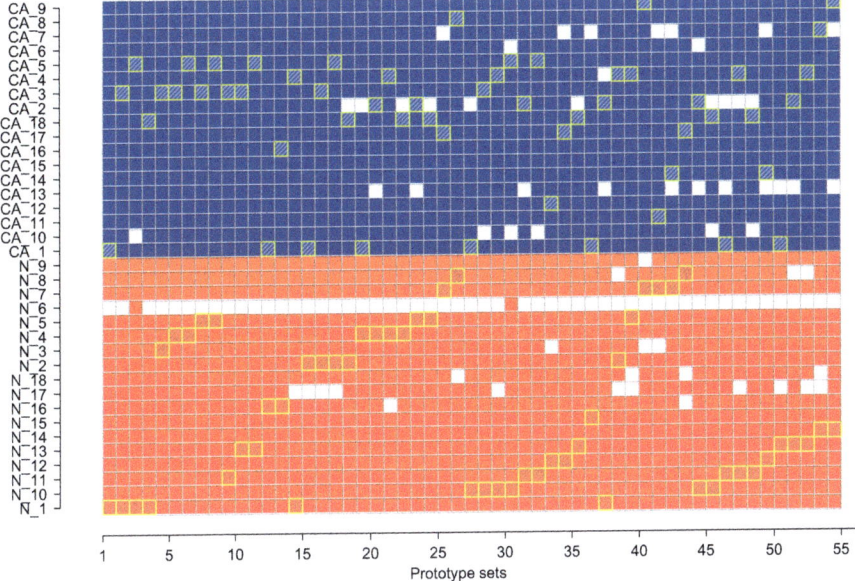

Fig. 2 Visualization of the predictions made by ensemble members chosen according to a ranking by accuracy on the training set. Here, *each row* corresponds to one sample in the data set, and *each column* corresponds to a single ensemble member. Cells are shown in *light grey* (class N) or *black* (class CA) respectively if they are predicted correctly by the corresponding ensemble member. They are shown in *white* if the ensemble member misclassifies the sample. For each ensemble member (*column*), the chosen prototypes for the two classes are highlighted as *shaded* cells

4 Discussion

A particular requirement in microarray classification is the interpretability of classifier models. Due to its inherent simplicity, the proposed RPS approach satisfies this need in many respects. For instance, one major advantage of RPS classifiers is their direct data dependency, which enables the analysis of data set characteristics through an exhaustive enumeration of RPS classifier models. Apart from descriptive analyses and visualizations, the complete set of classifiers can also serve as a basis for test statistics.

Here, we focus on the construction and analysis of ensemble methods based on RPS classifiers. We aim at constructing ensembles that preserve the interpretability of the single prototype sets. On the one hand, interpretability is supported by the use of simple fusion rules. This enables visual analyses that efficiently identify deficiencies of ensemble members (see Fig. 2). On the other hand, the selected prototypes of RPS classifiers may correspond to clinically relevant samples. It is therefore desirable to construct ensembles that use as few prototypes as possible. Our results generally indicate that the set of unique chosen prototypes remains small even for larger ensemble sizes. For example, the AdaBoost ensembles on the Notterman data set utilized no more than around eight different samples as prototypes even for large ensemble sizes. Such small sets of influential samples may be subjected to further analysis, as they may be characteristic for certain subgroups in the data.

Overall, the cross-validation performance of the ensembles can compete with state-of-the-art classification methods, such as support vector machines or the k-Nearest Neighbour classifier. AdaBoost and Bagging proved to be particularly suitable for the fusion of RPS. While RPS currently focuses on sample selection, the comparison with regularized support vector machines shows that feature reduction can bring an additional benefit for classification accuracy. Also, the Euclidean distance used here may suffer from influences of uninformative features. Further work will therefore cover the integration of feature selection into RPS classifiers and the evaluation of different distance measures.

Acknowledgements This work is supported by the German Science Foundation (SFB 1074, Project Z1) to HAK, and the Federal Ministry of Education and Research (BMBF, Gerontosys II, Forschungskern SyStaR, project ID 0315894A) to HAK.

References

Bittner, M., Meltzer, P., Chen, Y., Jiang, Y., Seftor, E., Hendrix, M., et al. (2000). Molecular classification of cutaneous malignant melanoma by gene expression profiling. *Nature, 406*, 536–540.

Breiman, L. (1996). Bagging predictors. *Machine Learning, 24*, 123–140.

Brighton, H., & Mellish, C. (2002). Advances in instance selection for instance-based learning algorithms. *Data Mining and Knowledge Discovery, 6*, 153–172.

Dasarathy, B. (1991). Nearest neighbor (NN) norms: NN pattern classification techniques. Los Alamitos: IEEE Computer Society.

Fix, E., & Hodges, J. (1951). Discriminatory analysis: Nonparametric discrimination: Consistency properties. Technical Report Project 21-49-004, Report Number 4, USAF School of Aviation Medicine, Randolf Field, TX.

Freund, Y., & Schapire, R. (1995). A decision-theoretic generalization of on-line learning and an application to boosting. In P. Vitányi (Ed.), *Computational learning theory. Lecture notes in artificial intelligence* (Vol. 904, pp. 23–37). Berlin: Springer.

Hart, P. (1968). The condensed nearest neighbor rule. *IEEE Transactions on Information Theory, 14*, 515–516.

Huang, Y., & Suen, C. (1995). A method of combining multiple experts for the recognition of unconstrained handwritten numerals. *IEEE Transactions on Pattern Analysis and Machine Intelligence, 17*, 90–94.

Kohonen, T. (1988). Learning vector quantization. *Neural Networks, 1*, 303.

Lausser, L., Müssel, C., & Kestler, H.A. (2012). Representative prototype sets for data characterization and classification. In N. Mana, F. Schwenker, & E. Trentin (Eds.), *Artificial neural networks in pattern recognition (ANNPR12). Lecture notes in artificial intelligence* (Vol. 7477, pp. 36–47). Berlin: Springer.

Lausser, L., Müssel, C., & Kestler, H. A. (2014). Identifying predictive hubs to condense the training set of k-nearest neighbour classifiers. *Computational Statistics, 29*(1–2), 81–95.

Müssel, C., Lausser, L., Maucher, M., & Kestler H. A. (2012). Multi-objective parameter selection for classifiers. *Journal of Statistical Software, 46*(5), 1–27.

Notterman, D., Alon, U., Sierk, A., & Levine, A. (2001). Transcriptional gene expression profiles of colorectal adenoma, adenocarcinoma, and normal tissue examined by oligonucleotide arrays. *Cancer Research, 61*, 3124–3130.

Schwenker, F., & Kestler, H. A. (2002) Analysis of support vectors helps to identify borderline patients in classification studies. In *Computers in cardiology* (pp. 305–308). Piscataway: IEEE Press.

Shipp, M., Ross, K., Tamayo, P., Weng, A., Kutok, J., Aguiar, R. C., et al. (2002). Diffuse large b-cell lymphoma outcome prediction by gene-expression profiling and supervised machine learning. *Nature Medicine, 8*, 68–74.

Tibshirani, R., Hastie, T., Narasimhan, B., & Chu, G. (2002). Diagnosis of multiple cancer types by shrunken centroids of gene expression. *Proceedings of the National Academy of Sciences, 99*, 6567–6572.

Vapnik, V. (1998). *Statistical learning theory*. New York: Wiley.

Wang, Q., Wen, Y.-G., Li, D.-P., Xia, J., Zhou, C.-Z., Yan, D.-W., et al. (2012). Upregulated INHBA expression is associated with poor survival in gastric cancer. *Medical Oncology, 29*, 77–83.

Webb, A. R. (2002). *Statistical pattern recognition* (2nd ed.). Chichester: Wiley.

Event Prediction in Pharyngeal High-Resolution Manometry

Nicolas Schilling, Andre Busche, Simone Miller, Michael Jungheim, Martin Ptok, and Lars Schmidt-Thieme

Abstract A prolonged phase of increased pressure in the upper esophageal sphincter (UES) after swallowing might result in globus sensation. Therefore, it is important to evaluate *restitution times* of the UES in order to distinguish physiologic from impaired swallow associated activities. Estimating the event $t^\star$ where the UES has returned to its resting pressure after swallowing can be accomplished by predicting if swallowing activities are present or not. While the problem, whether a certain swallow is pathologic or not, is approached in Mielens (J Speech Lang Hear Res 55:892–902, 2012), the analysis conducted in this paper advances the understanding of normal pharyngoesophageal activities.

From the machine learning perspective, the problem is treated as binary sequence labeling, aiming to find a sample $t^\star$ within the sequence obeying a certain characteristic: We strive for a best approximation of label transition which can be understood as a dissection of the sequence into individual parts. Whereas common models for sequence labeling are based on graphical models (Nguyen and Guo, Proceedings of the 24th International Conference on Machine Learning. ACM, New York, pp. 681–688, 2007), we approach the problem using a logistic regression as classifier, integrate sequential features by means of FFT-coefficients and a Laplacian regularizer in order to encourage a smooth classification due to the monotonicity of target labels.

N. Schilling (✉) • A. Busche • L. Schmidt-Thieme
Information Systems and Machine Learning Lab, University of Hildesheim, Hildesheim, Germany
e-mail: schilling@ismll.uni-hildesheim.de; busche@ismll.uni-hildesheim.de; schmidt-thieme@ismll.uni-hildesheim.de

S. Miller • M. Jungheim • M. Ptok
Hannover Medical High School, Klinik für Phoniatrie und Pädaudiologie, Hannover, Germany
e-mail: Miller.Simone@mh-hannover.de; Jungheim.Michael@mh-hannover.de; Ptok.Martin@mh-hannover.de

© Springer-Verlag Berlin Heidelberg 2015
B. Lausen et al. (eds.), *Data Science, Learning by Latent Structures, and Knowledge Discovery*, Studies in Classification, Data Analysis, and Knowledge Organization, DOI 10.1007/978-3-662-44983-7_30

1   Introduction

The work presented in this paper aims to support physicians analyzing high
resolution manometry data of regular swallowing activity. The medical problem
consists of estimating *restitution times* of the upper esophageal sphincter (UES)
which we define as *duration after swallowing* until swallow related activities
have subsided. We will denote this time stamp as the *restitution sample $t^\star$* from
which the restitution time can be inferred. Since estimation of *restitution times* is
time-consuming for physicians and not standardized yet, we develop a novel semi-
automatic application to assist the analysis.

1.1   Pharyngeal High-Resolution Manometry

In High-Resolution Manometry, pressures in the pharynx and UES are measured
using a probe containing k equidistantly aligned pressure sensors. This probe is
inserted through the patient's nose and stretches down to just below the UES.
Pressures are recorded at a predefined sample rate. Resulting data can be visualized
by a time-vs.-sensor plot as shown in Fig. 1 (Meyer et al. 2012).

1.2   Estimation of Restitution Times

Diagnostic and treatment of diseases such as dysphagia are expected to be enhanced
by a better understanding of human pharyngeal and especially upper esophageal
activities during swallowing. The UES maintains a basal pressure at rest to prevent
air from entering into the gastrointestinal tract during inspiration and to protect the
airways from material refluxing back from the esophagus into the pharynx. When

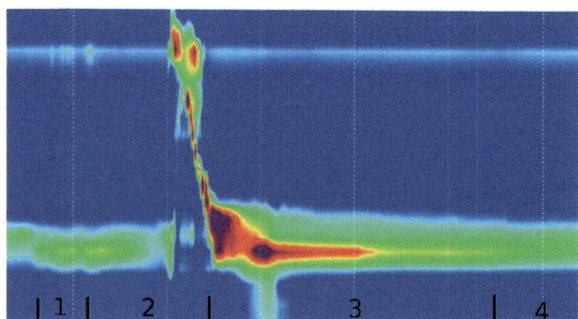

Fig. 1 Data is plotted as a time-vs.-pressure plot. Four different phases of the swallow can be
identified

swallowing, the UES relaxes, allowing a bolus to pass the sphincter region. This region can easily be depicted on the manometric recording (Fig. 1). The pressure decreases to a minimum and increases again to very high pressure values when the peristaltic wave passes the sphincter region during swallowing. After having reached the maximum pressure, sphincter pressures will eventually return to the resting pressure, this time interval is defined as *restitution time*. Since UES pressures decrease irregularly, it is difficult to define the exact $t^\star$.

For these reasons, it is important to develop a semi-automated method able to predict restitution times. The two main contributions of this paper are:

- To introduce the problem of estimating restitution times and regard it as a machine learning problem.
- To develop a method for estimating restitution times based on an annotated corpora of 100 swallows measured across 10 patients in two real-world medical application scenarios:

  1. *Intra-individual*: having learned the swallowing behavior of an individual patient, we apply the learned model to predict restitution times for the same patient.
  2. *Inter-individual*: having learned the swallowing behavior for several patients, we apply the learned model to predict restitution times of an unknown patient.

2   Related Work

2.1   Medical Background

Pressures in the pharyngeal and esophageal tract are typically measured by HRM (Fox and Bredenoord 2008). Several study groups have investigated pressure recordings associated with physiological as well as pathological swallowing (Mielens et al. 2012), analyzing among others minimal and maximal as well as resting pressures, time intervals (Meyer et al. 2012), integrals and velocities (Jungheim et al. 2013). In order to distinguish normal sphincter function from pathological swallowing activity, however, it is also necessary to evaluate the restitution time. Changes in restitution times could be indicative of, i.e. globus sensation, impaired bolus passage or regurgitation. Since the UES pressure decreases slowly but inconstantly (not in an asymptotic solution) after swallowing, it is difficult to define the exact point in time when the resting pressure is reached again. This is probably why restitution times have not been determined yet and why it is necessary to develop a computed model to determine restitution times of the UES.

2.2 Machine Learning Background

We will cast the problem as a sequence labeling problem and define the restitution sample $t^\star$ with the aid of a labeled sequence $\mathbf{y} = (y_1, \ldots, y_{|T|})$ where labels are categorical, i.e. $y_t \in \mathcal{Y}$ and $\mathcal{Y}$ is a finite set. Predicting structured output has been researched extensively throughout recent years (Nguyen and Guo 2007), especially since problems such as part-of-speech tagging, the process of categorizing words as *noun, verb, preposition* can be modeled. Earlier work Rabiner (1989) employs Hidden Markov Models on speech recognition. More recently, Lafferty et al. (2001) used Conditional Random Fields, a generalization of Markov Models for labeling sequence data.

Since $\mathbf{y} \in \mathcal{Y}^{|T|}$, the problem can also be understood as a multiclass problem and therefore be solved using multiclass Support Vector Machines. This approach is based on averaged perceptrons proposed in Collins (2002). The earliest approaches were reported by Altun et al. (2003) and Tashar et al. (2003). Nguyen and Guo (2007) found that SVM Struct Tsochantaridis et al. (2005) is among the most competitive methods and is therefore chosen as the competing method for the experiments.

3 Problem Definition

The data obtained by the High-Resolution Manometry forms a multivariate sequence of pressure values, as for each time stamp $t \in T$ an associated pressure vector $p_t \in \mathbb{R}^k$ exists, containing the recorded pressures by the sensors. Every swallow is assumed to have the same length for notation simplicity. Let $\mathcal{P}$ denote the set of all swallows of length $|T|$. Moreover, we obtain the *restitution samples* $t^\star \in T$ as ground truth information by two swallowing specialists. As these labels are subjective, we seek to learn an automated, more objective model.

3.1 Definition as Sequence Regression

One swallow consists of a sequence of associated pressure vectors and the restitution sample.

$$S^{\text{reg}} := \left(\{p_t\}_{t \in T} , t^\star \right) \in \mathcal{P} \times T. \tag{1}$$

The goal is to learn a classifier $\hat{f} : \mathcal{P} \to T$ that maps a swallow to a discrete time sample. However, one swallow corresponds to one instance, following this approach is therefore expensive since the model needs a large amount of training data which in this applicational scenario costs a lot of money and time to accumulate. Therefore, we aim to formulate the problem as sequence labeling.

3.2 Definition as Sequence Labeling

The formulation as sequence labeling involves two steps. The first step requires a definition of labels y_t (states) for each time sample t we employ binary labels $\mathcal{Y} = \{1, -1\}$.

$$S^{lab} := S := \{(p_t, y_t) \mid t \in T\}. \tag{2}$$

Consequently, we try to learn a classifier $\hat{y} : \mathbb{R}^k \to \mathcal{Y}$, which gives more instances to learn a model. One state contains all time samples t where swallow related activities are present, the other state forms the converse.

The second step consists of inferring $t^\star$ from a predicted sequence of labels. We segment the swallow according to these labels, which we accomplish through a derived maximum pressure p_{max} curve over the UES and additional knowledge in form of an annotated region of resting pressure before swallowing.

As the probe monitors the whole pharynx, the region of the UES is observed by a subset $I \subset \{1, \ldots, k\}$ of sensors. Over this subset, we compute a curve p_{max} of maximum pressure

$$p_{max}(t) := \max \{p_i(t) \mid i \in I\} \qquad t^{max} := \underset{t}{\operatorname{argmax}} \; p_{max}(t),$$

which assigns the maximum pressure in the UES region to every time sample t. Moreover, we denote its maximum position by t^{max}, using these properties enables us to segment a swallow into four different phases as can be seen in Fig. 1.

The first phase of the swallow describes the annotated resting pressure before swallowing. The state $y_t = -1$ is assigned. The second phase contains the beginning of activities in the velopharynx, until the t^{max} sample is reached. We will perform a supervised exclusion for phase two as it belongs to neither of the states. The third phase begins at the t^{max} sample, as from this point on swallow related activities of the UES are present we assign $y_t = 1$ for these time samples. The remaining samples are labeled $y_t = -1$, since by definition of $t^\star$ swallow related activities have subsided.

3.3 Deriving the Restitution Sample from a Labeled Sequence

For the sequence labeling techniques presented in the next section, we are required to derive the restitution sample if we are given a predicted labeled sequence of a test swallow. This procedure is given in Algorithm 1. Based on the segmentation presented in Fig. 1, going ahead in time from the t^{max} sample, we determine $t^\star$ as the first transition of states.

$$t^\star = \min \{i > t^{max} \mid y_i = 1 \land y_{i+1} = -1\}. \tag{3}$$

Algorithm 1 PredictRestitutionSample

1: **procedure** PREDICT
   input: $\mathcal{D}^{test}$, w

2:    $\forall\, x_t \in \mathcal{D}^{test}:$ $\hat{y}_t \leftarrow \hat{y}(x_t, \theta^*)$
3:    $\forall\, \hat{y}_t:$ $\overline{y}_t \leftarrow \frac{1}{2w+1} \sum_{i=t-w}^{t+w} \hat{y}_i$
4:    $t^* \leftarrow \min\{i > t^{max} \,|\, \overline{y_i} \geq 0 \wedge \overline{y_{i+1}} < 0\}$
5: **end procedure**

In an ideal case, we have a classification sequence where the labels change only once for all $t > t^{max}$, when the swallowing activities have subsided. Problems may arise when the classification is not perfectly smooth, consider a change of labels for only a short time period. We will overcome this issue by smoothing the final prediction with a predefined window size w representing an additional hyperparameter.

3.4 Preprocessing and Derivation of Additional Features

As a preprocessing step, we normalize the swallows individually, as to (a) account for possible probe calibration offsets, and to (b) roughly align the value ranges for swallows of different lengths. Normalizing the swallows patient-wise as opposed to normalizing each swallow individually has empirically proven to result in a weaker classification in our preliminary experiments.

Furthermore, since the UES region defined by I differs interindividually, we compute derived sphincter features as follows: we calculate local pressure maxima over the first, second, and last third of the UES region. We also repeat this process for the first and second half and for the overall UES region. Consequently, we obtain six additional sphincter features, denoted by s_t and discard the initial pressure features.

Moreover, from the maximum pressure curve p_{max}, we derive additional features by committing a discrete Fourier transformation (FFT) on every time stamp t of the p_{max} curve, using a constant window size $b = 128$. The obtained coefficients $c_t = (c_1^t, \ldots, c_b^t)$ of the Fourier transform, together with the current value $p_{max}(t)$, the extracted sphincter features s_t, and the label information are then concatenated to a vector $x_t \in \mathbb{R}^n$, which leads to the following swallow representation S and swallow data set $\mathcal{D}$.

$$S := \{(x_t, y_t) := (p_{max}(t), c_t, s_t, y_t) \,|\, t \in T\} \qquad \mathcal{D} = \{S_1, \ldots, S_{|\mathcal{D}|}\}. \qquad (4)$$

4 Proposed Methods

4.1 Prediction Based on Maximum Pressure Curve

The simplest prediction method is based only on the p_{max} curve. From the given resting pressure, we compute an average resting pressure p^{avg} over all resting pressure samples. Given these figures, we compute a labeled sequence as:

$$\hat{y}_t := \begin{cases} 1 & \text{if} \quad p_{max}(t) \geq p^{avg} \\ -1 & \text{else.} \end{cases} \tag{5}$$

Following Eq. (3), the restitution sample t^* is then estimated as the first sample $t > t^{max}$ where the p_{max} curve falls below the average resting pressure. This is a very simple method as it involves neither learning nor preprocessing besides computing p_{max}, and does not take sequential information into account. Its shortcomings will be discussed in the experiment section.

4.2 Logistic Regression

We perform a logistic regression, where model parameters $\theta \in \mathbb{R}^n$ are learned from the labeled data, optimized for logistic loss, and regularized by the common l^2-regularizer as to avoid overfitting:

$$\mathcal{L}_{Log}(\hat{y}(\theta), \mathcal{D}) = \sum_{x_i \in \mathcal{D}} \log\left(1 + e^{-y_i \langle \theta, x_i \rangle}\right) + \lambda \|\theta\|_2^2. \tag{6}$$

The model parameters θ are initialized randomly by a Gaussian $\mathcal{N}(0, \sigma^2)$ and iteratively optimized using gradient descent with a fixed step size η. Note that $\langle \cdot, \cdot \rangle$ denotes the scalar product in $\mathbb{R}^n$ and λ denotes a regularization parameter.

4.3 Laplacian Logistic Regression

We integrate the knowledge that a label change $(1 \rightarrow -1)$ is expected only once after t^{max}. To force learned parameters to adopt this property, we penalize label changes throughout the sequence by introducing a Laplacian regularizer $\mathcal{S}(\theta)$ on neighboring samples,

$$\mathcal{S}(\theta) := \sum_{x_i \in \mathcal{D}} \sum_{x_j \in w(x_i, s)} \frac{1}{2}\left(\sigma\left(\langle \theta, x_i \rangle\right) - \sigma\left(\langle \theta, x_j \rangle\right)\right)^2, \tag{7}$$

where $\sigma(\cdot)$ is a sigmoid function in order to make $S(\theta)$ differentiable, and $w(x_i, s) := \{x_{i-s}, \ldots, x_i, \ldots, x_{i+s}\}$ is the set of neighboring samples for a predefined window length s. This term will be added to the overall loss functional given in Eq. (6),

$$\mathcal{L}_{\text{Lap}}(\hat{y}(\theta), \mathcal{D}) := \mu \cdot \mathcal{L}_{\text{Log}}(\hat{y}(\theta), \mathcal{D}) + (1 - \mu) \cdot S(\theta) \tag{8}$$

using a weighting coefficient $\mu \in [0, 1]$ to capture the tradeoff between learning an accurate solution and learning a smooth solution.

5   Experiments

We are using SVM-HMM (Altun et al. 2003), as a competitor method for predicting structured output by converting our data as follows: One entire swallow is treated as a sentence, while one time stamp is treated as a token. As there are only two possible states, we convert these two states into two tags. More details concerning the implementation of SVM-HMM can be found in Joachims et al. (2009).

Regarding our initial questions on the performance of the proposed methods, we design two experiments as follows.

5.1   Experimental Setup

The dataset used in the experiments consists of ten patients, who have each conducted ten swallows, forming a dataset of 100 swallows to be used in our experiments, which is sufficient for a first feasibility analysis. The data is split per swallow into a training and a test set according to the use case, and determine optimal hyperparameters on a validation set using grid search.

Accuracy scores are reported for the predicted sequence as well as absolute time differences between the predicted restitution sample and the true restitution sample. For the Logistic Regression, we evaluated accuracies and sample differences after every iteration, while for the SVM-HMM, we employed the model learned after convergence of the algorithm. Note that, for an accurate prediction of the restitution sample t^*, the accuracy of the predicted binary sequence correlates only to a certain extent with finding the correct t^*, but does not necessarily lead to a correct prediction of t^*. As such, a model with a higher accuracy might predict an inaccurate t^*.

5.2 Hyperparameter Optimization

As stated earlier, the hyperparameters have been optimized using grid search. The step size η was searched in $10^{-5} \cdot \{1, 0.7, 0.5, 0.1\}$, the regularization constant λ and the window size w have been searched in $\{0.001, 0.01, 0.1, 1\}$ and $\{10, 75, 150\}$, respectively. For the Laplacian Logistic Regression, μ and the window extent s were searched in $\{0.01, 0.5, 1\}$ and $\{3, 5, 11, 31\}$.

The hyperparameters concerning the SVM-HMM were searched on the following grid. C was searched in $\{10^{-4}, \ldots, 10^3\}$, epsilon was set to 0.5 as a suggestion of the authors, the number of transitions was between 1 and 3, the number of emission was either 0 or 1.

5.3 Use Case 1: Intra-Individual

We train the classifiers individually in a leave-one-out cross validation, where we omit one swallow as test swallow for which we want to predict the restitution sample. Of the remaining swallows, two swallows are randomly picked as validation data. Thus, we have seven training swallows, two validation swallows and one test swallow.

Table 1 shows the results for all different methods. As can be seen clearly, SVM-HMM wins in accuracy, but the Laplacian Logistic Regression is best in predicting time differences. For all methods, we chose the model which gave the best accuracy on validation. Then, we apply the window extent that gives the best sample difference score on validation and apply the model with the chosen window extent. Note that the time differences are the target we are actually looking to optimize.

We can also see that the p_{max} method works well on Patient 1 and Patient 8. Nevertheless, the predicted t^* is inferior for very many other patients, such as 2, 5, 9, and 10.

Moreover, we observe that predicting restitution times seems to differ a lot regarding the individual patients as, for instance, for patient 2, the time differences are quite low, whereas especially for patient 10, the time differences are large. The confidence intervals reported are wide, since every patient conducted only ten swallows. Nevertheless, we see that our method outperforms the SVM-HMM for the majority of patients and we also see that including a Laplacian regularizer aids in finding t^*.

5.4 Use Case 2: Inter-Individual

We train the classifiers on nine of the ten patients (training patients) and predict the restitution times for the remaining test patient. We randomly leave two out of the

Table 1 Results for the Intra-Individual use case are shown

Patient	SVM-HMM	LogReg	LapLogReg	p_{max}	SVM-HMM	LogReg	LapLogReg
1	84.81 ± 6.3	82.75 ± 6.2	**85.49** ± 6.3	5.32 ± 3.65	9.72 ± 2.42	10.48 ± 2.83	**10.01** ± 2.45
2	81.45 ± 5.6	86.29 ± 9.0	**86.91** ± 9.2	22.84 ± 5.42	2.16 ± 1.03	1.41 ± 1.64	**0.58** ± 0.39
3	**83.20** ± 5.1	82.00 ± 4.4	80.96 ± 6.4	7.49 ± 3.42	**6.45** ± 2.74	6.66 ± 3.63	6.92 ± 3.44
4	85.86 ± 2.7	86.02 ± 3.1	**87.18** ± 4.5	5.08 ± 3.24	**3.20** ± 2.48	4.07 ± 3.78	4.12 ± 3.75
5	**88.52** ± 5.5	71.20 ± 4.1	87.20 ± 4.5	10.17 ± 4.74	2.69 ± 1.64	2.65 ± 1.34	**1.83** ± 1.27
6	**79.13** ± 5.1	68.29 ± 3.8	65.97 ± 3.2	7.19 ± 5.32	6.64 ± 1.93	3.26 ± 1.81	**2.50** ± 1.56
7	**86.51** ± 4.3	61.00 ± 1.9	61.64 ± 3.4	9.43 ± 6.05	3.90 ± 1.52	3.13 ± 1.30	**2.74** ± 1.12
8	**88.65** ± 9.4	66.85 ± 4.3	64.62 ± 3.8	3.47 ± 2.49	**4.18** ± 4.73	5.72 ± 4.92	6.46 ± 4.76
9	**78.38** ± 12.7	60.77 ± 6.2	62.24 ± 7.6	16.05 ± 5.54	8.48 ± 5.12	9.02 ± 4.58	**8.37** ± 4.82
10	**86.51** ± 4.0	63.34 ± 3.5	65.38 ± 4.6	18.12 ± 8.55	13.45 ± 9.10	**12.30** ± 10.69	14.16 ± 10.15

Average accuracies and 95 %-confidence intervals are reported on the left. Average sample differences converted to seconds and their confidence intervals are reported on the right

Table 2 Results for the inter-individual use case are shown

Patient	SVM-HMM	LogReg	LapLogReg	SVM-HMM	LogReg	LapLogReg
1	79.80 ± 8.2	83.76 ± 6.8	**83.79** ± 7.0	6.70 ± 6.70	**3.27** ± 2.32	4.24 ± 2.50
2	**87.45** ± 5.6	84.63 ± 8.2	85.10 ± 8.6	3.04 ± 1.03	1.02 ± 0.41	**0.97** ± 0.49
3	**78.48** ± 4.7	66.03 ± 9.8	65.64 ± 10.3	8.76 ± 3.17	**6.58** ± 3.12	7.19 ± 3.06
4	76.22 ± 7.3	80.02 ± 5.9	**81.29** ± 5.6	11.62 ± 2.41	5.37 ± 2.90	**4.56** ± 2.63
5	**90.26** ± 5.0	81.00 ± 8.5	80.67 ± 9.1	**2.36** ± 1.72	3.33 ± 1.79	3.29 ± 1.83
6	**88.08** ± 6.4	79.79 ± 8.1	79.99 ± 8.8	**1.87** ± 0.59	4.31 ± 3.47	4.17 ± 3.53
7	**71.29** ± 8.7	62.31 ± 8.4	68.70 ± 6.5	**4.62** ± 1.51	5.69 ± 1.53	6.11 ± 1.43
8	83.81 ± 13.4	87.94 ± 10.3	**88.47** ± 10.6	5.31 ± 4.34	3.81 ± 4.67	**3.00** ± 2.93
9	**84.20** ± 13.5	80.15 ± 8.5	80.31 ± 8.6	9.02 ± 5.54	6.17 ± 3.35	**6.13** ± 3.38
10	83.05 ± 6.3	84.10 ± 4.5	**84.78** ± 4.5	14.32 ± 8.61	10.47 ± 9.52	**10.12** ± 9.60

Average accuracies and 95 %-confidence intervals are reported on the left. Average sample differences converted to seconds and confidence intervals are reported on the right

remaining nine swallows per training patient as validation data. Thus, we build a training set of 72 swallows, a validation set of 18 swallows and a test set consisting of one swallow of the test patient. For each patient, ten different splits have been created.

Table 2 shows the results for the competing methods. Analogous to the first use case, we can observe that SVM-HMM outperforms our approaches with respect to accuracy, even though the margin is not that large anymore. In comparison with use case 1, predicting t^* for patient 1 seems to give better results, when the model did not learn on the same patient's swallows, which is rather surprising as for the majority of patients, the predicted t^* is worse. For this use case, adding a Laplacian regularizer seems to work best for some patients.

6 Conclusions and Future Work

We introduced the problem of estimating restitution times and formulated it as a machine learning problem suitable for semi-automation. For the learning process, an annotated corpora of swallows is required and parameters, such as the window size, have to be tuned manually. Furthermore, our initial study has shown that predicting restitution times is generally possible with acceptable accuracy. Among the tested methods, we empirically showed that the Laplacian Logistic Regression is the most promising method for predicting restitution times. However, as the swallow patterns differ inter-individually, we aim for a more stable and more accurate estimation in future work.

References

Altun, Y., Tsochantaridis, I., & Hoffmann, T. (2003). Hidden Markov support vector machines. *International Conference on Machine Learning, 3*, 3–10.

Collins, M. (2002). Discriminative training methods for hidden markov models: Theory and experiments with perceptron algorithms. In *EMNLP* (Vol. 10, pp. 1–8).

Fox, M. R., & Bredenoord, A. J. (2008). Oesophageal high-resolution manometry: Moving from research into clinical practice. *Gut, 75*, 405–423.

Joachims, T., Finley, T., & Chun-Nam, Y. (2009). Cutting-plane training of structured SVMs. *Machine Learning Journal, 72*(1), 27–59.

Jungheim, M., Miller, S., & Ptok, M. (2013). Methodologische Aspekte zur Hochauflösungsman-ometrie des Pharynx und des oberen Ösophagussphinkters. *Laryngo-Rhino-Otol, 92*, 158–164.

Lafferty, J., McCallum, A., & Pereira, F. (2001). Conditional random fields: Probabilistic models for segmenting and labeling sequence data. In *International Conference on Machine Learning* (pp. 282–289).

Meyer, S., Jungheim, M., & Ptok, M. (2012). High-resolution manometry of the upper esophageal sphincter. *HNO, 60*(4) , 318–326.

Mielens, J. D., Hoffman, M. R., Ciucci, M. R., McCulloch, T. M., & Jiang, J. J. (2012). Application of classification models to pharyngeal high-resultion manometry. *Journal of Speech, Language, and Hearing Research, 55*, 892–902.

Nguyen, N., & Guo, Y. (2007). Comparison of sequence labeling algorithms and extensions. In Z. Ghahramani (Ed.), *Proceedings of the 24th International Conference on Machine Learning* (pp. 681–688). New York, NY: ACM.

Rabiner, L. R. (1989). A tutorial on hidden markov models and selected applications in speech recognition. *Proceedings of the IEEE, 77*(2), 257–286.

Tashar, B., Guestrin, C., & Koller, D. (2003). Max-margin Markov networks. In *Advances in Neural Information Processing Systems* (Vol. 16).

Tsochantaridis, I., Joachims, T., Hofmann, T., & Altun, Y. (2005). Large margin methods for structured and interdependent output variables. *Journal of Machine Learning Research, 6*, 1453–1484.

Edge Selection in a Noisy Graph by Concept Analysis: Application to a Genomic Network

Valentin Wucher, Denis Tagu, and Jacques Nicolas

Abstract MicroRNAs (miRNAs) are small RNA molecules that bind messenger RNAs (mRNAs) to silence their expression. Understanding this regulation mechanism requires the study of the miRNA/mRNA interaction network. State-of-the-art methods for predicting interactions lead to a high level of false positive: the interaction score distribution may be roughly described as a mixture of two overlapping Gaussian laws that need to be discriminated with a threshold. In order to further improve the discrimination between true and false interactions, we present a method that considers the structure of the underlying graph. We assume that the graph is formed on a relatively simple structure of formal concepts (associated with regulation modules in the regulation mechanism). Specifically, the formal context topology of true edges is assumed to be less complex than in the case of a noisy graph including spurious interactions or missing interactions. Our approach consists thus in selecting edges below an edge score threshold and applying a repair process on the graph, adding or deleting edges to decrease the global concept complexity. To validate our hypothesis and method, we have extracted parameters from a real biological miRNA/mRNA network and used them to build random networks with fixed concept topology and true/false interaction ratio. Each repaired network can be evaluated with a score balancing the number of edge changes and the conceptual adequacy in the spirit of the minimum description length principle.

V. Wucher (✉)
INRA, UMR 1349 IGEPP, Le Rheu 35653, France

IRISA-INRIA, Campus de Beaulieu, 35042 Rennes cedex, France
e-mail: v.wucher@gmail.com

D. Tagu
INRA, UMR 1349 IGEPP, Le Rheu 35653, France
e-mail: denis.tagu@rennes.inra.fr

J. Nicolas
IRISA-INRIA, Campus de Beaulieu, 35042 Rennes cedex, France
e-mail: jacques.nicolas@inria.fr

© Springer-Verlag Berlin Heidelberg 2015
B. Lausen et al. (eds.), *Data Science, Learning by Latent Structures, and Knowledge Discovery*, Studies in Classification, Data Analysis, and Knowledge Organization, DOI 10.1007/978-3-662-44983-7_31

1 Introduction

MicroRNAs (miRNAs) are small RNA molecules that bind to and regulate the flow of messenger RNAs (mRNAs). They have a sequence of six nucleotides that bind to a complementary sequence, the binding site, of the target mRNA. Bound miRNAs repress the expression of their target mRNAs.

The interaction network created by miRNAs/mRNAs interactions is by definition a bipartite graph between miRNA nodes and mRNA nodes. Several bioinformatics methods can predict miRNAs/mRNAs interactions. The current state of the art offers only methods having a high level of false positive predictions (Chil et al. 2009; Reyes-Herrera et al. 2011). Even with scoring functions and a threshold, it is still hard to discriminate between true and false predictions.

Based on the biological function of miRNAs, i.e. repressing mRNAs translation, and their implication in many biological processes (Janga and Vallabhaneni 2011), authors have provided some evidence that miRNAs combine to regulate functional modules, i.e. clusters of mRNAs sharing similar functions (Bryan et al. 2014; Enright et al. 2003). This assumption is compatible with the observations of similar complexes for another major regulation actor, transcription factors. Thus true interactions could be distinguished in principle from false one on the basis of functional clusters (modules), i.e. set of miRNAs that regulate mRNAs with the same function.

Once a score threshold has been set, we intend to improve edge selection by detecting false negatives and false positives by taking into account the previous assumption in the framework of formal concept analysis.

2 Definition of Formal Concept Analysis

This section briefly recalls some notions of formal concept analysis as defined by Ganter and Wille (1999) and Klimushkin et al. (2010).

A *formal context* is a triplet $\mathbb{K} = (G, M, I)$ where G is the set of objects, M the set of attributes, and $I \subseteq G \times M$ is a binary relation between objects and attributes. The operator $(.)'$ is defined on $\mathbb{K}$ for $A \subseteq G$ and $B \subseteq M$ as: $A' = \{m \in M \,|\, \forall g \in A : gIm\}$ and $B' = \{g \in G \,|\, \forall m \in B : gIm\}$. A' is the set of common attributes to all objects in A and B' the set of common objects to all attributes in B.

A *formal concept* is a pair (A, B) defined on $\mathbb{K}$ with $A \subseteq G$ and $B \subseteq M$ where $A = B'$ and $B = A'$. Concept ordering can be based on set inclusion: For all formal concepts (A, B) and (C, D), let $(A, B) \leq (C, D)$ if $A \subseteq C$. If $(A, B) \leq (C, D)$ and there is no formal concept (E, F) such that $(A, B) < (E, F) < (C, D)$, then we write $(A, B) \prec (C, D)$.

The relation $<$ generates a *concept lattice* structure $\mathfrak{B}(\mathbb{K})$ on context $\mathbb{K}$. The order $\prec$ generates the edges in the covering graph of $\mathfrak{B}(\mathbb{K})$.

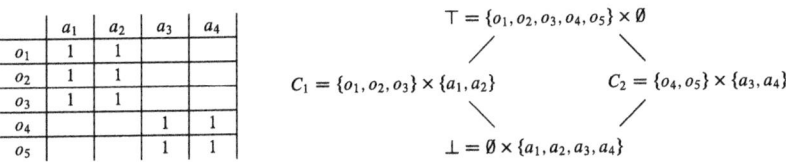

	a_1	a_2	a_3	a_4
o_1	1	1		
o_2	1	1		
o_3	1	1		
o_4			1	1
o_5			1	1

$\top = \{o_1, o_2, o_3, o_4, o_5\} \times \emptyset$

$C_1 = \{o_1, o_2, o_3\} \times \{a_1, a_2\}$ $C_2 = \{o_4, o_5\} \times \{a_3, a_4\}$

$\bot = \emptyset \times \{a_1, a_2, a_3, a_4\}$

Fig. 1 A formal context $\mathbb{K}_{ex}$ (*left*) with $G_{ex} = \{o_{1...5}\}$ the set of objects and $M_{ex} = \{a_{1...4}\}$ the set of attributes and the associated concept lattice $\underline{\mathfrak{B}}(\mathbb{K}_{ex})$ (*right*)

Figure 1 gives a small example of formal context and the associated concept lattice. It contains four formal concepts, namely C_1, C_2, the top concept $\top$, and the bottom concept $\bot$.

3 The Effect of Noise on Formal Concept Analysis

Formal concept analysis is a powerful method for binary data analysis because it extracts every complete group of related elements, i.e. such that every element from one set is related to every element in the second set. This advantage becomes a drawback in case of noisy data, because of its sensitivity to the presence of each relation.

Studies have already been conducted on fault-tolerant or approximated concepts analysis (Besson et al. 2005; Belohlavek and Vychodil 2006; Blachon et al. 2007). It consists mostly in retrieving dense rectangles of 1 in a binary matrix: a concept is indeed a submatrix filled with 1 values, up to line and column reordering. The constraint of requiring a complete set of 1 may be released by an optimization constraint requiring a maximal number of 1. Very few works exist aiming at retrieving original concepts from noisy formal concepts. One of the most advanced studies in this domain is due to Klimushkin et al. (2010), which showed that formal concepts can be recovered from a formal context including false relations and between 300 to 400 objects and 4 to 12 attributes. They used three statistical values on concepts to find the original concepts and concept lattice.

The next subsection introduces a toy example of noisy context in order to illustrate the effect of noise on the associated concept lattice. A more formal characterization of this effect is provided in a subsection.

3.1 Example of Noise Effect

In the context $\mathbb{K}_{noise}$ (Fig. 2), one spurious relation (o_5, a_2) has been added compared with Fig. 1 and a dissimilarity score is available for every relation. By setting a threshold of -0.2 and keeping every relation below this threshold, (o_3, a_2), an

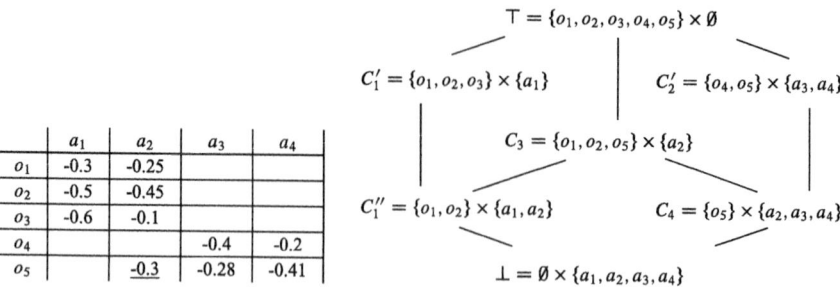

	a_1	a_2	a_3	a_4
o_1	-0.3	-0.25		
o_2	-0.5	-0.45		
o_3	-0.6	-0.1		
o_4			-0.4	-0.2
o_5		-0.3	-0.28	-0.41

Fig. 2 Formal context $\mathbb{K}_{\text{noise}}$ (*left*) with scores (spurious relation underlined) and the associated concept lattice $\underline{\mathfrak{B}}(\mathbb{K}_{\text{noise}})$ (*right*) obtained at threshold -0.2

original relation, is discarded while (o_5, a_2), a spurious relation, is kept. There are now seven concepts, three more than in $\mathbb{K}_{\text{ex}}$ (see Fig. 2). The deletion of element (o_3, a_2) has split concept C_1 into two concepts C_1' and C_1''. Concept C_2 still exists in $\mathbb{K}_{\text{noise}}$, renamed C_2' in the figure. Two new concepts, C_3 and C_4 have been created due to the addition of (o_5, a_2).

3.2 Spurious Relations

To better understand the local effect of spurious relations on a concept, we need to discriminate two types of relations: the original relations, $I^o \subseteq G \times M$ and the spurious relations, $I^s \subseteq G \times M$ with $I^o \cap I^s = \emptyset$. These two types of relations involve three types of contexts: the original context with no spurious relation $\mathbb{K}^o = (G, M, I)$, the context containing only the spurious relations $\mathbb{K}^s = (G, M, I^s)$, and the context with all the relations $\mathbb{K}^{os} = (G, M, (I \cup I^s))$. They generate three types of formal concepts, the set of original concepts $\mathfrak{C}^o$ defined on $\mathbb{K}^o$, the set of spurious concepts $\mathfrak{C}^s$ defined on $\mathbb{K}^s$, and the set of concepts $\mathfrak{C}^{os}$ defined on $\mathbb{K}^{os}$. The construction of $\mathfrak{C}^{os}$ from $\mathfrak{C}^o$ and $\mathfrak{C}^s$ depends on the contribution of each concept pair in $\mathfrak{C}^o \times \mathfrak{C}^s$.

Consider $C^o = (A^o, B^o) \in \mathfrak{C}^o$ and $C^s = (A^s, B^s) \in \mathfrak{C}^s$. Since I^o and I^s are exclusive, the concepts in $\mathfrak{C}^o$ and $\mathfrak{C}^s$ need to be disjoint. It means that either $A^o \cap A^s = \emptyset$ or $B^o \cap B^s = \emptyset$.

Assume with no lack of generality that $A^o \cap A^s \neq \emptyset$ and $B^o \cap B^s = \emptyset$. Then a new concept $C^{os} = (A^{os}, B^{os})$ may be created with $A^{os} = A^o \cap A^s$ and $B^{os} = B^o \cup B^s$. Note that if $A^s \subseteq A^o$ (resp. $A^o \subseteq A^s$), then C^s (resp. C^o) is not maximal in $\mathbb{K}^{os}$ since it is included in C^{os}.

Formally the contribution of two disjoint concepts to the set of extended concepts $\mathfrak{C}^{os}$ can be defined through the application of an inclusion operator:

Definition 1 The inclusion operator $i(.,.)$ is defined for a pair of disjoint concepts $(C^i, C^j) = ((A^i, B^i), (A^j, B^j))$ as $i(C^i, C^j) = \mathfrak{C}^{i \cup j}$ where $\mathfrak{C}^{i \cup j}$ is the set of concepts obtained on relation $\{(A^i \times B^i) \cup (A^j \times B^j)\}$.

The various types of results deriving from i application depending on the intersection between object or attribute sets are listed below:

$$i(C^i, C^j) = \{C^i, C^j\} \qquad\qquad \text{if } A^i \cap A^j = B^i \cap B^j = \emptyset; \quad (1)$$
$$= \{(A^i \cup A^j, B^i \cup B^j)\} \qquad \text{if } A^i = A^j \text{ or } B^i = B^j; \quad (2)$$
$$= \{C^j, (A^i \cup A^j, B^i \cup B^j)\} \qquad \text{if } A^i \subset A^j \text{ or } B^i \subset B^j; \quad (3)$$
$$= \{C^i, C^j, (A^i \cap A^j, B^i \cup B^j)\} \quad \text{if } A^i \cap A^j \nsubseteq \{\emptyset, A^i, A^j\}; \quad (4)$$
$$= \{C^i, C^j, (A^i \cup A^j, B^i \cap B^j)\} \quad \text{if } B^i \cap B^j \nsubseteq \{\emptyset, B^i, B^j\}. \quad (5)$$

$\mathfrak{C}^{os}$ can be defined using a fixpoint characterization: $\mathfrak{C}^{os}$ is the smallest set of concepts that cover the concepts of $\mathfrak{C}^s$ and $\mathfrak{C}^o$ and is closed under i. Concepts from $\mathfrak{C}^s$ and $\mathfrak{C}^o$ and concepts generated by operator i belong to $\mathfrak{C}^{os}$ if they are not covered by other concepts from $\mathfrak{C}^{os}$ as described above.

3.3 Missing Relations

As for spurious relations, one can proceed by distinguishing two types of relations: the original relations, $I^o \subseteq G \times M$ and the missing relations, $I^m \subseteq I^o$. They imply three types of contexts: the original context without missing relations $\mathbb{K}^o = (G, M, I^o)$, the context containing only the missing relations $\mathbb{K}^m = (G, M, I^m)$, and the context with all except the missing relations $\mathbb{K}^{om} = (G, M, (I^o \setminus I^m))$. These contexts entail three types of formal concepts, the set of original concepts $\mathfrak{C}^o$ defined on $\mathbb{K}^o$, the set of missing concepts $\mathfrak{C}^m$ defined on $\mathbb{K}^m$, and the set of concepts $\mathfrak{C}^{om}$ defined on $\mathbb{K}^{om}$. As for spurious relations, the objective is to describe how the sets $\mathfrak{C}^o$ and $\mathfrak{C}^m$ are combining in $\mathfrak{C}^{om}$. We first describe the general case where the relations of a concept in $\mathfrak{C}^m$ are overlapping those of a concept in $\mathfrak{C}^o$.

Consider $C^o = (A^o, B^o) \in \mathfrak{C}^o$ and $C^m = (A^m, B^m) \in \mathfrak{C}^m$, if $A^o \cap A^m \neq \emptyset$ and $B^o \cap B^m \neq \emptyset$, then the concept C^o cannot be in $\mathfrak{C}^{om}$ since it includes missing relations $A^m \times B^m$. Instead, two new concepts will be created in $\mathfrak{C}^{om}$, $C_1^{om} = (A^o, B^o \setminus B^m)$ and $C_2^{om} = (A^o \setminus A^m, B^o)$. Note that if $A^o \subseteq A^m$ (resp. $B^o \subseteq B^m$), then only the concept C_1^{om} is created (resp. C_2^{om}).

Formally the contribution of two overlapping concepts to the set of restricted concepts $\mathfrak{C}^{om}$ can be defined through the application of an exclusion operator:

Definition 2 The exclusion operator $e(.,.)$ is defined for a pair of overlapping concepts $(C^i, C^j) = ((A^i, B^i), (A^j, B^j))$ as $e(C^i, C^j) = \mathfrak{C}^{j \setminus i}$ where $\mathfrak{C}^{j \setminus i}$ is the set of concepts obtained on relation $\{(A^j \times B^j) \setminus (A^i \times B^i)\}$.

The various types of results deriving from e application depending on the intersection between object and attribute sets are listed below:

$$e(C^i, C^j) = C^j \qquad\qquad\qquad \text{if } A^j \cap A^i \text{ or } B^j \cap B^i = \emptyset; \quad (6)$$

$$= \{(A^j, B^j \setminus B^i), (A^j \setminus A^i, B^j)\} \quad \text{if } A^j \cap A^i \neq \emptyset, B^j \cap B^i \neq \emptyset; \quad (7)$$

$$= \{(A^j, B^j \setminus B^i)\} \qquad\qquad \text{if } A^j \subseteq A^i, B^j \not\subseteq B^i; \quad (8)$$

$$= \{(A^j \setminus A^i, B^j)\} \qquad\qquad \text{if } A^j \not\subseteq A^i, B^j \subseteq B^i; \quad (9)$$

$$= \emptyset \qquad\qquad\qquad\qquad \text{if } A^j \subseteq A^i, B^j \subseteq B^i. \quad (10)$$

$\mathfrak{C}^{om}$ can be defined using a fixpoint characterization: $\mathfrak{C}^{om}$ is the largest set of concepts which are included in the concepts of $\mathfrak{C}^o$ and is closed under e. Concepts from $\mathfrak{C}^o$ and concepts generated by operator e belong to $\mathfrak{C}^{om}$ if they do not contain a relation of I^m as described above.

3.4 Managing the Noise

The previous study points out that the number of concepts will increase depending on the type of noisy relations (spurious or missing) and the number of purely noisy concepts except for Eqs. (1) and (6) where no new concepts are created. For spurious relations, the number of new concepts in $\mathfrak{C}^{os}$ is bounded by the number n_s of disjoint concepts $C_j^s \in \mathfrak{C}^s$ with only one set that intersect with C and is bounded by n_s. For missing relations, the number of new concepts $C_i^{om} \in \mathfrak{C}^{om}$ locally created from a concept C depends on the number n_m of concepts $C_j^m \in \mathfrak{C}^m$ that overlap with C and is bounded by 2^{n_m}. Overall, the evolution of the number of new concepts is linear when adding spurious concepts and exponential when deleting missing concepts. To repair the context $\mathbb{K}^{osm}$ (the context with $I^{osm} = ((I^o \cup I^s) \setminus I^m)$) in order to retrieve $\mathbb{K}^o$, we need to define new operations that reverse the effect of operators i and e. These operations take advantage of the fact that most of the time, concepts resulting from the application of i or e are connected in the concept lattice by a direct relation or a sibling relation.

Concerning operator i, in Eq. (3) the two result concepts are ordered by $\prec$ in the concept lattice. As for Eqs. (4) and (5), the new concept is the direct precursor or the direct successor of C^i and C^j in the concept lattice. For operator e, in Eq. (7) the result concepts are ordered by $\prec$. The two sets A^j and B^j of the original concept can be easily recovered by crossing the noisy concepts.

4 Repair Process

4.1 Definition of Repair Operations

We have introduced two operations *delete* and *add* resp. defined from operators i and e, which select then suppress or insert relations based on concept lattice analysis. We assume in the following that these operations act on a pair of concepts (X, Y) with $X = (A, B)$ and $Y = (C, D)$.

Two types of selected (X, Y) pair selection exist, link pair if $X \prec Y$ and sibling pair, $X \approx Y$, if $\exists Z | X \prec Z$ and $Y \prec Z$. The following operations apply on these pairs:

$\forall (X, Y) \mid X \prec Y$ or $Y \prec X$ or $X \approx Y$:

$delete(X, Y) : \mathfrak{C} := \mathfrak{C} - Y;$ $\qquad\qquad\qquad$ $Noise := Noise \cup Y \setminus X;$

$\forall (X, Y) \mid X \prec Y$ or $Y \prec X$:

$add(X, Y) : \mathfrak{C} := \mathfrak{C} - X - Y + (C, B);$ $\quad$ $Noise := Noise \cup (C \setminus A) \times (B \setminus D);$

where $\mathfrak{C}$, initially the set of observed concepts, is the resulting set of concepts and *Noise* is the set of spurious or missing interactions.

In Fig. 2, $(C_2', C_4)_l$ and $delete(C_2', C_4)$ $=$ $delete((\{o_4, o_5\}, \{a_3, a_4\}),$ $(\{o_5\}, \{a_2, a_3, a_4\}))$ results in deleting a spurious relation: $\mathfrak{C} := \mathfrak{C} - C_4$ and $Noise := Noise \cup \{(o_5, a_2)\}$. The same way, $(C_1'', C_1')_l$ and $add(C_1'', C_1')$ $=$ $add((\{o_1, o_2\}, \{a_1, a_2\}), (\{o_1, o_2, o_3\}, \{a_1\}))$ results in adding a missing relation: $\mathfrak{C} := \mathfrak{C} - C_1'' - C_1' + (\{o_1, o_2, o_3\}, \{a_1, a_2\})$ and $Noise := Noise \cup \{(o_3, a_2)\}$

The whole repair process consists of the simultaneous application of a set of *delete/add* operations on a subset of pairs extracted from the initially observed set of concepts. The space of admissible pairs is naturally constrained: once a concept has been chosen for deletion for instance, it cannot be used for an *add* operation in another selection. This leads to a space of different subsets of concepts, induced by different repair alternatives. These alternatives have to be scored in order to keep the best one.

4.2 Minimum Description Length Optimization

In the spirit of the minimum description length principle, each set of concepts $\mathfrak{C}$ resulting from the application of *delete/add* operations on a subset of concept pairs $\mathfrak{C}$ gets a score defined as:

$$score(\mathfrak{C}) = \sum_{(A,B) \in \mathfrak{C}} (|A| + |B|) + \alpha \; card(Noise);$$

where α is an integer parameter set by default to 1. This score is minimized over all possible applications of *delete* and *add* operations on all admissible concept subsets $\mathfrak{C}$.

5 Experiments on Simulated Noisy Data

We have generated several random contexts with a fixed number of objects and attributes (from 20 to 40) to test our method for the detection of spurious interactions. For each context, five sets of interactions have been created, corresponding to five cross-products of a random number of objects and a random number of attributes each following a normal distribution (mean 5, standard deviation 2). The original concepts are obtained on these sets of interactions. The noisy concepts are obtained by adding a uniform noise with a fixed probability for each cell to be changed. For each set of parameters (number of objects/attributes, noise level, and weight α) 1,000 random contexts have been tried and the average ratio of original and spurious relations deleted has been computed. Results for the delete operation are shown in Table 1.

For all experiments in Table 1, the percentage of original and spurious relations deleted decreases when α increases. This observation is coherent with the defined score since α represents the relative minor importance of the number of deleted relations with respect to the description length of repaired concepts. In all cases,

Table 1 Mean and standard deviation (sd) on simulated noisy data of the percentage of original and spurious relations deleted by the repair process

Objects	Attributes	Noise	α	Original (%)		Spurious (%)	
				Mean	sd	Mean	sd
20	20	0.05	1	3.5	4.2	43.1	21.3
			2	3.3	4.1	43	21.3
			3	3.1	4	42.2	21.3
40	40	0.01	1	3.5	4.3	60.4	16
			2	2.3	3.1	60.6	15.9
			3	1.5	2.4	58.9	16.2
		0.05	1	0.4	1.5	29	11.2
			2	0.4	1.3	27.8	10.4
			3	0.3	0.1	24.2	9.1
		0.1	1	0.1	1.4	0.8	3.9
			2	0.1	1.4	0.8	3.9
			3	0	0	0.4	0.9
60	60	0.05	1	0.1	0.3	17.4	8.4
			2	0	0.3	15.7	6.8
			3	0	0.3	13.1	6.1

very few original concepts were affected by deletions. Half of spurious relations are detected for contexts that do not exceed 40 objects and 40 attributes and this rate decreases in line with the context size increase and noise level increase. These results seemed sufficient for real data management with a relatively stringent selection of interactions (a limited level of noise) and we have further experimented with more realistic data.

6 miRNAs/mRNAs Interaction Graph

The pea aphid (*Acyrthosiphon pisum*) is a crop pest that is a model for the study of phenotypic plasticity (The International Aphid Genomics Consortium 2010). During the warm seasons viviparous parthenogenetic females are produced whereas during autumn, sexual males and oviparous sexual females are produced. In order to understand the differences in the regulation between sexual and asexual embryogenesis, kinetic data for mRNA and miRNA expression have been collected in both contexts (Gallot et al. 2012). From these data, we have extracted 43 miRNAs and 2,033 mRNAs of interest exhibiting kinetics differences in the two embryogenesis.

To predict miRNAs/mRNAs interactions, we used TargetScan v5 (Grimson et al. 2007). TargetScan provides for each prediction a dissimilarity score, i.e. the lower the score, the stronger the interaction. This resulted in a scored interaction graph with 6,763 interactions between 41 miRNAs and 1,479 mRNAs. The prediction score distribution is shown Fig. 3.

The total distribution can be seen as a Gaussian mixture model (GMM, solid line) divided into two Gaussian curves, one centered around low values (dashed line) and the other centered around high values (spurious interactions, dotted lines). These curves are in agreement with the literature (Chil et al. 2009; Reyes-Herrera et al. 2011), which reveals a high false positive rate in the prediction methods. The TargetScan prediction forms thus a bipartite graph with a high level of noise,

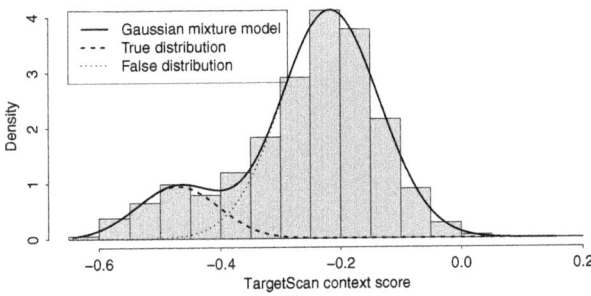

Fig. 3 TargetScan context score distribution (*solid line*: complete; *dashed line*: true prediction distribution; *dotted line*: spurious interactions)

something that does not allow to directly apply our method. Fortunately, it is possible to select the most interesting interactions by choosing a relatively stringent score threshold. Every true interactions above this threshold will be missing and all spurious interactions below this threshold will be retained.

7   Experiments on Simulated Biological Data

Since no miRNAs/mRNAs interactions dataset exist where actual interactions are known, we needed to simulate the interaction graphs in a controlled way to test our method. The scores were simulated by fitting a GMM to the data (solid line in Fig. 3). The degree of miRNAs and mRNAs vertices were determined using score dependent degree distributions for true interactions and spurious interactions.

We have generated 1,000 random miRNA/mRNA interactions graphs, keeping the number of miRNAs, mRNAs, and interactions in real data: 41 miRNAs, 1,479 mRNAs, and 6,763 interactions. Two thresholds have been tested, -0.3 and a more stringent one of -0.35, to restrict the number of spurious interactions while keeping a high number of true interactions (see Fig. 3). A threshold defines the set of original interactions (true interactions below the threshold), spurious interactions (false interactions below the threshold), and missing interactions (true interactions above the threshold).

For each graph and each threshold, the concept lattice has been computed and the ratio of deleted original and spurious relations has been obtained for $\alpha = 1$. Results for the *delete* operation are shown in Table 2.

In contrast to results on simulated noisy contexts, deletions affect both original and spurious relations. For both thresholds, the mean and standard deviation for spurious interactions is slightly higher than for the original interactions. A comparison between the two thresholds shows that the more stringent threshold has a higher mean and standard deviation. Interestingly, the same behavior is observed when comparing simulated noisy contexts of size 40×40 with a noise probability of 0.01 or 0.05 (Table 1).

Table 2 Mean and standard deviation (sd) on simulated data of the percentage of original and spurious relations deleted by the repair process with $\alpha = 1$

miRNAs	mRNAs	Interactions	Threshold	Original (%)		Spurious (%)	
				Mean	sd	Mean	sd
41	1,479	6,763	-0.3	5.6	3.2	8.7	4.5
			-0.35	22.5	4.9	34.8	7.6

8 Conclusion

We have formalized the effect of noise on a miRNAs/mRNAs interaction graph by considering it has a formal context. Two types of noise may occur, namely spurious and missing relations. We showed that noise has the effect of increasing the set of original concepts linearly or exponentially, respectively, with respect to purely spurious or missing concepts. In most cases, there exists some intersection/inclusion relation between noisy concepts observable as a direct or a sibling relation in the modified concept lattice, which allows to recover the original concepts.

Based on these observations, two repair operations: *delete* and *add* have been defined for spurious and missing relations. These operations are applied on subsets of concept pairs, looking for the optimization of a score based on minimum description length principle. We have shown on simulated noisy contexts that there exists a range of context sizes such that our method allows to increase the sensitivity of a highly specific prediction with the *delete* operation.

In order to test our method on more realistic data, we used a set of simulation parameters derived from a real miRNAs/mRNAs interaction graph. Unfortunately, the discriminative power of the repairing method on these data is insufficient as it deletes a significant number of true interactions.

Additional work is necessary to increase the deletion rate of spurious relations, to improve the discriminative power of the method for small and very large contexts and greater levels of noise. In the continuation of this work, we will also evaluate how well the *add* operation performs on the same data. Another perspective is to check how missing relations detected by our method compare to missing relations in approximated concepts, i.e. the 0 in dense rectangles of 1 (see Sect. 3).

Acknowledgements This work was founded by ANR project miRNAdapt and Région Bretagne. The authors thank R. Jullien, V. Picard, and C. Galiez for constructive remarks on the paper.

References

Belohlavek, R., & Vychodil, V. (2006). Replacing full rectangles by dense rectangles: Concept lattices and attribute implications. In *IEEE Information Reuse and Integration* (pp. 117–122). Hawaii: IEEE.

Besson, J., Robardet, C., & Boulicaut, J. F. (2005). Mining formal concepts with a bounded number of exceptions from transactional data. In *Knowledge discovery in inductive databases* (pp. 33–45). Berlin/Heidelberg: Springer.

Blachon, S., Pensa, R. G., Besson, J., Robartdet, C., Boulicaut, J. F., & Gandrillon, O. (2007). Clustering formal concepts to discover biologically relevant knowledge from gene expression data. *In Silico Biology, 7*(4), 467–483.

Bryan, K., Terrile, M., Bray, I. M., Domingo-Fernandéz, R., Watters, K. M., Koster, J., et al. (2014). Discovery and visualization of miRNA-mRNA functional modules within integrated data using bicluster analysis. *Nucleic Acids Research, 42*(3), e17.

Chil, S. W., Zang, J. B., Mele, A., & Darnell R. B. (2009). Argonaute HITS-CLIP decodes microRNA-mRNA interaction maps. *Nature, 460*, 479–486.

Enright, A. J., John, B., Gaul, U., Tuschl, T., Sander, C., & Marks D. S. (2003). MicroRNA targets in *Drosophila. Genome Biology, 5*(1), R1.

Gallot, A., Shigenobu, S., Hashiyama, T., Jaubert-Possamai, S., & Tagu, D. (2012). Sexual and asexual oogenesis require the expression of unique and shared sets of genes in the insect *Acyrthosiphon pisum. BMC Genomics, 13*(1), 76.

Ganter, B., & Wille, R. (1999). *Formal concept analysis: Mathematical foundations.* Berlin: Springer.

Grimson, A., Farh, K. K. H., Johnston, W. K., Garrett-Engele, P., Lim, L. P., & Bartel, D. P. (2007). MicroRNA targeting specificity in mammals: Determinants beyond seed pairing. *Molecular Cell, 27*(1), 91–105.

Janga, S. C., & Vallabhaneni, S. (2011). MicroRNAs as post-transcriptional machines and their interplay with cellular networks. In *RNA infrastructure and networks* (pp. 59–74). New York: Springer.

Klimushkin, M., Obiedkov, S., & Roth, C. (2010). Approaches to the selection of relevant concepts in the case of noisy data. In L. Kwuida & B. Sertkaya (Eds.), *Formal concept analysis* (pp. 255–266). Berlin/Heidelberg: Springer.

Reyes-Herrera, P. H., Ficarra, E., Acquaviva, A., & Macii, E. (2011). miREE: miRNA recognition elements ensemble. *BMC Bioinformatics, 12,* 454–473.

The International Aphid Genomics Consortium. (2010). Genome sequence of the pea aphid *Acyrthosiphon pisum. PLoS Biol, 8*(2), e1000313.

Part VI
Data Analysis in Education and Psychology

Linear Modelling of Differences in Teacher Judgment Formation of School Tracking Recommendations

Thomas Hörstermann and Sabine Krolak-Schwerdt

Abstract The present paper investigates the application of two regression-based approaches, individual multiple regression and hierarchical linear modelling, in modelling differences in judgment formation of primary school teachers' secondary school track recommendations. Both approaches share the same theoretical framework of judgment formation as a weighted linear information integration, but differ in their capacity to take differences in judgment formation into account. First, both approaches were applied to empirical data on teachers' track recommendations and led to deviating conclusions on differences in judgment formation. To investigate which approach results in more reliable representation of actual differences in judgment formation, both approaches were compared based on simulated data and hierarchical linear modelling performed slightly more accurate than individual regression. Thus, hierarchical linear modelling might be considered the preferable modelling approach in research on judgments on school tracking recommendations.

1 Introduction

Professional actors in educational fields, especially teachers, rely commonly on human judgments, and their competence in judgment formation is considered to be a component of their professional expertise (Ständige Konferenz der Kultusminister 2004). Teachers are mostly confronted with judgment situations in which multiple information about a student is available and has to be integrated into a judgment, but the consequences of teachers' judgments considerably vary throughout their professional activity. In educational systems with rather rigid separation of school tracks, like the German achievement-based differentiation of three secondary school tracks, teacher judgments influence students' academic career and later occupational possibilities, because primary school teachers' secondary school track recommendations are shown to be the key determinant of the attended secondary

T. Hörstermann (✉) • S. Krolak-Schwerdt
University of Luxembourg, ECCS Research Unit, Route de Diekirch, L-7220 Walferdange, Luxembourg
e-mail: thomas.hoerstermann@uni.lu; sabine.krolak@uni.lu

© Springer-Verlag Berlin Heidelberg 2015
B. Lausen et al. (eds.), *Data Science, Learning by Latent Structures, and Knowledge Discovery*, Studies in Classification, Data Analysis, and Knowledge Organization, DOI 10.1007/978-3-662-44983-7_32

367

school track (Bos et al. 2003). Much effort was invested by research to investigate the judgment formation of teachers' track recommendations, either to identify the student attributes the recommendations are based on (Glock et al. 2013; Stubbe and Bos 2008), the social disparities caused by tracking recommendations (Maaz et al. 2010), or to reveal systematic differences in teachers' formation of tracking decisions (van Ophuysen 2006).

1.1 Linear Modelling of Judgment Formation

Especially the investigation of differences in judgment formation requires reliable models of the judgment formation process. A prominent theoretical approach in modelling judgment formation is to consider judgment formation as weighted linear integration of information, as stated in Anderson's (1974) *theory of cognitive algebra*. Empirical studies on judgment formation in various domains implemented this theoretical framework of weighted linear integration by modelling judgment processes in form of *individual multiple regression* (IR) (e.g., Dawes and Corrigan 1974; Dhami and Harries 2001; Hunter et al. 2003) and interpreted the regression slopes as the impact of the corresponding piece of information on judgment. Equation (1) shows how the judgment J of a teacher t for a student s is modelled via individual regression as the sum of the weighted pieces of information I_{ks}. The regression slopes β_{kt} represent the relevance of a piece of information k for teacher t, the intercept α_t represents the teacher's general mildness in judgment, and the error component ϵ_{st} refers to the variance in judgment that cannot be explained by the teacher's judgment model.

$$J_{st} = \alpha_t + \sum_{k=1}^{K} \beta_{kt} I_{ks} + \epsilon_{st} \tag{1}$$

Main interest of previous studies applying multiple individual regression in judgment modelling was to compare weighted linear models of judgment with simplified heuristic or nonlinear judgment models (e.g., Dhami and Harries 2001; Hunter et al. 2003). The question whether individual multiple regression is the optimal approach of weighted linear judgment models has not been directly addressed. In the case of modelling differences in judgment formation, the question arises because individual multiple regression results only in individual judgment models for each judge, and modelling differences rely on the a posteriori comparison of the distribution of individual regression coefficients or simply the number of significant coefficients per judge (Dhami and Harries 2001). A possible alternate approach of modelling differences in judgment formation, which can integrate differences of judgment formation between groups of judges into the judgment model, is *hierarchical linear modelling* (HLM) (Aitkin et al. 1981; Raudenbush and Bryk 1986). In hierarchical linear modelling, judgment models are not estimated individually for each judge, but each judge's information weighting is composed of a constant β_k, the influence γ_{kt} of the group of judges η_t the judge belongs to, and an error component ζ_{kt} specific to each judge [see Eq. (2)]. For interpretation, γ_{kt} is of special interest, because a significant γ_{kt} reflects differences in the weighting of a piece of information between groups of judges assumed to differ in their judgment formation process.

$$J_{st} = (\alpha + \gamma_0 \eta_t + \zeta_{t0}) + \sum_{k=1}^{K} (\beta_k + \gamma_k \eta_t + \zeta_{kt}) I_{ks} + \epsilon_{st} \qquad (2)$$

Although hierarchical linear modelling is commonly applied in various domains of educational research, its application in judgment formation research is rather limited. The aim of this study is to compare both approaches described in modelling differences in teachers' secondary school track recommendation judgments and to investigate whether one of the approaches more adequately reflects differences in teachers' judgment formation processes. First, both approaches are applied to empirical data from educational research on track recommendations to illustrate whether the choice of a modelling approach may lead to deviating conclusions on differences in teachers' judgment formation. Second, both approaches are compared based on simulated data with a priori known differences in underlying judgment formation processes. The simulation-based comparison shows which approach more reliably detects systematic differences in judgment formation and therefore, in case of deviating conclusions between the modelling approaches, which judgment model is to be preferred in interpretation.

2 Empirical Comparison of Modelling Approaches

Both modelling approaches are applied to empirical data on secondary school track recommendations of 21 female and 4 male primary school teachers from the German federal state of *Nordrhein-Westfalen* with a mean teaching experience of 23.6 years. The data set stems from a study investigating the *ecological validity* of educational judgment formation research based on student case vignettes, that means if judgment formation processes observed for student case vignettes can be generalized to teachers' judgments on actual students. In the study, teachers in charge of a fourth grade primary school class assessed their actual students on school achievement related information and indicated which secondary school track they recommended to each student. After the teachers assessed their actual students, they received 24 student case vignettes and were again asked to indicate an adequate secondary school track for these students. The distribution and intercorrelation of the student information in the case vignettes matched the distribution and intercorrelations observed for the actual students. In total, 1,148 school track recommendations were recorded, of which 549 were recommendations for the teachers' actual students.

Four pieces of information are included in the analysis: (a) the student's average grade in German, math, and science, ranging from 1 (*very good*) to 5 (*to be improved*), (b) the student's average of working and learning behavior ratings, ranging from 1 (*very good*) to 5 (*very bad*), (c) the student's available parental support on a five-point scale from 1 (*very good*) to 5 (*very bad*), and (d) the student's migration background (0—no migration background, 1—migration background). Two teachers have to be excluded from the analysis, because they either did not report assessments of their students' parental support or showed no variance in the migration background of their students. The teachers' school track recommendations is split into two binary judgments, lower track (0) vs. intermediate or higher track (1) and lower or intermediate track (0) vs. higher track (1), according to usual procedures in previous studies on tracking recommendations (e.g., Kristen 2006). The judgment formation is then modelled in terms of individual logistic regression equations for each teacher, as shown in Eq. (3). The β-distributions for each predictor are compared between actual students and case vignettes and significant differences are interpreted as indicators of differences between the judgment formation processes. The estimation of individual regression is done in the R 2.10.1 software. Due to the low number of recommendations per teacher, *separation* problems might arise (Albert and Anderson 1984), that means a nearly perfect prediction by one predictor and a respective infinite β. To avoid separation, a penalized maximum likelihood estimation (Heinze and Schemper 2002) is done using the logistf function of the logistf package.

$$\ln\left(\frac{P(X_{\text{higher track}} = 1)}{1 - P(X_{\text{higher track}} = 1)}\right) = \beta_0 + \beta_1 I_{\text{grades}} + \beta_2 I_{\text{behavior}} + \beta_3 I_{\text{support}} + \beta_4 I_{\text{migration}} \quad (3)$$

For the hierarchical linear modelling of teachers' track recommendations, a factor η_t is included, indicating whether the recommendation was done for an actual student (0) or a case vignette (1). Thus, the weighting of each piece of information k is a composite of the constant β_k, the influence γ_k of the vignette factor η_t, and a teacher-specific weighting error ζ_{kt} [see Eq. (4)].

$$\ln\left(\frac{P(X_{\text{higher track}_{st}} = 1)}{1 - P(X_{\text{higher track}_{st}} = 1)}\right) = (\alpha + \gamma_0 \eta_{\text{vignette}_t} + \zeta_{0t})$$

$$+ (\beta_1 + \gamma_1 \eta_{\text{vignette}_t} + \zeta_{1t}) I_{\text{grades}_{st}}$$
$$+ (\beta_2 + \gamma_2 \eta_{\text{vignette}_t} + \zeta_{2t}) I_{\text{behavior}_{st}} \quad (4)$$
$$+ (\beta_3 + \gamma_3 \eta_{\text{vignette}_t} + \zeta_{3t}) I_{\text{support}_{st}}$$
$$+ (\beta_4 + \gamma_4 \eta_{\text{vignette}_t} + \zeta_{4t}) I_{\text{migration}_{st}}$$

β_k can be interpreted as the information weighting for the actual students, and significant γ_k coefficients indicate systematic differences in weighting the corresponding piece of information between real students and case vignettes. The estimation is done via the `glmer` function of the `R 2.10.1 - lme4` package.

2.1 Results

When inspecting the regression coefficients of both approaches shown in Table 1, the judgment models reflect a mainly similar pattern of coefficients, thus depicting a rather high correspondence between the judgment models. However, following the common conventions in educational research to take only significant coefficients into account for interpretation, the modelling approaches lead to substantially deviating conclusions on differences between teachers' judgment formation for actual students and case vignettes. Considering the recommendation to the higher track, the individual regression judgment model shows that the track recommendations for actual students are mainly based on student's grades ($\bar{\beta}_1 = -1.50$, $SE = 0.35$), working and learning behavior ($\bar{\beta}_2 = -1.32$, $SE = 0.53$), and partly on the parental support ($\bar{\beta}_3 = -0.54$, $SE = 0.18$). The migration background has no significant influence on the higher track recommendation ($\bar{\beta}_4 = 0.30$, $SE = 0.15$). In contrast to actual students, the individual regression judgment model indicates significant differences in judgment formation for case vignettes throughout all student information. Students' grades show a higher impact on the recommendation ($\Delta\bar{\beta}_1 = -1.69$, $t_{22} = 4.66$, $p < 0.01$), whereas the influence of behavioral information diminishes ($\Delta\bar{\beta}_2 = 1.33$, $t_{22} = 2.30$, $p = 0.03$). The influence of parental support changes its direction ($\Delta\bar{\beta}_3 = 0.96$, $t_{22} = 4.96$, $p < 0.01$), and a

Table 1 Regression coefficients of judgment models from individual regression (IR) and hierarchical-linear modelling (HLM), separated by type of case and recommendation

Information	IR			HLM	
	$\bar{\beta}_{student}$	$\bar{\beta}_{vignette}$	$\Delta\bar{\beta}$	$\beta_{student}$	γ
Higher track vs intermediate/lower track					
Grades	−1.50*	−3.19*	−1.69*	−7.38*	−5.69*
Working and learning behavior	−1.32*	0.01	1.33*	−1.71*	3.38
Parental support	−0.54*	0.42*	0.96*	−1.52*	2.91
Migration background	0.30	−0.54*	−0.84*	1.00	−2.44
Higher/intermediate track vs lower track					
Grades	−2.05*	−1.76*	0.29	−6.66*	0.88
Working and learning behavior	−0.05	−1.07*	−1.02*	−0.41	−2.18*
Parental support	−0.18	−0.52*	−0.34	−1.31*	0.35
Migration background	−0.25	−0.19	0.06	0.31	−1.10*

* $p < 0.05$

migration background tends to lead to lower chances for a recommendation to the higher track ($\Delta \bar{\beta}_4 = -0.84$, $t_{22} = 5.70$, $p < 0.01$).

In sum, the individual regression judgment model of higher track recommendations indicates severe differences between the judgment formation for actual students and case vignettes, which can be readily interpreted in line with the educational theories and previous research. In daily classroom, working and learning behavior is assessed by longtime continuous interaction with the student. In case vignettes, this experience is missing, thus the behavioral information might be not as salient. The observed effect of migration background can be explained in line with the Maaz et al.'s (2010) findings that a migration background might not only directly influence school tracking recommendations, but also earlier assessments, e.g., prior grading, and thus influences tracking recommendations via its impact on grades as well as assessments of behavior and parental support. In case vignettes, the possibility of prior mediating effects does not exist, and therefore information on migration background might be directly taken into account for judgment formation.

Noteworthy, the conclusions stated above about teachers' judgment formation are fully dependent on the individual regression judgment model. If the hierarchical linear model is taken into account, neither significant differences for working and learning behavior ($z = 1.94$, $p = 0.05$), parental support ($z = 1.58$, $p = 0.11$), or migration background ($z = 1.75$, $p = 0.08$) are observed. In line with the slightly less deviating findings for lower track recommendations, the comparison of modelling approaches shows the relevance of the choice of a modelling approach in applied research on teachers' judgment formation, at least for the school tracking recommendations observed in this sample.

3   Simulation-Based Comparison of Modelling Approaches

Although demonstrating possible differences in judgment models between the modelling approaches, the comparison based on empirical judgment data cannot reveal which of the approaches should be preferred in the interpretation of teachers judgment formation of track recommendations. To identify which modelling approach can be considered as more accurate in reflecting the true underlying differences in judgment formation, a simulation-based comparison of modelling approaches is performed and the approaches can be evaluated in terms of their capacity of correctly detecting a priori defined differences.

3.1   Data Simulation

The aim of the data simulation is to create data sets that resemble the situation of school tracking recommendations, as observed in the empirical comparison of modelling approaches. Corresponding to findings that school tracking recommendations rely on three domains of student information (Glock et al. 2013), the number of information underlying the simulated judgments is set to three. Each piece of information I_k is a composite of an information-specific factor F_k and a general factor G common to all pieces of information, thus resulting in intercorrelated pieces of information [see Eq. (5)].

$$I_k = w_G G + w_F F_k, \quad k = 1, \ldots, 3, \quad F, G, I \sim N(0, 1) \tag{5}$$

By choosing appropriate values for the weights w_F and w_G, the amount of intercorrelation between the information is varied from low ($r = 0.30$) to moderate ($r = 0.50$) to high intercorrelation ($r = 0.70$), resembling the range of intercorrelation of student information observed in the empirical comparison. Furthermore, the number of judges $T = \{40, 100\}$ and the number of judgments per judge $S = \{25, 10\}$ is varied to reflect usual class sizes and sample sizes in research on school track recommendations.

The simulated judgment is constructed according to the theoretical assumption of a weighted linear integration of information. The judges are divided into two equally sized groups r with different underlying judgment formation processes. The first group of judges weights information I_1 stronger than the second group, which, in turn, weights information I_3 stronger. For information I_2, no differences between the groups are assumed. Thus, the simulated judgment J of a judge t for a student s is simulated as the sum of the pieces of information I_1 to I_3, weighted by the corresponding weights v_{kr} of the judgment formation process of the teacher's group [see Eq. (6)]. The size of difference between the groups' judgment formation processes is varied, with $v_{10} = \{1.00, 0.80, 0.65, 0.60, 0.55\}$ and $v_{11} = \{0.00, 0.20, 0.35, 0.40, 0.45\}$. For v_{30} and v_{31}, the pattern is mirrored, and v_{20} and v_{21} are equally set to 0.50.

$$J_{st} = \sum_{k=1}^{3} (v_{kr} + E_{V_{kt}}) I_{kst} + E_{J_{st}} \quad t = 1, \ldots, t, \quad s = 1, \ldots, S, \quad r = 0, 1 \tag{6}$$

E_V reflects a teacher-specific weighting error, and E_J is a judgment-specific error component, both accounting for 20 % of the variance of judgments resulting from $\sum_{k=1}^{3} v_{kr} I_{kst}$. To reflect the situation of school tracking recommendations, in which the judgment is binary and the pieces of information are discrete, the simulated judgment J is transformed into a binary judgment J_D, according to the observed distribution function Φ of higher track recommendations in the empirical comparison [see Eq. (7)]. The pieces of information I_k are transformed accordingly into discrete variables I_{D_k}, matching the distribution function of observed grades [see Eq. (8)].

$$J_D = \begin{cases} 0 & \text{for } \Phi(J) < .654 \\ 1 & \text{else} \end{cases} \tag{7}$$

$$I_D = \begin{cases} 1 & \text{for } \Phi(I) < .080 \\ 2 & \text{for } .080 < \Phi(I) < .476 \\ 3 & \text{for } .476 < \Phi(I) < .825 \\ 4 & \text{else} \end{cases} \tag{8}$$

3.2 Results

For each combination of varied factors in the simulation, 50 data sets are computed and both modelling approaches are applied to the simulated data sets as described for the empirical comparison of approaches. For the individual regression judgment model, a correctly signed significant difference of the distributions of $\bar{\beta}$ coefficients is counted as a correct detection of the differences in judgment formation for the corresponding piece of information. For the hierarchical linear model, a correctly signed significant γ coefficient indicates correct detection.

When considering only the pieces of information I_{D_1} and I_{D_3}, for which differences in the underlying judgment formation processes of the groups of judges exist, slightly higher detection rates are observed for the hierarchical linear model, rather independent of the size of differences in judgment formation and the amount of intercorrelation (see Table 2). For the data sets containing 100 judges and 25 judgments per judge, HLM shows higher detection rates in 57 % of the simulation conditions, compared to 7 % for the individual regression approach. If the number of judges is lowered to 40, higher detection rates for the HLM are observed in 70 % and for the individual regression in 3 % of the cases. If the number of judgments per judge is lowered to 10, the advantage of HLM becomes more pronounced, showing 90 % (100 judges) and 83 % (40 judges) differences in detection rates in favor of HLM, compared to 7 % in favor of individual regression. If only significant differences in detection rates are considered, all differences throughout the simulation conditions are in favor of the HLM approach.

In sum, despite the usually only slight improvement in detection rates, hierarchical linear modelling tends to be the more reliable approach in detecting differences between judgment formation processes. Although the simulation is designed to reflect common empirical data in research on school tracking recommendations and hierarchical linear modelling shows advantages throughout all simulated conditions, the generalizability of the results is limited to empirical judgment data in line with the simulated conditions. Furthermore, the underlying true judgment formation is designed according to the theoretical framework of weighted linear integration of information, and the advantages of HLM might not exist if this assumption is violated in empirical research.

Table 2 Percentage differences of detection rates between individual regression (IR) and hierarchical linear modelling (HLM), separated by judgment rule, intercorrelation and number of judgments and judges

	100 judges				40 judges			
	25 judgments		10 judgments		25 judgments		10 judgments	
Judgment rule	I_{D_1} (%)	I_{D_3} (%)	I_{D_1} (%)	I_{D_3} (%)	I_{D_1} (%)	I_{D_3} (%)	I_{D_1} (%)	I_{D_3} (%)
$r = 0.30$								
1.00 vs. 0.00	2	8	10	14	0	0	20*	12
0.80 vs. 0.20	8	30*	6	−6	4	12	22*	22*
0.65 vs. 0.35	0	0	18*	2	6	6	16	2
0.60 vs. 0.40	16	20	6	16*	0	4	0	2
0.55 vs. 0.45	0	0	2	0	0	14	6	8
$r = 0.50$								
1.00 vs. 0.00	4	8	2	2	0	4	16*	10
0.80 vs. 0.20	20	8	12	22*	18*	6	22*	16
0.65 vs. 0.35	0	0	14	8	4	10	4	10
0.60 vs. 0.40	2	0	6	12	0	0	4	0
0.55 vs. 0.45	8	18	−4	6	0	2	−4	12
$r = 0.70$								
1.00 vs. 0.00	0	0	6	4	8	4	30*	32*
0.80 vs. 0.20	0	0	8	10	10	12	20	22*
0.65 vs. 0.35	10	−6	10	16	8	4	8	14
0.60 vs. 0.40	2	−2	24*	26*	−2	4	2	−4
0.55 vs. 0.45	6	20	10	10	2	10	0	4

Note. Positive values indicate higher detection rates of HLM approach
*Fishers' test $p < 0.05$

4 Discussion

The study compares two approaches of modelling judgment formation, individual multiple regression and hierarchical linear modelling, regarding their accuracy in detecting differences in the judgment formation of primary school teachers for their secondary school track recommendations. When applied to empirical data on school track recommendations, each approach indicates differences in teachers' judgment formation, which are only partly replicated by the alternate approach. Because the differences revealed by only one approach can be readily interpreted in the framework of theories of education and psychology, the comparison demonstrates the importance of the choice of a reliable modelling approach especially in educational research, as research findings on teacher judgments may influence later decisions of educational policy makers and are readily incorporated into perpetuating public discussions on the quality of and social disparities within the educational system. Reliable judgment models may support valid insights, for example, if a change in administrative regulations of school tracking changed teachers' judgment formation of track recommendations or if training courses for ongoing teachers led to a more professional and competent judgment formation. To identify which of the approaches is more reliable in detecting differences in teachers' judgment formation of track recommendations, both approaches are compared in detecting underlying judgment differences in simulated data sets, resembling track recommendation data observed in the empirical comparison. Hierarchical linear modelling performs better than individual regression throughout the simulated data sets. Given the limitations of the simulation, hierarchical linear modelling tends to be the more adequate approach in modelling empirical judgments on school tracking recommendations. Regarding the mainly only slight advantages of hierarchical linear modelling in detection rates and the possible consequences of empirical findings on school track recommendations, applying the more reliable modelling approach should only assist other common approaches for reliable findings on differences in teachers' judgment formation, e.g., metaanalytic techniques trying to unify findings of separate empirical studies.

References

Aitkin, M., Anderson, D., & Hinde, J. (1981). Statistical modelling of data on teaching styles. *Journal of the Royal Statistical Society. Series A, 144*, 419–461.

Albert, A., & Anderson, J. A. (1984). On the existence of maximum likelihood estimates in logistic regression models. *Biometrika, 71*, 1–10.

Anderson, N. H. (1974). Cognitive algebra: Integration theory applied to social attribution. In L. Berkowitz (Ed.), *Advances in experimental social psychology* (Vol. 7, pp. 1–100). New York: Academic Press.

Bos, W., Lankes, E.-M., Prenzel, M., Schwippert, K., Wather, G., & Valtin, R. (2003). *Erste Ergebnisse aus IGLU: Schülerleistungen am Ende der vierten Jahrgangsstufe im internationalen Vergleich.* Münster: Waxmann.

Dawes, R. M., & Corrigan, B. (1974). Linear models in decision making. *Psychological Bulletin, 81*, 95–106.

Dhami, M. K., & Harries, C. (2001). Fast and frugal versus regression models of human judgment. *Thinking and Reasoning, 7*, 5–27.

Glock, S., Krolak-Schwerdt, S., Klapproth, F., & Böhmer, M. (2013). Prädiktoren der Schullaufbahnempfehlung für die Schulzweige des Sekundarbereichs I. *Pädagogische Rundschau, 67*, 349–367.

Heinze, G., & Schemper, M. (2002). A solution to the problem of separation in logistic regression. *Statistics in Medicine, 21*, 2409–2419.

Hunter, D. R., Martinussen, M., & Wiggins, M. (2003): Understanding how pilots make weather-related decisions. *International Journal of Aviation Psychology, 13*, 73–87.

Kristen, C. (2006). Ethnische Diskriminierung in der Grundschule? Die Vergabe von Noten und Bildungsempfehlungen. *Kölner Zeitschrift für Soziologie und Sozialpsychologie, 58*, 79–97.

Maaz, K., Baumert, J., Gresch, C., & Mcelvany, N. (2010). *Der Übergang von der Grundschule in die weiterführende Schule. Leistungsgerechtigkeit und regionale, soziale und ethnisch-kulturelle Disparitäten.* Bonn: Bundesministerium für Bildung und Forschung.

Raudenbush, S. W., & Bryk, A. S. (1986). A hierarchical model for studying school effects. *Sociology of Education, 59*, 1–17.

Ständige Konferenz der Kultusminister der Länder in der Bundesrepublik Deutschland (KMK) (2004). *Standards für die Lehrerbildung: Bildungswissenschaften.* Bonn: KMK.

Stubbe, T. C., & Bos, W. (2008). Schullaufbahnempfehlungen von Lehrkräften und Schullaufbahnentscheidungen von Eltern am Ende der vierten Jahrgangsstufe. *Empirische Pädagogik, 22,* 49–63.

Van Ophuysen, S. (2006). Vergleich diagnostischer Entscheidungen von Novizen und Experten am Beispiel der Schullaufbahnempfehlungen. *Zeitschrift für Entwicklungspsychologie und Pädagogische Psychologie, 38,* 154–161.

Psychometric Challenges in Modeling Scientific Problem-Solving Competency: An Item Response Theory Approach

Ronny Scherer

Abstract The ability to solve complex problems is one of the key competencies in science. In previous research, modeling scientific problem solving has mainly focused on the dimensionality of the construct, but rarely addressed psychometric test characteristics such as local item dependencies which could occur, especially in computer-based assessments. The present study consequently aims to model scientific problem solving by taking into account four components of the construct and dependencies among items within these components. Based on a data set of 1,487 German high-school students of different grade levels, who worked on computer-based assessments of problem solving, local item dependencies were quantified by using testlet models and Q_3 statistics. The results revealed that a model differentiating testlets of cognitive processes and virtual systems fitted the data best and remained invariant across grades.

1 Introduction

1.1 Problem Statement

Solving complex problems is regarded as a multidimensional ability, which enables students to cope with real-life situations. As many processes are involved in problem solving, psychological research has focused on the dimensionality of the construct by emphasizing how students acquire and apply knowledge in interactive environments such as virtual laboratories (Funke 2010; Wirth and Klieme 2004). In the last years, many computer-based assessment approaches have been developed in order to capture students' competencies of exploring unknown systems, building mental models of the system's structure, controlling variables within a system, and, finally, evaluating potential solutions to the problem (Koppelt 2011; Wüstenberg et al.

R. Scherer (✉)
Faculty of Educational Sciences, Centre for Educational Measurement (CEMO), University of Oslo, P.O. 1161, Blindern, 0318 Oslo, Norway
e-mail: ronny.scherer@cemo.uio.no

© Springer-Verlag Berlin Heidelberg 2015
B. Lausen et al. (eds.), *Data Science, Learning by Latent Structures, and Knowledge Discovery*, Studies in Classification, Data Analysis, and Knowledge Organization, DOI 10.1007/978-3-662-44983-7_33

379

2012). Although these assessments provide meaningful and reliable measurement tools, many psychometric problems occur, especially when using item response theory to model the structure and dimensionality of the underlying constructs. One major problem, that has already been extensively discussed on a theoretical level, refers to local item dependencies (Ragni and Löffler 2010; Wüstenberg et al. 2012). As models of item response theory such as the Rasch model are based on the assumption of independent student responses given their ability (De Ayala 2009), using these models for describing cognitive processes within computer-based assessments is compromised. Until now, this problem has not yet been addressed in research on problem solving and has only been discussed theoretically. The present study consequently aims to analyze the effects of item dependencies within virtual environments and thereby combine a theoretical framework of problem solving with psychometric modeling approaches. For the first time, testlet models are applied to describe dependencies among items which are assigned to different cognitive processes.

1.2   The Present Study

In light of the shortcomings in modeling students' problem-solving competencies, the present study aims to

1. Model the structure of scientific problem solving by combining theoretical assumptions on cognitive processes with statistical approaches of item response theory (IRT).
2. Investigate the effects of local item dependencies within virtual problem-solving environments by using testlet models.

2   Theoretical Background

2.1   A Theoretical Model of Scientific Problem Solving

In 2011, Koppelt proposed a theoretical framework of complex, scientific problem solving which systematically combined previous approaches from the fields of intelligence research, large-scale assessments such as the PISA study (OECD 2004), and standards of science education. In this framework, four cognitive components are distinguished: *Understanding & Characterizing the problem (PUC)*, *Representing the problem (PR)*, *Solving the problem (PS)*, and *Reflecting & Communicating the*

Table 1 Detailed description of the four problem-solving factors

Dimension	Description: students are asked to …
Understanding & characterizing the problem (PUC)	Understand the problem situation
	Identify relevant information
	Extract information from scientific sources
Representing the problem (PR)	Build representations of the problem
	Shift between different representations
Solving the problem (PS)	Perform systematic and strategic methods
	Achieve a goal state meeting specific criteria
	Plan and execute scientific investigations
Reflecting & communicating the solution (SRC)	Evaluate and reflect solutions
	Find alternative solutions
	Communicate solutions towards an audience
	Use scientific language

solution (SRC). These processes cover a broad range of constructs involved in problem solving and could be used to describe students' performance in different grades (Scherer 2012). Additionally, they provide guidelines for developing items and tasks which are intended to measure the construct of scientific problem solving. In Table 1, a more detailed description of these four cognitive processes is given.

2.2 Psychometric Shortcomings of Modeling Problem Solving

From a psychometric perspective, researchers have to ensure that measurement models of problem solving are identified and show reasonable goodness-of-fit statistics (Ragni and Löffler 2010). Especially in the process of modeling complex problem solving, research faces various problems of scalability because of the effects of item dependencies, task interactivity, and interferences with covariates such as system knowledge and intelligence (Funke 2010; Wirth and Klieme 2004). For instance, as Wüstenberg et al. (2012) argued, dependencies among items within a virtual problem-solving environment are large and, thus, do not permit to apply traditional IRT scaling approaches. In their argumentation, they claimed that different virtual scenarios (i.e., "systems") with varying numbers of variables and relationships among these variables should be used to assess problem-solving competency appropriately. However, these dependencies have not yet been explicitly quantified or modeled. This shortcoming is of major importance as it addresses the question of how to assess problem-solving processes with appropriate environments.

3 Methodology

3.1 Sample and Design

In this study, the data set of $N = 1,487$ students in chemistry was used, who participated in the *EnkoPro* study on the development of students' complex problem-solving competencies in chemistry (see Scherer 2012). The total sample consisted of 61 science classes from 17 German high schools in the federal states Berlin and Brandenburg. These classes belonged to grade levels 8 ($n_8 = 506$), 10 ($n_{10} = 476$), and 12 ($n_{12} = 505$). 51.2 % of the students were female. Students' mean age was 15.6 years ($SD = 1.7$ years) and ranged between 13 and 21 years. In this cross-sectional study, the four problem-solving dimensions were linked between adjacent grades by at least 60 anchor items within six virtual scenarios (two per grade). These scenarios contained complex problems from different topics of the chemistry curricula in Germany. For instance, tenth graders were asked to identify unknown chemicals, which represented simple organic (e.g., alcohols) and inorganic (e.g., metal oxides) compounds. For each of the three grade levels, two complex problems were developed, of which one was used as a link to the adjacent grade level. For details, please review Scherer (2012) or Scherer and Tiemann (2014). Finally, the assessments for each grade contained between 127 and 154 items which were assigned to one of the four cognitive processes. Table 2 contains the number of items which were used to link two adjacent grades (anchor items) and items which were used as grade-specific items. Based on this common-item design, a concurrent calibration showed that measurement invariance was present (Kolen and Brennan 2004; for details, see Scherer 2012). This characteristic can be regarded as a prerequisite of comparing students' performance across grade levels (Millsap 2011).

Table 2 Numbers of grade-specific and anchor items

Grade	8	8–10	10	10–12	12
Dimension	Grade-specific	Anchor	Grade-specific	Anchor	Grade-specific
PUC	14	17	6	11	14
PR	16	21	2	31	32
PS	15	18	1	16	16
SRC	14	12	3	16	14
Total	59	68	12	74	76

3.2 Modeling Local Stochastic Item Dependencies

Models of item response theory such as the one-parametric logistic model (Rasch model) are commonly used in educational measurement, as they provide tools to establish a common scale for equated tests and for evaluating students' abilities $(\theta_v, v = 1,\ldots, N)$ in relation to item difficulties $(\beta_i, i = 1,\ldots, I)$ (Bond and Fox 2007; De Ayala 2009). In these models, conditional independence of students' binary responses X_{vi} ($x_{vi} = 1$ for correct response of student v on item i) given their abilities is assumed. The resulting model and its likelihood function $\mathcal{L}$ for N students and I items are given as follows:

$$\text{logit}\{P(X_{vi} = 1|\theta_v, \beta_i)\} = \theta_v - \beta_i \tag{1}$$

$$P(X_{v1} = x_{v1},\ldots, X_{vI} = x_{vI}|\theta_v; \beta_1,\ldots, \beta_I) = \prod_{i=1}^{I} P(X_{vi} = x_{vi}|\theta_v, \beta_i), \forall v \tag{2}$$

$$\mathcal{L} = P(\mathbf{X}|\theta_1,\ldots, \theta_N; \beta_1,\ldots, \beta_I) = \prod_{v=1}^{N}\prod_{i=1}^{I} P(X_{vi} = x_{vi}|\theta_v, \beta_i) \tag{3}$$

On the basis of this model, Yen (1993) defined Q_{3ij} statistics as residual correlations between two items i and j for all persons v ($v = 1,\ldots,N$):

$$e_{vi} = x_{vi} - P(X_{vi} = 1|\theta_v, \beta_i) \tag{4}$$

$$Q_{3ij} = r(e_i, e_j) \tag{5}$$

The resulting correlations are comparable to residual correlations in confirmatory factor analyses (De Ayala 2009). These statistics reveal information on the effects of item dependencies. However, they do not provide a modeling approach which could be applied in order to handle these dependencies within an IRT framework. As an alternative, researchers proposed the Rasch testlet model, which contains a general latent trait θ_v for a person v, the difficulty β_i of an item i, and the interaction $\gamma_{vd(i)}$ between a person and a testlet $d(i)$ ($i = 1,\ldots, I$) (e.g., Brandt 2012; Ip 2010).

$$\text{logit}\{P(X_{vi} = 1|\theta_v, \beta_i)\} = \theta_v - \beta_i + \gamma_{vd(i)} \tag{6}$$

In this model, further restrictions are introduced:

1. Testlet factors are uncorrelated.
2. There is no covariance between testlet factors and the trait θ.
3. The sum of testlet effects equals zero: $\sum_d \gamma_{vd(i)} = 0$.

As Robitzsch et al. (2011) suggested, these models could be used to quantify the effects of testlet structures on students' performance and item difficulties. But the resulting estimates of person abilities (θ) are difficult to interpret since they are based on a different scale than the unidimensional Rasch model. However, in order to investigate whether or not testlet effects are present, model fit criteria can be used to compare models with and without testlet structures (Brandt 2012; DeMars 2012). In these models, testlet variances indicate the effects of dependencies within a testlet or an item bundle (De Boeck and Wilson 2004).

In the present study, item dependencies were first quantified by using Q_3 statistics for each virtual laboratory (system) and each grade level. In a second step, unidimensional Rasch models and Rasch testlet models have been specified and compared by using information criteria and the likelihood ratio test, which are based on the loglikelihood values $\mathrm{Log}\mathcal{L}$ (DeMars 2012). In this step, systems, cognitive dimensions, and their interactions were regarded as testlet factors. In order to specify these models, the maximum likelihood procedures implemented in ConQuest 2.0 (Wu et al. 2007) and the R package sirt (Robitzsch 2013) have been applied.

4 Results

In a first step, item dependencies were quantified as Q_3 statistics within each grade level and each problem-solving environment. The resulting statistics are shown in Table 3. According to the argumentation of Lucke (2005) who showed that if the mean of item dependencies is approximately zero, an unbiased estimate of reliability is still obtained. In light of the results in this study, these dependencies were quite low and showed negligible effects.

Table 3 Q_3 statistics for virtual laboratories (systems) in each grade level

Grade	Systems	$M(Q_3)$	$SD(Q_3)$	N_{Items}
8	Pigments	0.003	0.129	54
	Titanium(IV) oxide	−0.002	0.111	59
10	Polyester	−0.003	0.133	67
	Titanium(IV) oxide	0.004	0.106	60
12	CNTs	−0.002	0.132	70
	Polyester	−0.009	0.129	66

However, due to large standard deviations, there is evidence of specific patterns (testlets) within these dependencies. Thus, the second step of analyses focused on these testlets. By applying different testlet models which were compared to the unidimensional Rasch model (Ip 2010), further information on the variance explained by testlet factors was obtained. In a series of nested models, which assume no testlet factors (1dim RM), two testlets representing the two different systems (2dim TM), four testlet factors representing the cognitive processes (4dim TM), and eight testlet factors with an interaction between systems and cognitive processes (8dim TM), information criteria and the likelihood ratio test were used to compare these models (see Table 4). Based on the resulting data, the eight-dimensional testlet models were empirically preferred in each grade, and were thus accepted.

Table 5 contains the variances within the testlets obtained by the interaction between systems and cognitive processes. These variances show a broad range

Table 4 Comparisons between Rasch and Rasch testlet models

Grade	Model	Log$\mathcal{L}$ (df)	AIC	BIC	Likelihood ratio test
8	1dim RM	−18,956(115)	38,142	39,122	−
	2dim TM	−18,888(117)	38,011	38,506	$\Delta\chi^2(2) = 136.5^{***}$
	4dim TM	−18,539(119)	37,316	38,330	$\Delta\chi^2(2) = 699.1^{***}$
	8dim TM	−18,319(123)	36,884	37,932	$\Delta\chi^2(4) = 440.5^{***}$
10	1dim RM	−18,100(128)	36,455	37,537	−
	2dim TM	−17,970(130)	36,200	36,741	$\Delta\chi^2(2) = 256.7^{***}$
	4dim TM	−17,867(132)	35,998	37,113	$\Delta\chi^2(2) = 205.7^{***}$
	8dim TM	−17,625(136)	35,522	36,671	$\Delta\chi^2(4) = 484.0^{***}$
12	1dim RM	−16,496(137)	33,267	34,412	−
	2dim TM	−16,430(139)	33,139	33,726	$\Delta\chi^2(2) = 129.2^{***}$
	4dim TM	−16,103(141)	32,487	33,666	$\Delta\chi^2(2) = 654.9^{***}$
	8dim TM	−15,957(145)	32,203	33,416	$\Delta\chi^2(4) = 291.9^{***}$

$^{***}\,p < .001$

Table 5 Testlet variances σ_γ^2 of the eight-dimensional testlet models

Testlet d	Grade 8		Grade 10		Grade 12	
	$\sigma_{\gamma d}^2$	$SD(\sigma_{\gamma d}^2)$	$\sigma_{\gamma d}^2$	$SD(\sigma_{\gamma d}^2)$	$\sigma_{\gamma d}^2$	$SD(\sigma_{\gamma d}^2)$
Intercept (Gf)	0.662	0.813	0.527	0.726	0.626	0.791
PUC×S1	0.824	0.907	0.666	0.816	0.769	0.877
PR×S1	2.361	1.537	1.060	1.030	1.649	1.284
PS×S1	0.156	0.395	0.344	0.586	0.047	0.218
SRC×S1	0.321	0.566	0.803	0.896	0.021	0.143
PUC×S2	0.392	0.626	0.499	0.707	1.140	1.068
PR×S2	1.623	1.274	1.461	1.209	1.507	1.228
PS×S2	0.291	0.540	0.214	0.463	0.169	0.411
SRC×S2	0.187	0.432	0.165	0.407	0.108	0.328

Notes. S1, S2 systems 1 and 2, *Gf* general factor

and differ across testlet factors. Especially for the testlets of *Representing the problem (PR)*, large values were obtained which explain a substantial amount of variance in the model. In contrast, testlets of *Solving the problem (PS)* and *Reflecting & Communicating the solution (SRC)* revealed the lowest values in each grade, indicating lower effects of dependencies.

5 Discussion and Conclusion

The present study primarily focused on psychometric modeling approaches of scientific problem solving, based on a theoretical model with four cognitive dimensions. By using IRT models, the following results were obtained:

1. In each grade level, the theoretically implied model with four testlet dimensions fitted the data better than the unidimensional model.
2. Item dependencies for virtual environments (systems) were small.
3. Testlet models which specify the interaction between systems and cognitive processes were empirically preferred.

In light of previous discussions on the structure of scientific problem solving, it was possible to show that the theoretically implied framework of four cognitive processes could be linked to models of item response theory with four dimensions. As this result held across grade levels and sets of anchor items, measurement invariance was present (for further details, see Scherer 2012). Consequently, the computer-based assessments are able to reflect students' progressions within the four dimensions and could, thus, be used to model competence development (De Ayala 2009; Kolen and Brennan 2004; Millsap 2011). However, Köller and Parchmann (2012) pointed out that it is also important to capture students' individual development rather than focusing on comparisons across cohorts of different grade levels. Future research could, therefore, assess scientific problem-solving competencies within longitudinal settings. Moreover, the findings presented in this study could be validated by examining how students' problem-solving competencies are related to the development of their general abilities of reasoning. Such an external validation may provide further insights into the developmental mechanisms of problem solving (Scherer and Tiemann 2014).

For the first time, item dependencies were quantified for the construct of complex scientific problem solving. On average, these dependencies revealed small effects for the different systems. Following the argumentation of DeMars (2012), modeling dependencies for systems would not be necessary within further analyses. However, it was found that testlet models which do not only take into account dependencies for *systems*, but also for *cognitive factors* (PUC, PR, PS, and SRC) and their interactions significantly outperformed models without testlets. These results indicate that item dependencies should not be neglected in problem-solving environments and, thus, support the theoretical argumentation of Wüstenberg et al. (2012). As the 8-dimensional testlet model fitted the data best, the present study extends the argumentation of Wüstenberg et al. (2012), who only focused on *systems*. Rather than assuming dependencies within a virtual assessment or system, item dependencies should be modeled for cognitive dimensions *and* systems (Scherer 2012). However, they often cannot be avoided, but, instead of creating polytomously scored "super items," testlet models provide an approach to quantify these dependencies (Ip 2010). Especially in educational measurement, meaningful tasks require a certain number of items that refer to one stimulus in order to represent the underlying construct adequately. To some extent, the validity of test scores is thereby ensured.

In this study, item dependencies were modeled as uncorrelated testlet factors which represented *trait dependence*. Other approaches also take into account that students' responses on previous items affect the probability of solving subsequent items correctly. These models refer to *task dependence* (Marais and Andrich 2008) and were not modeled in the present study because items were developed in such a way that they could be solved independently (Koppelt 2011; Scherer 2012).

Taken together, the present study addressed major psychometric concepts of modeling scientific problem-solving competency. First, it was shown that the theoretically implied structure of cognitive processes involved in problem solving held across grades and, second, item dependencies were modeled for virtual scenarios. The importance of these test characteristics is indisputable, but these have rarely been taken into account (Köller and Parchmann 2012).

Acknowledgements The author wishes to thank Professor Dr. Rüdiger Tiemann (Humboldt-Universität zu Berlin, Germany) for his conceptual support in conducting the proposed study. This research has been partly funded by a grant of the German Academic Exchange Service (DAAD).

References

Bond, T. G., & Fox, C. M. (2007). *Applying the Rasch model: Fundamental measurement in the human sciences* (2nd cd.). Mahwah, NJ: Lawrence Erlbaum.

Brandt, S. (2012). Robustness of multidimensional analyses against local item dependence. *Psychological Test and Assessment Modeling, 54*, 36–53.

De Ayala, R. J. (2009). *The theory and practice of item response theory*. New York, NY: The Guilford Press.

De Boeck, P., & Wilson, M. (2004). *Explanatory item response models*. New York, NY: Springer.

DeMars, C. E. (2012). Confirming testlet effects. *Applied Psychological Measurement, 36*, 104–121.

Funke, J. (2010). Complex problem solving: A case for complex cognition? *Cognitive Processing, 11*, 133–142.

Ip, E. H. (2010). Empirically indistinguishable multidimensional IRT and locally dependent unidimensional item response models. *British Journal of Mathematical and Statistical Psychology, 63*, 395–416.

Köller, O., & Parchmann, I. (2012). Competencies: The German notion of learning outcomes. In S. Bernholt, K. Neumann, & P. Nentwig (Eds.), *Making it tangible – Learning outcomes in science education* (pp. 165–185). Münster: Waxmann.

Kolen, M. J., & Brennan, R. L. (2004). *Test equating, scaling, and linking*. New York, NY: Springer.

Koppelt, J. (2011). *Modellierung dynamischer Problemlösekompetenz im Chemieunterricht.* Berlin: Mensch & Buch.

Lucke, J. F. (2005). "Rassling the Hog": The influence of correlated item error on internal consistency, classical reliability, and congeneric reliability. *Applied Psychological Measurement, 29,* 106–125.

Marais, I., & Andrich, D. (2008). Formalizing dimension and response violations of local independence in the unidimensional Rasch model. *Journal of Applied Measurement, 9,* 200–215.

Millsap, R. E. (2011). *Statistical approaches to measurement invariance.* New York, NY: Taylor & Francis.

OECD. (2004). *Problem solving for tomorrow's world.* Paris: OECD.

Ragni, M., & Löffler, C. M. (2010). Complex problem solving: another test case? *Cognitive Processing, 11,* 159–170.

Robitzsch, A. (2013). *R package sirt – Supplementary functions for item response theory.* Salzburg: Bifie.

Robitzsch, A., Dörfler, T., Pfost, M., & Artelt, C. (2011). Die Bedeutung der Itemauswahl und die Modellwahl für die längsschnittliche Erfassung von Kompetenzen. *Zeitschrift für Entwicklungspsychologie und Pädagogische Psychologie, 43,* 213–227.

Scherer, R. (2012). *Analyse der Struktur, Ausprägung und Messinvarianz komplexer Problemlösekompetenz im Fach Chemie.* Berlin: Logos.

Scherer, R., & Tiemann, R. (2014). Evidence on the effects of task interactivity and grade level on thinking skills involved in complex problem solving. *Thinking Skills and Creativity, 11,* 48–64.

Wirth, J., & Klieme, E. (2004). Computer-based assessment of problem solving competence. *Assessment in Education: Principles, Policy and Practice, 10,* 329–345.

Wu, M., Adams, R., Wilson, M., & Haldane, S. (2007). *ACER ConQuest 2.0.* Camberwell: ACER.

Wüstenberg, S., Greiff, S., & Funke, J. (2012). Complex problem solving – more than reasoning? *Intelligence, 40,* 1–14.

Yen, W. M. (1993). Scaling performance assessments: Strategies for managing local item dependence. *Journal of Educational Measurement, 30,* 187–213.

The Luxembourg Teacher Databank 1845–1939. Academic Research into the Social History of the Luxembourg Primary School Teaching Staff

Peter Voss and Etienne Le Bihan

Abstract From 1845 to 1939 the pedagogical journal *Der Luxemburger Schulbote* published a comprehensive annual directory of the primary school teaching staff of the Grand Duchy. On the basis of this directory, we have established a databank encompassing 75,000 entries relating to a total of approx. 4,700 primary school teachers, both male and female, who taught in the Grand Duchy during this period. With the assistance of IBM SPSS Statistics, we have been able to process the data and compile a collective biography or prosopography that provides a profound insight into the development of an occupational group over a period of nearly 100 years at a local, regional and national level. This paper presents an analysis of initial research findings relating to the number of teaching staff, length of service and the level of qualification and mobility among teaching staff for the first half of this period from 1845 to 1895.

1 Introduction

After gaining independence in 1839, the government of the Grand Duchy of Luxembourg made the establishment of an operational school system one of its top priorities. At that time, Luxembourg was still suffering from a lack of institutional and economic development due to the disruptive influence of the many political regime changes that had occurred in the period since the French revolution. The first major step in fulfilling this goal was the introduction of the Primary School Act of 1843, which laid the regulatory foundations for the future primary school system as well as setting up a hierarchical school administration. These administrative and institutional reforms were accompanied by the establishment

P. Voss (✉)
University of Luxembourg, ECCS, 7201 Walferdange, Luxembourg
e-mail: peter.voss@uni.lu

E. Le Bihan
University of Luxembourg, INSIDE, 7201 Walferdange, Luxembourg
e-mail: etienne.lebihan@uni.lu

© Springer-Verlag Berlin Heidelberg 2015
B. Lausen et al. (eds.), *Data Science, Learning by Latent Structures, and Knowledge Discovery*, Studies in Classification, Data Analysis, and Knowledge Organization, DOI 10.1007/978-3-662-44983-7_34

of a pedagogical journal, *Der Luxemburger Schulbote* (i.e., "School Herald"), and the regular publication of school statistics. These descriptive statistics performed an important role which went far beyond their basic purpose, providing not just numbers but a gauge of the success of the newly created system, signalling the competitiveness of the newly founded nation state to both to its own population and on the international stage (Voss 2012).

As the most important instrument of social "self-monitoring" and as "a form of political rhetoric" (Osterhammel 2009), these statistics offered reassurance to the school administration, acted as a basis for planning and raised the level of uniformity across the school system. This applies in particular to one part of the statistics, namely the teacher directory. From 1845 onwards, one volume of the *Schulbote* contained a comprehensive annual record of the teaching staff—*Liste générale des personnes qui exercent dans le Grand-Duché de Luxembourg les fonctions d'instituteur*—listing the educators of the Grand Duchy in a kind of "Who's who?" with numbered entries classified according to school district, detailing full name, place of work and rank. These basic entries were supplemented in later editions by information on earnings, bonuses, length of service and class sizes. Publication of the directory continued uninterrupted for the entire period from 1845 to 1939 without any major changes in the document. In fact, the directory provides a consistent data source that covers the best part of a century.

2 The Luxembourg Teacher Databank: Research Scope

The Teacher Directory can be considered an instrument of classification and standardization. Only those persons mentioned in this directory had the right to practice officially as primary school teachers in the country. As certified educators, they were appointed to a particular school and performed their duties under the authority of a school inspector. In effect, these records created a national professional corps of Luxembourg primary school teachers, placing them under the scrutiny of both the public and their own colleagues. These records, therefore, provide their own unique insight into the gradual transformation that took place within this developing profession, charting the progress from early village teachers left largely to their own devices to more modern instructors (*instituteurs*) equipped with a code of professional ethics and expected to be both model citizens and loyal representatives of the national education programme. As part of the "bureaucratic imposition of uniform standards and measures" (Porter 1995), the directory also performed a disciplinary function for the new occupational group of certified primary school teachers. According to an official statement from the state in 1828, schools were considered "the cradle of the new citizen". The importance the state placed on the development of these future citizens is clearly reflected in the fundamental re-evaluation of the status of the teachers entrusted with their education. The annual teachers directory, therefore, can be regarded as one indicator of the new esteem afforded to the Luxembourg teaching staff.

According to "longue durée", the teacher directory reflects the evolution of an occupational group from the mid-nineteenth to the mid-twentieth century. Charting several generations of Luxembourg teachers practising between 1845 and 1939, the directory offers the opportunity to create a complete collective biographical analysis of a profession over a period of nearly 100 years.

Luxembourg is a very small nation. In 1839, the Grand Duchy had only 175,000 inhabitants, rising to 236,000 in 1900 and reaching 300,000 in 1935. In 1845, there were 430 educators working in the primary schools of the Grand Duchy. This number had risen to 790 by 1895, rising further to 950 in 1912 and to 1,100 in 1939. Compared to other nation states, these are of course very modest figures. From the perspective of a macro analysis, however, the fact that Luxembourg is such a small nation state is a distinct advantage because it enables us to conduct an extensive analysis of an occupational group at a national level. Research of this kind would simply not be logistically possible for larger states such as England, France or Germany. At the same time, the directory also allows a detailed microanalysis of the teaching staff and the development of the school infrastructure of every municipality even of every single village school throughout the Grand Duchy. Overall, therefore, the main advantage of this case study is the ability to conduct a comprehensive analysis at local, regional and national level.

3   Methodology

The entries in the teacher directory published in the *Luxemburger Schulbote* contain the following items: school year, school district, school inspector, municipality, section of the municipality (administrative subdivision), full name of teacher, level of qualification (provisional authorization, Brevet Level 4, 3, 2 or 1), years of service, salary and bonuses (from 1895 to 1915), type of school (boys school, girls school, mixed school), pupils per teacher (from 1902 to 1919). In addition, it is possible to derive the gender of the teacher from the first name. Also, the prefix to the name, for example "sister" or "soeur" allows us to identify how many of the teaching staff were Catholic nuns or so-called teaching sisters.

The first step in this research project was to collate individual annual data using Microsoft EXCEL, a program that is not only user friendly, but also compatible with a wide range of other statistical software allowing for easy data transfer. The data pool entered in this step comprises 75,000 entries relating to a total of 4,600 individual primary school teachers, both male and female, who taught in the Grand Duchy during the period from 1845 to 1939. In the second stage, the data set was imported into IBM SPSS statistical software where more complex statistical analyses were conducted.

Before the analysis stage could begin, however, it was necessary to review and revise the historical data. One major stumbling block encountered was the identification of individual teachers listed in the annual directories, which proved far more difficult than initially anticipated. First of all, there were many cases of staff bearing the same name. For example, there were a total of eight different Nicolas Müllers and six different Jean Schmits over the period. Another problem concerned first names, which were not always printed in full. A teacher entered as Johann-Peter Schmit in 1870 may well have been referred to as Johann Schmit in 1871 or as Peter Schmit in 1872 or vice versa. Finally, the fact that Luxembourg is a bilingual country also led to confusion because the spelling of Christian names varied depending on the official language during the period of entry. For example, the same person entered in German as "Johann" one year may appear in French as "Jean" in another.

This meant that identification often had to be verified by the researcher on a case by case basis using a combination of the different data categories available, such as work place, level of qualification and so on. Although the procedure inevitably involved a degree of subjectivity, we believe the findings ascertained were of significant value. To the best of our knowledge, it would not have been possible to automate this review and revision process. There are other comparable sources, for example matriculation registers of the teaching staff in the national archives of Luxembourg, which could be used to verify and further revise the data.

4   Initial Findings for the Period from 1845 to 1895

This paper presents initial findings relating to the number of Luxembourg primary school teaching staff, length of service and level of professional qualification and mobility for the first half of the period, the years 1845 to 1895.

4.1   The Number of the Luxembourg Primary School Teaching Staff, 1845–1895

Figure 1 shows the increase in the number of teachers working in the Grand Duchy of Luxembourg from 1845 to 1895. As the population grew, new primary schools were established resulting in the creation of new teaching posts. Within the first 50 years of records, the number of teachers doubled, rising gradually from 400 to 800 staff. Many of these new positions were occupied by female candidates. In fact, from the 1860s onwards, the proportion of women entering the profession increased markedly. Until this point, the occupation of primary school teacher in Luxembourg had been almost exclusively a male domain. By around 1890, however, more than 40 % of the teaching staff was female. In contrast, the number of male teachers remained relatively stable, fluctuating around 400.

It is possible that the feminization of the teaching profession during this period caused the occupation to become less attractive to male candidates and may have resulted in men leaving teaching posts to seek work in alternative professions. Another possible contributory factor to this shift was the expansion of the administration that resulted from increasing industrialization and the construction of rail

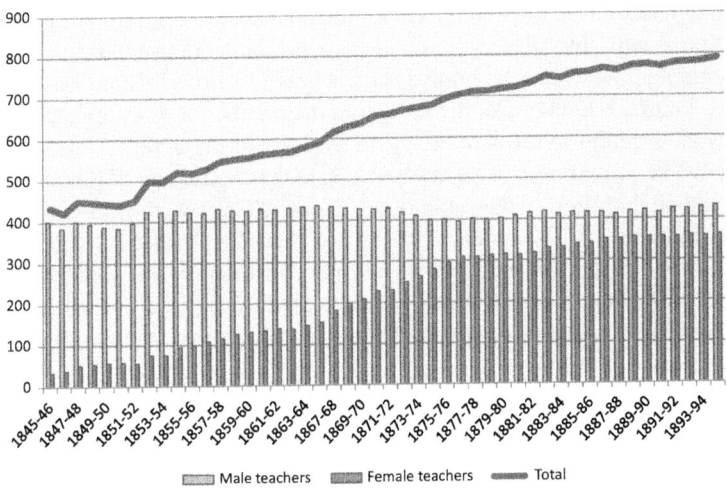

Fig. 1 Number of primary school teachers in Luxembourg, 1845–1895

networks during the second half of the nineteenth century. These developments opened up alternative career paths for men that were often both more prestigious and better paid than a teaching post. Another cause of the decline in the number of male teaching staff in Luxembourg in the period from 1871 to approximately 1880 was that many teachers trained in the state took up better renumerated posts in the neighbouring Elsass-Lothringen region, which found itself under German administration following the Franco-German War of 1870–1871.

In the 1840s, Catholic congregations from France, in particular the *Soeurs de la Doctrine chrétienne* from Nancy, settled in Luxembourg and started entering the primary school teaching profession. From the outset, the teaching staff at Luxembourg's primary schools included both lay female teachers and Catholic nuns. However, the teacher directory did not include entries relating to this religious status until 1861. It was therefore necessary to ascertain this information from other sources (Bombardier and Lepage 1999). From the late 1860s on, Catholic nuns became more and more prominent in the Luxembourg primary education system, representing approximately 40 % of the total female teaching staff. The increase of female teachers in the Grand Duchy is, therefore, to a great extent due to the "forgotten contribution of the teaching sisters", as Marc Depaepe pointed out in the case of neighbouring Belgium (Hellinckx et al. 2009).

In the second half of the nineteenth century, the high demand for teaching staff in girls' schools could not have been met without calling upon these "teaching sisters". Another attraction to the authorities of recruiting catholic nuns was that their salaries were considerably lower than those of their secular colleagues. During this period, the teaching profession certainly took on the character of a sacerdotal activity. For example, celibacy was expected from all female staff, even those from secular backgrounds.

Although the set-up of primary schooling in Luxembourg followed more or less population trends throughout the country, approximately doubling overall, there were some regions where the number of teaching positions created was influenced by other factors. For example, the growth in the number of teaching jobs in certain parishes in agricultural areas in the north and east of the country was slower than average, while growth was particularly strong in the capital city of Luxembourg and in the industrial regions in the south of the country. There were three times as many teachers in the city of Luxemburg in 1895 as there had been 50 years previously. In Hollerich, situated on the edge of railroad construction and industrial regions, the number of teachers rose fourfold and in the industrial town of Esch-sur-Alzette, in the heart of Luxembourgs iron and steel producing area, the number even rose fivefold.

4.2 Length of Service

In the first part of our sample, covering 1845 to 1895, we identified a total of 2,700 individual teaching staff, for 1,152 of whom we were unable to reconstruct a complete professional biography. This group either began teaching before 1845, meaning we had no record of their commencement or took up a position during the period but continued to teach beyond 1895. In total, there are therefore comprehensive records for 1,548 teachers, comprising all those teachers who first appear in our data from the year 1846 at the earliest, leaving the profession by 1894 at the latest.

Figure 2 shows that more than half of these teachers (52%) remained in the profession for 5 years at best, while 20% worked for up to 10 years. Thus more

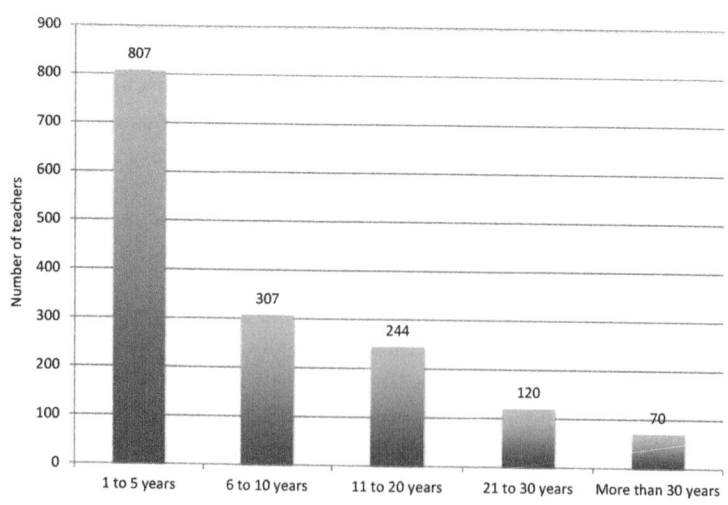

Fig. 2 Primary school teachers: length of service, 1845–1895

than half of the total teaching staff dropped out within 5 years; nearly two thirds left their jobs before reaching 10 years of service, and only 28 % spent more than 10 years in the profession. This "brain drain" was probably largely attributable to relatively low wages and poor working conditions. Primary school teaching seems to have been seen by many as an entry job which candidates used as a stepping stone to move on to another profession as quickly as possible. It should be noted that the data for the period 1845 to 1895 reflect a school system going through a transitional period of initial development. It is predicted that when the second part of the data set for the period from 1895 onward is analysed, the increased continuity and stability in the system during the second half of the nineteenth century will coincide with increasing lengths of service among teaching staff.

In general, female teachers dropped out earlier than their male colleagues and their average careers were shorter than those of male teachers (7.9 years vs. 9.3 years). Among female teachers, Catholic teachers or "teaching sisters" retired on average a little later than the lay female teachers. In the period from 1845 to 1895, they generally spent 8.2 years at work vs. 7.8 for the lay female teachers.

4.3   Occupational Qualification

In the 1840s, the level of formal occupational qualifications held by Luxembourg primary school teachers was quite low. Almost half of the teaching staff (49 %) had the lowest recognized level of qualification, the Brevet 4, while 29 % even began work on the basis of "provisional authorizations", a status not initially foreseen by school officials or local authorities. Twenty percent of staff held the next qualification, a Level 3 degree, while teachers holding higher, level 2 or level 1 degrees, represented an insignificant minority (Fig. 3). Until at least the 1860s, there was a lack of qualified candidates to take up available teaching positions. The Luxembourg teacher training college, *Ecole Normale*, started taking in male candidates in 1845; only 10 years later, in 1855 were the first female trainees admitted. In its first few years, the *Ecole Normale* simply could not keep pace with the rapidly rising demand for well-qualified teachers.

As the overall number of teachers increased, so too did the proportion of teachers holding the lowest possible level 4 degree because this is the level at which most teachers took up their first post if they did not begin on a provisional basis. It was not until the 1890s that this trend began to change with teachers increasingly obtaining a Brevet level 3 or level 2. In 1895, for example, more than half of the teaching staff (53,5 %) held at least a Level 3 degree. In general, male teachers were better qualified than female teachers. Women did, however, make up ground quite rapidly, especially in the decade from 1885 to 1895. In 1895, female teachers were more or less nearly as qualified as their male colleagues. While male teachers were generally better qualified than female teachers, lay female teachers were normally better qualified than the teaching sisters. In 1895, 61 % of the lay female educators but only 38 % of the teaching sisters held a degree better than level 4.

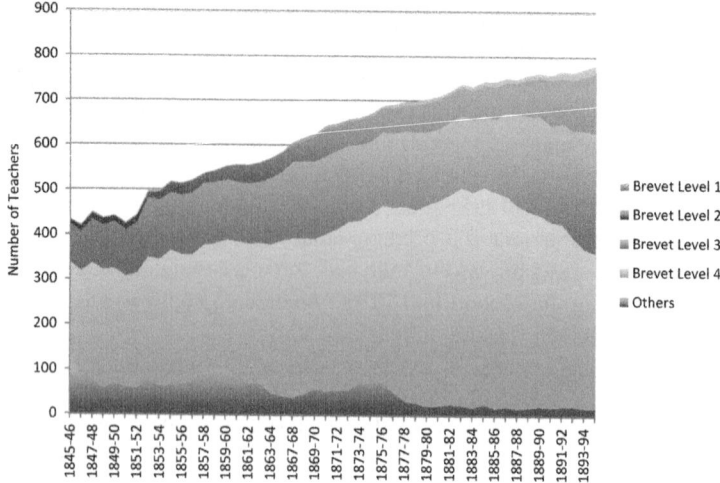

Fig. 3 Occupational qualifications of primary school teachers, 1845–1895

Although the Luxembourg school authorities demanded professional development, two thirds of the teaching staff failed to improve their level of qualification at all in the period from 1845 to 1895, finishing their careers holding the same level of qualification with which they had started, and this was generally the minimum level 4 degree. Only one third of teachers gained a higher level of qualification during their careers, obtaining, for example, a level 4 degree after having started teaching on the basis of a "provisional authorization" or obtaining a Level 3 degree after having started at Level 4. Upward mobility was stronger among male than among female teachers. One possible explanation is that, numerically speaking, it was not until the 1860s that women made up a significant proportion of the teaching staff. Male teachers, therefore, had a head start when it came to gaining qualifications. Nevertheless, these figures reveal a relative stagnation in the level of qualification among teaching staff in the sample. We anticipate that the figures for the period 1896–1939 will display a higher level of upward mobility in terms of qualification.

One example of a complex statistical model we applied during our analysis of this data set is a logistic model, whereby it was used in conjunction with a GEE marginal model in order to take into account the non-independence of observations (Fitzmaurice et al. 2011). The dependent variable in the analysis was the level of degree held by a teacher. We modeled the probability of advancing beyond the basic level four qualification. The explanatory variables were year of commencement of employment (as a continuous variable), gender and, in the case of female staff, religious status. The working correlation matrix structure chosen was autoregressive order 1. The analysis was performed using SPSS statistical software.

Table 1 shows that female teachers had a lower probability of advancing beyond a level 4 degree, especially if they were religious. It can also be seen from Table 1

Table 1 Parameter estimates of a logistic GEE model for the probability of advancement beyond a basic level 4 teaching qualification

Parameter	B	Std. error	95 % CI		p
			Lower	Upper	
Intercept	−2.197	.0848	−2.364	−2.031	0.000
Gender	−1.094	.1343	−1.357	− .830	0.000
	0				
Year of commencement	.034	.0027	.029	.040	0.000
Women					
Not religious	.721	.1477	.432	1.011	0.000
Religious	0				

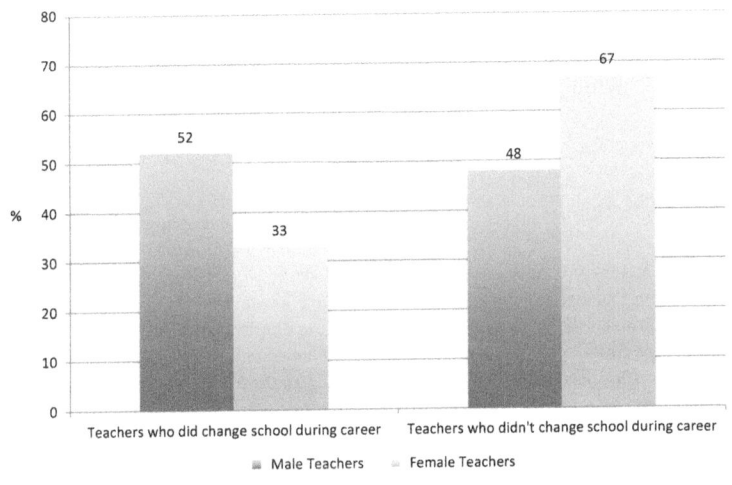

Fig. 4 Geographical mobility of teaching staff, 1845–1895

that the later employment commenced, the greater the probability of advancement beyond the basic level four teaching qualification.

4.4 Geographical Mobility

Geographical mobility was high in the newly established school system with 44 % of primary school teachers changing their workplace at least once during their career (Fig. 4). Female teachers were more settled than their male counterparts with 48 % of the male staff moving at least once while 66 % of female staff did not change their workplace at all during their careers.

5 Conclusion

These initial findings for the period from 1845 to 1895 offer an insight into the establishment of a professional teaching corps in an emerging nation state in the nineteenth century. Primary school teachers played a central role in the nation building process in Luxembourg, acting as educators in terms of both general schooling and civil education for the future citizens of this new nation state. The increases evident in terms of the number of teachers, length of service, level of qualification and level of mobility depict the gradual professionalization of teaching during this period. In the next stage of research, the same analytical approaches will be applied to the entire database spanning the period of 1845 to 1939. Among many other things, this analysis will allow us to track local, regional and national trends in levels of teaching qualification and salaries, relative proportions of male and female teaching staff, teacher mobility, teaching careers among different age groups and the formation of "teaching dynasties" over a period of almost an entire century.

References

Bombardier, J., & Lepage, A.-M. (1999). *Les soeurs de la Doctrine chrétienne de Nancy durant le dix-neuvième siècle: L'expansion au Luxembourg et en Algérie.* (Histoire des soeurs de la Doctrine chrétienne de Nancy, t. 4). Nancy: Doctrine chrétienne.

Der Luxemburger Schulbote (1846 to 1939). *Le Courrier des Ecoles dans le Grand-Duché de Luxembourg. Eine Zeitschrift zunächst für die Schullehrer des Großherzogtums Luxemburg bestimmt* (Vols. 3 and 96). Luxembourg: Lamort.

Fitzmaurice, G. M., Laird, N. M., & Ware, J. H. (2011). *Applied longitudinal analysis* (2nd ed.). Hoboken, NJ: Wiley.

Hellinckx, B., Simon, F., & Depaepe, M. (2009). *The forgotten contribution of the teaching sisters. A historiographical essay on the educational work of catholic women religious in the 19th and 20th centuries.* Leuven: Leuven University Press.

Osterhammel, J. (2009). *Die Entdeckung der Welt. Eine Geschichte des 19. Jahrhunderts.* München: C. H. Beck.

Porter, T. M. (1995). *Trust in numbers: The pursuit of objectivity in science and public life.* Princeton, NJ: Princeton University Press.

Voss, P. (2012). Der bürokratische Wendepunkt von 1843: Die Primärschule im Prozess der Luxemburger Nationalstaatsbildung des 19. Jahrhunderts. In M. Geiss, & A. de Vincenti (Eds.), *Verwaltete Schule. Geschichte und Gegenwart* (pp. 53–82). Wiesbaden: Springer.

Part VII
Data Analysis in Musicology

Correspondence Analysis, Cross-Autocorrelation and Clustering in Polyphonic Music

Christelle Cocco and François Bavaud

Abstract This paper proposes to represent symbolic polyphonic musical data as contingency tables based upon the duration of each pitch for each time interval. Exploratory data analytic methods involve weighted multidimensional scaling, correspondence analysis, hierarchical clustering, and general autocorrelation indices constructed from temporal neighborhoods. Beyond the analysis of single polyphonic musical scores, the methods sustain inter-voices as well as inter-scores comparisons, through the introduction of ad hoc measures of configuration similarity and cross-autocorrelation. Rich musical patterns emerge in the related applications, and preliminary results are encouraging for clustering tasks.

1 Introduction

This paper aims to produce an exploratory data analysis of symbolic polyphonic musical data represented as contingency tables, which count the duration of each pitch for each time interval, given a predefined partition of the musical score into equal durations. This representation, not so far from the piano-roll representation or from the Chroma representation for audio files (see, e.g., Müller and Ewert 2011 or Ellis and Poliner 2007), has the advantage of representing digital polyphonic musical scores, being usable with common data analytic methods, such as correspondence analysis and being aggregation-invariant (Sect. 2).

In Sect. 3.1, analyses of whole music pieces are proposed, by means of correspondence analysis and a flexible autocorrelation index able to deal with general neighborhoods. Both methods grasp intrinsic structures of musical scores and provide pattern visualizations. Multiple voices within a single musical score are analyzed through soft multiple correspondence analysis and a cross-autocorrelation index (Sect. 3.2). Finally, based on the choice of the contingency table, a similarity

C. Cocco (✉) • F. Bavaud
University of Lausanne, Lausanne, Switzerland
e-mail: christelle.cocco@unil.ch; francois.bavaud@unil.ch

© Springer-Verlag Berlin Heidelberg 2015
B. Lausen et al. (eds.), *Data Science, Learning by Latent Structures,
and Knowledge Discovery*, Studies in Classification, Data Analysis,
and Knowledge Organization, DOI 10.1007/978-3-662-44983-7_35

measure, aimed to cluster music pieces according to composers, is proposed and illustrated (Sect. 3.3).

2 Data Representation

In this contribution, symbolic music files are used, and especially files in *Humdrum* **kern format, as they are well structured with all voices, independent of the performer and freely available on the web (http://kern.ccarh.org/). Moreover, *Humdrum extras* (http://extra.humdrum.org/) are used when modifications, such as transposition, are needed, as well as to transform **kern files in *Melisma* format (http://www.link.cs.cmu.edu/music-analysis/), easily handleable for the representation proposed in this paper. Note that the representation proposed in the followings could have also been obtained with other digital score formats, such as ABC or MIDI files, especially if the latter is performed with a constant tempo.

Each musical score is represented, with all repeated passages, as a contingency table $X = (x_{tj})$ crossing *pitches* ($j = 0, \ldots, m$) and *time intervals* ($t = 1, \ldots, n$). The table gives the duration of each pitch in each time interval. Notice that the repetition of notes of the same pitch within a time interval is not coded. In more detail, MIDI note numbers (0 to 127) are transformed in a 12-note octave-equivalent pitch set using a modulo 12, where 0 stands for C; 1, for C♯ or D♭; 2, for D; etc. Moreover, a *true rest z* is added whenever no note is played. Thus, j can take on 13 different values: 0 to 11 and z. Regarding time intervals, each one has a constant *duration* of τ which can take any value, such as a 16th note, a measure or a number of milliseconds. Consequently, the total duration of the musical score is $\tau_{\text{tot}} = n\tau$. An example of the transposed contingency table is given in Fig. 1 for two different values of τ.

Besides the advantage to deal with polyphonic music, this representation is aggregation-invariant in the sense that doubling τ amounts to summing counts within two consecutive parts. So, considering an interval T made out of smaller intervals t, the new counts are $\tilde{x}_{Tj} = \sum_{t \in T} x_{tj}$. Lavrenko and Pickens (2003) and Morando (1981) use a quite similar representation, except that the former do not take into account the duration of and between notes and the latter bases his representation upon the succession of chords. However, in contrast to the present representation, theirs are not aggregation-invariant.

Then, as a second step, the contingency table $X = (x_{tj})$ is normalized to $\Xi = (\xi_{tj})$ in order that the sum of each row $\sum_j \xi_{tj} = \xi_{t\bullet}$ equals to 1, that is $\xi_{tj} = \frac{x_{tj}}{x_{t\bullet}}$. Thus, the same importance is given to each time interval, regardless of the duration and the number of pitches.

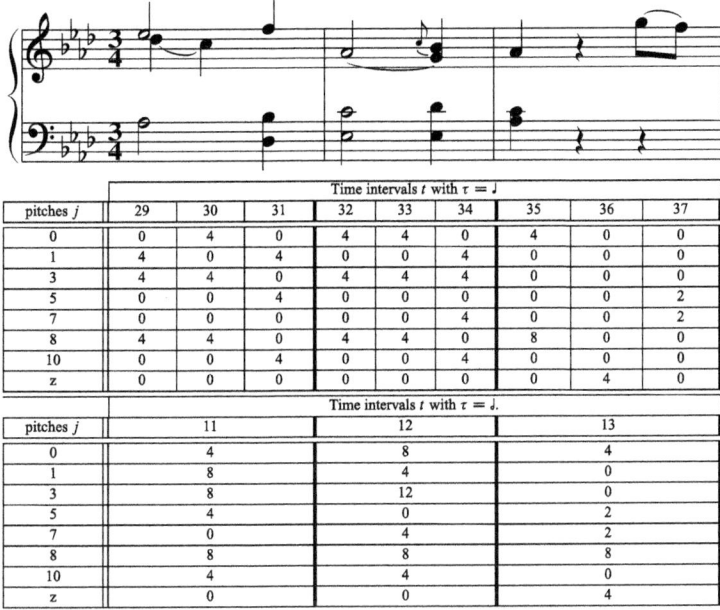

pitches j	Time intervals t with $\tau =$ ♩								
	29	30	31	32	33	34	35	36	37
0	0	4	0	4	4	0	4	0	0
1	4	0	4	0	0	4	0	0	0
3	4	4	0	4	4	4	0	0	0
5	0	0	4	0	0	0	0	0	2
7	0	0	0	0	0	4	0	0	2
8	4	4	0	4	4	0	8	0	0
10	0	0	4	0	0	4	0	0	0
z	0	0	0	0	0	0	0	4	0

pitches j	Time intervals t with $\tau =$ ♩.		
	11	12	13
0	4	8	4
1	8	4	0
3	8	12	0
5	4	0	2
7	0	4	2
8	8	8	8
10	4	4	0
z	0	0	4

Fig. 1 Extract of the *third movement of the Beethoven's Piano Sonata No. 1 in F minor, Op. 2, No. 1*. Transposed display of the contingency table $X = (x_{tj})$, giving the duration of each pitch (in units of 16th note), for τ equal to a quarter note (*top*) and to a dotted half note (*bottom*)

3 Methods and Applications

3.1 Single Score Analysis

3.1.1 Correspondence Analysis

To perform the correspondence analysis (CA) on the $\varXi$ matrix, an equivalent method is used which consists in applying a weighted multidimensional scaling on the chi-squared dissimilarities between time intervals $\hat{D} = (\hat{D}_{st})$ and between pitches $\check{D} = (\check{D}_{ij})$:

$$\hat{D}_{st} = \sum_{j} \rho_j \, (q_{sj} - q_{tj})^2 \qquad \check{D}_{ij} = \sum_{t} f_t (q_{ti} - q_{tj})^2 \qquad (1)$$

where $f_t = 1/n$ is the relative weight of time intervals, $\rho_j = \xi_{\bullet j}/n$ is the relative weight of pitches, and $q_{tj} = \xi_{tj} n / \xi_{\bullet j}$ is the independence ratio.

In a nutshell, scalar products between time intervals $\hat{B} = (\hat{b}_{st})$ and between pitches $\check{B} = (\check{b}_{ij})$ are computed from the dissimilarity matrices as:

$$\hat{B} = -\frac{1}{2} H^f \hat{D} (H^f)' \qquad \check{B} = -\frac{1}{2} H^\rho \check{D} (H^\rho)'$$

where $H^f = I - 1f'$, $H^\rho = I - 1\rho'$ are the corresponding centering matrices. Then, weighted scalar products $\hat{K} = (\hat{k}_{st})$ and $\check{K} = (\check{k}_{ij})$ are defined as:

$$\hat{k}_{st} = \sqrt{f_s f_t}\,\hat{b}_{st} \qquad \check{k}_{ij} = \sqrt{\rho_i \rho_j}\,\check{b}_{ij} \qquad (2)$$

The spectral decomposition of the matrix $\hat{K}$ (respectively $\check{K}$) provides the eigenvectors $u_{t\alpha}$ (resp. $v_{j\alpha}$) and the corresponding eigenvalues λ_α (identical for both matrices) from which stem the factor coordinates for time intervals ($x_{t\alpha}$) and for pitches ($y_{j\alpha}$):

$$x_{t\alpha} = \frac{\sqrt{\lambda_\alpha}}{\sqrt{f_t}}u_{t\alpha} = \frac{1}{\sqrt{\lambda_\alpha}}\sum_{j=1}^{m}\rho_j q_{tj}y_{j\alpha} \qquad y_{j\alpha} = \frac{\sqrt{\lambda_\alpha}}{\sqrt{\rho_j}}v_{j\alpha} = \frac{1}{\sqrt{\lambda_\alpha}}\sum_{t=1}^{n}f_t q_{tj}x_{t\alpha}$$

An example of this formalism for the well-known French monophonic nursery melody *Frère Jacques* (*Are you sleeping?* in English) is given in Fig. 2. The graph on the left shows the result obtained with τ equal to an eighth note, which means that no more than one pitch is played during each time interval, i.e. the representation is totally monophonic. In that case, chi-squared dissimilarities between time intervals are "star-like", i.e. of the form $\hat{D}_{st} = a_s + a_t$ (see, e.g., Critchley and Fichet 1994). Consequently all λ_α are equal and data are difficult to compress by factor analysis. When τ is equal to a measure (graph in the middle), the graph reveals the structure of the music piece, with each measure played two times. Note the "horseshoe effect" resulting from the temporal ordering of time intervals. The right graph highlights that when increasing the duration τ, the percentage of explained inertia climbs, except when τ is smaller than or equal to a eighth note, the smallest duration of a note, and between τ equal to a whole note (corresponding to a measure) and equal to two whole notes, due to the repeated structure of the piece.

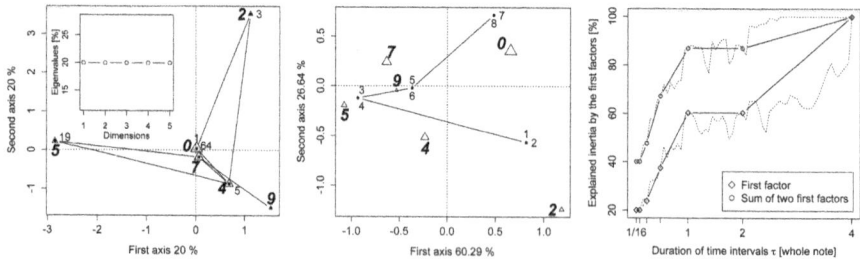

Fig. 2 CA on *Are you sleeping?* in *C major*. *Left*: scree graph and biplot with τ equal to an eighth note. Triangles with large-sized figures in italic stand for pitches (triangle size is proportional to the quantity of the pitch in the music piece) and full circles, sometimes with small-sized figures, represent time intervals and are linked in consecutive order according to the time progression. *Middle*: biplot with τ equal to a measure. *Right*: explained inertia by the (two) first factors according to τ. *Dotted lines* represent results for all durations and *solid lines* stand for results for integer divisors of τ_{tot}

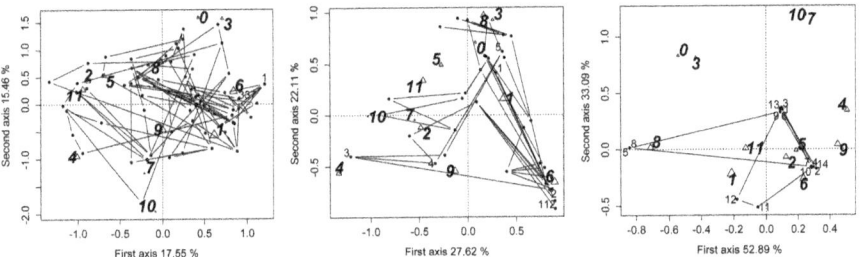

Fig. 3 CA on *Mazurka Op. 6, No. 1 in F♯ minor by Chopin*. Biplots with τ, respectively, equal to a quarter note (*left*), a measure (*middle*), and eight measures (*right*)

Another example is given in Fig. 3 for a Mazurka by Chopin, with three different interval durations. The structure emerges more clearly for large values of τ. In particular, the right graph, with τ equal to eight measures, reveals the similar (e.g., 1, 3, 6, 9 and 13) and different passages (e.g., 2 against 3).

While these two examples clearly highlight the structure of the piece, results are less comprehensible when a motif is transposed in the same piece or when a true rest appears. In fact, in the latter case, the first factor often exclusively expresses the contrast between true rests and pitches.

3.1.2 Autocorrelation Index

Consider now the neighborhood analysis between ordered time intervals, represented by the rows of Ξ. Temporal neighborhoods can be defined by a non-negative symmetric *exchange matrix* $E = (e_{st})$ obeying $e_{t\bullet} = e_{\bullet t} = f_t = 1/n$. The associated *autocorrelation index* (Bavaud et al. 2012) is calculated as:

$$\delta := \frac{\Delta - \Delta_{\text{loc}}}{\Delta} \in [-1, 1] \tag{3}$$

where Δ is the (global) inertia and Δ_{loc} is the local inertia:

$$\Delta := \frac{1}{2} \sum_{st} f_s f_t \hat{D}_{st} = \frac{1}{2n^2} \sum_{st} \hat{D}_{st} \qquad \Delta_{\text{loc}} := \frac{1}{2} \sum_{st} e_{st} \hat{D}_{st} \tag{4}$$

Thus, the autocorrelation index measures the difference between the overall variability of chi-squared interval dissimilarities and the local variability within some neighborhood defined by E, generalizing the usual "immediate left-right neighborhood" (see, e.g., Morando (1981) for a musical data-analytic approach). A large positive (resp. negative) autocorrelation means that the pitches distributions are more (resp. less) similar in the neighborhood than in randomly chosen intervals.

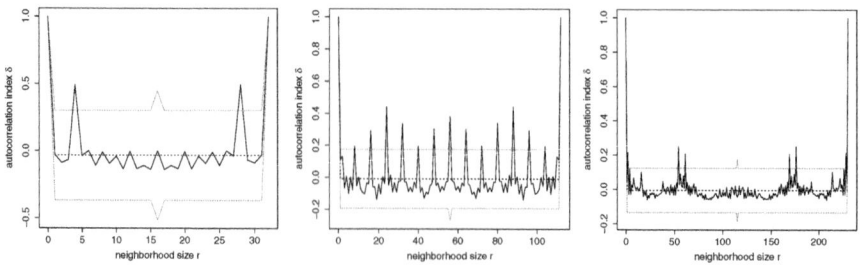

Fig. 4 Autocorrelation index according to the lag r varying from 0 to n (*solid line*), its expected value (*dashed line*) plus or minus $u_{0.975} = 1.96$ times the standard deviation (*dotted line*). *Left: Are you sleeping?* with τ equal to a quarter note. *Middle: Mazurka Op. 6, No. 1 by Chopin* with τ equal to a measure. *Right: sonata L. 12 (K. 478) by Scarlatti* with τ equal to a measure

Among all possible exchange matrices, it turns out to be convenient to define a *periodic* exchange matrix, with a neighborhood at temporal *distance* (or *lag*) r (right and left) of the current interval, $E^{(r)}$:

$$e_{st}^{(r)} = \frac{1}{2n}[1(t = (s \pm r) \bmod n) + 1((s \pm r) \bmod n = 0) \cdot 1(t = n)]$$

For statistical testing of the autocorrelation index, see, e.g., Cliff and Ord (1981) and Bavaud (2013). Note that in contrast to the usual autocorrelation function in time series analysis (see, e.g., Box and Jenkins 1976) which considers a single numerical variable, the autocorrelation index can deal with multiple simultaneous categorical variables.

The autocorrelation index is computed on three musical scores (Fig. 4). As expected, $\delta = 1$ for $r = 0$ and the figures are symmetric, since the neighborhood is periodic ($E^{(r)} = E^{(n-r)}$). Moreover, noticeable peaks appear in all graphs. For the monophonic music piece *Are you sleeping?*, the highest value ($\delta = 0.495$) appears for $r = 4$ which corresponds to the duration of a measure. In fact, due to the systematic repetition of each measure, at each point the same pitches are played at a distance equal to four, sometimes on the left, sometimes on the right. For Chopin's piece, peaks occur each eight measures as expected by the results obtained in Fig. 3. Finally, for Scarlatti's sonata, there are two remarkable peaks ($\delta = 0.25$ and $\delta = 0.21$), for $r = 54$ and $r = 61$ measures, corresponding to the length of the two repeated parts of the piece, which compose the whole piece. In conclusion, peaks of δ appear to detect strict or approximate repetitions, but do not detect transposed repetitions.

3.2 Between Voices Analysis

3.2.1 Soft Multiple Correspondence Analysis

Let Ξ^v denote the row-normalized contingency table for *voice* $v = 1, \ldots, V$ occurring in a music piece. The complete contingency table of the musical score

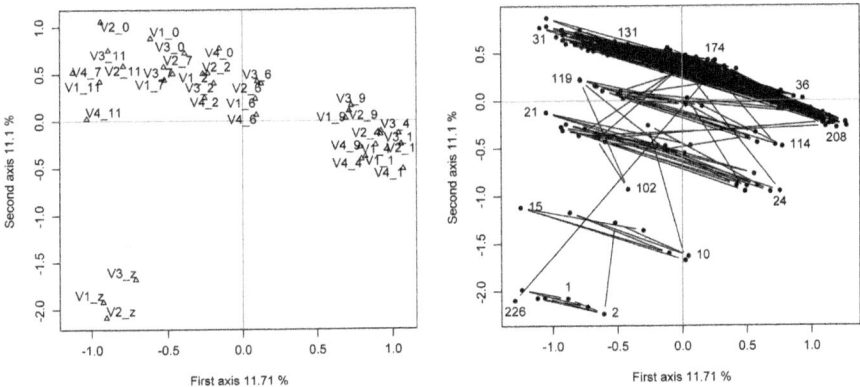

Fig. 5 Soft MCA on the *Canon in D Major by Pachelbel* with τ equal to a quarter note. *Left*: factor coordinates for the pitches, whose names are preceded by V1 for violin I, V2 for violin II, V3 for violin III, and V4 for Harpischord. *Right*: factor coordinates for the time intervals

is given as $\varXi^{\text{COMP}} = (\varXi^1 | \varXi^2 | \dots | \varXi^V)$, on which a CA is carried out. Whereas an usual multiple correspondence analysis (MCA) is computed on a disjunctive table, the present procedure is applied to row cells containing, due to row-normalization, the pitch *proportions* of the voice during a given t, and hence constitutes a soft variant of MCA.

Figure 5 shows the results obtained for the Pachelbel's canon. On the right graph, different zones appear depending on the number of instruments which are playing. For instance, in the bottom zone, only the harpsichord is playing, and so there are true rests for the three violins. Again, as for CA, the clarity of pattern representation largely depends upon the value of τ.

3.2.2   Cross-Autocorrelation Index

Define the "raw" coordinates of the voice $\varXi^v$ as $^*\xi_{tj}^v = \sqrt{\rho_j^v}(q_{tj}^v - 1)$, with the property that the associated squared Euclidean distances $D_{st} = \sum_j (^*\xi_{sj}^v - {}^*\xi_{tj}^v)^2$ are equal to the chi-squared distances $\hat{D}_{st}$ of Eq. (1).

To extend the autocorrelation index to two voices (α and β), one proposes a *cross-autocorrelation* index for multidimensional variables $\varXi^\alpha$ and $\varXi^\beta$, which measures the similarity between the pitch distribution of α and the pitch distribution of β within a fixed lag or, more generally, a defined neighborhood, namely:

$$\delta(\varXi^\alpha, \varXi^\beta) := \frac{\Delta(\varXi^\alpha, \varXi^\beta) - \Delta_{\text{loc}}(\varXi^\alpha, \varXi^\beta)}{\sqrt{\Delta(\varXi^\alpha)\Delta(\varXi^\beta)}} \in [-1, 1]$$

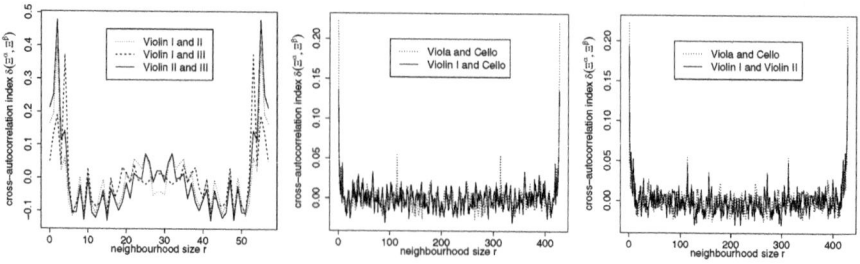

Fig. 6 Cross-autocorrelation index according to the lag r varying from 0 to n. *Left: Canon in D Major by Pachelbel* with τ equal to a measure. *Middle and right: first movement of the String Quartet No. 1 in F major, Op. 18 by Beethoven* with τ equal to a measure

In the latter, $\Delta(\Xi^v)$ is the inertia of the voice v [see the first part of (4)], $\Delta(\Xi^\alpha, \Xi^\beta) = \frac{1}{2} \sum_{st} f_s f_t D_{st}^{\alpha\beta} = \sum_s f_s \sum_j {}^*\xi_{sj}^\alpha {}^*\xi_{sj}^\beta - \sum_j {}^*\bar{\xi}_j^\alpha {}^*\bar{\xi}_j^\beta$ is the cross-inertia between the voice α and the voice β, where $D_{st}^{\alpha\beta} = \sum_j ({}^*\xi_{sj}^\alpha - {}^*\xi_{tj}^\alpha)({}^*\xi_{sj}^\beta - {}^*\xi_{tj}^\beta)$ is the cross-dissimilarity between two time intervals of two voices, and finally $\Delta_{\mathrm{loc}}(\Xi^\alpha, \Xi^\beta) = \frac{1}{2} \sum_{st} e_{st} D_{st}^{\alpha\beta} = \sum_s f_s \sum_j {}^*\xi_{sj}^\alpha {}^*\xi_{sj}^\beta - \sum_{st} e_{st} \sum_j {}^*\xi_{sj}^\alpha \xi_{tj}^\beta$ is the local cross-inertia between voices α and β.

In particular, $\Delta(\Xi, \Xi) = \Delta(\Xi)$ and $\Delta_{\mathrm{loc}}(\Xi, \Xi) = \Delta_{\mathrm{loc}}(\Xi)$, so $\delta(\Xi, \Xi) = \delta(\Xi) = \delta$ given in (3). It must be noticed that this formalism works in this specific context because $f_t^\alpha = f_t^\beta = f_t = \frac{1}{n}$ due to the normalization of Ξ or Ξ^v and since all voices have the same number of time intervals.

This cross-correlation index is computed on two multiple-voice music pieces with the same exchange matrix as the one proposed for the autocorrelation index (Fig. 6). For Pachelbel's canon, highest peaks on the left graph appear at $r = 2$ for the cross-autocorrelation between violins I and II and between violins II and III, and at $r = 4$ between violins I and III, corresponding to the lag of two or four measures between the starts of each violin. For Beethoven's string quartet (center and right graphs), peaks at $r = 0$ reveal largest melodic similarities between violin I and violin II on the one hand, and between viola and cello on the other hand. Moreover, both graphs exhibit large peaks at $r = 114$ measures, corresponding to a repetition in the music piece.

Thus, the cross-autocorrelation index allows the comparison of different voices of a music piece. It can also be implemented to compare two music piece variants. See, e.g., Ellis and Poliner (2007), who apply cross-correlation on audio files.

3.3 Between Scores Analysis

To measure the configuration similarity between two musical scores a and b, a weighted dual version of the RV-Coefficient proposed by Robert and Escoufier

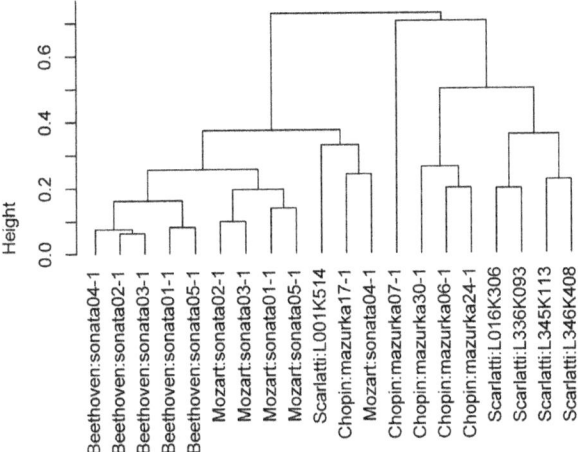

Fig. 7 Hierarchical clustering of 20 music pieces with the Ward aggregation method

(1976) is computed:

$$CS_{ab} = \frac{Tr(\check{K}^a \check{K}^b)}{\sqrt{Tr((\check{K}^a)^2)Tr((\check{K}^b)^2)}}$$

where $\check{K}^a$ (resp. $\check{K}^b$) is the weighted scalar product between pitches of the musical score a (resp. b) as defined in the second part of the Eq. (2). By construction, the components of $\check{K}^a$ (or $\check{K}^b$) are zero for a pitch absent in the corresponding musical score. Both $\check{K}^a$ and $\check{K}^b$ depend upon the reference duration τ, chosen as identical for both music pieces.

Define the dissimilarity between two musical scores as $D_{ab} = 1 - CS_{ab}$. This dissimilarity can be seen as a generalization of the well-known cosine distance (see, e.g., Weihs et al. 2007), and turns out to be squared Euclidean. Usual clustering methods between musical scores, based upon D_{ab}, can in turn be applied.

Figure 7 presents the results obtained with an agglomerative hierarchical clustering on a dataset made up of 20 music pieces written by four composers:

- **Scarlatti:** Sonatas L. 1 (K. 514), L. 16 (K. 306), L. 336 (K. 93), L. 345 (K. 113), and L. 346 (K. 408). They all have a 2/2 time signature.
- **Mozart:** First movement of piano sonatas n°1, 2, 3, 4, and 5.
- **Beethoven:** First movement of piano sonatas n°1, 2, 3, 4, and 5.
- **Chopin:** Mazurkas Op. 6 (No. 1), Op. 7 (No. 1), Op. 17 (No. 1), Op. 24 (No. 1), and Op. 30 (No. 1).

For comparison sake, the 20 music pieces are all transposed in C, with a common τ value of one measure. Although the dataset is small, this first result is encouraging, producing well-grouped music pieces with respect to each composer, especially for Beethoven.

4 Conclusion

The present data-analytic treatment of musical scores is based upon two primitives, namely a dissimilarity matrix and a neighborhood matrix between time intervals, defined with respect to a reference duration. It covers and generalizes well-known multi-categorical, factorial and time-series techniques, and is able to treat polyphonic pieces, as well as performing between-voices and between-scores analyses, with encouraging clustering results. Its modest computational cost makes it amenable to the automatic treatment of large symbolic musical data sets. Furthermore, it allows the consideration of flexible alternatives, both for the dissimilarity matrix (other than the chi-square) and for the exchange matrix (other than periodic neighborhood), deserving further investigation.

So far, exploratory analyses are interpretable in a fairly satisfactory way, although the complex factorial structures exhibited by rich music pieces certainly deserve further attention. In the near-future agenda, within the present formalism, we hope to progress in the automatic detection of τ, motif recognition and large dataset clustering or classification.

References

Bavaud, F. (2013). Testing spatial autocorrelation in weighted networks: The modes permutation test. *Journal of Geographical Systems, 15*, 233–247.

Bavaud, F., Cocco, C., & Xanthos, A. (2012). Textual autocorrelation: Formalism and illustrations. In *11èmes Journées internationales d'analyse statistique des données textuelles* (pp. 109–120). Liège: Université de Liège.

Box, G. E. P., & Jenkins, G. M. (1976). *Time series analysis: Forecasting and control.* San Francisco: Holden-Day.

Cliff, A. D., & Ord, J. K. (1981). *Spatial processes: Models and applications.* London: Pion.

Critchley, F., & Fichet, B. (1994). The partial order by inclusion of the principal classes of dissimilarity on a finite set, and some of their basic properties. In: B. Van Cutsem (Ed.), *Classification and dissimilarity analysis* (pp. 5–65). New York: Springer.

Ellis, D. P. W., & Poliner, G. E. (2007). Identifying 'Cover Songs' with chroma features and dynamic programming beat tracking. In *IEEE International Conference on Acoustics, Speech and Signal Processing, 2007* (pp. IV-1429–IV-1432).

Lavrenko, V., & Pickens, J. (2003). Polyphonic music modeling with random fields. In *Proceedings of the Eleventh ACM International Conference on Multimedia* (pp. 120–129), Berkeley, CA.

Morando, M. (1981). L'analyse statistique des partitions de musique. In J.-P. Benzécri, et al. (Eds.), *Pratique de l'analyse des données, tome 3: Linguistique et lexicologie* (pp. 507–522). Paris: Dunod.

Müller, M., & Ewert, S. (2011). Chroma toolbox: Matlab implementations for extracting variants of chroma-based audio features. In *Proceedings of the 12th International Conference on Music Information Retrieval* (pp. 215–220).

Robert, P., & Escoufier, Y. (1976). A unifying tool for linear multivariate statistical methods: The RV-coefficient. *Journal of the Royal Statistical Society. Series C (Applied Statistics), 25*, 257–265.

Weihs, C., Ligges, U., Mörchen, F., & Müllensiefen, D. (2007). Classification in music research. *Advances in Data Analysis and Classification, 1*, 255–291.

Impact of Frame Size and Instrumentation on Chroma-Based Automatic Chord Recognition

Daniel Stoller, Matthias Mauch, Igor Vatolkin, and Claus Weihs

Abstract This paper presents a comparative study of classification performance in automatic audio chord recognition based on three chroma feature implementations, with the aim of distinguishing effects of frame size, instrumentation, and choice of chroma feature. Until recently, research in automatic chord recognition has focused on the development of complete systems. While results have remarkably improved, the understanding of the error sources remains lacking. In order to isolate sources of chord recognition error, we create a corpus of artificial instrument mixtures and investigate (a) the influence of different chroma frame sizes and (b) the impact of instrumentation and pitch height. We show that recognition performance is significantly affected not only by the method used, but also by the nature of the audio input. We compare these results to those obtained from a corpus of more than 200 real-world pop songs from The Beatles and other artists for the case in which chord boundaries are known in advance.

1 Introduction

A chord is defined as "three or more pitches sounded simultaneously or functioning as if sounded simultaneously" (Randel 1999, p. 135) . In Western music, chords form the building blocks of harmony, the "relationship of tones considered as they sound simultaneously, and the way such relationships are organised in time" (Randel 1999, p. 286). This paper deals with pitch class profiles (PCP; Fujishima 1999)

D. Stoller (✉) • I. Vatolkin
TU Dortmund, Chair of Algorithm Engineering, 44227 Dortmund, Germany
e-mail: daniel.stoller@tu-dortmund.de; igor.vatolkin@tu-dortmund.de

M. Mauch
Queen Mary University of London, Centre for Digital Music, London E1 4NS, UK
e-mail: matthias.mauch@eecs.qmul.ac.uk

C. Weihs
TU Dortmund, Chair of Computational Statistics, 44227 Dortmund, Germany
e-mail: claus.weihs@tu-dortmund.de

© Springer-Verlag Berlin Heidelberg 2015
B. Lausen et al. (eds.), *Data Science, Learning by Latent Structures, and Knowledge Discovery*, Studies in Classification, Data Analysis, and Knowledge Organization, DOI 10.1007/978-3-662-44983-7_36

411

or chroma vectors (computational representations of harmony and chords in audio recordings of music).

Chroma vectors provide a summary of the energy (or activation) of the 12 pitch classes C, C# ,D , ..., B in a piece of audio. Pitch height (octave information) is discarded. Chroma implementations differ in precisely how this is achieved, but two steps are generally applied: (1) mapping a spectral representation to a semitone-based representation and (2) wrapping the semitone spectrum into one octave. Spectral frequencies are mapped onto semitones as follows:

$$p(f) = \left[12 \log_2 \frac{f}{440}\right] + 69, \tag{1}$$

where f is the frequency and $p(f)$ is the corresponding position in the semitone spectrum. Once a vector $A[p]$ of semitone activation for all semitones p is available, it is mapped to pitch class activations

$$\text{PCP}(p_w) = \sum_{(p \bmod 12) = p_w} A[p], \tag{2}$$

where $p_w \in \{0, 1, .., 11\}$ is the pitch class of the corresponding semitone.

In order to deal with time-varying harmonic content, all chroma implementations divide the time-domain audio signal into T short time frames which chroma is calculated on. This results in the chromagram, a $(12 \times T)$-matrix of chroma vectors.

A robust computational representation of chord progressions and other harmonic properties is valuable for both scholarly and commercial applications such as the analysis of rules and characteristics of genre or style (Anglade et al. 2009), recognition of moods (Baume 2013) and music recommendation (Celma 2010). Chroma profiles are often used if score information is not available. Yet the extraction of harmonic audio properties is not without error, especially due to the varying properties of instruments with overlapping harmonic partials playing simultaneously.

In recent years, chroma features have been used as a basis for chord recognition systems. An established method is to model chords as variables in hidden Markov models (HMM) with chroma emissions (e.g. Sheh and Ellis 2003). A structured overview of chord recognition systems used up until 2010 is provided in Mauch (2010). Several implementations of chroma combined with different approaches for chord recognition were evaluated for songs from The Beatles in Jiang et al. (2011).

The evaluation of complete chord recognition systems usually focuses on overall performance, and little room remains for the extended statistical analysis of possible sources for errors in subsystems. In previous work, we examined the impact of instrumentation on the recognition of intervals (Mattern et al. 2013). In this paper, we concentrate on a very basic aspect for chord recognition, specifically, the frame size used during the feature extraction process. We use three different chroma vector implementations and a simple chord prediction method in mixtures of up to four tones and short sequences of tones in popular songs.

This paper is structured as follows. Section 2 describes the parameters that we vary to learn about the properties of chroma extraction. Section 3 describes the data and evaluation used in our experiments, and the results are discussed in Sect. 4. We conclude with a summary of results and an outlook.

2 Parameters Under Investigation

Instead of trying to optimise the configurations of a complete chord recognition system, in this study we focus on the impact of different parameters on classification performance, including the chroma implementation, the analysis window size and the feature processing method.

2.1 Chroma Implementations

We compare three different chroma implementations with default settings on audio data sampled at 22,050 Hz: NNLS Chroma, MIR Toolbox Chroma and Yale Valueseries Chroma. NNLS chroma (Mauch and Dixon 2010) is based on the enhanced model of note profiles using overtone distribution after Gómez (2006) and analysis of tones with fundamental frequencies between 27.5 and 3,322 Hz. The MIR Toolbox chroma implementation (Lartillot and Toivainen 2007) is based on the work of Gómez (2006), taking into account the frequencies between 500 and 6,400 Hz. The Yale Valueseries chroma plugin (Mierswa et al. 2006) is based on the findings of Goto (2003).

2.2 Window Size

In order to investigate the effects of different frame sizes we used 512, 1,024, 2,048 and 4,098 samples as frame sizes. Because each music signal is downsampled to 22,050 Hz, these settings approximately correspond to the frame sizes of 23, 46, 93 and 186 ms. Frames of 8,192 and more samples would lead not only to a more exact frequency resolution, but also to an increased danger of a possible overlap of subsequent notes, specifically, if the beats are uniformly distributed, the tempo of approximately 161 beats per minute is enough for a gap of 372 ms (8,192 samples) between the beats.

2.3 Feature Processing

In the feature processing stage, the $(12 \times T)$-matrix containing the chroma vectors of each frame is processed and reduced to a single chroma vector for later classification. A very simple feature processing method is to calculate the average of all the per-frame chroma vectors. We call this method "FP1". A more complex method ("FP2") is based upon the idea of filtering possibly existing non-harmonic components at the beginnings and the ends of notes. We estimate the attack and release phase of the onset envelopes using the MIR Toolbox and restrict our analysis to the frames between the onset event and the middle of the release phase. If two or more attack/release pairs are estimated, we default to defining the attack phase as the interval between the beginning of the recording and the frame with the highest root mean square energy. Likewise, the release phase is between the frame with the highest root mean square energy and the final frame of the recording.

2.4 Classification

After obtaining the chroma vector from the feature processing stage, the actual chord recognition stage involves classifying the given chroma vector. In this study, a chord class is defined as the set of pitch classes that the chord contains. We adopt a simple classifier which, for a given chroma vector and the number n of notes in the chord, returns the pitch classes with the n highest values in the chroma vector. We deliberately avoid using complex, possibly trained classification methods capable of correcting errors from earlier processing stages in order to be able to clearly identify error sources.

3 Experimental Set-up

This section describes the two datasets used in the experiments (one artificial and one consisting of recorded pop music) and the metrics used for evaluation.

3.1 Datasets

The first dataset ("DS1") was built from artificially created polyphonic mixtures of single tones from different instruments from McGill University samples,[1] RWC

---

[1]http://www.music.mcgill.ca/resources/mums/html.

database (Goto et al. 2003) and the University of Iowa samples.[2] It does not include sounds such as background noises or additional instruments playing at the same time, resulting in a dataset for evaluating chord recognition performance under close to ideal conditions not often found in the real world. Each recording contains exactly one chord which is either played by one instrument alone or for each of the corresponding notes a random instrument is selected. This splits the dataset into nine different subsets: a set of chords with exactly one instrument for each of eight instrument classes (acoustic guitar, cello, electric guitar, flute, piano, trumpet, viola and violin) and a set of chords where the notes are shared across at least two instruments. Generating 50 triads and 50 tetrachords for each of those nine subsets results in 18 subsets comprised of 900 chords in total, enabling a detailed analysis of classification performance depending on the type of chord and the instruments that were used.

The second dataset ("DS2") acts as an additional challenge for the system and contains 225 real-world songs from Queen, The Beatles, Carole King and Zweieck, including 19,162 segments with chords and thereby covering 544 unique chords (Mauch et al. 2009). The annotations belonging to this dataset provide exact information about the start and end times of each of these chords. However, precise onset and offset time events of each individual instrument tone are not available. Each recording contains the audio signal of a segment with unchanging chords. We removed chords consisting of more than four or less than three notes and with a duration shorter than 0.5 s.

3.2 Metrics for Classification Performance

In order to evaluate the accuracy of the chord recognition system when used with a specific configuration and to compare the performance between different configurations, we define two metrics.

The first metric, called "relative success", measures the performance of the classifier described in Sect. 2.4. Given the ground truth as the set of played notes $\mathcal{P}$ in the input signal and the set of recognised notes $\mathcal{C}$ obtained from the classifier, it is defined as follows:

$$s_{\text{rel}} = \frac{|(\mathcal{P} \cap \mathcal{C})|}{|\mathcal{P}|} \tag{3}$$

The resulting values in the interval [0, 1] represent the fraction of notes correctly recognised by the chord recognition system.

Our second approach is based upon the idea of calculating the error in the chroma vector obtained from the feature processing stage by comparing it to an "ideal

---

[2]http://www.theremin.music.uiowa.edu.

chroma vector", thus circumventing the classification stage with its associated error sources and focusing on the errors introduced by earlier chord recognition stages. Intuitively, the ideal chroma vector for a given chord has the same, positive values for every entry corresponding to a note contained in the chord and thus in the set of played notes $\mathcal{P}$, while all the other entries are zero. Applying the cosine similarity to the ideal chroma vector **a** and the chroma vector from feature processing stage **b** yields an indicator for the amount of deviation from the ideal chroma vector. This method is also referred to as the binary template-based approach (Jiang et al. 2011).

More precisely, the cosine similarity represents the angle θ between the non-negative 12-dimensional chroma vectors with values in the range of $[0, 1]$:

$$\text{cosine similarity} = \cos(\theta) = \frac{\sum_{i=1}^{12} a_i \cdot b_i}{\sqrt{\sum_{i=1}^{12} (a_i)^2} \cdot \sqrt{\sum_{i=1}^{12} (b_i)^2}} \tag{4}$$

4 Discussion of Results

Figure 1 shows the average relative success values for increasing frame sizes for 12 different set-ups: three chroma vectors (MIR Toolbox, Yale, NNLS), two processing methods (FP1, FP2), and two kinds of chords (triad, tetrachord) from dataset DS1. In all cases, larger frame sizes correspond to a higher performance of chord recognition, however this progress slows down as the frame sizes increase. The

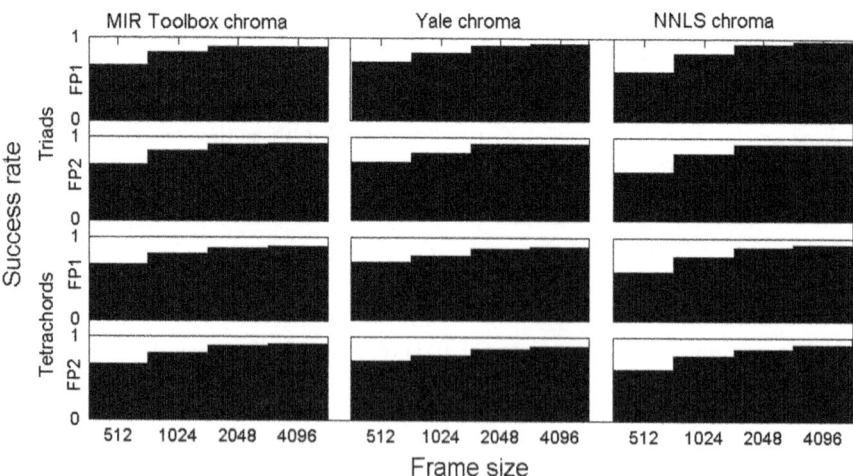

Fig. 1 Mean relative success for triads and tetrachords from DS1 and different feature extraction and processing methods

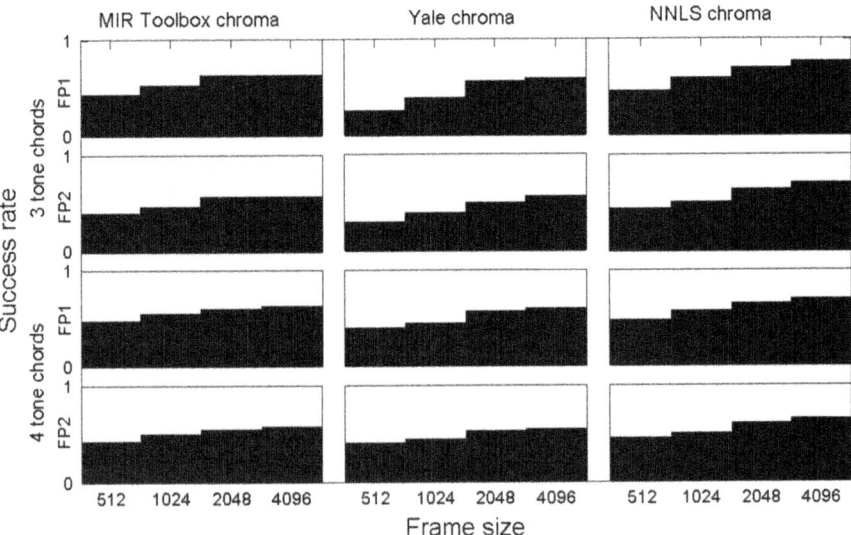

Fig. 2 Mean relative success for chords of 3 and 4 tones from DS2 and different feature extraction and processing methods

application of the Wilcoxon signed rank test confirms this increase in performance to be significant in all cases except for the triads when changing frame size from 2,048 to 4,096 samples and using MIR Toolbox chroma.

Similar results can be observed for chords from real-world songs (dataset DS2) and are illustrated in Fig. 2. Here, the increase of performance of chord recognition is significant with respect to Wilcoxon signed rank test in all combinations of settings of parameters, even if this improvement is rather marginal for MIR Toolbox, triads, and the change of 2,048–4,096 samples. As expected, the overall performance is lower than for DS1 because of background noise like additional instruments in the real-world song segments. The best mean relative success is achieved by NNLS chroma with FP1 and 4,096 samples (77 %).

Because the mean relative success of chord recognition increases with higher frame sizes for all chroma vectors, FP1 and FP2 feature processing methods, and also both datasets (artificial mixtures and real-world recordings), we recommend to use the frame size of 4,096 samples as default value for chroma extraction. Larger frame sizes may provide a further slight increase in performance, but are not suited for music with tempo above 160 beats per minute (see Sect. 2.2). Another possibility is to first estimate the tempo of a music piece and then set the frame size of chroma vectors to the maximum reasonable value according to the mean gap between two beats.

We use the results obtained with a frame size of 4,096 samples for an investigation into instrument effects. Figure 3 shows the mean relative success by instrument and chroma feature. Chords played by the acoustic guitar are recognised consistently

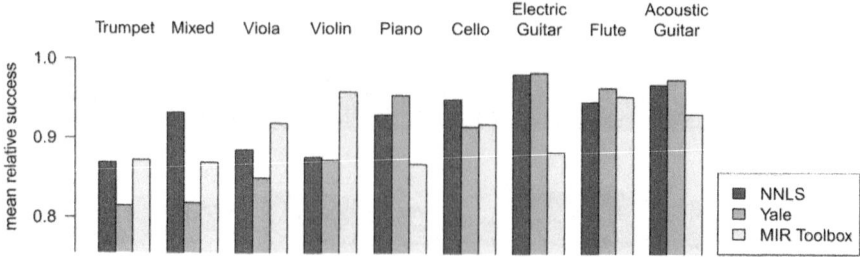

Fig. 3 Mean relative success with different instruments and feature extraction methods from DS1 and a frame size of 4,096 samples. Instruments are ordered by mean relative success across chroma methods

well overall ($s_{rel} > 0.93$ for all chroma types), while trumpet chords are recognised worst on average (mean $s_{rel} \approx 0.85$), even worse than mixed instrumentation. This can be explained by the underlying database of trumpet sounds: a part of tones contained slight vibrato, and many trumpet spectra were characterised by a rich number of harmonic components of similar amplitudes. Perhaps most striking is the large difference between chroma methods for some instruments. For example, electric guitar chords are almost perfectly classified using NNLS and Yale chroma ($s_{rel} > 0.98$), but substantially worse by the MIR Toolbox chroma ($s_{rel} \approx 0.88$). In the case of violin chords, the roles are reversed, with classification based on the MIR Toolbox at $s_{rel} \approx 0.96$, outperforming the other two by more than 0.08. Further study is needed to understand what properties of the instruments were responsible for the differences and how significant the differences are on a larger dataset.

Particularly for small frame sizes, we may expect worse performance for lower frequencies due to an insufficient resolution of frequency bins. This hypothesis can be tested by comparing the estimated chroma to the ideal chroma as discussed in Sect. 3.2. A look at the resulting cosine similarity values reveals that this assumption is not true in all cases and its correctness depends on the instruments used, as illustrated in the example of piano and viola in Fig. 4. Here, the individual chords are marked with small circles. The general trends are outlined by regression lines. The strong variance of performance for the same mean frequencies is reasoned by varying properties of tones: for each pitch and instrument class, the original databases contained tones with slightly different playing techniques or different bodies of the corresponding instruments. Even if the similarity to the ideal chroma vector significantly increases for lower frequencies when larger frame sizes are used (columns from left to right), for viola, the chords with a lower mean pitch tend to have a higher cosine similarity to the ideal chroma than chords with a higher mean pitch. For more robust statistics, a larger number of chords should be analysed, which was beyond of the scope of this study.

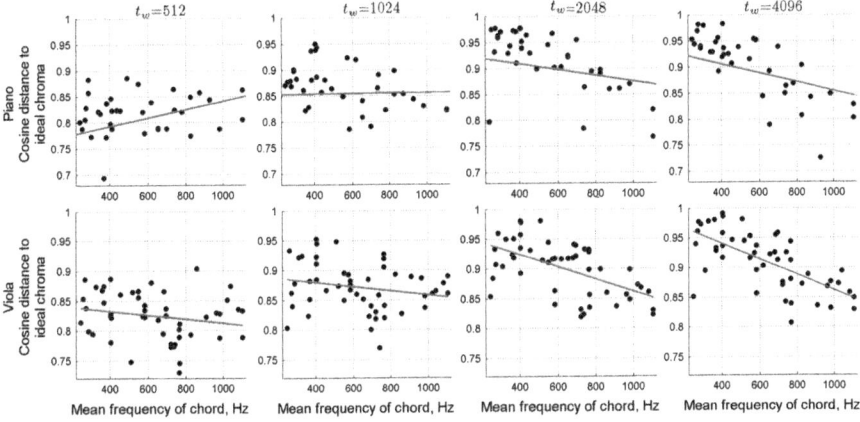

Fig. 4 Impact of mean frequency of chords on cosine distance to ideal chroma for piano (*upper row*) and viola (*bottom row*)

5 Conclusions

In our study, we examined the impact of frame size and instrumentation on the recognition of chords based on pitch class profiles (chroma vectors). To provide a higher general interpretability of results, we used two datasets: the first one with a precise ground truth available for randomly generated chords from single tones, and the second one containing longer segments of real-world songs with unchanging chords. To limit the impact of further parameters, we used a simple classification method based on the selection of highest peaks and did not change the default parameters for the three tested chroma vectors.

In almost all cases, an increase of extraction frame size for chroma vectors leads to a significant increase of classification performance. However, we would not recommend using higher frame sizes than 4,096 samples (for the sampling rate of 22,050 Hz) except for slow music pieces, because larger frame sizes may contain the signal of more than one chord. Further, the performance depends on the instruments played: acoustic guitar and flute perform best in general, while the chords from trumpet tones seem to be the most challenging. The performance sometimes varies for different chroma vectors, so that no particular chroma implementation can be explicitly recommended. Finally, we compared the complexity of chord recognition to the so-called ideal chroma depending on the mean frequency of the chord. The performance again depends on the instrument due to the varying distribution of overtones and non-harmonic frequencies.

Summarising the results of our study, we express that the recognition of chords may be improved if more high-level information about the tempo and instrumentation of songs would be available: for slower songs it is possible to further increase the frame size of the chroma vector, and the knowledge of instrumentation may be integrated into a machine learning approach for chord prediction. Because automatic instrument and tempo recognition is not straightforward for polyphonic audio recordings, a possible compromise is to gather statistics of user tags and text descriptions from the Internet: where the precise annotation of chords for a concrete audio recording is only seldom available, the information about instruments and tempo may be derived from such metadata.

For further studies we plan to analyse the performance of chord recognition systems on real-world songs for which no information about chord boundaries is known and also to integrate more complex classification methods using further high-level features (instrumentation, genres, tempo) as input. For songs with changing tempo or a strongly varying number of notes per bar it is possible to extract the chroma vector with varying frame sizes.

References

Anglade, A., Ramirez, R., & Dixon, S. (2009). Genre classification using harmony rules induced from automatic chord transcriptions. In *Proceedings of the 10th International Conference on Music Information Retrieval (ISMIR)* (pp. 669–674).

Baume, C. (2013). Evaluation of acoustic features for music emotion recognition. In *Proceedings of Audio Engineering Society Convention 134*.

Celma, O. (2010). *Music recommendation and discovery*. Berlin/Heidelberg: Springer.

Fujishima, T. (1999). Realtime chord recognition of musical sound: A system using common LISP music. In *Proceedings of the International Computer Music Conference (ICMC)* (pp. 464–467).

Gómez, E. (2006). Tonal description of music audio signals. PhD thesis, Department of Information and Communication Technologies, Universitat Pompeu Fabra, Barcelona.

Goto, M. (2003). A chorus-section detecting method for musical Audio signals. In *Proceedings of the 2003 IEEE International Conference on Acoustics, Speech, and Signal Processing (ICASSP)* (pp. 437–440).

Goto, M., Hasiguchi, H., Nishimura, T., & Oka, R. (2003). RWC music database: Music genre database and musical instrument sound database. In *Proceedings of the 4th International Conference on Music Information Retrieval (ISMIR)* (pp. 229–230).

Lartillot, O., & Toivainen, P. (2007). MIR in matlab (II): A toolbox for musical feature extraction from audio. In *Proceedings of the 8th International Conference on Music Information Retrieval (ISMIR)* (pp. 127–130).

Mattern, V., Vatolkin, I., & Rudolph, G. (2013). A case study about the effort to classify music intervals by chroma and spectrum analysis. In B. Lausen, D. Van den Poel, & A. Ultsch (Eds.), *Algorithms from and for nature and life* (pp. 519–528). Berlin/Heidelberg: Springer

Mauch, M. (2010). *Automatic chord transcription from audio using computational models of musical context*. PhD thesis, School of Electronic Engineering and Computer Science Queen Mary, University of London.

Mauch, M., & Dixon, S. (2010). Approximate note transcription for the improved identification of difficult chords. In *Proceedings of the 11th International Society for Music Information Retrieval Conference (ISMIR)* (pp. 135–140).

Mauch, M., Cannam, C., Davies, M., Dixon, S., Harte, C., Kolozali, S., Tidhar, D., & Sandler, M. (2009). OMRAS2 metadata project 2009. In *Late-Breaking Session at the 10th International Conference on Music Information Retrieval (ISMIR)*.

Mierswa, I., Wurst, M., Klinkenberg, R., Scholz, M., & Euler, T. (2006) YALE: Rapid prototyping for complex data mining tasks. In *Proceedings of the 12th ACM SIGKDD International Conference on Knowledge Discovery and Data Mining (KDD)* (pp. 935–940).

Jiang, N., Grosche, P., Konz, V., & Müller, M. (2011). Analyzing chroma feature types for automated chord recognition. In *Proceedings of the 42nd AES Conference* (pp. 285–294).

Randel, D. M. (1999). *The Harvard Concise Dictionary of Music and Musicians*. Cambridge: Harvard University Press.

Sheh, A., & Ellis, D. P. W. (2003). Chord segmentation and recognition using EM-trained hidden markov models. In *Proceedings of the 4th International Conference on Music Information Retrieval (ISMIR)*. (pp. 135–141).

Interpretable Music Categorisation Based on Fuzzy Rules and High-Level Audio Features

Igor Vatolkin and Günter Rudolph

Abstract Music classification helps to manage song collections, recommend new music, or understand properties of genres and substyles. Until now, the corresponding approaches are mostly based on less interpretable low-level characteristics of the audio signal, or on metadata, which are not always available and require high efforts for filtering the relevant information. A listener-friendly approach may rather benefit from high-level and meaningful characteristics. Therefore, we have designed a set of high-level audio features, which is capable to replace the baseline low-level feature set without a significant decrease of classification performance. However, many common classification methods change the original feature dimensions or create complex models with lower interpretability. The advantage of the fuzzy classification is that it describes the properties of music categories in an intuitive, natural way. In this work, we explore the ability of a simple fuzzy classifier based on high-level features to predict six music genres and eight styles from our previous studies.

1 Towards a Higher Interpretability of Classification Models

Recognition of high-level music categories such as genres and styles provides an efficient and automatic way to organise large music collections and recommend new songs. A large number of past and recent studies have addressed this task: Sturm (2012) lists almost 500 references related to genre recognition. The development of new features and complex classification techniques led to significant improvements in the quality of music classification systems. However, in many cases user-centred evaluation criteria as discussed in Hu and Liu (2010) remain completely untouched and are not integrated into the optimisation of parameter settings. One of these criteria is the interpretability of classification models: if some rules are trained to predict music categories, it may be useful for both music scientists and listeners to better understand and interpret their properties.

I. Vatolkin (✉) • G. Rudolph
TU Dortmund, Chair of Algorithm Engineering, 44227 Dortmund, Germany
e-mail: igor.vatolkin@tu-dortmund.de; guenter.rudolph@tu-dortmund.de

© Springer-Verlag Berlin Heidelberg 2015
B. Lausen et al. (eds.), *Data Science, Learning by Latent Structures,
and Knowledge Discovery*, Studies in Classification, Data Analysis,
and Knowledge Organization, DOI 10.1007/978-3-662-44983-7_37

423

A basic chain of algorithms for classification consists of three steps: feature extraction, feature processing, and the training of classification models. Each of these steps may be designed either to facilitate highly comprehensible models as the final output or ignore the request for interpretability.

The first step towards an interpretable classification model is to start with a set of high-level features which are related to music theory and are understood by a music listener rather than by a signal processing expert. The difference between several levels of interpretability of features from the perspective of a listener is very well illustrated in Celma and Serra (2008): They distinguish between three abstraction layers: low-level features which describe the audio signal and physical properties of the sound, mid-level features which correspond to musical characteristics, such as key and mode, and high-level features which are very close to a listener: moods, opinions, memories, etc.

The goal of feature processing is at first to prepare feature vectors for classification, but also to reduce the dimensionality of the original feature matrix, because it may be very large, in particular for short-frame features. Very popular statistical feature processing methods such as principal component analysis are especially dangerous for the keeping of the interpretability because they transform the original feature space (Essid et al. 2006). A suitable solution to strongly reduce the feature matrix keeping the original feature space is to apply selection both on time and feature dimension: to select a limited amount of time frames according to events related to music structure, such as onsets, beats, and tatums (Vatolkin et al. 2012), and apply feature selection for the identification of the most relevant features (Guyon et al. 2006).

The final step is to train classification models. Again, many methods, such as well-established support vector machines, estimate linear combinations of original features or even transform them to higher dimensional spaces. Other successful methods combine the results of many classifiers, e.g., by stacking as proposed in Wolpert (1992) or building ensembles (Zhou 2012). These approaches often lead to a high classification quality, but the models are not comprehensible any more. One of the possibilities to address interpretability is to build classification rules with linguistic variables using fuzzy controllers (Zhang and Liu 2006), or optimise fuzzy controllers with genetic algorithms (Geyer-Schulz 1998). Fuzzy classification was recommended in Mckay and Fujinaga (2006) as the method which "would significantly improve the quality of ground truth, and would make the evaluation of systems more realistic" but still plays a minor role in most music classification applications. In particular, we are not aware of any work which applies fuzzy classification based on a large set of high-level audio features. To name a few related publications, in Friberg (2005) the prediction of emotional expressions was done using fuzzy modelling of so-called cues (tempo, sound level, and articulation). Application of a fuzzy classifier to predict emotions was reported in Yang et al. (2006). In Fernández and Chávez (2012), a fuzzy-rule based system is optimised with the help of evolutionary algorithms to distinguish between classical and jazz recordings using several features which describe frequencies with the strongest amplitudes, and Abeßer et al. (2009) introduced a rule-based framework for genre classification.

2 High-Level Audio Features

In this study we concentrate on audio features, which can be extracted independently of the popularity of songs or the availability of the score. Even if the estimation of some complex signal characteristics is time intensive, this can be done offline only once for the building of the feature set. However, the main challenge is that it is very hard to robustly extract high-level features from the polyphonic signal. This task can be solved to a certain level by machine learning approaches. Probably the first work which integrated this approach for the extraction of high-level features, so-called anchors, is Pachet and Zils (2003). In Vatolkin (2013), we have proposed a novel optimisation approach called sliding feature selection (SFS), where high-level features are iteratively predicted from other high-level and low-level characteristics, and the building of classification models is optimised by means of multi-objective evolutionary feature selection.

Table 1 lists high-level features, which are used in this work as the basis for further fuzzy recognition of genres and styles, providing some examples (second column) and the overall number of feature dimensions. These features can be roughly distinguished in three groups. The first one contains directly implemented mostly harmonic and short-frame characteristics. The second one corresponds to features derived with the help of an SFS-optimised machine learning approach. The

Table 1 Groups of high-level audio features with examples. Dim.: overall number of dimensions of the corresponding group

Group	Examples	Dim.
Directly implemented features		
Chroma and harmony	Tonal centroid, key, strengths of intervals	129
Chord statistics	Number of different chords in 10 s	5
Tempo, rhythm, and structure	Duration, beats per minute	9
SFS-optimised high-level features		
Instruments and effects	Guitar, piano, effects distortion	48
Singing characteristics	Singing solo rough, singing solo polyphonic	56
Harmony	Major, minor	16
Melody	Melodic range > octave, melody linear	32
Moods	Earnest, energetic	72
Structural complexity		
Chord, harmony, instruments, tempo and rhythm complexity		70

last group of features describes structural changes of several high-level groups as introduced in Mauch and Levy (2011). Because for the first group of short-framed harmony characteristics we have estimated the mean and standard deviation values for larger music intervals, the overall number of dimensions of features is 566.

3   Measuring of Feature Relevance with Linguistic Terms

A preliminary step in the design of a fuzzy classifier is to describe the values of features with linguistic terms. Usually not more than 5–7 terms are used, such as *very low*, *low*, *moderate*, *high*, *very high*. The values of features can then be mapped to a membership function which estimates the relationship grade to a category.

Figure 1 provides an example of a relevant feature (top) and not relevant feature (bottom) for the prediction of the category Pop. The segments of songs of the training set which are used as classification instances are marked with small circles. Because in our scenario the songs are assigned as either belonging to a category (positive examples) or not belonging to it (negative examples), the corresponding membership functions are equal to 0 resp. 1. In the upper subfigure it can be observed that songs which belong to the category Pop have the values of the feature "Drum recognition share" always equal or greater than 0.4. On the other side, songs

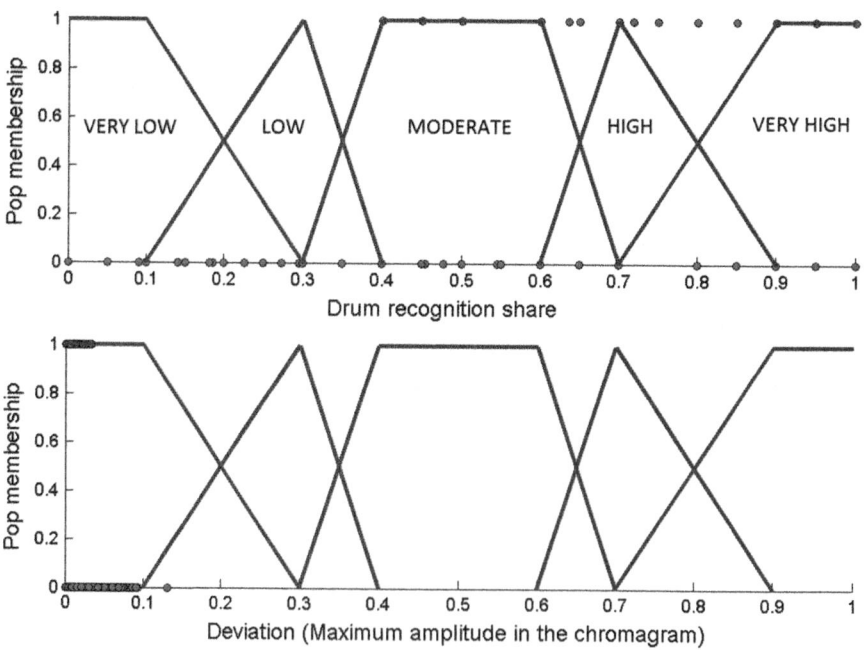

Fig. 1 Features with high (*top subfigure*) and low (*bottom subfigure*) relevance

which do not belong to Pop may contain very different shares of drums between 0 and 1. In other words, if the share of drums is below 0.4 in a song, it is very probable that this song does not belong to the category Pop—at least according to the previously selected songs for the training set.

Figure 1, lower subfigure, provides an example of a feature which is not well suited for the recognition of Pop songs. The deviation of the maximum amplitude of the chromagram is almost always below 0.1 for both positive and negative training songs.

The estimation of individual relevance of features may be used for the prediction of the membership function. If C_k would denote the kth category to predict (e.g., Pop or Classic), X_i would denote the feature i (e.g., share of drum recognitions), and T_j would describe a linguistic term (*very low, low, moderate, high, very high*), the membership grade can be calculated using the Bayes theorem as the conditional probability:

$$P(C_k|\text{feature } X_i \text{ is } T_j) = \frac{P(\text{feature } X_i \text{ is } T_j|C_k) \cdot P(C_k)}{P(\text{feature } X_i \text{ is } T_j)}, \tag{1}$$

where the terms in the right half of the equation are estimated from the training data.

Because we use approximately the same number of positive and negative songs in training sets for a better balance, $P(C_k) \approx 0.5$. The non-relevant feature from the bottom subfigure of Fig. 1 outlines a problematic issue of the application of the Eq. (1). For $T_j = very\ low$, $P(\text{feature } X_i \text{ is } T_j) = 1$ (this feature has always *very low* values). Further, $P(\text{feature } X_i \text{ is } T_j|C_k) = 1$, so that $P(C_k|\text{feature } X_i \text{ is } T_j) \approx 0.5$. However, it is better to set a significantly lower value for the relevance of the rule "IF deviation of the maximum amplitude of the chromagram is *very low* THEN Pop". Therefore, the features which almost always belong to a certain linguistic term may be penalised using the following formula for the estimation of the relevance of a rule "IF feature X_i is T_j THEN category C_k":

$$R(C_k, X_i, T_j) = P(\text{feature } X_i \text{ is } T_j|C_k) \cdot (1 - P(\text{feature } X_i \text{ is } T_j)). \tag{2}$$

Table 2 lists the most relevant rules for the prediction of three music genres (Classic, Pop, Rap) and music styles (ClubDance, HeavyMetal, ProgRock). For simplicity reasons, we omit some details of the feature estimation, such as the underlying supervised classification method. The features in these rules provide a comprehensible description of the properties of the tested categories, compared to low-level characteristics of the audio signal. The linguistic terms *very high* and *very low* belong to rules with highest relevance values. For example, the rule "IF structural complexity of harmony is *moderate* THEN Classic" has the position 383 ($R(C_k, X_i, T_j) = 0.1441$) in the list of rules sorted according to their relevance, and "IF structural complexity of harmony is *high* THEN Classic" has the position 1,251 ($R(C_k, X_i, T_j) = 0.0401$).

Table 2 The most relevant rules for the recognition of three genres and three styles

Rule	$R(C_k, X_i, T_j)$
Genre classic	
IF structural complexity of harmony is *very high* THEN Classic	0.4030
IF melodic range greater than octave is *very high* THEN Classic	0.3821
IF mood Earnest is *very high* is THEN Classic	0.3816
IF mood Stylish is *very low* is THEN Classic	0.3813
Genre pop	
IF singing solo rough is *very high* THEN Pop	0.3844
IF key major is *very low* THEN Pop	0.3498
IF key minor is *very high* THEN Pop	0.3342
IF number of segment changes is *very high* THEN Pop	0.3277
Genre Rap	
IF mood PartyCelebratory is *very high* THEN Rap	0.4895
IF melodic range less than octave is *very high* THEN Rap	0.4699
IF mood Sentimental is *very low* THEN Rap	0.4502
IF singing solo position medium is *very high* THEN Rap	0.4225
Style ClubDance	
IF mood Energetic is *very high* THEN ClubDance	0.3421
IF mood PartyCelebratory is *very high* THEN ClubDance	0.3420
IF melodic range greater than octave is *very low* THEN ClubDance	0.3348
IF singing solo clear is *very high* THEN ClubDance	0.3288
Style HeavyMetal	
IF mood Aggressive is *very high* THEN HeavyMetal	0.3861
IF effects distortion is *very high* THEN HeavyMetal	0.3850
IF singing solo rough is *very high* THEN HeavyMetal	0.3514
IF singing solo clear is *very low* THEN HeavyMetal	0.3505
Style ProgRock	
IF singing solo rough is *very high* THEN ProgRock	0.3142
IF mood Stylish is *very low* THEN ProgRock	0.2910
IF number of segment changes is *very high* THEN ProgRock	0.2904
IF melodic range greater than octave is *very high* THEN ProgRock	0.2867

A further possibility which was not examined for this paper but is promising for future studies is the combination of rules using fuzzy operators for "and" and "or", e.g. "IF structural complexity of harmony is *very high* AND structural complexity of harmony is *high* THEN Classic". However, the number of possible rules to analyse may explode: for simple rules based on a single feature the number of possible rules is already 2,830 (five linguistic terms for 566 high-level audio features). Not only the combination of two and more rules may be helpful for fuzzy classification, but also the optimisation of the definition areas of the linguistic terms, as done in Fernández and Chávez (2012) with the help of evolutionary algorithms.

4   Application of a Simple Fuzzy Classifier

A simple multi-class fuzzy classifier may estimate the average relevance of M most relevant rules and select the genre with a highest value:

$$\hat{C}_k = \max_{k \in \{1,...,C\}} \left(\frac{1}{M} \sum_{m=1}^{M} R\left(C_k, X(m), T(m)\right) \right),\qquad (3)$$

where C is the number of (exclusive) genres to predict, $X(m)$ is the feature from the mth rule for the genre C_k, $T(m)$ is the linguistic term from the mth rule for the genre C_k, and the rules are strictly ordered by decreasing $R(C_k, X_i, T_j)$ as defined in Eq. (2). If there are several equal maximum values, ties are broken at random.

The basic challenge of this method is to find the optimal value for M. The one extreme is to apply only the most relevant rule. Because only one high-level feature is used for the prediction of the category in that case, the classification quality may be often too low. For example, we compared the most relevant rules for six genres (Classic, Electronic, Jazz, Pop, Rap, R'n'B) for the identification of the genre of the song "Let Me Put My Love Into You" from AC/DC. The rule with the highest relationship grade classifies this song to R'n'B, the next one to Jazz, and the third one to Pop. However, if we average $R(C_k, X_i, T_j)$ for 50 rules as described in Eq. (3), the song is correctly predicted as belonging to the category Pop.

Another extreme is to apply a very large number of rules. In that case the classification performance can be significantly increased, as illustrated in Fig. 2. Here, the averaged $R(C_k, X_i, T_j)$ of up to the 300 most relevant rules is estimated for the classification of 120 test songs. However, if a high number of rules are used for a genre prediction, too many high-level music features contribute to the final decision and the interpretability decreases. A compromise solution would be to

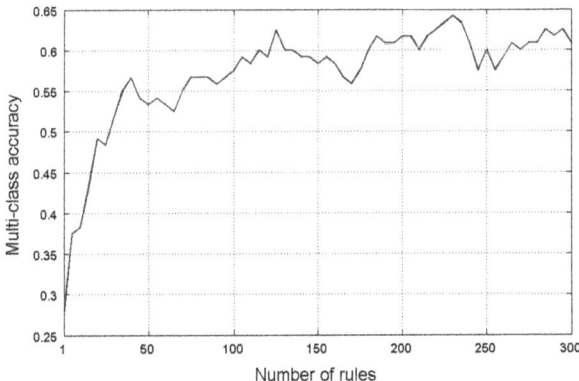

Fig. 2 Multi-label accuracy for the prediction of six music genres using 1–300 most relevant fuzzy rules (with the step size of five rules)

apply 20 rules: this method has the multi-class accuracy of 49.17 %, which has a potential to be improved, but is already significantly above the performance of a random classifier for six genres which would achieve an expected probability of 16.67 %. Another local optimum has the accuracy of 56.67 % (40 rules).

The aggregation of rules may also be applied for binary classification. Here we compare the results to Vatolkin (2013), where the same set of high-level features was used, however four different supervised classification methods and evolutionary multi-objective feature selection were applied for the recognition of genres and styles. In spite of significant improvements of the classification performance, the complex methods developed for the aforementioned study have also some drawbacks: the optimisation requires large computing times (up to more than 24 h if combined with support vector machines), and the interpretability of original high-level features suffers if a complex classification method (support vector machine or ensemble of many decision trees) provides the best classification model. If a simple fuzzy classification model (as discussed above) contains a limited number of rules M, other advantages are that the models have very small storage demands and the classification of new songs can be done very fast.

Table 3 compares the results from both studies. The column "HL-all" lists the classification errors if complete feature set is used for the classification with four tested methods, and the column "HL-FS" describes the errors after the optimisation by means of multi-objective evolutionary feature selection. All models were trained

Table 3 Balanced classification error for experiments with several classification methods (columns HL-all, HL-FS) and fuzzy rules (columns HL-fz10, HL-fz50, HL-fz100)

Task	HL-all	HL-FS	HL-fz10	HL-fz50	HL-fz100
Recognition of genres					
Classic	0.0365	0.0137	0.0524	0.0524	0.0238
Electronic	0.2010	0.1191	0.2571	0.2524	0.2524
Jazz	0.0866	0.0605	0.1904	0.1904	0.1904
Pop	0.2890	0.1270	0.4156	0.3444	0.3244
Rap	0.0852	0.0650	0.1143	0.1143	0.1143
R & B	0.1931	0.1484	0.3000	0.2762	0.2762
Recognition of styles					
AdultContemporary	0.2358	0.1344	0.2909	0.2227	0.2227
AlbumRock	0.2084	0.1066	0.3500	0.3500	0.3500
AlternativePopRock	0.2015	0.1092	0.1875	0.1875	0.1875
ClubDance	0.2484	0.1389	0.2500	0.2500	0.2500
HeavyMetal	0.1384	0.0778	0.1024	0.1024	0.1024
ProgRock	0.1818	0.0973	0.3963	0.3963	0.3963
SoftRock	0.2253	0.1197	0.3003	0.2147	0.2147
Urban	0.1467	0.0837	0.2273	0.2273	0.2273

For details see the text

from small training sets of ten positive and ten negative music pieces for each category. The balanced classification error was estimated for the independent validation set of songs not involved into the learning procedure. For details, see Vatolkin (2013).

The third column ('HL-fz10') contains the smallest errors from the combinations of 1 to 10 most relevant rules after the simple fuzzy classification discussed above (Eq. (3), $M = 1, \ldots, 10$). Similarly, for columns "HL-fz50" and "HL-fz100" the aggregation of up to 50 and up to 100 most relevant rules was estimated. The performance of a simple classifier using up to 100 fuzzy rules depends on the category: e.g., for Electronic and Pop the error is always higher than using four methods and all features (the column "HL-all"), and on the other side for AlternativePopRock and HeavyMetal the aggregation of up to ten rules performs already better than the classification with four methods using all high-level features. In all cases the classification based on fuzzy rules leads to significantly higher errors than the classification with four methods using the optimised feature set ('HL-FS'). However, the fuzzy approach has still enough room for optimisation without losing the interpretability.

5   Summary and Outlook

In this paper we have discussed a basic approach to music classification where each algorithm step is designed to provide as interpretable outputs as possible, from comprehensible high-level audio characteristics to a simple fuzzy classifier, which aggregates a limited number of categorisation rules which describe well the most important musical properties of music genres and styles. The results show that although the classification performance is inferior to the approach where several classification methods and feature selection are applied, they are still significantly better than for a random classifier, the method is very fast, and the classification models are comprehensible for listeners and music scientists.

There exist several possibilities to improve our basic implementation with extended techniques keeping the high interpretability of classification models. In particular, the combination of different high-level features using fuzzy "and" and "or" operators for the corresponding linguistic terms is promising, among others because the features are then not treated independently and may be relevant for a certain category only in their combination. The adaptation of the definition areas of linguistic terms as proposed in Fernández and Chávez (2012) and the application of rule selection similar to feature selection as done in Vatolkin (2013) are other starting points for further investigations.

Acknowledgements We thank the Klaus Tschira Foundation for the financial support.

References

Abeßer, J., Lukashevich, H., Dittmar, C., & Schuller, G. (2009). Genre classification using bass-related high-Level features and playing styles. In *Proceedings of the 10th Int'L Conference on Music Information Retrieval (ISMIR)* (pp. 453–458).

Celma, Ò., & Serra, X. (2008). FOAFing the music: Bridging the semantic gap in music recommendation. *Journal of Web Semantics: Science, Services and Agents on the World Wide Web, 6*(4), 250–256.

Essid, S., Richard, G., & David, B. (2006). Musical instrument recognition by pairwise classification strategies. *IEEE Transactions on Audio, Speech, and Language Processing, 14*(4), 1401–1412.

Geyer-Schulz, A. (1998). Fuzzy genetic algorithms. In H. T. Ngyen & M. Sugeno (Eds.), *Fuzzy systems*. Boston: Kluwer Academic Publishers.

Guyon, I., Nirkavesh, M., Gunn, S., & Zadeh, L. A. (2006). *Feature extraction. foundations and applications*. Berlin/Heidelberg: Springer.

Fernández, F., & Chávez, F. (2012). Fuzzy rule based system ensemble for music genre classification. In *Proceedings of the 1st International Conference on Evolutionary and Biologically Inspired Music, Sound, Art and Design (EvoMUSART)* (pp. 84–95). Berlin: Springer.

Friberg, A. (2005). A fuzzy analyzer of emotional expression in music performance and body motion. In J. Sundberg & B. Brunson (Eds.), *Proceedings of Music and Music Science*.

Hu, X., & Liu, J. (2010). User-centered music information retrieval evaluation. In *Proceedings of the Joint Conference on Digital Libraries (JCDL) Workshop: Music Information Retrieval for the Masses*.

Mauch, M., & Levy, M. (2011). Structural change on multiple time scales as a correlate of musical complexity. In *Proceedings of the 12th International Society for Music Information Retrieval Conference (ISMIR)* (pp. 489–494).

Mckay, C., & Fujinaga, I. (2006). Musical genre classification: Is it worth pursuing and how can it be improved? In *Proceedings of the 7th International Conference on Music Information Retrieval (ISMIR)* (pp. 101–106).

Pachet, F., & Zils, A. (2003). Evolving automatically high-level music descriptors from acoustic signals. In *Proceedings of the 1st International Symposium on Computer Music Modeling and Retrieval (CMMR)* (pp. 42–53).

Sturm, B. (2012). A survey of evaluation in music genre recognition. In *Proceedings of the 10th International Workshop on Adaptive Multimedia Retrieval (AMR)*.

Vatolkin, I., Theimer, W., & Botteck, M. (2012). Partition based feature processing for improved music classification. In W. A. Gaul, A. Geyer-Schulz, L. Schmidt-Thieme, & J. Kunze (Eds.), *Challenges at the interface of data analysis, computer science, and optimization* (pp. 411–419). Berlin: Springer.

Vatolkin, I. (2013). Improving supervised music classification by means of multi-objective evolutionary feature selection. PhD thesis, Department of Computer Science, TU Dortmund, 2013.

Yang, Y. -H., Liu, C. -C., & Chen, H. H. (2006). Music emotion classification: A fuzzy approach. In: K. Nahrstedt, M. Turk, Y. Rui, W. Klas, & K. Mayer-Patel (Eds.), *Proceedings of the 14th ACM International Conference on Multimedia* (pp. 81–84).

Wolpert, D. (1992). Stacked generalization. *Neural Networks, 5*(2), 241–260.

Zhang, H., & Liu, D. (2006). *Fuzzy modeling and fuzzy control*. Boston/Basel/Berlin: Birkhäuser.

Zhou, Z. -H. (2012). *Ensemble methods: Foundations and algorithms*. Boca Raton: CRC Press.

Part VIII
Data Analysis in Communication and Technology

What Is in a Like? Preference Aggregation on the Social Web

Adrian Giurca, Daniel Baier, and Ingo Schmitt

Abstract The Social Web is dominated by rating systems such as the ones of Facebook (only *"Like"*), YouTube (both *"Like"* and *"Dislike"*), or the Amazon product review *5-star rating*. All these systems try to answer on *How should a social application pool the preferences of different agents so as to best reflect the wishes of the population as a whole?* The main framework is the theory of *social choice* (Arrow, Social choice and individual values, Wiley, New York, 1963; Fishburn, The theory of social choice, Princeton University Press, Princeton, 1973) i.e., agents have preferences, and do not try to camouflage them in order to manipulate the outcome to their personal advantage (moreover, manipulation is quite difficult when interactions take place at the Web scale). Our approach uses a combination between the *Like/Dislike* system and a *5-star satisfaction* system to achieve local preference ranks and a global partial ranking on the outcomes set. Moreover, the actual data collection can support other preference learning techniques such as the ones introduced by Baier and Gaul (J. Econ. 89:365–392, 1999), Cohen et al. (J. Artif. Intel. Res. 10:213–270, 1999), Fürnkranz and Hüllermeier (Künstliche Intelligenz 19(1):60–61, 2005), and Hüllermeier et al. (Artif. Intel. 172(16–17):1897–1916, 2008).

A. Giurca (✉)
Brandenburgische Technische Universität Cottbus, Platz der Deutschen Einheit 1, 03046 Cottbus, Germany

Binarypark, Erich-Weinert-Str. 1, 03046 Cottbus, Germany
e-mail: giurca@tu-cottbus.de

D. Baier • I. Schmitt
Brandenburgische Technische Universität Cottbus, Platz der Deutschen Einheit 1, 03046 Cottbus, Germany
e-mail: daniel.baier@tu-cottbus.de; schmitt@tu-cottbus.de

1 Introduction and Motivation

Nowadays world is dominated by the online activities: we see a great movement of the retail commerce on e-commerce platforms, we see the retail and B2B becoming more social as more and more niche social networks develop along with specific interests of their members. All major players move fast to e-commerce as they discover that the young generation of consumers perform most of their activities on the Web. Recently a survey (CMO Council 2011) of more than 1,300 consumers and 132 senior marketers discovered a discrepancy between the consumer intentions and actions on the Social Web and the marketers beliefs about that. The survey concludes that consumers want more experiences, more engagement, more rewards, and more reasons to connect but the brands do not fully understand how to obtain this. Various scientists look nowadays to develop feasible models to derive user preferences from their online social activities—some develop voting systems by processing graph models (Boldi et al. 2009), others use agent based models (Walter et al. 2008; Giurca et al. 2012a,b). Recently the Semantic Web community starts being more and more involved in social network analysis (Mika 2005; Matsuo and Yamamoto 2009; Symeonidis et al. 2011). Other recent research tries to understand if the traditional survey method can be replaced by the online review analysis (Rese et al. 2014). However, many of these models have a large complexity, therefore subject of expensive implementations. We adapt the well-known model of pair comparison to the actual model of social web endorsements i.e., likes, dislikes, $+1$, and so on.

2 Preference Units and the Voting Model

Let $\mathcal{P}$ denote a set of n groups of agents, each with an arbitrarily number of agents and the group of agents are not disjoint, i.e., each agent can participate in one or more groups. Let $\mathcal{O}$ a finite set of outcomes (or alternatives) subject of preference voting. An agent i can state partial weighted preference on various outcomes from $\mathcal{O}$, e.g., $(a \succeq b) : w$ (*weak weighted preference*), respectively $(a \succ b) : w$ and $(a \sim b) : w$ for *strict weighted preference* and *weighted indifference* and each agent preference is a transitive relation. The weight meaning is the agent's satisfaction on the specific preference.

Definition 1 (Preference Unit) Let $a, b \in \mathcal{O}$ be two outcomes. A *unit* U is a tuple $U = (\{a, b\}, \text{like}, w)$ where $\text{like} : \{a, b\} \longrightarrow \{0, 1\}$ are ratings and $w \in [0, 1]$ is the agent's satisfaction with the current vote. Overloading the set membership notation we denote $a \in U$. When there is no confusion we denote $U = \{a, b\}$. A *survey* S is a set of units such that $\forall o \in \mathcal{O} \, \exists U \in S, \, o \in U$. S_i denotes a survey against the group of agents i. In other words, a preference encodes a block of two outcomes. Each outcome can be liked/disliked independently. However an outcome cannot be in the same time liked and disliked, i.e., the user judgment is consistent with law of non-contradiction. The agent is provided with many such units covering a desired set of outcomes.

Everywhere in this paper we use the usual notation for preferences, i.e., $a \prec b$ (b is preferred to a), $a \sim b$ (a and b are equal preferred, indifference). The weight factor w is the overall satisfaction, $(a \prec b) : w$ according to the modified variant of the logistic function

$$w(x) = \begin{cases} \frac{1}{1+e^{\alpha(3-x)}} & x \neq 1, \; x \neq 5 \\ 0 & x = 1 \\ 1 & x = 5 \end{cases}$$

where α is the function growth rate.

We denote $\perp$ the agent *non action* (not voting, not rating) , i.e., like(a) $= \perp$ means the agent did not vote the outcome a nor like (1) or dislike (0). $w = \perp$ denotes that the agent did not assign a satisfaction weight to the unit. These weights are not probabilities, as they do not represent the absolute satisfaction of the agent but just its local behavior against the presented outcomes. Therefore the satisfaction weights can change during any kind of process of computing agent preferences.

Example 1 An entertainment company is looking to decide on the optimal activities that can be done on its campus. They are able to organize the following activities (the outcomes): going to the video arcade (a); playing basketball (b); driving in a cart (c) and make a dance contest (d). They have three consumer interest groups (the agents groups) and of course, each consumer has a different preference over these activities, which is represented as a vote on a specific unit. A survey on these outcomes is: $\{\{a,b\},\{a,c\},\{a,d\},\{b,c\},\{b,d\},\{c,d\}\}$ and the goal is to understand what are the most desired activities, per each interest group as well as in the entire community.

3   Combining Likes and Satisfaction: Constructing Weighted Preference from Choice

Let $U = (\{a,b\}, \text{like}, w)$ be a survey unit. The elicitation of preferences through choices is of particular importance in our use case, where choices of large groups of agents are the only available empirical data. Recall, our model considers preferences as states of mind whereas choices are actions. Before describing any mapping rule from choices to weighted preferences the following axioms are introduced as part of the model:

1. *Non-contradiction of agent's satisfaction.* $(a \succ b) : w \Leftrightarrow (b \succ a) : \bar{w}$
2. *Inconsistency under indifference.*
   $(a \sim b) : w \Leftrightarrow (a \succ b) : w \wedge (b \succ a) : \bar{w} \Leftrightarrow (b \succ a) : w \wedge (a \succ b) : \bar{w}$, where $\bar{w}$ is the complement of w in a complemented lattice $[0, 1]$.

An agent performs actions (like/dislike/rate satisfaction) against a presentation unit. Missing of satisfaction rating is mapped the same as the neutral rating (3-star). For example, we obtain the same indifference when (a) an agent likes both outcomes and w-rate them or (b) an agent dislikes both outcomes and w-rate them. While this solution may not be satisfactory for a general model, our work is interested to capture not all agent preferences but to understand the positive ones and that's the reason of assuming complete satisfaction on fully supported outcomes (the case when the agent likes both outcomes but does not assign a satisfaction weight.)

Agent's *complete choice* is described by the following rules:

1. IF like(a) > like$(b) \wedge w \neq \bot$, THEN $(a \succ b) : w$.
   The agent likes the outcome a, dislikes the outcome b, and gives a weight to the unit.
2. IF like(a) > like$(b) \wedge w = \bot$, THEN $(a \succ b) : 1/2$.
   The agent doesn't assign a weight to the unit but likes one outcome while disliking the other one, neutrally supports this unit.
3. IF like(a) = like$(b) = 1 \wedge w \neq \bot$ THEN $(a \sim b) : w$
   The agent likes both outcomes and assigns a weight to the unit. This rule derives an indifference on outcomes
4. IF like(a) = like$(b) = 1 \wedge w = \bot$, THEN $(a \sim b) : 1$.
   Any agent not weighting a unit but liking both outcomes fully supports this unit. While both outcomes were liked the rule translates these actions into an indifference (both outcomes are equally supported).
5. IF like(a) = like$(b) = 0 \wedge w \neq \bot$, THEN $(a \sim b) : w$
   The agent dislikes both outcomes and assign a weight to the unit.
6. IF like(a) = like$(b) = 0 \wedge w = \bot$, THEN $(a \sim b) : 0$ Any agent not weighting a unit but disliking both outcomes does not support this unit.

Below we describe the agent's *partial choice* (derived from incomplete votes). Absence of voting information is interpreted as similar negative voting because Web users are lazy voters. Absence of satisfaction rating is interpreted as neutral.

1. IF like(a) = $\bot$ $\wedge$like$(b) = 0 \wedge w \neq \bot$, THEN $(a \succ b) : w$.
2. IF like(a) = $\bot$ $\wedge$like$(b) = 1 \wedge w \neq \bot$, THEN $(b \succ a) : w$.
3. IF like(a) = $\bot$ $\wedge$like$(b) = 0 \wedge w = \bot$, THEN $(a \succ b) : 1/2$.
4. IF like(a) = $\bot$ $\wedge$like$(b) = 1 \wedge w = \bot$, THEN $(b \succ a) : 1$.

Agent's *missing choice* is mapped into possible weighted indifference:

1. IF like(a) = $\bot$ $\wedge$like(b) = $\bot$ $\wedge w \neq \bot$, THEN $(a \sim b) : w$.
2. IF like(a) = $\bot$ $\wedge$like(b) = $\bot$ $\wedge w = \bot$, THEN $(a \sim b) : 1/2$.

This model does not support *negative indifference* $(a \sim b) : 0$ and *positive indifference* $(a \sim b) : 1$ therefore all such votes will be discarded to the next step in evaluation.

4 Computing the Scores of Preferences

Let U be a preference unit and let j be a group of agents. This model is agnostic with respect to the classification criteria used on clustering agents into groups. Such criteria depend heavily on the nature of the presented outcomes as well as on the agent attributes such as age group, geographical location, income, and so on. Each agent can participate in an unlimited number of groups.[1] $\{a \succ b\}_j$ (or U_j) denotes the distribution of the preferences of all group agents against U. Let i be an agent member of the group j. Then, $[(a \succ b) : w]_{i,j}$ denotes the preference of the agent i when in group j on the unit $\{a, b\}$.

Definition 2 (Raw Frequency) We denote by $|(a \succ b) : w|_j$ the total number of agents voted the unit $\{a, b\}$ in group j and call this number the *raw frequency of* $(a \succ b) : w$ *in the group* j.

Example 2 (Distribution of Preferences) As in Example 1, the agent group 1 (A_1) may have the following distribution of preference $a \succ b$:

$$\{a \succ b\}_1 = \{((a \succ b) : 0.15; 20), ((a \succ b) : 0.25; 40), ((a \succ b) : 1.0; 60)\}$$

that is 20 agents voted $(a \succ b) : 0.15$; 40 agents voted $(a \succ b) : 0.25$ and 60 agents voted $(a \succ b) : 1.0$.

Then the raw frequency of the preference $a \succ b$ in group 1 is $|(a \succ b) : 0.15|_1 = 20$, while $|(a \succ b) : 0.35|_1 = 0$

The weighted preferences such as $(a \succ b) : w_i$ represent variations of the same preference but tailored with the agent's satisfaction degree. As discussed by Thomas and McFayden (1995), people express degrees of satisfaction proportional to the certainty with which they hold those beliefs and, therefore, the evaluation models should tend to judge the reliability of the communicated information according to the confidence with which it is expressed. In addition, we have little knowledge about the agents to base a decision, therefore this model assumes that the most satisfied agents are the most likely to be correct. Accordingly, we introduce the *raw preference frequency* as an additive aggregation of these variations.

Definition 3 (Raw Preference Frequency) Let $a \succ b$ be a preference. The *raw preference frequency with respect to agents group* j is

$$\| a \succ b \|_j = \sum_{i \in j} w_i |(a \succ b) : w_i|_j$$

---

[1] The typical use case of groups of agents that are not partitions is when users visit *different* sites containing presentation units part of the same survey.

Table 1 Preferences per agent group

$\| a \succ b \|_j$	A_1	A_2	A_3	$pf(\alpha, A_j)$	A_1	A_2	A_3
$a \succ b$	73	25	54	$a \succ b$	1.1186	0.6786	0.8857
$a \succ c$	15	0	0	$a \succ c$	0.6027	0.5000	0.5000
$a \succ d$	0	70	15	$a \succ d$	0.5000	1.0303	0.6071
$b \succ c$	47	40	38	$b \succ c$	0.8219	0.7857	0.7714
$b \succ d$	59	66	70	$b \succ d$	0.9041	0.9714	1.1481
$c \succ d$	0	35	20	$c \succ d$	0.5000	0.7500	0.6429

Example 3 Considering the preference distribution from Example 2 then

$$\| a \succ b \|_j = 0.15 \times 20 + 0.25 \times 40 + 1.0 \times 60 = 73$$

Table 1 depicts the result of the overall voting process. The reader shall observe that there are different total number of votes for each preference ($a \succ b$ was voted by 152 agents while $c \succ d$ by only 55) and different agent groups may have different total number of votes (A_1 voted 194 times and A_2 voted 236 times).

Because the group of agents are not disjoint there is a clear dependency between various votes on the outcomes in different groups, therefore characterizing a preference $a \succ b$ by only its raw preference frequency in a group is not an overall measure. Moreover, it is unlikely that a large number of the same vote in a group can carry additively the significance of a single occurrence, that is summing over all groups may not be enough to get an overall raw frequency. As such we consider a model inspired from the Tf-idf [see Manning et al. (2008), Schmitt (2005)], document classification in information retrieval.[2] Such method does not consider only the *raw preference frequency* but also the *inverse group frequency* as a weight on influence of the voting in different groups.

Definition 4 (Preference Score) Let j be a group of agents and $a \succ b$ a preference. The *preference frequency* of $a \succ b$ inside j is

$$\mathrm{pf}(a \succ b, j) = \begin{cases} 0.5 + \dfrac{0.5 \times \|a \succ b\|_j}{\max_\alpha \{\|\alpha\|_j, \, |\alpha \neq (a \succ b)\}} \\ 1, \text{ otherwise} \end{cases}$$

Note that $\mathrm{pf}(a \succ b, j) = 1$ when preference $a \succ b$ is the only one rated in the group j.

---

[2]There are many developed ranking functions. This work does not intend to compare all these various solutions.

Table 2 Preference scores
per each agent group

Score(α, j)	A_1	A_2	A_3
$a \succ b$	1.1186	0.6786	0.8857
$a \succ c$	0.3151	0.2614	0.2614
$a \succ d$	0.4120	0.8489	0.5002
$b \succ c$	0.8220	0.7857	0.7714
$b \succ d$	0.9041	0.9714	1.1481
$c \succ d$	0.4119	0.6179	0.5297

The *inverse raw group frequency* is:

$$\text{gf}(a \succ b) = |\{j \in N \; : \| a \succ b \|_j \neq 0\}|$$

i.e., the number of groups that voted the preference $a \succ b$.

Then, the *inverse group frequency* is defined as:

$$\text{igf}(a \succ b) = 1 - \log \frac{|N|}{\text{gf}(a \succ b)}$$

The *preference score* is:

$$\text{score}(a \succ b, j) = \text{pf}(a \succ b, j) \times \text{igf}(a \succ b)$$

Following the data from the Table 1 the final preference scores are shown in Table 2.

The interpretation of the preference score inside of each group is the relevance of that preference inside of the group, the greater the score the greater the relevance a preference has. The preference score allows us to construct an evaluation scenario centered on the weight learning algorithm, Cohen et al. (1999) as an adaptation of the weighted majority algorithm proposed by Littlestone and Warmuth (1994) under the main assumption that overall preference is a weighted sum of individual preferences. The main differences with the original algorithm is the generalization to sets of preferences, obtaining a global outcome rank derived from N preference sets, one for each group of agents.

The weighting learning algorithm computes a rank function by considering weighted binary preferences. It uses the preferences as a directed weighted graph where the initial set of vertices V is equal of the set of outcomes $\mathcal{O}$ and each edge $a \rightarrow b$ has weights $\delta = \text{score}(a \succ b, j)$ (a is preferred to b with score δ in group j). Each vertex v gets a potential $\pi(v)$ which is the sum of the incoming edges scores minus the sum of outgoing edges scores. Algorithm 1 (see Cohen et al. (1999) for a proof on optimality) picks some node t that has a maximum potential, assigns it a rank $\rho(t) = |V|$ and then ordering in the same way the remaining nodes, after updating the nodes potentials. The potentials define a stratification of the directed weighted graph into node sets of the same potential. The stratum with the

Data: $\mathcal{O}$
Result: ρ
$V = \mathcal{O};$
for $v \in V$ **do**
  for $j \in N$ **do**
    | $\pi_j(v) = \sum_{x \in V} score(v \succ x, j) - \sum_{x \in V} score(x \succ v, j);$
  end
  $\pi(v) = \sum_j \pi_j(v);$
end
while $V \neq \emptyset$ **do**
  $t = argmax_{x \in V} \pi(x);$
  $\rho(t) = |V|;$
  $V = V - \{t\};$
  for $x \in V$ **do**
    | $\pi(x) = \pi(x) + \sum_{j,t}(score(x \succ t, j) - score(t \succ x, j));$
  end
end

highest potential is processed first. When processing the nodes in the same stratum the algorithm has a choice between them, i.e., the output ordering depends on the $argmax$ implementation. The overall preference rank obtained by Algorithm 1 on the data provided in Table 2 is $a \succ b \succ c \succ d$.

5 Conclusion and Future Work

This work describes the initial process and its first outcomes of a voting aggregation mechanism towards understanding group preferences. This procedure applies exclusively on information collected from agent activities on web pages embedding presentation units as well from the agent activities when using dedicated mobile applications. Our data collection can support other preference learning techniques such as the ones introduced by Baier and Gaul (1999), Cohen et al. (1999), Fürnkranz and Hüllermeier (2005), and Hüllermeier et al. (2008). The presentation of units is compliant with the main criteria of online surveys developed by the marketing research community such as Brusch et al. (2002) and Brusch and Baier (2005).

The voting system is compliant with the survey Web design criteria agreed by many communities such as Göritz et al. (2000), Theobald (2000), Görts and Behringer (2003), Brusch and Baier (2005), and possibly others:

1. *anonymity of the survey*—all information is collected according to the legal matter rules. We do not collect and store any information towards identifying the agent.
2. *self-selection and initiative of respondents*—any agent can like/dislike/rate an item according to his/her free choice. There is no direct request on rating a specific presentation unit or outcome.

3. *asynchrony of the survey*—presentation units (parts of a survey) are shown to the agent independently and at different time moments and application sessions.
4. *length of the survey*—the results of our procedure do not depend on a complete survey (showing all possible two-pair outcomes from *n* available). We obtain good results when all possible outcomes appear in at least one presentation unit.
5. *availability of specific target groups*—presentation units can be freely embedded on any Web site. However, there are dedicated Web sites and mobile applications where they are embedded too.
6. *the drop-out phenomenon*—presentation units are shown to all visitors of dedicated web sites as well to all users of mobile applications. Our voting system does not consider such phenomena because it does not request the user to perform a complete vote on a specific survey.

Along with the traditional raw number of votes (no. of likes/dislikes) we introduced the inverse group frequency as a weight on influence of the voting in different groups, a much more sensible mechanism and including a weighted preference model representing variations of the same preference but tailored with the agent's satisfaction degree (computed from agent ratings).

Our approach is completely setup on user freedom of choice and we applied it to a specific kind of presentation units, i.e. time limited offers on specific events such as music festival tickets. The preliminary tests had good results even in presence of a limited number of votes and when the agent voting is incomplete.

The future work is concerned with the development of an adequate mechanism to understand surveys from various presentation units, according to the criteria discussed by Brusch et al. (2002) and others. When it comes to apply established marketing techniques such as conjoint analysis on Web data, the reality of Web 2.0 applications is different from the traditional centric expert-based approach. While traditional conjoint analysis is based on surveys developed by experts the Web 2.0 case involves pair comparisons developed by any trading actor on the Web.

References

Arrow, K. J. (1963). *Social choice and individual values* (2nd ed.). New York: Wiley.
Baier, D., & Gaul, W. (1999). Optimal product positioning based on paired comparison data. *Journal of Econometrics, 89*, 365–392.
Boldi, P., Bonchi, F., Castillo, C., & Vigna, S. (2009, 2–6 November). Voting in social networks. In *Proc. of The 18th ACM Conference on Information and Knowledge Management* (pp. 777–786), Hong Kong.
Brusch, M., & Baier, D. (2005). Vergleich von persönlich-computergestützten und webbasierten Erhebungsformen in der Marktforschung am Beispiel der Conjointanalyse. *Forum der Forschung 18/2005: 161-166*, BTU Cottbus, Eigenverlag, ISSN-Nr.: 0947 - 6989.
Brusch, M., Baier, D., & Treppa, A. (2002). Conjoint analysis and stimulus presentation. A comparison of alternative methods. In K. Jajuga, A. Sokolowski, & H. H. Bock (Eds.), *Classification, clustering, and analysis* (pp. 203–210). Berlin: Springer.

CMO Council (2011). Variance in the Social Brand Experience, http://www.cmocouncil.org/images/uploads/216.pdf.

Cohen, W., Schapire, R. E., & Singer, Y. (1999). Learning to order things. *Journal of Artificial Intelligence Research, 10*, 213–270.

Fishburn, P. C. (1973). *The theory of social choice*. Princeton: Princeton University Press.

Fürnkranz, J., & Hüllermeier, E. (2005). Preference learning. *Künstliche Intelligenz, 19*(1), 60–61.

Giurca, A., Schmitt, I., & Baier, D. (2012a). Adaptive conjoint analysis. Training data: Knowledge or beliefs? A logical perspective of preferences as beliefs. In *Proceedings of the Federated Conference on Computer Science and Information Systems* (pp. 1127–1133), ISBN 978-83-60810-51-4.

Giurca, A., Schmitt, I., & Baier, D. (2012b, 28 August). Can adaptive conjoint analysis perform in a preference logic framework? In G. Nalepa, J. Canadas, & J. Baumeister (Eds.), *Proceedings of 8th Workshop on Knowledge Engineering and Software Engineering* (KESE8), at the 20th Biennial European Conference on Artificial Intelligence (ECAI 2012) (Vol. 949), Montpellier, CEUR-WS.

Göritz, A. S.,Batinic, B., & Moser, K. (2000). Online marktforschung. In W. Scheffler, K.-I. Voigt (Eds.), *Entwicklungsperspektiven im Electronic Business. Grundlagen – Strukturen – Anwendungsfelder* (pp. 187–204). Wiesbaden: Gabler.

Görts, T., & Behringer, T. (2003). Online-Conjoint – Chancen und Grenzen: Ein Fallbeispiel aus dem Telekommunikationsmarkt. In A. Theobald, M. Dreyer, & T. Starsetzki (Eds.), *Online-Marktforschung. Theoretische Grundlagen und praktische Erfahrungen* (pp. 283–296). Wiesbaden: Gabler.

Hüllermeier, E., Fürnkranz, J., Cheng, W., & Brinker, K. (2008). Label ranking by learning pairwise preferences. *Artificial Intelligence, 172*(16–17), 1897–1916.

Littlestone, N., & Warmuth, M. (1994). The weighted majority algorithm. *Information and Computation, 108*(2), 212–261.

Manning, C., Raghavan, P., & Schütze, H. (2008). *Introduction to information retrieval*. Cambridge: Cambridge University Press.

Matsuo, Y., & Yamamoto, H. (2009, 20–24 April). Community gravity: Measuring bidirectional effects by trust and rating on online social networks. In *Proceedings of WWW 2009*, Madrid. ACM available at http://www2009.org/proceedings/pdf/p751.pdf.

Mika, P. (2005). Flink: Semantic web technology for the extraction and analysis of social networks. *Journal of Web Semantics, 3*, 2–3.

Rese, A., Schreiber, S., & Baier, D. (2014). Technology acceptance modeling of augmented reality at the point of sale: Can surveys be replaced by an analysis of online reviews? *Journal of Retailing and Consumer Services, 21*, 869–876.

Schmitt, I. (2005). *Ähnlichkeitssuche in Multimedia-Datenbanken-Retrieval, Suchalgorithmen und Anfragebehandlung*. München: Oldenbourg Wissenschaftsverlag GmbH, ISBN 3-486-57907-X.

Symeonidis, P., Tiakas, E., & Manolopoulos, Y. (2011, 23–27 October). Product recommendation and rating prediction based on multi-modal social networks. In *In Proceedings of ACM Recommender Systems 2011, RecSys'11*, Chicago, IL.

Theobald, A. (2000). *Das Word Wide Web als Befragungsinstrument*. Wiesbaden: DUV.

Thomas, J. P., & McFayden, R. G. (1995). The confidence heuristic: A game-theoretic analysis. *Journal of Economic Psychology, 16*, 97–113.

Walter, F. E., Battiston, S., & Schweitzer, F. (2008, February). A model of a trust-based recommendation system on a social network. *Autonomous Agents and Multi-Agent Systems, 16*(1), 57–74.

Predicting Micro-Level Behavior in Online Communities for Risk Management

Philippa A. Hiscock, Athanassios N. Avramidis, and Jörg Fliege

Abstract Online communities amass vast quantities of valuable knowledge and thus generate major value to their owners. Where these communities are incorporated in a business as the main means of sharing ideas and issues regarding products produced by the business, it is important that the value of this knowledge endures and is easily recognized. For good management of such a business, risk analysis of the integrated online community is required.

We choose to focus on the process of knowledge creation rather than the knowledge gained from individual messages isolated from context. Consequently, we model collections of messages, linked via tree-like structures; these message collections we call threads. Here we suggest a risk framework aimed at managing micro-level thread related risks. Specifically, we target the risk that there is no satisfactory response to the original message after a period of time. Risks are considered as binary events; the event can therefore be flagged when it is predicted to occur for the attention of the community manager. To predict such a binary response, we use several methods, including a Bayesian probit regression estimated via Gibbs sampling; results indicate this model to be suitable for classification tasks such as those considered.

1 Introduction

Online communities have evolved at an ever-increasing rate in the recent past and continue to grow steadily. Their use is not limited to domestic domains, being widespread in various business, scientific and public service domains. Likewise, substantial economic value is no longer only generated by high profile public communities, e.g. Twitter and Facebook, but also by business communities, such as the SAP Community Network (SCN) (http://scn.sap.com/) and IBM's Connections (http://www-03.ibm.com/software/products/us/en/conn). Online communities are now pivotal elements in corporate management and marketing, product support,

P.A. Hiscock (✉) • A.N. Avramidis • J. Fliege
University of Southampton, Southampton SO17 1BJ, UK
e-mail: P.A.Hiscock@soton.ac.uk; aa1w07@soton.ac.uk; J.Fliege@soton.ac.uk

© Springer-Verlag Berlin Heidelberg 2015
B. Lausen et al. (eds.), *Data Science, Learning by Latent Structures, and Knowledge Discovery*, Studies in Classification, Data Analysis, and Knowledge Organization, DOI 10.1007/978-3-662-44983-7_39

customer relations management, product innovation, and targeted advertising. Members of such communities are connected in a way that opinion, knowledge, and ideas may be shared to facilitate collaboration.

Each online community is a valuable ecosystem that is full of information, the micro- and macro-dynamics (i.e., structure, behavior, and economics) of which are unclear. It is obvious that risks and overlooked emerging opportunities present threats to the health of such an ecosystem. Techniques that enable the health in online communities to be measured, managed, analyzed, protected, and optimized are therefore invaluable. This paper outlines tools utilized and developed to enable timely analysis and decision support of micro-level risks and/or opportunities in the SCN.

In the following, we consider a classification prediction task based around a thread-level opportunity relevant to managers of online communities (Sect. 2). Anderson et al. (2012) study binary classification on thread-based events; the set of features they used inspired and informed our choice (Sect. 2.2.1). However, we consider a different and broader set of methods described shortly in Sect. 3: Bayesian probit; generalized linear model with probit link and with logit link; linear discriminant analysis. Finally, we discuss results obtained (Sect. 4) and draw corresponding conclusions (Section "Conclusion").

2 The Online Community Considered and Problem Definition

2.1 SCN: The SAP Community Network

SAPs community network (SCN) is a business community platform where any uniquely registered person, referred to as *user*, may discuss and share their ideas and issues regarding SAP products. This community mainly consists of a number of fora, each relating to a unique product or topic. A user may post a message in any forum and a collection of messages form a *thread*. The first message in a thread is the parent message (i.e., "question") and subsequent messages are linked via a tree-like structure. As messages are linked, they are given a *time rank* and *wall clock*, that is an arrival order and minutes since thread creation. The user who posts the parent message is known as the *original poster* (OP). A user who makes a post in response to the parent message is called a *respondent*. Within the tree-like message structure of a thread, the *most responded to message* (MRTM) is that with greatest number of messages posted in direct response. Similarly, the *most responded to user* (MRTU) (including the OP) is the user to whom the greatest number of direct responses is made of all users to post in the thread.

The OP is the only user capable of making certain actions with respect to their thread. Each respondent may be awarded *points* by the OP based on the quality of their response, see Table 1. The SCN places light restrictions on the way an OP

Table 1 SCN's point awarding system via the original poster

Original poster's view of respondent's post	Points awarded
Respondent "solved" the issue of the parent post	10
Respondent was "very helpful" towards the issue of the parent post	6
Respondent was "helpful" towards the issue of the parent post	2

awards points in a thread such that only one 10 and two 6 point scores may be awarded. Consequently we define a thread to be *solved* only if the OP has awarded a 10 point score to a response; the associated respondent is known as the *thread solver* (TS). A more relaxed version of the TS is the *highest point scorer* (HPS). Where the HPS is not the TS, there may be more than one HPS. In the SCN, points awarded are connected to the corresponding respondents message allowing respondents to increase their *reputation*. A user's reputation is the total points accumulated over time. We view respondent reputation to be forum-specific due to forum topic inhomogeneity. Assuming a thread has at least one respondent, the *most reputable respondent* (MRR) is that with greatest lifetime reputation.

The OP in addition can change the *status* of a thread from the default "Unanswered" to "Answered." However, there are no restrictions on when an OP may change the status of a thread. For example, a thread does not have to be solved to have status "Answered"—the converse also holds true.

SAP made available a complete trace of actions of 95 fora (a third of the total byte size of the SCN) from February 2004 to July 2011. We select three for a showing variance in micro-level activity during the period analyzed: forum 50 spiking; forum 142 staying mostly level; and forum 246 decreasing.

2.2 Problem Definition

Problem motivation arose after observing only 23.26% of threads created within the dataset to be solved. Of the unsolved threads, approximately 12% are never responded to. We consider a classification event which may be viewed as either a risk or an opportunity. That is, after a time threshold, t_s minutes, of creation, the thread is solved (opportunity) or unsolved (risk).

Assuming the ith thread to be eventually solved, we note the wall clock time (minutes since thread creation) of this event as w_i. The default value of w_i is ∞. Thus the binary response observed, y_i, for the ith thread, is

$$y_i = \begin{cases} 1 & \text{if } 0 \leq w_i \leq t_s, \\ 0 & \text{otherwise,} \end{cases} \tag{1}$$

where $i = 1, \ldots, n_o$ and n_o is the number of threads observed within the sample population.

Fig. 1 Percentage of threads solved by hours since thread creation for all threads within dataset created between 2008 and 2010

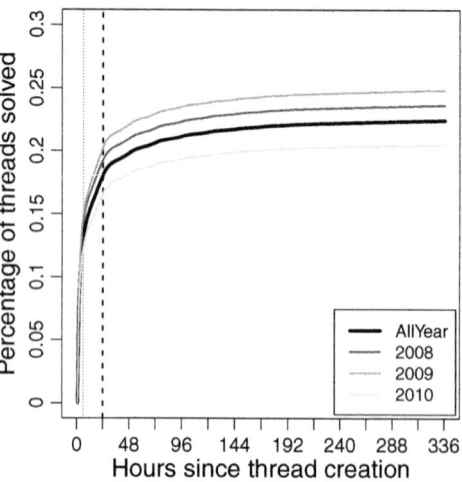

Considering only those threads created at least one year prior to our last observation and having at least one respondent, 13.76 % are not responded to within the first 24 h and 0.56 % are only responded to a year after thread creation. This highlights that the vast majority of threads receive greatest attention within the first 24 h after creation. Of the threads responded to within the first 24 h, 28.56 % are subsequently solved and of these 74.57 % are solved within the first 24 h. Within the threads solved in the first 24 h, 62.64 % are solved by the first response. In comparison, of the threads responded to only after the initial 24 h period, 15.18 % are eventually solved. However, within these latter solved threads, 64.77 % are solved by the first response. Thus, although a thread seems less likely to be solved if not responded to within the first 24 h of creation, it is still most likely to be solved by the first respondent.

Figure 1 illustrates the percentage of threads solved across all fora, grouped by year, over hours since thread creation. In all cases, the curve incline begins to reduce 6 h after thread creation and starts to level off 24 h after thread creation. We therefore take t_s in (1) to be 1,440 min (24 h).

2.2.1 Features Available for Prediction

The full set of features available for prediction following t minutes since thread creation is given below. The choice of t affects feature inclusion. For t sufficiently close to t_s, one could argue prediction is made too closely to the occurrence of the event. In addition, for t close to zero (thread creation) there exist uninformative features where all observations hold the same value. We trialed $t \in \{30, 60, 180, 360\}$ minutes; here, we report on $t = 30$ min due to space constraints and lack of improvement for larger t. Those features marked by an asterisk are included in our feature space for modeling y_i in (1) given t.

- **OP features:** OP reputation*; OP reputation in past year*; # thread OP partici-pated*; # thread OP created*; # thread OP created subsequently solved; # thread OP solved*; # messages OP posted*; # messages OP posted in thread*; # days since OP registration (first appearance in relevant forum)*.
- **TS features:** TS reputation; TS reputation in past year; # thread TS participated; # thread TS created # messages TS posted in thread.
- **MRR features:** MRR reputation*; MRR reputation in past year*; # messages MRR posted in thread*.
- **HPS features:** # HPSs*; mean HPS reputation*; # responses to HPS*.
- **MRTM features:** # MRTM*; # responses to MRTM*; mean MRTM reputation*; mean MRTM points earned*.
- **MRTU features:** # MRTU*; # responses to MRTU*; mean MRTU reputation*; mean MRTU points earned*; mean # thread MRTU solved; mean # thread MRTU created*; mean # messages MRTU posted*.
- **Temporal features:** minutes till first reply*; mean minutes till respondents first message*; mean minutes between messages*; median minutes between messages*; minimum minutes between messages*; TS time rank; TS wall clock; MRR time rank*; MRR wall clock*; mean HPS time rank*; mean HPS wall clock*; minimum MRTM time rank*; minimum MRTM wall clock*; minimum MRTU time rank*; minimum MRTU wall clock*.
- **Thread summary features:** indicator for TS is MRR; indicator of thread status*; indicator of thread solved; # users to participate*; sum points awarded*; # messages posted*; mean respondent reputation*; median respondent reputation*; mean respondent reputation in past year*.

3 Classification Methods Applied

We apply four linear methods for classification against the baseline model of ran-domized prediction (RAND) informed by observed class proportion in the training set. These models are: Bayesian probit (BP) model; generalized linear model with probit link (GLMP); generalized linear model with logit link (GLML); and linear discriminant analysis (LDA). Here, only the first model is non-standard, being taken from (Albert and Chib 1993), as such, some detail is given below. Details of GLM and LDA methods fitted are described in Venables and Ripley (2002) Chapters 7 and 12, respectively [function names glm() and lda()]. For a comprehensive guide to generalized linear models, see Mccullagh and Nelder (1989). A more general introduction to linear models for classification, including LDA, is provided by Hastie et al. (2011). In Sect. 4 we compare quality characteristics of classification predictions made with respect to the problem defined in Sect. 2.2.

First, we introduce some general notation. Let X be the normalized column matrix with rows $x_i^T = [x_{i,1}, \ldots, x_{i,p}]$ where $x_{i,j}$ is the ith observation of the jth feature; and $i = 1, \ldots, n_o$, for n_o the number of observations (here the number of threads) within the sample population. To avoid identifiability or non-integrability issues later on, we assume that $X^T X$ is non-singular. Given that X has full column-rank, this assumption is always satisfied. In addition let β be the corresponding p length vector of (elasticity) coefficients. For all four methods considered, the ith binary response Y_i is modeled via the corresponding latent variable Z_i; in the methods taking the probit link function

$$Z_i \overset{\text{i.d.}}{\sim} N\left(x_i^T \beta, \sigma\right), \tag{2}$$

$$Y_i = \begin{cases} 1 & \text{if } Z_i > 0, \\ 0 & \text{if } Z_i \leq 0. \end{cases}$$

where "$\overset{\text{i.d.}}{\sim}$" means "independently distributed" and $N(\mu, \sigma)$ denotes a normal distribution with mean μ and variance σ. Note that we set $\sigma = 1$ in (2) such that the distribution on the error terms is the standard normal.

Given a sample $y = [y_1, \ldots, y_{n_o}]^T$ and an associated column matrix X, a statistical inference problem about the coefficients β arises. Frequentist treatment of the above model, assuming probit link, leads to the generalized linear model with probit link. Here, one maximizes the β-likelihood; integrating out the latent variables analytically to give $L(\beta) = \prod_{i:y_i=1} \mathbb{P}(Z_i > 0) \prod_{i:y_i=0} \mathbb{P}(Z_i \leq 0) = \prod_{i:y_i=1} \Phi(x_i^T \beta) \prod_{i:y_i=0}(1 - \Phi(x_i^T \beta))$, where Φ is the standard normal (cumulative) distribution function. This can be similarly shown for the generalized model with logit link.

The Bayesian probit model takes a Bayesian approach to inference, following Albert and Chib (1993). We use the notation (β, Z) to denote the $(p + n_o)$-dimensional random vector consisting of the elements of β and of $Z = [Z_1, Z_2, \ldots, Z_{n_o}]$ and the symbol "$\propto$" as "is proportional to." Let β have prior probability density function, $\pi_0()$, then the posterior of (β, Z) is

$$\pi(\beta, Z) \propto \pi_0(\beta) \prod_{i=1}^{n_o} [1(y_i = 1)1(Z_i > 0) + 1(y_i = 0)1(Z_i \leq 0)] \times \phi\left(Z_i; x_i^T \beta\right) \tag{3}$$

assuming this is integrable. In (3), $\phi\left(Z_i; x_i^T \beta\right) \propto \exp\left(-\left(Z_i - x_i^T \beta\right)^2 / 2\right)$ is the normal (Gaussian) density with mean $x_i^T \beta$ and variance 1; and $1()$ is the indicator function. More concretely, letting $x = (\beta, Z)$ and taking μ as the $(p + n_o)$-dimensional Lebesgue measure, the function in (3) is (a version of) the density of a probability measure on $\mathbb{R}^{(p+n_o)}$ with respect to μ only where $C \overset{\text{def}}{=} \int_{\mathbb{R}^{p+n_o}} \pi(x)\mu(dx)$ is finite.

We take a flat prior for π_0, meaning that all points in $\mathbb{R}^p$ are, essentially, "equally likely," as is later mentioned, other choices are available. As the support is unbounded and the intended "density" is a positive constant, this π_0 does not give a probability measure on $\mathbb{R}^p$ and is hence *improper*. This is not a problem where (3) defines a probability measure. Thus the target of inference is the resulting β-marginal of (3). This target is denoted π_β.

The Bayesian probit method for binary response data as prescribed by Albert and Chib (1993) utilizes Gibbs sampling. Gibbs sampling is a particular method of Markov Chain Monte Carlo class and works by sampling from conditional distributions of the target probability measure, the (3) here. See Casella and George (1992) for an introduction; for a thorough treatment, see Chapters 9 and 10 of Robert and Casella (2004). Whilst the conditional distributions of the target (3) are easy to sample from Albert and Chib (1993), the level of ease depends on the choice of π_0. For conditionals with uniform π_0, see Albert and Chib (1993). In addition, Albert and Chib (1993) give integrable, that is *proper*, possibilities for the prior π_0 which directly enable Gibbs sampling. Given that we assume a uniform prior distribution for the regression coefficients, related issues of the propriety of the posterior distribution are studied by Chen and Shao (1999). In practice, the initial state of the Markov chain for β is taken to be the maximum likelihood (ML) estimate, $\tilde{\beta}_{ML} = \left(X^T X\right)^{-1} X^T y$.

Given the predictors x_i^T, the posterior mean of Y_i is $\mathbb{P}_{\pi_\beta}(Y_i = 1) = \mathbb{P}_{\pi_\beta}(Z_i > 0) = \mathbb{E}_{\pi_\beta}[\Phi(x_i^T \beta)]$, where $\mathbb{P}_{\pi_\beta}$ and $\mathbb{E}_{\pi_\beta}$ denote the probability and expected value with respect to π_β. Assuming certain conditions, a consistent estimator of this mean is the corresponding sample average of the Gibbs sample; whereby consistency we mean convergence with probability one as the sample size tends to infinity (Robert and Casella 2004, Theorem 6.63; Cappé et al. 2005, Theorem 14.2.53). Hence, given the sample $\{\beta^{(1)}, \beta^{(2)}, \ldots, \beta^{(M)}\}$, with M sufficiently large, $M^{-1} \sum_{m=1}^{M} \Phi(x_i^T \beta^{(m)})$ is an appropriate estimator of $\mathbb{P}_{\pi_\beta}(Y_i = 1)$.

4   Results

We implement our methods in the language and environment R (version 3.0.1) (R Core Team 2013) on a stand-alone computer with 64-bit operating system and 16 GB of memory. With regard to the Bayesian probit model (Sect. 3), ad-hoc experimentation led us to believe a "burn-in" period of $t_b = 90,000$ and subsequent sample of $t_r = 10,000$ to result in estimates accurate for our purpose. In all instances, we implement 10-fold cross-validation.

To assess the quality of our classification predictions, we consider the receiver operating characteristic (ROC). This characterizes true positive rate (TPR) and false positive rate (FPR) as the discrimination threshold (d) is varied, where $\hat{y}_i = 1$ if and only if the posterior probability $\mathbb{P}(Y_i = 1) > d$ and

$$\text{TPR} = \frac{\sum_{i=1}^{n_o} 1(\hat{y}_i = 1, y_i = 1)}{\sum_{i=1}^{n_o} 1(y_i = 1)}, \quad \text{FPR} = \frac{\sum_{i=1}^{n_o} 1(\hat{y}_i = 1, y_i = 0)}{\sum_{i=1}^{n_o} 1(y_i = 0)} \quad (4)$$

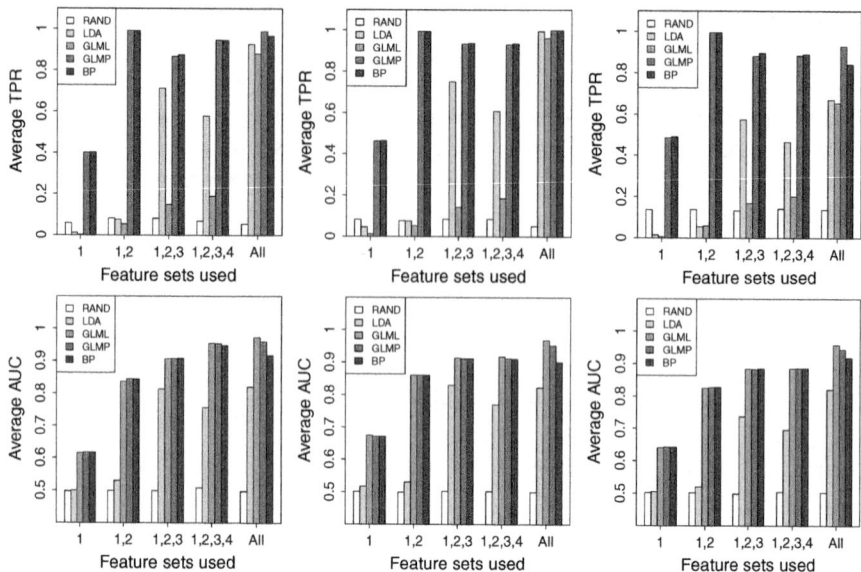

Fig. 2 Classification quality characteristics, true positive rate with discrimination threshold 0.5 (*top*) and area under ROC curve (*bottom*), averaged over cross-validation sets; given the event in Sect. 2.2 for for a 50, 142, and 246 (*left to right*) with 13,236, 35,933, and 34,102 unique threads respectively

(Fawcett 2006). In (4), $1()$ is the indicator function, $\hat{y}_i$ and y_i are the predicted and observed classifications for the ith thread where $i = 1, \ldots, n_o$. The area under ROC curve (AUC) is used to summarize our observations of the ROC curves, calculated using the R package ROCR (Sing et al. 2005).

Predictions for the event of Sect. 2 are made both for the entire thread population and thread subpopulations, segregated by forum. As stated (Sect. 2.1), we discuss here only three fora of the SCN. We started with our full feature set, described in Sect. 2.2.1, and performed classification with subsets of these, partitioned by feature type (indicated in bold feature within the feature list). The complete set of features is noted S_{All}; the subset of original poster features, S_1; the subset of MRR features, S_2; the subset of HPS features, S_3; the subset of both MRTM and MRTU features, S_4; and the subset of temporal and thread summary features, S_5.

Results for the application of the methods in Sect. 3 are reported in Fig. 2. Observe that the original poster features (S_1) have good predictive power across all fora. As expected, the most reputable respondent features (S_2) are very useful—increasing the AUC by almost 20 points and doubling the TPR for BP, GLMP, and GLML methods in every fora. The quality of the LDA method classifications is greatly improved by including features for the highest point scorer (S_3). Here, the AUC improves for all methods, although the TPR dips for those methods involving probit link. By including the features regarding both the most responded to message and user (S_4), very little appears to be gained. When all features are included

(S_{All}), the quality of method classifications is high, with regard to both AUC and TPR measures. The GLML method classification quality sees a substantial increase in TPR, more than trebling with regard to all fora. However, the GLML method consistently has lowest TPR of all methods across all fora (excluding the random baseline method). On the other hand, considering only the AUC, the quality of the classifications is lowest for the LDA method.

Thus we see that incorporating the rich micro-level community dynamics surrounding an original post significantly aids in determining whether a satisfactory response will be made in good time—no matter the method. In addition, we stress that these features are extracted only 30 min after the original post was made and are predicting whether a satisfactory response will arrive during the subsequent 1,410 min. We find it promising towards real-time application that after a mere 30 min there is sufficient information to predict, comparatively long-term, whether a thread will be solved. In addition, that the main value is from those features which are not direct evaluations of the original post.

5 Conclusion

Given the question-answer nature of the online community considered and the ever increasing complexity of community dynamics, it is valuable to think of each "question" related set of messages as a series of connected information. We have demonstrated how the rich structure of the SAP community network can be used to identify important characteristics of linked messages such that original posts needing additional help via the manager of the community can be identified. In our ongoing work, we found Bayesian probit models to be promising tools for predicting such binary classification risk events. We see our approach to be promising for question-answer communities in general.

Our Bayesian probit model incorporating a Gibbs sampler does require care when implementing. First, one must ensure that the multivariate Markov chain for β has converged to the desired target, regardless of the initial state of the chain. This typically involves verifying conditions of irreducibility, aperiodicity, positivity, and Harris recurrence; see Robert and Casella (2004, Chapter 6), for example. Second, selecting an appropriate burn-in and retained sample size tends to be challenging; see Robert and Casella (2004, Chapter 12).

Further investigation into quality characteristics for comparing binary classifiers is required. This is motivated by the discussion in the literature on the validity of AUC as a stand-alone measure of classification performance occurring primarily between Flach and Hand (Berrar and Flach 2012; Flach 2010; Hand 2009, 2006). Consequently, with increased automation, classifying streaming data would become more flexible.

Acknowledgements We thank Dr. Adrian Mocan of SAP for his contribution to defining the risk events. We also thank Edwin Tye, School of Mathematics, University of Southampton, for his assistance in processing the data. This work has been supported by the EU FP7 project ROBUST, EC Project Number 257859.

References

Albert, J. H., & Chib, S. (1993). Bayesian analysis of binary and polychotomous response data. *Journal of the American Statistical Association, 88*(422), 669–679.

Anderson, A., Huttenlocher, D., Kleinberg, J., & Leskovec, J. (2012). Discovering value from community activity on focused question answering sites: A case study of stack overflow. In *KDD* (pp. 850–858). New York, NY: ACM Press.

Berrar, D., & Flach, P. A. (2012). Caveats and pitfalls of ROC analysis in clinical microarray research (and how to avoid them). *Briefings in Bioinformatics, 13*, 83–97.

Cappé, O., Moulines, E., & Rydén, T. (2005). *Inference in hidden Markov models.* Springer series in statistics. New York: Springer.

Casella, G., & George, E. I. (1992). Explaining the Gibbs sampler. *The American Statistician, 46*(3), 167–174.

Chen, M. H., & Shao, Q. M. (1999). Properties of prior and posterior distributions for multivariate categorical response data models. *Journal of Multivariate Analysis, 71*(2), 277–296.

Fawcett, T. (2006). An introduction to ROC analysis. *Pattern Recognition Letters, 27*(8), 861–874.

Flach, P. A. (2010): ROC analysis. In C. Sammut, & G. I. Webb (Eds.), *Encyclopedia of machine learning* (pp. 869–875). Boston, MA: Springer.

Hand, D. J. (2006). Classifier technology and the illusion of progress (with discussion). *Statistical Science, 21*(1), 1–34.

Hand, D. J. (2009). Measuring classifier performance: a coherent alternative to the area under the ROC curve. *Machine Learning, 77*(1), 103–123.

Hastie, T., Tibshirani, R., & Friedman, J. (2011). *The Elements of statistical learning: Data mining, inference, and prediction* (2nd ed.). New York: Springer.

McCullagh, P., & Nelder, J. A. (1989). *Generalized linear models* (2nd ed.). London: Chapman and Hall/CRC.

Robert, C. P., & Casella, G. (2004). *Monte carlo statistical methods* (2nd ed.). New York: Springer.

R Core Team. (2013). *R: A language and environment for statistical computing.* Vienna: R Foundation for Statistical Computing. http://www.R-project.org/.

Sing, T., Sander, O., Beerenwinkel, N., & Lengauer, T. (2005). ROCR: Visualizing classifier performance in R. *Bioinformatics, 21*(20), 3940–3941. http://rocr.bioinf.mpi-sb.mpg.de.

Venables, W., & Ripley, B. (2002). *Modern applied statistics with S* (4th ed.). New York: Springer.

Human Performance Profiling While Driving a Sidestick-Controlled Car

Ljubo Mercep, Gernot Spiegelberg, and Alois Knoll

Abstract We have established a metric for measuring human performance while operating a sidestick-controlled car and have used it in conjunction with a known environment type to identify unusual steering trends. We focused on the analysis of the vehicle's offset from the lane center in the time domain and identified a set of this signal's features shared by all test drivers. The distribution of these features identifies a specific driving environment type and represents the essence of the proposed metric. We assumed that the driver performance, while operating a sidestick-controlled car, is determined by the environment type on one side and the driver's own mental state on the other. The goal is to detect the mismatch of the assumed driving environment, gained from the introduced metric, and a ground truth about the actual environmental type, which can be obtained through map and GPS data, in order to identify unusual steering trend possibly caused by a change in driver fitness.

1 Introduction

The most recent basic guidelines for the considerations on the driving context data were provided by the European AIDA project. The main identified context features were:

- Goal of the current voyage as provided by the navigational component
- Basic traffic information extended with car to infrastructure and car-to-car communication

L. Mercep (✉) • A. Knoll
Institute for Informatics, Chair for Robotics and Embedded Systems, Technische Universität München, München, Germany
e-mail: ljubo.mercep@tum.de; knoll@in.tum.de

G. Spiegelberg
Institute for Advanced Study der Technischen Universität München / Siemens AG, München, Germany
e-mail: gernot.spiegelberg@siemens.com

© Springer-Verlag Berlin Heidelberg 2015
B. Lausen et al. (eds.), *Data Science, Learning by Latent Structures, and Knowledge Discovery*, Studies in Classification, Data Analysis, and Knowledge Organization, DOI 10.1007/978-3-662-44983-7_40

- Assessment of driver's current state, both mental and physical
- Assessment of vehicle's current state

This work represents the effort to indirectly assess the driver's current mental state, by measuring his driving performance when operating a sidestick-controlled vehicle. The optimal input device for the primary driving task in road vehicles is a very debatable subject. In the scope of the project Diesel Reloaded, sidestick has been proposed as the future input modality in the automotive domain (Mercep et al. 2013). As compared to a central stick, where the input device is located between the driver's or pilot's legs, a sidestick is located to the left or to the right (or both) of the driver. One advantage is the integration of longitudinal and lateral vehicle dynamics' control in one single physical device, saving space and reducing the amount of physical force necessary to operate the vehicle. Another advantage is accessibility, since the device can be operated by people with a wide range of physical impediments. However, one of the key assumptions for the acceptance of sidestick-controlled vehicles is a reliable and affordable drive-by-wire system (Spiegelberg 2005). Therefore, the acceptance of the new input device might not be a question of ergonomics, but rather of engineering and regulatory changes taking place in other vehicle subsystems. Vehicle information and communication architecture is one of the key enabling technologies for innovation in the area of human–machine interaction and driver assistance (Buckl et al. 2012).

In this work, we propose analyzing the lane keeping task as the primary factor describing the successful performance of the driving task. A blind analysis of the vehicle's offset from the lane center over the course of time is performed. The goal is to find a lane offset-based metric which describes the driver's performance in a specific environment. The focus is on the definition and the validation of the metric through experimental data. Once the driver performance in a specific environment is sufficiently described by the metric, we assume that any sudden change in this description directly relates to a new and unusual steering trend in a specific environment. The fact that the driver suddenly altered his driving performance is therefore directly attributed to the change in its mental state. This result can be used as an input for other driver assistance systems. It should be noted that this work remains plagued by the absence of any related research, since the lane following task has mostly been analyzed from the driver intention, collision avoidance, or autonomous driving point of view. The assessment of the driver performance for sidestick-controlled vehicles seems to be a novel domain, what is not surprising considering the non-existing market share of such vehicles. Nevertheless, the more general task of target following with a sidestick represents a very interesting field of research for different vehicle types and different lower-level applications.

This work is organized as follows. In Sect. 2 we describe the proposed method for obtaining the necessary metric. The experiment design is described in Sect. 3. Pre-processing methods used on the data gathered during the experiment are described in Sect. 4. Results are presented in Sect. 5. Finally, we conclude and elaborate on future work in Sect. 6.

2 Method

In this section, we describe the method used to obtain the metric for the driver's performance. Based on the previous work in the area of driver steering prediction, we opted for an approach which reduces the driving task to a lane following task, in which the driver uses his previous knowledge, current state and future predictions to keep the vehicle from leaving the road margins (MacAdam and Charles 1981). We assume that a perfect lane detection exists and that it provides the lateral vehicle offset from the middle of the lane. The lateral offset was taken as the sole input of the method. The driver's sidestick input directly changes the lane offset, but the key difference between analyzing the sidestick input and the lane offset is the suppression of the influence of the road profile. A driver following a very dynamic road at high speeds produces a relatively large amount of lateral sidestick activity, but if he still manages to follow the road profile, the activity of the lane offset will be reduced to under- and over-steering. Driver's input for the longitudinal vehicle control, i.e. throttle and brake, is only used as an additional parameter in the further analysis.

Let offset(n) represent the time series containing the lateral vehicle offset from the middle of the lane. Let $\delta(n)$ be the first differential of the function offset(n). The set Δ_0 can be defined as:

$$\Delta_0 := \{\delta_x \in \Delta \mid |\delta_x| < \epsilon_0\}. \tag{1}$$

meaning that Δ_0 contains the segments of $\delta(n)$ where the lane offset signal underwent a trend change with a magnitude described by ϵ_0. Δ_0 is a set of subsegments of $\delta(n)$ of various lengths, in which the differential fell beneath the ϵ_0. Let us now define a so-called trigger set T as a set containing the first and last element of every subsegment in Δ_0. In a case of a subsegment which is one point wide, meaning that the first and the last element are the same, the trigger set T includes it only once. An example of the trigger set is given in Fig. 1.

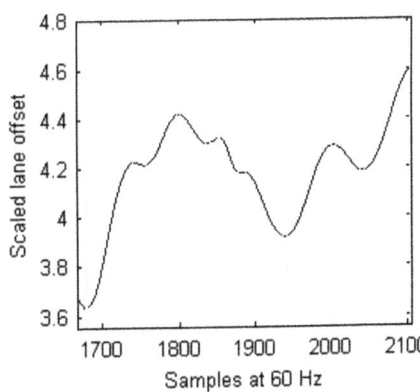

Fig. 1 Trigger set contains points in which signal started to rapidly change, here denoted with *vertical lines*

In the next step, we generate an alternative description of the trigger set, based on average densities of the triggers in a fixed time window. Inside of a larger time window, we iterate a smaller N-points time window, in which the number of triggers is counted. The counter value is added to an appropriate bin, i.e. N-point window containing four triggers increases the counter value of the bin number four. After the entire larger window has been processed, it is fully described by the final value of all the bins. Our hypothesis is that for a fixed sidestick sensitivity and a fixed sidestick sampling rate, a fixed number of bins will take on a typical average value for a specific driver and environment. The shapes of the bins' values and their relation to each other might also prove advantageous in the driver performance assessment. We propose that each environment will impose upon (or require from) the driver a specific behavior of the road offset signal, which we try to capture with the proposed metric. The static values of the bins as well as the perturbations between the bins should behave in a same manner for the same environment and for the same driver. Such perturbations can also be imagined as spectral shifts of the road offset in the frequency domain, even though we did not engage in spectral analysis in the scope of this work.

2.1 Trivial Solutions

There are, of course, trivial ways of identifying the environment based on the lateral component of the sidestick input. Long and extreme turning will signify an urban environment. Average number of sidestick corrections can trivially differentiate between inside and outside of city. The problem with these "summarizing" approaches is that they do not provide any possibility of further analysis, since most of the useful data is discarded in the averaging process.

3 Experiment Design

A total of 23 participants, all in possession of a valid driver license inside the European Union, took part in the experiment (19 male and 4 female). Mean age was 26.48, minimal 18 and maximal 36 years. A pre-experiment survey was filled out in order to determine possible alcohol or caffeine intake. The Virtual Test Drive (VTD) software from the company VIRES was used for the data collection. It was integrated into an automobile mock-up, a complete chassis of a Smart automobile, as shown in Fig. 2. A sidestick was mounted on the right of the driver, at the location usually taken by the gear shift. The sidestick did not provide force feedback. A simulation of the driving environment was shown on a large screen in front of the vehicle mock-up. Simulated vehicle dynamics were those of a typical personal automobile.

Fig. 2 Virtual Test Drive driving simulator with a complete vehicle mock-up

The experiment started with a target following game which was played with the sidestick inside the vehicle simulator. The goal was to learn the sensitivity and the behavior of the sidestick prior to the driving phase. Even though the sidestick is almost completely absent from the current road vehicles, all the participants possessed experience of using a common joystick, which lessened the learning curve. The participants were required to keep an object shaped as a circle in the middle of a large moving rectangular target for as long as possible. Penalty points were gathered when the circle failed to keep up with the rectangle. A randomly generated target following scenario was executed in each 30-s run. The game lasted no more than 3 min.

In the next step, the driving simulation was started. This step consisted of a new learning phase and, finally, the real driving phase. The learning phase lasted no more than 5 min. Participants were able to explore the simulation and further increase their grip on the sidestick skills. In the second phase, all the participants started from the same position inside the simulated world and the data was collected using the VTD RDB interface shown in Fig. 3. The participants started the drive on the outskirts of a virtual city and proceeded to drive towards and finally into the city, continuing on the city roads. Most of the participants chose to take the same route out of the city and back to the original starting position, but this was not strictly required in order to complete this phase. The real driving phase and the respective data collected lasted around 7 min.

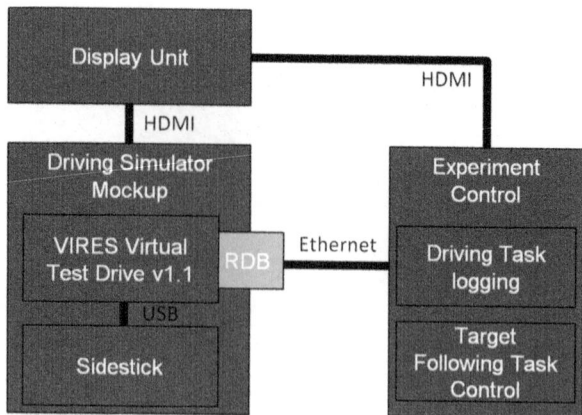

Fig. 3 Data flow between software and hardware components used in the experiment

4 Preprocessing Collected Data

The lane offset data collected in the experiment had to be pre-processed before being fed to the previously described method. Additionally, the window sizes and the ϵ_0 had to be defined. All of these values are directly dependent on the sidestick device and had to be derived from the data.

A value of 15-s has been chosen for the previously defined larger sliding window, while the smaller sliding window was fixed to 200-points (about 3 s). This has been chosen by a brute force analysis of the impact different window sizes have on the observed features and remains in direct connection with the sidestick sensitivity and sample rate. These parameters and their further refinement remain an open question and were not covered further the scope of this work.

A value of ϵ_0 of 0.3 was chosen on the same terms.

In order to eliminate the bouncing artifacts of the collected lane offset signal, present when the sidestick is switching from one discrete position to another we iterate a 3-phase 15-point moving average smoothing over the signal. The artifacts removed are rapid oscillations around a stable or steadily transient (ramp) sidestick position. They can be removed with a low-pass frequency filter, but the result has proven to be generally worse during the experiment: As the smoothing effect approaches the level of a simple average smoothing, the filter progressively removes more of the important signal features.

5 Results

After applying the binning procedure, two bins started to contain relatively large and stable signal features which stayed similar for all participants. These bins were bin number 4 and bin number 5, which count the number of 200-point windows containing, respectively, 4 and 5 triggers in larger 15-s time window. Lower bins have not been deemed useful for classification and started to fill bottom-up only during long steering maneuvers. The bins higher than 5 were almost always empty and would appear only in the most erratic and non-realistic driving situations, when the participants opted for a short chase through the streets (even though they were advised not to beforehand). The emerging signal features in bins 4 and 5 differed in two ways throughout a course of every experiment.

The first difference was the relative difference of the same bin value between different environments. Driving inside the city, as well as driving outside the city as higher speeds, trivially raised the value of bins 4 and 5 throughout all test subjects. In addition, any sudden increase in speed was intuitively countered with over-steering in the following curves, which would create significant spikes in the bin 5. This type of differences was only marginally useful for classifying environments, since the average value can drift through a large value range inside the same environment without being classified as another environment, but still denoting a change in driver performance. In other words, too much data about the driver performance is discarded by only focusing on the values of bins. This is, in fact, a version of the previously mentioned trivial solution.

The second type of differences focuses on the shapes of the bins 4 and 5 and their mutual ratio. This has proven to be the most valuable approach and it mostly tied to the surges in value of the bins 4 and 5. There were four identified sub-types, presented in Figs. 4 and 5, which are further denoted as F0, F1, F2, and SW.

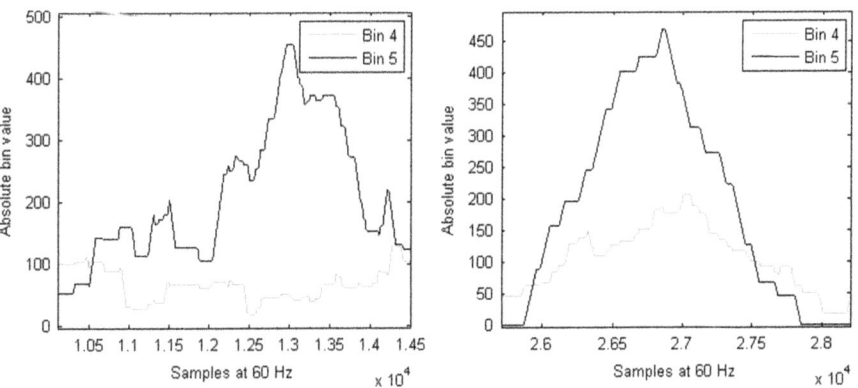

Fig. 4 Sub-type F0 on the *left*, sub-type F1 on the *right*

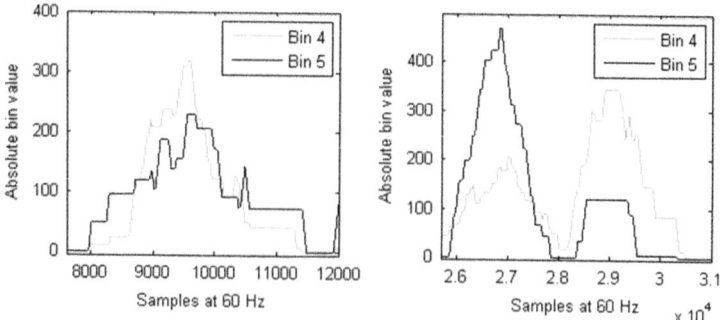

Fig. 5 Sub-type F2 on the *left*, sub-type SW on the *right*

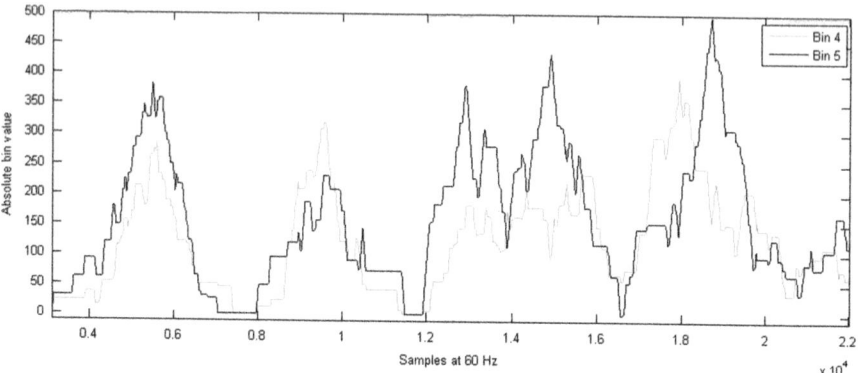

Fig. 6 Typical form of the bin 4 and bin 5 signals during the drive

Table 1 Occurrence of sub-types for different environments

Environment	SW	F0	F1	F2
Inside the city	3	3	67	14
Transition between environments	18	0	4	2
Outside the city	1	0	13	29

The SW sub-type represents a switch of absolute values between the bins 4 and 5 and was mostly observed on the borders of two environments and during a change of driving style inside a certain environment. The F0 sub-type represents a surge of bin 5 which is not followed by the bin 4. The F1 sub-type represents a surge of bin 5 which is moderately followed by the bin 4. The F2 sub-type represents a surge of bin 5 followed by a same or similar surge by the bin 4.

Figure 6 demonstrates the appearance of the sub-types during a drive in which the participant first drove outside the city (F2), than inside (F1) and than began leaving the city (SW).

Table 1 demonstrates the occurrence of sub-types in different environments for different drivers.

The relatively large amount of collected data (160 min of driving sampled at 60 Hz) resulted in a relatively low amount of detected sub-types, due to their size (some are formed over a period of 60 s) and due to the presence of other signal forms, which did not take a stable form. Nevertheless, the data clearly shows a correlation between the environment type and the signal features based on the proposed metric.

6 Conclusion

A metric for measuring driver performance for a sidestick-operated road vehicle was proposed. The driving task was first reduced to lane following task, taking the lane offset as the main element of the metric. Binning of the average number of trend changes inside the road offset signal produced several signal features which can be used for classification of the driving environment. The assumption is that each environment requires a specific performance of the lane keeping task. In this sense, we have identified the signal features which correlate with the performance and performance changes of the lane keeping task.

Future work involves classifying additional environment types and comparison with other sidestick devices, with their own sensitivity and sampling rates. Additionally, we will compare how the metric fares in more generic target following tasks, joystick being the main input method.

References

Buckl, C., Camek, A., Kainz, G., Simon, C., Mercep, Lj., Staehle, H., et al. (2012). The software car: Building ICT architectures for future electric vehicles. In *IEEE Electric Vehicle Conference (IEVC) 2012*, Greenville.

MacAdam, C. C., & Charles, C. (1981). Application of an optimal preview control for simulation of closed-loop automobile driving. *IEEE Transactions on Systems, Man and Cybernetics, 11*(6), 393–399.

Mercep, L., Spiegelberg, G., & Knoll, A. (2013). A case study on implementing future human-machine interfaces. In *IEEE Intelligent Vehicles Symposium*, Gold Coast.

Spiegelberg, G. (2005). Advantage in integration and active safety by using drive-by-wire technologies. In *Results of EU-Project PEIT and SPARC*, Autotec 2005, Stuttgart.

Multivariate Landing Page Optimization Using Hierarchical Bayes Choice-Based Conjoint

Stefanie Schreiber and Daniel Baier

Abstract Landing pages are defined to be the home page of a website (e.g., an online shop) or a specific webpage that appears in response to an ad. Their design plays an important role in decreasing the number of visitors leaving the website without any activity (e.g., clicking a banner, purchasing a product). For improving landing pages, the traditional A/B testing approach offers a simple but limited solution to evaluate two different variants. However, recently, new approaches have been introduced. Webpages with multiple variations of website elements (e.g., navigation menu, advertising banners) generated through experimental designs are rated by customers (Gofman et al., J. Consum. Mark. 26(4):286–298, 2009).

The paper explores a new approach for multivariate landing page optimization using hierarchical Bayes choice-based conjoint analysis (CBC/HB) that combines the potential to test a large number of variants with a short survey. The new approach is discussed and applied to improve the online shop of a popular German Internet pharmacy. Choice data are collected from a large sample of customers. From the results an optimal landing page is derived and implemented.

1 Introduction

In the context of e-commerce, companies need to attract consumers in the first seconds they enter their online shopping website. Home pages of a website or specific landing pages are more and more designed to increase conversion rates (e.g., clicking a banner, purchasing a product) as well as to decrease the number of visitors leaving the website without any activity. Altogether, the companies aim is to convert visitors into buyers to gain more revenues per site visit. Hence, it is becoming increasingly important to establish continuous landing page optimization to be successful in capturing the customer's attention (Ash et al. 2011).

S. Schreiber (✉) • D. Baier
Institute of Business Administration and Economics, Brandenburg University of Technology Cottbus-Senftenberg, Cottbus, Germany
e-mail: stefanie.schreiber@tu-cottbus.de; daniel.baier@tu-cottbus.de

© Springer-Verlag Berlin Heidelberg 2015
B. Lausen et al. (eds.), *Data Science, Learning by Latent Structures, and Knowledge Discovery*, Studies in Classification, Data Analysis, and Knowledge Organization, DOI 10.1007/978-3-662-44983-7_41

465

In recent years several approaches have been suggested in the literature to solve the problem of landing page optimization. However, far too little attention has been paid to the methods' validation. Hence, our research question focuses on developing a valid method for website optimization based on consumer perceptions and preferences. The study explores a new approach for multivariate landing page optimization using hierarchical Bayes choice-based conjoint analysis (CBC/HB).

This paper is structured as follows. Section 2 introduces traditional approaches of landing page optimization and briefly discusses the state of the art. Section 3 describes the theoretical framework and the procedure of our new landing page optimization method. An online questionnaire where the customer had to choose between randomly created landing page concepts is set up to investigate how attributes and corresponding levels affect customer preferences for an online-shop landing page in Sect. 4. In the following Sect. 5 we perform choice modeling at both aggregate and individual level to analyze the data. This research also focuses on the question whether the predictions made by using a conjoint analysis for landing pages are valid. In Sect. 6 we conclude with a discussion of our findings.

2　Landing Page Optimization: Traditional Approaches

For years improving a landing page has only depended on the web designer's subjective preferences or was based on targets. With experimentation based landing page optimization which is not founded on prior knowledge of the target audience the methods became more consumer-oriented. The approaches try to capture consumer's preferences through closed or open-ended experiments. The A/B split test is a really simple approach which involves only a very limited set of alternatives. In the majority of cases only two variations of a landing page are examined. A/B tests often measure conversion or bounce rates. Finally, the results of the standard and improved version are compared (Gofman et al. 2009).

Multivariate landing page optimization is an advanced form of landing page optimization. The papers of Gofman (2007) and Gofman et al. (2009) were the first major work on multivariate landing page optimization (MVLPO). The approach used conjoint analysis to test multiple variations of a landing page created through experimentation with different website elements. In their study 127 members from a web panel participated in a questionnaire and had the task to rate 27 webpages according to their preference. These were unique sets of different landing pages designed by a method called rule-developing experimentation (RDE). Four attributes (shipment options, promotions, featured items, main pictures) with three levels each were used to build varying attribute-level-combinations (Gofman et al. 2009). RDE generates these prototypes by arranging the attribute levels according to isomorphic permuted experimental designs. After rating the prototypes RDE allows an individual modeling of utilities for each participant, an identification of latent segments and interactions between the attribute levels (Gofman et al. 2010).

3   Landing Page Optimization: The New Approach

The entire presented research on landing page optimization in Sect. 2 did not cover the topic of methods validity. Furthermore the A/B test yields limited findings and the MVLPO introduced by Gofman (2007) with its 27 required webpage ratings is very demanding for respondents. Our research employs hierarchical Bayes choice-based conjoint analysis to examine the preferences for each attribute and its corresponding levels to take this criticism into account.

To determine the customer's preference structure this decompositional approach uses an overall evaluation of a product or service profile that consists of different attributes and attribute levels. Compared with the approach of rating or ranking a particular concept (e.g., Green et al. 1981) in choice-based conjoint analysis a set of product or service alternatives is presented to the respondents and they are asked to indicate which concept they would choose (Louviere and Woodworth 1983).

Choice-based conjoint analysis is the most widely used method for preference measurement for pricing, product development, and market segmentation purposes (Sattler and Hartmann 2008). To choose a product or a service is equal to the decision process customers make in the market place and provides a realistic situation (Cohen 1997). With regard to this simple choice procedure the application of a choice-based conjoint analysis for landing page optimization is easily conceivable. As choice-based conjoint analysis typically presents three or more alternatives in each choice task, more product or service concepts are seen by each respondent which also leads to a reduction of survey length (Sawtooth Software 2013).

In choice-based conjoint analysis for each attribute level a numerical part-worth utility value is estimated using a multinomial logit model. The highest part-worth utility value is associated with the most preferred attribute level. The applied multinomial logit model is a specific type of discrete choice models (see, e.g., McFadden 1974; Train 2009 for a detailed mathematical foundation) where the utility U_{ij} of a decision maker i to choose the jth alternative is defined by the equation:

$$U_{ij} = V_{ij} + \epsilon_{ij} \quad \forall ij. \tag{1}$$

V_{ij} is a systematic component and observable by the researcher. This representative utility is usually specified as $V_{ij} = \beta_i' x_{ij}$ where x_{ij} represents the observed variables concerning the jth alternative. ϵ_{ij} is a random term to capture uncertainty associated with the jth alternative. Furthermore ϵ_{ij} is distributed independently, identically type I extreme value, a so-called Gumbel distribution (McFadden 1974).

The probability that the decision maker i chooses the kth alternative with

$$p_{ik} = \text{Prob}(V_{ik} + \epsilon_{ik} > V_{ij} + \epsilon_{ij} \quad \forall j \neq k) \tag{2}$$

leads to the derived logit choice probabilities

$$p_{ik} = \frac{e^{V_{ik}}}{\sum_j e^{V_{ij}}} = \frac{e^{\beta_i' x_{ik}}}{\sum_j e^{\beta_i' x_{ij}}}. \tag{3}$$

For an estimation of β_i maximum likelihood procedures can be applied (Train 2009).

In contrast to rating- or ranking-based conjoint analysis results have traditionally been analyzed at the aggregate or group level (Louviere and Woodworth 1983), but by the application of hierarchical Bayes (HB) estimation methods, unique model parameters can be calculated for each individual customer. The first applications of Bayesian methods to choice-based conjoint analysis data were made almost 20 years ago by e.g., Allenby et al. (1995), Allenby and Ginter (1995) and Lenk et al. (1996).

A Bayesian analysis combines prior information about model parameters with information about model parameters contained in the observed data. The Sawtooth CBC/HB system was used for parameter estimation in this study. The iterative estimation procedure uses a Monte Carlo Markov chain (MCMC) method which is a combination of Metropolis-Hastings algorithm and Gibbs sampling. Further information concerning the estimation are provided by Sawtooth Software (2009). The applied hierarchical Bayes model has two levels. At the higher level, individuals' part-worths β_i are expected to have a multivariate normal distribution which is characterized by a vector α of means of the distribution of individuals' part-worths and a matrix $\mathbf{D}$ of variances and covariances of the distribution of part-worths across individuals (Sawtooth Software 2009).

$$\beta_i \sim MNV(\alpha, \mathbf{D}) \tag{4}$$

At the lower level, as in Eq. (3), given an individual's part-worths, the probability p_{ik} of an ith individual choosing the kth concept in a choice task is controlled by a multinomial logit model (Sawtooth Software 2009).

4 Empirical Test: Design

4.1 Survey

First of all it is necessary to investigate which website elements need to be considered in landing page design. For this reason we conducted an extensive review of literature covering the topics design and usability of websites. We include current insights from, e.g., the Software Usability Research Laboratory (http://www.surl.org/), the Nielsen Norman Group (http://www.nngroup.com/topic/web-usability/), and the eResult GmbH (http://www.eresult.de/studien_artikel/forschungsbeitraege.

html). The experimental data concerning the examined website elements are rather controversial, and there is no general agreement about design and positioning of elements such as navigation, font as well as button and advertisements.

Based on the knowledge that a large number of attributes is difficult to handle in choice-based conjoint analysis with regard to the respondents' cognitive efforts it is essential to reduce the number of attributes to be examined (Sawtooth Software 2013). Therefore, discussions with substantial customers and employees working in the field of web design as well as several pre-test questionnaires were conducted to generate relevant key attributes and levels.

As a consequence, the examined online shop home pages consist of (1) navigation menu: at the top, at the left, or in the middle; (2) search and shopping basket area: static or flexible; (3) button size: small or large; (4) font size: small (12pt) or large (14pt); and (5) advertising banner: big size banner, junior page, full banner, skyscraper banner or button. The five banner advertising variants were suggested to be the most accepted by a study from Maxl and Fahrleitner (2007). The navigation menu positioning in the middle differs from the current practice to place it at the top or at the left. We propose the positioning in the center of the page as eye-tracking studies assume that people tend to fixate this position first (Tatler 2007).

As a consequence, the examined online shop home pages consist of (1) navigation menu: at the top, at the left, or in the middle; (2) search and shopping basket area: static or flexible; (3) button size: small or large; (4) font size: small (12pt) or large (14pt); and (5) advertising banner: big size banner, junior page, full banner, skyscraper banner, or button. The five banner advertising variants were suggested to be the most accepted by a study from Maxl and Fahrleitner (2007). The navigation menu positioning in the middle differs from the current practice to place it at the top or at the left. We propose the positioning in the center of the page as eye-tracking studies assume that people tend to fixate this position first (Tatler 2007).

Before answering the online questionnaire, the respondents received a detailed and illustrated description of the core landing page attributes and levels. For example:

Search and shopping basket area.

a. The search and shopping basket area is static. The area disappears if the customer uses the browsers scrolling down function.
b. The search and shopping basket area is flexible. The area remains visible at any time even the customer uses the browsers scrolling down function.

We constructed a questionnaire with 12 choice-based conjoint tasks. Each choice task included three concepts without a "None" alternative. The choice tasks were created randomly using the "Balanced Overlap" design methodology since it is efficient in measuring main effects and, in addition, considers interactions. Therefore, the method allows a certain overlap of attribute levels in each choice task but forbids the occurrence of duplicate concepts (Sawtooth Software 2013).

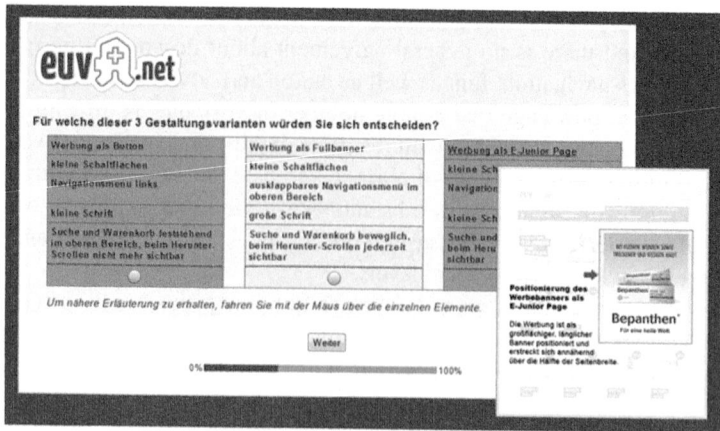

Fig. 1 Example of a choice task with mouse over function: Respondents received a detailed and illustrated description of every attribute level

A sample choice task with a mouse over function for further information is shown in Fig. 1.

For the holdout questions four different landing pages were designed which were shown after the choice-based conjoint task. Holdout tasks are not used for utility estimation but they are necessary to assess the predictive validity of the estimated models. Furthermore holdout tasks allow to test configurations of particular interest (Johnson and Orme 2010). Based on the former elaborations holdout question one with Small font size, Small button size, Static search and shopping basket area, Navigation menu at the left and a Button for banner advertising was expected to be the worst and holdout question three with Large font size, Large button size, Flexible search and shopping basket area, Navigation menu at top and a Skyscraper banner was expected to be the best home page concept. The other two concepts were prototypes which the pharmacy wanted to be evaluated. The respondents had to rate each concept on a 1 to 11 scale dependent on their preference. Figure 2 presents the first out of four evaluated landing page concepts.

4.2 Sample and Data Collection

In our sample application choice-based conjoint data were collected from actual customers of the Internet pharmacy EU-Versandapotheke (www.euva.net). The GfK Consumer Panel ranked the pharmacy seventh in Germany in terms of awareness. Additionally, the German Institute for Service Quality honored the company 2013 for being the best German Internet pharmacy in terms of service quality.

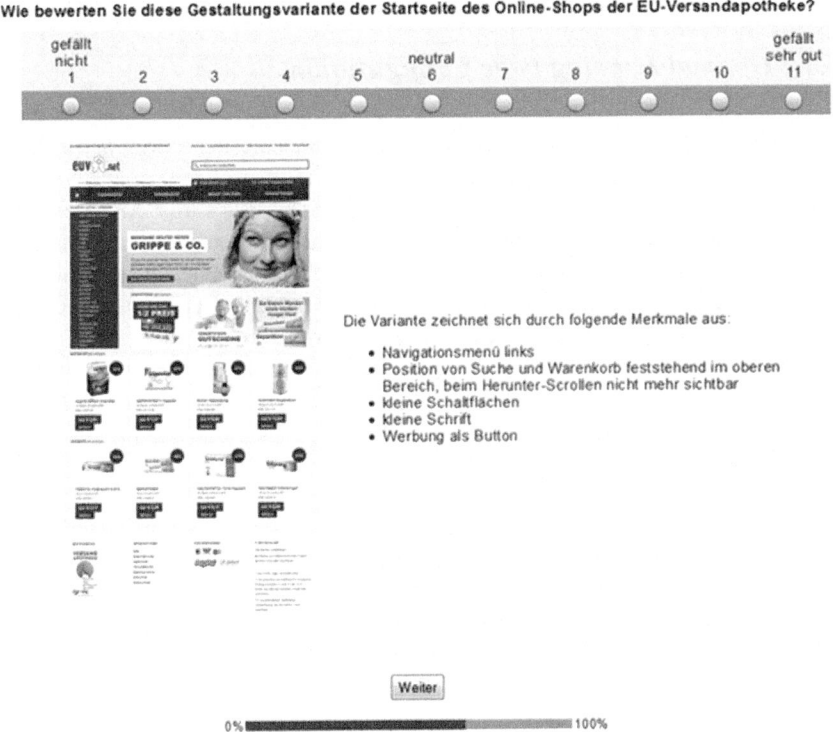

Fig. 2 Holdout questions 1: Respondents were asked to rate four concepts according to their preference

The questionnaire was distributed via the pharmacy's newsletter and enabled 36,231 customers to participate in the survey. The participants had access to the questionnaire from January 20th until February 3rd 2013. A total of 2,489 customers completed the questionnaire. 54 % of those were female and 46 % male. The group aged 45 years and above had the largest number of respondents (61 %). The majority (72 %) has been a customer for more than 1 year. 33.5 % of the respondents uses the online shop once in month and 40.3 % purchase a product once every 2 to 3 month. The sample characteristics imply that our study is biased towards older and experienced internet pharmacy customers. Additionally, the newsletter recipients select themselves into the sample which might induce less transferability of the outcomes. However, our sample matches the age distribution (age above 45: 58.5 %) of a leading marketing research company's online consumer panel that interviewed 20,054 participants concerning the purchase in Internet pharmacies (GfK 2011).

5 Empirical Test: Results

5.1 Optimal Landing Page Configuration

An aggregate and an individual analysis were conducted using Sawtooth Software's CBC/HB module. For the hierarchical Bayes estimation 10,000 draws were executed in the burn-in phase with 10,000 used iterations afterwards. An extension of 190,000 draws for the burn-in phase obtained no improvement of estimation quality.

Our study suggests the landing page concept with the attribute levels "Navigation menu at the left," "Search and shopping basket flexible," "Large button size" and "Large font size" as well as a "Button" for banner advertising as the configuration with the highest utility values. Table 1 presents the normalized part-worth values.

5.2 Validations

Additionally, the validity of the predictions was assessed. The root likelihood value (RLH) is a typical internal validity value in the context of conjoint analysis and measures how well the solution fits the data. In this study the aggregate RLH is with 0.35 relatively low concerning the possible range of 1/3 to 1 (Sawtooth Software 2013). On the individual level the mean RLH value is 0.54 which is an improvement compared to the aggregate level. For measuring predictive validity first choice hit rate (FCHR) and Spearman's rank correlation coefficient were used to determine the predictive validity at aggregate level as well as on individual level. The Spearman's

Table 1 Part-worth values of key attributes

Attributes	Levels	Part-worth
Navigation menu	Navigation menu at top	0.1121
	Navigation menu at the left	0.2289
	Navigation menu in the middle	0.0000
Search and shopping basket area	Static	0.0000
	Flexible	0.1723
Button size	Small	0.0000
	Large	0.1349
Font size	Small (12pt)	0.0000
	Large (14pt)	0.3118
Advertising Banner	Big size banner	0.0226
	Junior page	0.0000
	Full banner	0.0479
	Skyscraper banner	0.1170
	Button	0.1520

Table 2 Internal and predictive validity

	Internal validity	Predictive validity (%)	
Aggregate	RLH = 0.35	FCHR = 0.55	$r_s = 0.800$
Individual	RLH = 0.54	FCHR = 0.66	$r_s = 0.371$

rank correlation coefficient measures the correlation between the observed rankings in the holdouts tasks and the ranking of the predicted utility values. The calculated Spearman's rank correlation coefficient for the aggregate data is $r_s = 0.8$.

Moreover the FCHR measures the choice probability of the rank 1 concept in the holdout task. At the aggregate level the FCHR has the value 0.55. The individual Spearman's rank correlation coefficient is with $r_s = 0.371$ slightly worse than in the aggregate analysis but we can determine an improvement of FCHR to 0.66. A summary of all validity measurements is represented in Table 2. Selka et al. (2014) and Selka and Baier (2014) analyzed the validity of 1,777 commercial application of CBC and figured out a mean FCHR with 0.557 and a mean RLH with 0.533. Our research achieves these values on the individual level.

6 Conclusion and Discussion

The purpose of the current study was to determine a better approach for landing page optimization. We introduced an innovative approach using hierarchical Bayes Choice-Based conjoint analysis. The results of this study indicate that hierarchical Bayes Choice-Based conjoint analysis is an adequate method for improving landing pages. This new approach shows some theoretical and practical advantages over traditional landing page optimization methods. We can evaluate more product concepts with a shorter survey. We gained improved internal validity through hierarchical Bayes estimation and an appropriate predictive validity. As the examined group of customers (e.g., more than 50 % over 45 years old) is very unique, further research should be done to investigate whether the results are also valid for other customer structures and industries. Furthermore our research did not include all possible website elements. Further research might explore other attributes and levels that can potentially influence consumer's preference for an online shop landing page. Advanced research options are given for individual conjoint data in the case the customer's part-worths are very heterogeneous. By the application of segmentation methods, e.g., cluster analysis, the individual preference values can be used to synthesize specific landing pages optimized for subgroups of customers. Additionally, it would be interesting to set up experiments with eye-tracking systems to have a closer look at customer behavior to validate our findings. Moreover it would be helpful to analyze the protocols of landing page usage to measure and compare key figures, e.g. conversion rate, bounce rate.

References

Allenby, G. M., Arora, N., & Ginter, J. L. (1995). Incorporating prior knowledge into the analysis of conjoint studies. *Journal of Marketing Research, 32*(2), 152–162.

Allenby, G. M., & Ginter, J. L. (1995). Using extremes to design products and segment markets. *Journal of Marketing Research, 32*(4), 392–403.

Ash, T., Ginty, M., & Page, R. (2011). *Landing page optimization: The definitive guide to testing and tuning for conversions.* Hoboken, NJ: Wiley.

Cohen, S. H. (1997). Perfect union. *Marketing Research, 9*(1), 12–17.

GfK (2011). *CP Panel Internetapotheke 2011.* Nuremberg: GfK - Gesellschaft für Konsumforschung.

Gofman, A. (2007). Consumer driven multivariate landing page optimization: Overview, issues, and outlook. *Transactions on Internet Research, 3*(2), 7–9.

Gofman, A., Moskowitz, H. R., Bevolo, M., & Mets, T. (2010). Decoding consumers perceptions of premium products with rule-developing experimentation. *Journal of Consumer Marketing, 27*(5), 425–436.

Gofman, A., Moskowitz, H. R., & Mets, T. (2009). Integrating science into web design: Consumer driven website optimization. *Journal of Consumer Marketing, 26*(4), 286–298.

Green, P. E., Carroll, J. D., & Goldberg, S. M. (1981). A general approach to product design optimization via conjoint analysis. *Journal of Marketing, 45*(1), 17–37.

Johnson, R., & Orme, B. (2010). *Including holdout choice tasks in conjoint studies.* Sawtooth software research paper series (pp. 1–3). Sequim, WA: Sawtooth Software

Lenk, P. J., Desarbo, W. S., Green, P. E., & Young, M. R. (1996). Hierarchical Bayes conjoint analysis: Recovery of part-worth heterogeneity from reduced experimental designs. *Marketing Science, 15*(2), 173–191.

Louviere, J. J., & Woodworth, G. G. (1983). Design and analysis of simulated choice and allocation experiments: An approach based on aggregate date. *Journal of Marketing Research, 20*(4), 350–367.

Maxl, E., & Fahrleitner, P. (2007). Werbewirkung von Onlinewerbeformen. *Transfer, Werbeforschung & Praxis, 2*(3), 36–41.

Mcfadden, D. (1974). Conditional logit analysis of qualitative choice behaviour. In: P. Zarembka (Ed.), *Frontiers in econometrics* (pp. 105–142). New York: Academic Press.

Sattler, H., & Hartmann, A. (2008). Commercial use of conjoint analysis. In M. Höck, & K.-I. Voigt (Eds.), *Operations management in theorie und praxis: Aktuelle entwicklungen des industriellen managements* (pp. 103–119). Wiesbaden: Gabler.

Sawtooth Software (2009). *The CBC/HB system for hierarchical Bayes estimation version 5.0 technical paper.* Orem, UT: Sawtooth Software.

Sawtooth Software (2013). *The CBC system for choice-based conjoint analysis version 8.* Orem, UT: Sawtooth Software.

Selka, S., & Baier, D. (2014). Kommerzielle Anwendung auswahlbasierter Verfahren der Conjointanalyse: Eine empirische Untersuchung zur Validitätsentwicklung. *Marketing Zeitschrift für Forschung und Praxis, 36*(1), 54–64.

Selka, S., Baier, D., & Kurz, P. (2014). The validity of conjoint analysis: An investigation of commercial studies over time. *Studies in Classification, Data Analysis, and Knowledge Organization, 48*, 227–234.

Tatler, B. W. (2007). The central fixation bias in scene viewing: Selecting an optimal viewing position independently of motor biases and image feature distributions. *Journal of Vision, 7*(14), 4, 1–17.

Train, K. (2009). *Discrete choice methods with simulation.* Cambridge: Cambridge University Press.

Distance Based Feature Construction
in a Setting of Astronomy

Tobias Voigt and Roland Fried

Abstract The MAGIC and FACT telescopes on the Canary Island of La Palma
are both imaging Cherenkov telescopes. Their purpose is to detect highly energetic
gamma particles sent out by various astrophysical sources. Due to characteristics
of the detection process not only gamma particles are recorded, but also other
particles summarized as hadrons. For further analysis the gamma ray signal has
to be separated from the hadronic background. So far, so-called Hillas parameters
(Hillas, Proceedings of the 19th International Cosmic Ray Conference ICRC, San
Diego, 1985) are used as features in a classification algorithm for the separation.
These parameters are only a first heuristic approach to describe signal events, so that
it is desirable to find better features for the classification. We construct new features
by using distance measures between the observed Cherenkov light distribution in the
telescope camera and an idealized model distribution for the signal events, which
we deduce from simulations and which takes, for example, the alignment and shape
of an event into account. The new features added to the Hillas parameters lead to
substantial gains in terms of classification.

1 Introduction

In very high energy (VHE) gamma-ray astronomy, so-called Hillas parameters
are used as features for classification to separate a gamma-ray signal from a
hadronic background. These variables, introduced by Hillas (1985), are based on
fitting ellipses to images recorded by Cherenkov telescopes and using the ellipse's
parameters such as length and width as features. A so-called shower image and
the fitted ellipse can be seen in Fig. 1. The approach we are following in this
paper is to extend the idea of fitting an ellipse to the shower image, but instead
of fitting an ellipse we fit a bivariate, possibly elliptic-symmetric density to the
image and use features calculated from the fitted distribution. Since the contour lines

T. Voigt (✉) • R. Fried
Department of Statistics, TU Dortmund University, Vogelpothsweg 87, 44227 Dortmund,
Germany
e-mail: voigt@statistik.tu-dortmund.de; fried@statistik.tu-dortmund.de

© Springer-Verlag Berlin Heidelberg 2015   475
B. Lausen et al. (eds.), *Data Science, Learning by Latent Structures,
and Knowledge Discovery*, Studies in Classification, Data Analysis,
and Knowledge Organization, DOI 10.1007/978-3-662-44983-7_42

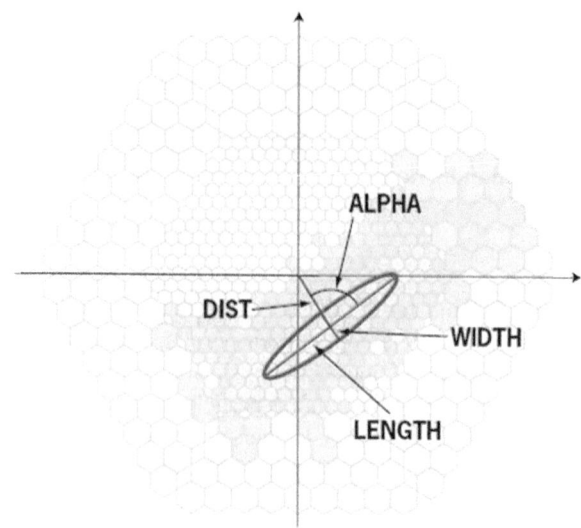

Fig. 1 Some Hillas parameters from fitting an ellipse (source: http://ihp-lx.ethz.ch/Stamet/magic/parameters.html)

of an elliptic-symmetric density form ellipses, or ellipsoids in higher dimensions, our approach implies a promising generalization of the Hillas parameters and additionally allows to incorporate information on the height and the tail behavior of the shower. Our approach corresponds to the fact that a shower consists of a number of Cherenkov light photons under the assumption of a bivariate distribution for the arrival coordinates of each photon. In Fig. 2 a bivariate normal distribution is fitted to a shower image. This means an intuitive first extension of the idea of fitting an ellipse, as the bivariate normal distribution is the best known and most popular elliptical distribution. When fitting a normal distribution instead of an ellipse additional information is implied about the shape of the shower inside and outside the ellipse. As we see in Fig. 2, the fit is not very good since the underlying shower image is obviously skewed. A skewed distribution thus might give a better fit. This is the case not only in this example, but for gamma ray images in general, which are known to be skewed. This information cannot be used by an elliptical-symmetric distribution like the bivariate normal. A skewed extension of the normal is the bivariate skew-normal distribution (Azzalini and Dalla Valle 1996). It has seven parameters instead of five of the normal distribution. With these two additional parameters the skewness of the distribution is described. A skew-normal distribution fitted to the same shower image as before can also be seen in Fig. 2.

We can use further information when fitting a distribution to a shower. We know, for example, that signal showers are always aligned to the center of the camera. This can be seen on the left side of Fig. 3, where a simulated signal event is displayed. To incorporate this information, we force a fitted distribution to be aligned to the center of the camera. For this we rotate the camera image so that the center of gravity of the shower lies on the x-axis of the camera as seen in Fig. 4. We separately fit

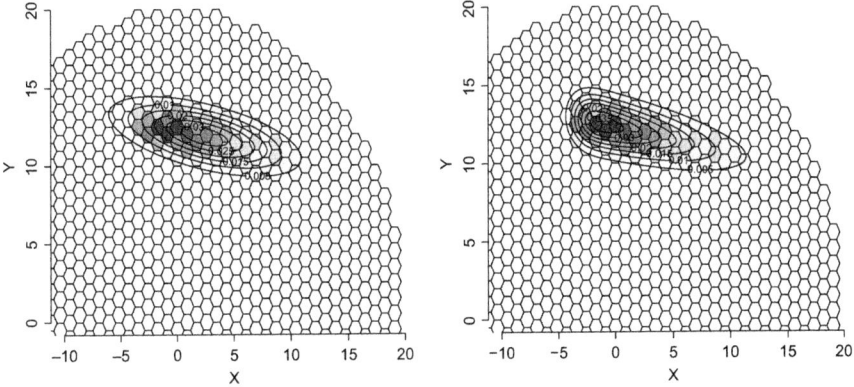

Fig. 2 Contour lines of a normal (*left*) and skew-normal (*right*) distribution fitted to a random shower image. Only a part of the camera is shown

two one-dimensional distributions to the *x*- and *y*-directions of the shower. For now we fit two univariate normal distributions. Jointly they represent a bivariate normal distribution with zero covariance, so that we get a bivariate normal with one direction being aligned to the center, when we rotate the image and the fitted distribution back.

To summarize we try three different approaches of fitting a bivariate distribution:

- Normal
- Skew-normal
- Aligned normal: Two univariate normals fitted to a rotated image

In the following we use the terms signal event and background event for shower images induced by a gamma-ray signal or hadronic background particle, respectively.

2 Model Fitting

One possibility for fitting a density to the shower images is to use maximum likelihood estimators (MLEs) for the parameters of the distribution. For observations from normal distributions this is easy, as MLEs are known and simple to calculate, namely the arithmetic mean and the empirical covariance matrix. In our case of discretized data on a crude hexagonal pixel structure, however, these estimators have to be adjusted for this discretization. A well-known possibility to adjust estimators for binned data are Sheppard's corrections (see, for example, Kendall and Stuart 1963). These are, however, not applicable in our case, as they require equidistant (and therefore rectangular) class limits. As no corrections for hexagonal binning is known to us we use intuitive, heuristical estimators. The principle underlying these is to use the center point of each bin weighted by the relative number of events in this bin. The estimators are weighted means, variances, and covariance given by:

$$\hat{\mu}_x = \sum_i m_{xi} \frac{M_i}{M},$$

$$\hat{\mu}_y = \sum_i m_{yi} \frac{M_i}{M},$$

$$\hat{\sigma}_x^2 = \sum_i \frac{M_i}{M} (m_{xi} - \hat{\mu}_x)^2,$$

$$\hat{\sigma}_y^2 = \sum_i \frac{M_i}{M} (m_{yi} - \hat{\mu}_y)^2,$$

$$\hat{\sigma}_{xy}^2 = \sum_i \frac{M_i}{M} (m_{xi} - \hat{\mu}_x) (m_{yi} - \hat{\mu}_y),$$

where M_i is the observed intensity in the ith pixel, M is the sum of all intensities in the shower, and $m_i = (m_{xi}, m_{yi})^T$ is the vector of the center coordinates of the ith pixel.

We use these estimators to fit the normal and bivariate normal distributions. For the skew-normal distribution the fit is not that simple, but can be obtained numerically making use of the R package sn (Azzalini 2013).

3 Distance Measures

We use the fitted distributions to construct new features for classification. Here, the idea is that we expect the fitted model to approximate gamma showers well, so that the fit should be better for signal events than for background events. The reason for this is that signal events are known to have a more regular shape than background events. This is illustrated in Fig. 3, where side views of sample signal and background showers can be seen. The shape seen in the side views directly translates into the camera image where we fit distributions.

We thus measure the distance between the fitted and the empirical distribution of each observed event. There are many distance measures between observed and fitted distributions, for example goodness of fit measures like the Chi-Squared (e.g., Greenwood and Nikulin 1996) or the Kolmogorov–Smirnov distance (e.g., Massey 1951), which can both be extended to the bivariate case. There are also distance measures for densities like the Kullback–Leibler divergence (Kullback and Leibler 1951) or the Hellinger distance (Nikulin 2001). We consider the Chi-Squared distance, the Kullback–Leibler divergence, and the Hellinger distance in

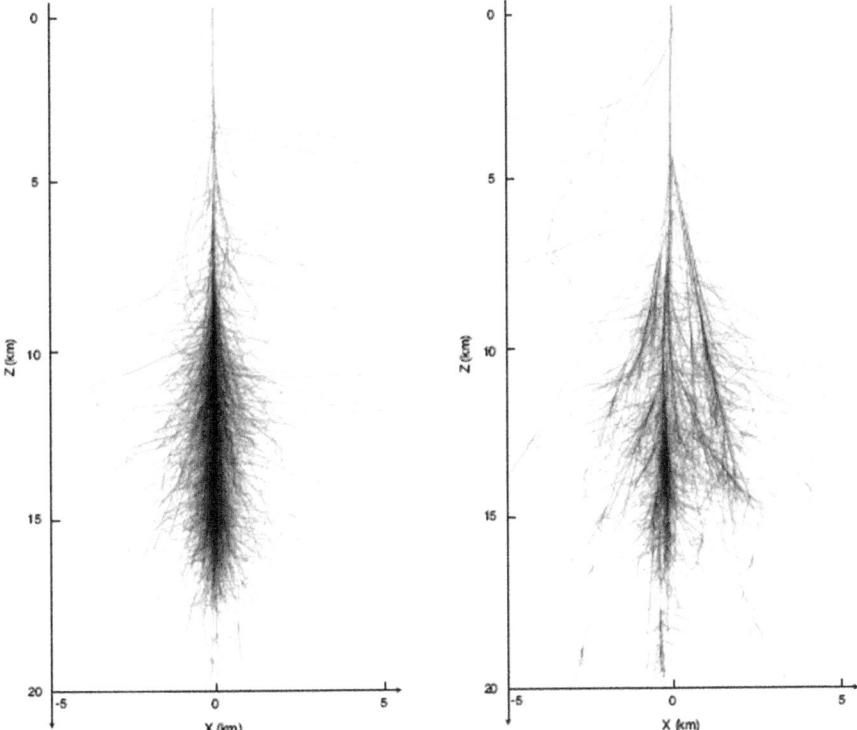

Fig. 3 Simulated side views of a signal (*left*) and background (*right*) event. It can be seen that the background event spreads much wider than the signal event. This irregular shape translates directly into the camera of Cherenkov telescopes

the following. We do not use the Kolmogorov–Smirnov distance because we had to adapt it to our binned data.

3.1 Chi-Squared Distance

An advantage of the Chi-Squared distance, underlying the corresponding test, in our situation is that it is based on binned data as given here. Hence, we can directly apply it to our data without the need of further modification.

The resulting Chi-Square distance is

$$Q_n = \sum_{i=1}^{m} \frac{(n_i - np_i)^2}{np_i},$$

where n_i is the observed intensity in pixel i, $n = \sum_{i=1}^{m} n_i$, p_i the probability that a gamma-induced photon arrives at pixel i, and m the number of pixels in the camera. Because we have discrete (binned) observed values we discretize the fitted distribution and directly calculate the p_i.

Large values of Q_n indicate that the observed event is likely to be a background event. That means we expect signal events to have a smaller Q_n than background events.

3.2 Kullback–Leibler Divergence

Another possibility is to use distance measures for densities. The best known of these is the Kullback–Leibler divergence (Kullback and Leibler 1951).

Let P and Q be two distributions and p and q their corresponding probability density functions (pdf). The asymmetric Kullback–Leibler Divergence between p and q is defined as

$$D_k(P, Q) = \sum_x p(x) \log\left(\frac{p(x)}{q(x)}\right)$$

for discrete distributions. The Kullback–Leibler divergence is not symmetric, so in general $D_k(P, Q) \neq D_k(Q, P)$.

In our case, we want to measure the distance between the empirical distribution with cdf F_n and pdf f_n and the fitted distribution with cdf F_0 and pdf f_0. As said in the previous paragraph, we discretize the fitted distribution leading to the discrete empirical density given by $\frac{n_i}{n}$ with n and n_i the same as above. The values of the discretized fitted distribution are given by $p_1, \ldots, p_m$, so that the Kullback–Leibler divergence reduces in our case to

$$D_k(F_0, F_n) = \sum_{i=1}^m p_i \log\left(\frac{np_i}{n_i}\right)$$

and accordingly

$$D_k(F_n, F_0) = \sum_{i=1}^m \frac{n_i}{n} \log\left(\frac{n_i}{np_i}\right).$$

3.3 Hellinger Distance

The Hellinger distance is another distance measure between densities. For discrete distributions P and Q with pdfs p and q it is in general given by

$$D_h(P, Q) = 1 - \sum_x \sqrt{p(x)q(x)}.$$

The Hellinger distance is symmetric and bounded by 0 and 1, i.e. $D_h(P, Q) = D_h(Q, P)$ and $0 \leq D_h(P, Q) \leq 1$.

With the same reasoning as above the Hellinger distance simplifies in our case to

$$D_h(F_n, F_0) = 1 - \sum_{i=1}^{m} \sqrt{p_i \frac{n_i}{n}}.$$

3.4 Additional Feature

Besides the distance measures and distributions described above, we use one further feature, which we construct from the rotated shower image described in Sect. 1. It is the quotient of the variances in y- and x-directions of the rotated image (Fig. 4). It is not a distance measure, but we use this feature because we expect that this variance ratio is usually less than 1 for signal events, and about 1 for background events.

4 Application to FACT Data

In this section, we apply the feature construction described above to simulated data from the FACT telescope. This allows us to assess the quality of the classification with and without the new features.

New features can be constructed by combining different distributions to be fitted to the shower images and distance measures. Because of limited space we do not consider all combinations of distributions and measures. We focus on combinations

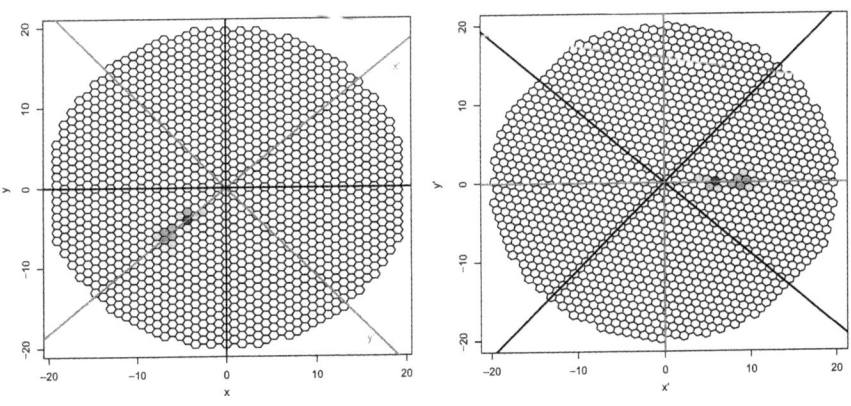

Fig. 4 Rotating an image to fit two univariate normal distributions to the new x- and y-directions

Table 1 Newly constructed features for classification

Variable	Explanation
normhell	$D_h(F_n, F_0)$ with F_0: cdf of bivariate normal
skewhell	$D_h(F_n, F_0)$ with F_0: cdf of bivariate skew-normal
alignnormhell	$D_h(F_n, F_0)$ with F_0: cdf of bivariate normal aligned to source
normkullobs	$D_k(F_n, F_0)$ with F_0: cdf of bivariate normal
normkulltheor	$D_k(F_0, F_n)$ with F_0: cdf of bivariate normal
normchi	Q_n of χ^2-test for bivariate normal
varquot	Ratio of variances in the rotated shower picture (VarY/VarX)

of the bivariate normal distribution with all distance measures mentioned in Sect. 3 and all combinations of the Hellinger distance with all distributions outlined in Sect. 2. An overview of the constructed features can be seen in Table 1.

The FACT data set we use consists of 13,185 events, 6,478 of which are simulated signal events and the other 6,707 are real events. As we did not have pure background data at the time of the writing of this paper, we took real events as background of which we know they also contain signal events. Since the signal-to-background ratio is about 1:1,000 in real data, we expect only about 6–7 signal events in our background data. This should not influence the outcome very much, but needs to be kept in mind.

The data set includes 14 standard Hillas-Parameters as well as the new features explained in Table 1.

From this data set we draw subsamples with different signal-to-background ratios. As stated above we expect a ratio of 1:1,000 in real data, so we would like to consider such unfavorable ratios. However, the sample size of about 13,000 events is too small to include such small ratios in our study. Instead we look at ratios 1:1, 1:10, 1:20, ..., 1:100. We randomly draw 10 subsamples per ratio, train a random forest with the remaining events and let it classify the drawn sample. The number of background events in each subsample drawn is 1,000 with a number of signal events corresponding to the signal-to-background ratio.

4.1 Results

A part of the results is displayed in Table 2, where the error rates of signal and background events can be seen, depending on the signal-to-background ratio. For the signal we see that the error is increasing, while it is decreasing for background.

Table 2 Error rates depending on the signal-to-background ratio

Gamma-error												
Ratio	1:1	1:10	1:20	1:30	1:40	1:50	1:60	1:70	1:80	1:90	1:100
Hillas	0.097	0.501	0.675	0.775	0.866	0.872	0.941	0.942	0.960	0.970	0.980
Hillas + new	0.090	0.444	0.503	0.708	0.812	0.824	0.888	0.886	0.911	0.936	0.948
Hadron-error												
Ratio	1:1	1:10	1:20	1:30	1:40	1:50	1:60	1:70	1:80	1:90	1:100
Hillas	0.082	0.007	0.002	0.001	<0.001	<0.001	<0.001	<0.001	<0.001	<0.001	<0.001
Hillas + new	0.054	0.003	0.001	<0.001	<0.001	<0.001	<0.001	<0.001	<0.001	<0.001	<0.001

The signal error is the relative frequency of signal events classified as background. The background error is the relative frequency of background events classified as signal. The error of background cannot be improved, but there is a relevant gain for signal

In background the error decreases very fast, so that it approaches 0 very quickly, and cannot be improved much by other features. So what is desirable is to improve the signal error, while maintaining the good background error. We see in Table 2 that we accomplish this. Although with and without the new features the error increases as a function of the ratio, the error is smaller when we use the additional features.

We also look at the individual features to find out how important each of them is for the classification. A Random Forests use the Mean Gini Decrease as an importance measure, which is provided in Table 3 for different ratios. We observe that one of the new features, *varquot*, is overall very important, while most other new features have a small importance. Note that with decreasing signal-to-background ratio the importance of all the new features gets larger. For a ratio of 1:100 two of the new features are among the five most important ones, while the others also have a rather high importance.

Table 3 The Hillas-parameters and the new features and their importance in a random forest measured by the mean Gini decrease for different signal-to-background ratios

Variable	Ratio 1:1	1:50	1:100
Hillas-parameters			
Size	181.6	15.37	7.67
Width	262.1	10.73	5.64
Length	172.7	9.29	4.97
Area	154.4	9.11	4.78
Delta	82.1	8.43	4.69
Alpha	674.7	14.97	5.83
Conc1	166.0	8.74	4.96
Conc	181.8	8.85	5.06
CogX	406.5	8.57	3.63
CogY	297.9	7.74	3.40
NumberIslands	5.4	0.42	0.21
NumberShowerPixel	62.3	5.13	2.90
Leakage1	111.1	1.29	0.58
Leakage2	287.0	2.79	1.20
New features			
alignnormhell	312.1	12.38	6.36
normhell	98.9	8.05	4.82
skewhell	104.7	8.71	4.69
normkullobs	105.3	8.29	4.67
normkulltheor	91.2	8.25	4.60
normchi	127.4	10.53	5.17
varquot	1,113.2	27.79	13.21

5   Conclusions and Outlook

We have introduced features for classification in the FACT experiment, which are based on fitting a bivariate distribution to an image and measuring the distance of the fitted distribution to the observed one. This led to better classification results than the old features alone. As a side-product of our modeling approach, we obtained a new feature, *varquot*, which incorporates additional physical information about signal showers and improves the classification further. However, the results presented here are preliminary and have to be further validated. Smaller signal-to-background ratios and more combinations of measures and distributions have to be investigated. Other measures and distributions could also be investigated as well as classification algorithms different from random forests. A feature selection may make sense, because the newly constructed features may be highly correlated and thus redundant. Still, the results presented look rather promising.

Acknowledgements Part of the work on this paper has been supported by Deutsche Forschungs-gemeinschaft (DFG) within the Collaborative Research Center SFB 876 "Providing Information by Resource-Constrained Analysis," project C3. We gratefully thank the FACT collaboration for supplying us with the test data sets and the ITMC at TU Dortmund University for providing computer resources on LiDO. We are grateful to the reviewers for their helpful comments.

References

Azzalini, A. (2013). R package 'sn': The skew-normal and skew-t distributions (version 0.4–18). http://www.azzalini.stat.unipd.it/SN.

Azzalini, A., & Dalla Valle, A. (1996). The multivariate skew-normal distribution. *Biometrika*, *83*, 4.

Greenwood, P. E. & Nikulin, M. S. (1996). *A guide to chi-squared testing*. Wiley Series in Probability and Statistics. New York: Wiley.

Hillas, A. M. (1985). Cherenkov light images of EAS produced by primary gamma. In *Proceedings of the 19th International Cosmic Ray Conference ICRC* (Vol. 3, p. 445), San Diego.

Kendall, M. G., & Stuart, A. (1963). *The advanced theory of statistics* (Vol. 1). London: Griffin.

Kullback, S., & Leibler, R. A. (1951). On information and sufficiency. *Annals of Mathematical Statistics*, *22*, 79–86.

Massey, F. J. (1951). The Kolmogorov-Smirnov test for goodness of fit. *Journal of the American Statistical Association*, *46*, 68–78.

Nikulin, M. S. (2001). Hellinger distance. In M. Hazewinkel (Ed.), *Encycolpedia of mathematics*. Berlin: Springer.

Part IX
Data Analysis in Administration and Spatial Planning

Hough Transform and Kirchhoff Migration for Supervised GPR Data Analysis

Andre Busche, Daniel Seyfried, and Lars Schmidt-Thieme

Abstract Ground penetrating radar (GPR) is a widely used technology for detecting buried objects in the subsoil. Radar measurements are usually depicted as radargram images, including distorted hyperbola-like shapes representing pipes running non-parallel to the measurement trace. Also because of the heterogeneity of subsoil, human experts are usually analysing radargrams only in a semi-automatic way by adjusting parameters of the detection models (exposed by the software used) to get best detection results.

To gain a set of approximate hyperbola apex positions, unsupervised methods such as the Hough transform (HT) or Kirchhoff Migration are often used. By having high-quality, large-scale real-world measurement data collected on a specialized test site at hand, we both (a) analyse differences and similarities of the HT and Kirchhoff Migration quantitatively and analytically with respect to different preprocessing techniques, and (b) embed results from either technique into a supervised framework. The primary contribution of this paper is the conduction of an exhaustive experiment, not only showing their equivalence, but also showing that their application for the automated analysis of GPR data, unlike it is currently assumes, does not improve the detection performance significantly.

1 Introduction

Ground penetrating radar (GPR) is used to investigate the shallow surface, e.g., to find buried landmines or pipes and cables underneath (road) surfaces (Daniels 2004). Our current data is measured using an on-site vehicle and illuminates

A. Busche (✉) • L. Schmidt-Thieme
Information Systems and Machine Learning Lab, University of Hildesheim, Hildesheim, Germany
e-mail: busche@ismll.uni-hildesheim.de; schmidt-thieme@ismll.uni-hildesheim.de

D. Seyfried
Technische Universität Braunschweig, Institut für Hochfrequenztechnik, Braunschweig, Germany
e-mail: daniel.seyfried@ihf.tu-bs.de

© Springer-Verlag Berlin Heidelberg 2015
B. Lausen et al. (eds.), *Data Science, Learning by Latent Structures, and Knowledge Discovery*, Studies in Classification, Data Analysis, and Knowledge Organization, DOI 10.1007/978-3-662-44983-7_43

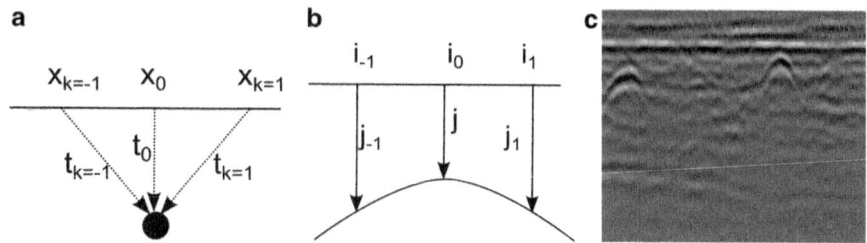

Fig. 1 Geometric approach for the introduction of the GPR analysis problem: (**a**) Wave travel times t_k differ at different positions x_k when hitting the same object. (**b**) Depicting wave travel times in radargram coordinates (i, j) results in reflection hyperbolas. (**c**) real-world images also contain clutter

structures in subsoil down to about 3–4 m in depth. We are aiming at assisting the GPR analysis process by means of probabilistic methods, while a special focus is put on the identification of pipes and cables of various types (e.g., PE, metal, stoneware) which are represented as hyperbola-like structures on measured radargram images (see Fig. 1).

One ultimate goal of GPR data analysis is the derivation of supply maps, that is, maps of buried objects of a certain kind. The creation of those maps is required, e.g., as municipalities, according to one of our project partners and at least in Germany, seldom have a single map of their buried structures. Instead, maps and plans of buried pipes and cables are cluttered and only available in a distributed manner, and can only be partially gathered and combined when requested by all parties owning buried objects, such as water supply companies, power supply companies, and telecommunication companies. In any case, existing maps may be inaccurate and not recent, causing additional problems when highly accurate maps are required.

This paper represents a first step towards (semi-)automated creation of such supply maps, by means of developing supervised Machine Learning methods for an automated detection of such buried objects. The overall pipe detection process can be split up into two distinct aspects: (a) the detection of individual objects in radargram images, and (b) the creating of supply maps out of individually detected object locations. We focus on the first aspect here.

For detecting individual hyperbolas in radargram images, we investigate both the Hough Transform (HT; Simi et al. 2008) and the Kirchhoff Migration (Daniels 2004). Though we will be using GPR as an application scenario for our analysis, any other application domain involving the detection of hyperbolas may be used as well. Our main reason for using GPR is the fact that the Kirchhoff Migration stems from the Radar Analysis Community (Hagedoorn 1954), whereas the Hough Transform is a general technique stemming from the Image Analysis community (Hough 1962).

In light of this application setting, this paper makes the following contributions:

- We show the equivalence of the Hough Transform and Kirchhoff Migration for the analysis of GPR data both analytically and empirically.
- We show in an exhaustive experiment empirically that their embedding in a state-of-the-art supervised framework does not make any statistically significant difference.

2 Data Collection Process and Geometric Relations

The collection of individual radargram images is done as follows: A specialized measurement vehicle drives at a constant speed along the x-axis (cf. Fig. 1a) and measures a radargram image. This image visualizes reflected energies/intensities at discrete time points (j-axis in Fig. 1b). While in theory, one is able to *induce* the appearance of a radargram image out of known subsoil structures (e.g., by means of a numerical simulation software), the reverse action—the *deduction of subsoil structures* out of radargram images—is a highly non-trivial task in real-world situations for a variety of reasons, including: (a) hyperbola reflections get distorted by supplementary reflections from horizontal layer breaks, (b) the signals' energy decays with increasing depth, resulting in lower (visual) contrasts, (c) the pipes' type and fillings causes multiple vertical reflections.

This paper first presents the Hough Transform in Sect. 4, and thereafter the Kirchhoff Migration in Sect. 5 as two unsupervised methods. We will be integrating both into a supervised framework by means of a Convolutional Neural Network by treating the hyperbola estimation task as a patch-based classification problem. This requires us of having an a priori labeled dataset containing patches (fractions of a radargram image) being labeled positive (patches containing hyperbola shapes) and negative (clutter; background noise). We refer to our work in Busche et al. (2013) on how to obtain such a dataset.

According to the analysis of Kempen and Sahli (1999), the reflection hyperbola may be parameterized by three parameters according to

$$h(x) = \sqrt{\alpha^2 + \beta(x - x_0)^2} \qquad (1)$$

for which x_0 denotes the horizontal position, α denotes the apex depth index position, and β describes the curvature of the reflection hyperbola. We will be analysing both, the Hough Transform and the Kirchhoff Transform, for their relation to this equation.

From the physical fact that reflection hyperbolas are only visible to a certain extent to the left and right branch of the apex due to antenna radiation pattern, we will be increasing the overall performance of both the Hough Transform and the

Kirchhoff Migration by using a triangular weight

$$w(i\,;x_0,e) := \begin{cases} 1 - \frac{|i-x_0|}{e}, & \text{if } |i - x_0| < e \\ 0, & \text{else} \end{cases} \qquad (2)$$

for which $i - x_0$ denotes the horizontal offset from an (assumed) hyperbola apex position x_0. We use e to denote the "width" of the triangular weighting. For our current data at hand, setting $e = 50$ pixels is a good choice.

3 Related Work

The ultimate goal of GPR analysis is the derivation of complete and accurate supply maps based on usually just a small set of radargrams (Simi et al. 2008 Chen and Cohn 2011). Before radargram images are fed into an automated algorithm, they are usually preprocessed. This process is usually visually (Busche et al. 2012) or methodologically (Chen and Cohn 2010, 2011) driven.

One out of three different approaches for hyperbola detection in radargram images can be distinguished: (a) Estimation from sparse data (Chen and Cohn 2010), (b) the usage of Hough Transform, see Simi et al. (2008) and Pochanin (2009) for an application in GPR and Forsyth and Ponce (2004) for an introduction for image analysis, (c) the usage of migration techniques, see Scheers (2001) for an overview and related work. If embedded into supervised machine-learning methods such as Neural Networks for patch-based classification (AL-Nuaimy et al. 2000, Sermanet et al. 2013), training data needs to be carefully collected beforehand, e.g., as done in Busche et al. (2013).

4 Hough Transform for Hyperbola Detection

The general idea behind the Hough Transform is simply stated as inverting any functional dependency against the pixel coordinates within an image. For every position (i, j) within a (radargram) image, the parameters of a target function crossing that point are calculated and accumulated into a so-called accumulator array h, also being called the model space. The accumulator array is an n-dimensional array, n being the number of parameters of the function, while each element therein denotes a certain parameter combination. Depending on how the accumulator array is designed, its size may become excessive due to the Hough Transform having a memory complexity of $O(\Pi_i^n p_i)$, with p_i being the amount of parameter bins for parameter i.

Input : A radargram image r^k, x_{min}, x_{max}, β_{min}, β_{max}, β_{inc}
Output: An accumulator array h

```
forall the i ∈ I^k do                // iterate over all horizontal positions
    forall the j ∈ J do              // iterate over all vertical positions
        forall the x_i ∈ [x_min, x_max] do              // assumed apexes
            forall the β ∈ [β_min, β_max]_{β_inc} do    // discretized curvatures
                α² = y² − β ∘(x − x_0)²;
                if 1 < α/β < 10 then                    // restrict curvature
                |   increment(α,β,x_i);
                end
            end
        end
    end
end
```

Fig. 2 A Hough transform implementation for detecting hyperbolic shapes using the three parametric hyperbola equation $h(x) = \sqrt{\alpha + \beta(x - x_0)^2}$

Performing a Hough Transform for detecting hyperbolas requires the inversion of the three parametric hyperbola equation (1), and either α, β, or x_0 needs to be calculated, while the other parameters are iterated on a fixed grid. Figure 2 gives a pseudocode of the Hough Transform if the α parameter is calculated, and different values for the curvature β are iterated. Calculating the α value actually represents the *calculation of the depth* of a reflection hyperbola apex.

Instead of inverting the three-parametric hyperbola equation for calculating the α values, one may also iterate over the α values on a fixed interval at fixed increments, to calculate the β values. Doing so changes both, (a) the loop over the β to be $a_i \in [a_{min}, a_{max}]_{a_{inc}}$ and (b) the calculus of the $\beta = \frac{y^2 - \alpha^2}{(x - x_0)^2}$, while all other parts in the algorithm stay the same.

The most significant/visible hyperbola reflection in an image is determined by searching the parameter combination having the most "votes" within the accumulator array:

$$(\alpha, \beta, x_0) = \arg\max_{\alpha, \beta, x_0} h. \tag{3}$$

For the detection of multiple hyperbolas within images, one needs to apply one or multiple heuristics to avoid collocations of hyperbola apex positions. One example is given by using the "detect-and-remove" heuristic. Alternatively, we could also use a 95 % quantile of the accumulator array to obtain suitable hyperbola apex positions or embed the Hough Transform into a probabilistic framework, as done in Barinova et al. (2010).

5 Kirchhoff Migration

The Kirchhoff Migration corresponds to a baseline technique which is widely used in the GPR community as it (a) is easily implementable, as we will show below, and (b) generally results in good estimation results. It is usually introduced by considering an "exploding point reflector", for which the explosion propagates as a semicircle in subsoil. As time progresses, the half-circle increases. If the half circle anywhere hits the object to be detected (in a binary world), the corresponding time index $t_{k=-1}$ is taken as a depth index j_{-1}, see also the dissection in Fig. 1. For each location at each point in time, a certain signal strength can be measured. Claerbout (2009) claims of having "best tutorial Kirchhoff migration-modeling program [he] could devise", which is also given in Fig. 3 below using our terminology.

The Kirchhoff Migration assumes a homogeneous medium and thus a constant propagation velocity of the wave in subsoil. Such a velocity has to be chosen by an engineer and is usually based on visual investigations.

Assume that any arbitrary object is located at (x_0, t_0), as depicted in Fig. 1a, which is inspired by Fig. 7.32 in Daniels (2004). If the measurement vehicle is exactly located at x_0, the travel time of the wave until it hits the object exactly equals $t_0 = j$. Let the movement of the measurement vehicle as it moves forward or backward be encoded by relative positions $x_{k=1}$, resp. $x_{k=-1}$. The wave travel times until the wave hits the object may now be calculated using the plain Pythagorean

Input : A radargram image r^k, a model mdl, a time discretization dt, a constant velocity v, a boolean fwd

```
forall the i₁ ∈ 𝓘ᵏ do                    // fix an apex location
    forall the i₂ ∈ 𝓘ᵏ do                // assume a horizontal position
        forall the j ∈ [0, 80]₅₁₂⁸⁰ do    // assume a depth index
            hₛ = (i₁−i₂)/v;                // horizontal offset
            t = √(hₛ² + (0.5 ∘ j)²);       // wave travel times - Pythagorean
            distance
            t* = 1.5 + t/dt;              // discrete depth; value is casted
            if fwd then
                /* copies reflection intensity to migrated image
                */
                mdl[i₁][t*] += rᵢ₂,ⱼ;
            else
                rᵢ₂,ⱼ += mdl[i₁][t*];      // reconstructs radargram image
            end
        end
    end
end
```

Fig. 3 Kirchhoff migration tutorial code, conceptually taken from Claerbout (2009)

distance, as

$$t_k = \sqrt{\left(\frac{i - x_k}{c_{\text{medium}}}\right)^2 + t_0^2} \tag{4}$$

while the denominator at the first term under the square root scales for the propagation speed of the wave in the medium. Equation (4) can be reformulated by its relationships according to the actual measurement locations (i, j) in radargram images. In this scenario, we have reflection intensities at (i, j) in the radargram image, but the apex position $(x_0, t_0 = j)$ is to be determined.

The pseudocode in Fig. 3 adapts the formalism of Claerbout (2009) and shows the relationship between continuous real-world coordinates (x_0, t_0) and pixel coordinates (i, j) by taking into consideration the mapping of continuous real-world wave travel times into discrete pixel coordinates. In the pseudocode, the geometric relationships as shown in Fig. 1 are directly embedded into an algorithm for migrating the energy of a hyperbola.

The pseudocode in Fig. 3 iterates over pairs of (horizontal) measurement locations i_1 and i_2 and over the individual one-way wave travel times j between 0 and 80 ns (recording time) at $\frac{80}{512}$ increments. h_s denotes the horizontal offset $(x_0 - x_i)$ by scaling the value according to the wave propagation velocity. The wave travel time calculation exactly corresponds to Eq. (4). Having calculated the "diagonal" travel time t_k, all geometrically relevant factors are computed. We may either copy the reflection intensity from the radargram image to its transformation (*fwd* = true). Vice versa, setting *fwd* = false allows for a transformation from a Kirchhoff migrated image back into a radargram image, and thus allows a reconstruction of the hyperbolic reflection patterns from a migrated image. In the algorithm, "mdl" corresponds to the model which is created, being the migrated image of a radargram image, showing the energy of the hyperbola focused in its apex.

The derivation of hyperbola apex positions is similar to the derivation of apex positions when performing a Hough Transform: given that the assumed propagation velocity of the subsoil is correct, the Kirchhoff Migration focuses the hyperbolic reflection intensity in the apex of the migrated image. To this end, one may again apply a 95 % quantile to get hyperbola apex positions.

6 Comparison to the Hough Transform

When comparing both pseudocodes from Figs. 2 and 3, their equivalences is clearly visible, if we (a) assume the curvature β within the Hough Transform implementation is given priori, and (b) the mapping from the continuous values to discrete ones is removed. The assumption of a given curvature corresponds to the assumed, fixed and constant propagation velocity being a prerequisite of the Kirchhoff Migration. The Kirchhoff Migration tutorial code furthermore adds the aspect of the two-way conversion, either from the raw radargram image to the migrated one, and vice versa back. This, however, can also be integrated into the Hough Transform.

Having identified its equivalence, we now need to discuss the consequences and possibilities of the fixed β parameter for any generic Hough Transform implementation. As it has also been analysed, e.g., by Pochanin (2009), the β parameter has a focusing aspect within the Hough space, which can visually be validated by considering the fact that focusing the energy of a hyperbola corresponds to a summation of its individual reflection intensities along the correct curvature, while this value is maximal if the curvature is correct, and decreasing if it is either too high or too low. If we ran a Hough Transform while β is iterated in a small interval around the actual corresponding propagation velocity, we would be able to take into consideration local variances in subsoil velocities.

Both algorithms are inherently slow for their multiple loops over the data, but can be accelerated and improved in various ways (cf. Claerbout 2009), e.g., by (a) reordering terms and (b) early stopping, if the calculation of t^* exceeds defined coordinates. An improvement can be found by considering the triangular weighting as described when introducing the Hough Transform.

7 A Supervised Framework for Hough Transform and Kirchhoff Migration

After having shown the equivalence between the Hough Transform and the Kirchhoff Migration, we embed these unsupervised GPR analysis techniques into a supervised, patch-based classification framework. Our aim is to estimate the location of reflection hyperbola apex positions based on a binary classification setting on extracted patches from radargram images. Patches of size 32×32 pixels are extracted from each radargram image and classified according to whether or not they contain a hyperbolic reflection pattern.

It is our assumption that migration techniques improve the detectability of hyperbolas in the setting of supervised Machine Learning, as we are integrating domain knowledge to the analysis pipeline. For running the experiments, we will be using a state-of-the art-classifier for patch-based classification, a Convolutional Neural Network implemented in the eblearn library (Sermanet et al. 2013). This has already shown superior performance on a variety of application scenarios in the last years.

Besides other performance scores, the eblearn library reports a class-normalized accuracy score, which weights the accuracy according to the class imbalance. We will be using this score to report experimental results.

8   Experiments

Our real-world dataset contains 13 and 6 radargram images of different subsoil types $t = 1$ and $t = 3$ and is split each such that a whole radargram image is contained in the validation and the test set, resp. Labels and known subsoil velocities were obtained according to our work in Busche et al. (2013). The following preprocessed dataset variants were created:

- The *accurate* dataset $\mathcal{D}^{\mathrm{accu}}$ contains unprocessed radargram images.
- The *Kirchhoff* datasets are preprocessed according to the known propagation velocities. We are using relative propagation velocity variants (multipliers) 1.0, 0.9, 1.1, resp. $\mathcal{D}^{k=1.0}$, $\mathcal{D}^{k=0.9}$, $\mathcal{D}^{k=1.1}$ to study the effect of wrong manual estimations. These values are considered to be still a good estimate.

Positive patches denoting hyperbolic reflection patterns were extracted according to annotated hyperbola apex position, as well as its direct neighbouring patches. We discarded potential patches in a $70\% \cdot 32\,px = 23\,px$ radius for not getting obfuscated class boundaries. Negating patches denoting background noise were randomly sampled from all remaining image positions (i, j) at a 2% rate. The final class imbalance was about $1 : 8$. The experiments reported here were obtained on a 492 CPU compute cluster and took about 60 days to compute.

Hyperparameters ranges were initially determined on a smaller data subset and afterwards used to obtain the best model on the following grid for the full dataset as follows: The learn rate was tested for $\eta \in \{5, 1, 0.5, 0.1, 0.01\} \cdot 10^4$. Regularization parameters for both $l1$ and $l2$ regularization were tested for values $\{0.01, 0.001, 0.0001, 0.0\}$. An undocumented "reg"[ularization] configuration parameter was both tested for a values of 0.0 and 0.001. A learn rate decay of 0.00005 after the first iteration was tested alternatively as well.

Table 1 shows experimental results for class-normalized accuracy scores along with 95% confidence bounds for our experiments on two subsoil types on the datasets as presented above. As it can be seen, though the performance for subsoil type $t = 1$ on the migrated datasets is better on average, no statistical difference could be identified. For subsoil type $t = 3$, the unprocessed dataset is even better, compared to all migrated dataset variants.

Table 1 Embedding the Hough transform, also called Kirchhoff migration with known subsoil propagation velocity, into a supervised machine learning framework shows no significant improvement when the propagation velocity matches exactly ($\mathcal{D}^{k=1.0}$), or either over-, or underestimates ($\mathcal{D}^{k=0.9}$ and $\mathcal{D}^{k=1.1}$, resp.), nor shows an improvement over the unmigrated dataset $\mathcal{D}^{\mathrm{accu}}$

Subsoil type	$\mathcal{D}^{\mathrm{accu}}$	$\mathcal{D}^{k=0.9}$,	$\mathcal{D}^{k=1.1}$	$\mathcal{D}^{k=1.0}$
$t = 1$	92.91 (2.57)	94.34 (2.59)	94.12 (2.19)	94.04 (3.00)
$t = 3$	86.41 (2.97)	82.83 (5.95)	85.19 (6.38)	81.23 (6.40)

95% confidence bounds in brackets

9 Conclusion and Future work

In this paper, we showed the equivalence of the Hough Transform and the Kirchhoff Migration for detecting hyperbola apex positions in image data. Embedding those techniques into a supervised Machine Learning scenario by trying to detect hyperbola apex positions by means of patch-based classification techniques could not show a significant improvement when applying either technique as a preprocessing step beforehand.

For validating our experiments, we will be running an even larger experiment by integrating more background/noise patches. We hope that the shown effect is an artefact of our data preparation phase.

Acknowledgements This work is co-funded by the European Regional Development Fund project AcoGPR (http://acogpr.ismll.de) under the grant agreement no. WA3 80122470.

References

Al-Nuaimy, W., Huang, Y., Nakhkash, M., Fang, M., Nguyen, V., & Eriksen, A. (2000). Automatic detection of buried utilities and solid objects with gpr using neural networks and pattern recognition. *Journal of Applied Geophysics, 43*, 157–165.

Barinova, O., Lempitsky, V., & Kohli, P. (2010). On the detection of multiple object instances using Hough transforms. In *IEEE Conference on Computer Vision and Pattern Recognition*.

Busche, A., Janning, R., Horváth, T., & Schmidt-Thieme, L. (2012). A unifying framework for GPR image reconstruction. In *Proceedings of the 36th Annual Conference of the Gesellschaft für Klassifikation (GfKl)*.

Busche, A., Janning, R., & Schmidt-Thieme, L. (2013). Analysing the potential impact of labeling disagreements for engineering sensor data. In *Workshop on Knowledge Discovery, Data Mining and Machine Learning (KDML)*.

Chen, H., & Cohn, A. G. (2010). Probabilistic robust hyperbola mixture model for interpreting ground penetrating radar data. In *IEEE World Congress on Computational Intelligence* (pp. 3367–3374).

Chen, H., & Cohn, A. G. (2011). Buried utility pipeline mapping based on multiple spatial data sources: a bayesian data fusion approach. In *IJCAI* (pp. 2411–2417).

Claerbout, J. (2009). Zero-offset migration/tutorial kirchhoff code. http://www.reproducibility.org/RSF/book/bei/krch/paper_html/node8.html. Accessed 05.09.2013.

Daniels, D. (2004). *Ground penetrating radar* (2nd ed.). London: The Institution of Engineering and Technology.

Forsyth, D.A. & Ponce, J. (2004). Computer Vision: A Modern Approach. Prentice Hall Series in Arti cial Intelligence, (1st edn.).

Hagedoorn, J. G. (1954). A process of seismic reflection interpretation*. *Geophysical Prospecting, 2*, 85–127.

Hough, P. (1962). Method and means for recognizing complex patterns. US Patent No. 3069654.

Kempen, L. V., & Sahli, H. (1999). Ground penetrating radar data processing: a selective survey of the state of the art literature. Technical Report tr-0060, IRIS VUB-ETRO Department.

Pochanin, G. (2009). Some advances in UWB GPR. In *Unexploded Ordnance Detection and Mitigation NATO Science for Peace and Security Series B: Physics and Biophysics* (pp. 223–233).

Scheers, B. (2001). Ultra-wideband ground penetrating radar, with application to the detection of anti personnel landmines. Ph.D. thesis, Université Catholique de Loucain Laboratoire d'Hyperfréquences.

Sermanet, P., Kavukcuoglu, K., & Lecun, Y. (2013). Eblearn: open-source energy-based learning in C++. http://www.eblearn.cs.nyu.edu:21991/. Accessed 30.05.2013.

Simi, A., Bracciali, S., & Manacorda, G. (2008). Hough transform based automatic pipe detection for array GPR: algorithm development and on-site tests. In *Radar Conference, 2008, RADAR '08, IEEE* (pp. 1–6).

Application of Hedonic Methods in Modelling Real Estate Prices in Poland

Anna Król

Abstract This paper concentrates on empirical and methodological issues of the application of econometric methods to modelling real estate market. The presented hedonic analysis of apartments' prices in Wrocław is based on the dataset consisting of over ten thousands offers from the secondary real estate market. The models estimated as the result of the research allow for pricing the apartments, as well as its characteristics.

The foundations of hedonic methods are formed by the so-called hedonic hypothesis which states that heterogeneous commodities are characterized by a set of attributes relevant both from the point of view of the customer and the producer. As a consequence, the price of a commodity is determined as an aggregate of values estimated for each significant characteristic of this commodity. The hedonic model allows to price the commodity as well as to identify and estimate the prices of respective attributes, including the prices which are not directly observable on the market. The latter is particularly useful for the real estate market as it enables pricing location-related, neighbourhood-related and structure-related characteristics of housing whose values cannot be obtained otherwise.

1 Introduction

This paper presents the results of hedonic analysis of apartments' prices in Wrocław, with special attention paid to the problems of data gathering and model specification. The primary research objective was to investigate the relationship between the price of an apartment and its significant characteristics, and to create an efficient tool for pricing the apartments on Wrocław real estate market.

Hedonic methods were invented in the early twentieth century (first attempts at hedonic analyses were made by Haas 1922, Wallace 1926, Waugh 1929, Court 1939), and revisited at the end of the century when the rapid progression of technology started (cf. Boskin et al. 1996; Berndt et al. 1995). The sudden

A. Król (✉)
Wrocław University of Economics, Komandorska 118/120, Wrocław, Poland
e-mail: anna.krol@ue.wroc.pl

© Springer-Verlag Berlin Heidelberg 2015
B. Lausen et al. (eds.), *Data Science, Learning by Latent Structures, and Knowledge Discovery*, Studies in Classification, Data Analysis, and Knowledge Organization, DOI 10.1007/978-3-662-44983-7_44

501

development of production engineering and the emergence of completely new, advanced technologies caused acceleration in changes of the quality of goods. This shortening of products' life cycles has led to the situation in which the goods present on the market were not comparable with the goods whose prices were observed in the past. As a result, standard methods of price changes measurement have yielded biased results, usually an overestimation of price growth rates.

The problem of quantification of the so-called true price change affects two classes of heterogeneous goods:

- goods which undergo very rapid technological development (e.g. consumer electronics, household appliances, cars, IT and ICT devices),
- goods which are strictly heterogeneous. For heterogeneous goods it is highly unlikely to encounter two identical specimens whose prices could be compared without the risk of quality-related biases (e.g. apartments, houses, land parcels).

For the first class of goods the difficulties in comparing prices from periods t and $t + 1$ arise either from a significant change in characteristics of a commodity (e.g. when a given computer is endowed with the hard drive of a considerably larger capacity), or from the final withdrawal of a certain good from the market (e.g. withdrawal of obsolete, dangerous, or otherwise needless good whose price has been measured in the period t), or from development of completely new technology (e.g. replacement of CD drives by DVD drives). In the class of strictly heterogeneous goods the price of a good observed in the period t may only be compared with the price in period $t + 1$ of a "similar" good. Because of the above-mentioned fact the problem of quality difference is immanent feature of this measurement process.

Other application areas of hedonic methods include (cf. Dziechciarz 2004; Triplett 1986) facilitating the development of a pricing strategy, pricing the commodities, as well as estimating the prices of respective attributes, including the prices which are not directly observable on the market. The latter is particularly useful for the real estate market (cf. Sheppard 1999; Can 1992), as it enables pricing location-related, neighbourhood-related and structure-related characteristics of housing whose values cannot be obtained otherwise (e.g. the distance from city centre to the apartment, the quality of air in the vicinity of the housing, the age of the building).

2 Hedonic Models

2.1 Basic Concepts

The foundations of hedonic methods are formed by the so-called hedonic hypothesis which states that heterogeneous commodities are characterized by a set of relatively homogeneous attributes (characteristics) relevant both from the point of view of the customer and the producer (cf. Brachinger 2002; Triplett 2006). For a given good described by m attributes this set may be formally written as a vector X:

$$X^T = [x_1 \ x_2 \ \ldots \ x_m] . \tag{1}$$

Moreover it is assumed that there exists a relationship between the price of the good and its significant characteristics which may be described by a certain function f, called the hedonic function. As a consequence, the price of a commodity is determined as an aggregate of values estimated for each significant characteristic of this commodity. A commonly used method for obtaining the hedonic function is the application of regression model described in the following general notation:

$$P = f (X; \beta; \epsilon) , \tag{2}$$

where P is the commodity price, β a vector of parameters, and ϵ the error term.

The estimate of the vector of parameters β, obtained by estimation of a correctly specified hedonic regression model using a dataset, allows to calculate the theoretical price of a given good with a specified set of significant characteristics. This property of hedonic models is crucial for their application in revealing the "true" price change because of a change in characteristics. In general, implementation of hedonic methods to make adjustments for quality changes consists in incorporating results obtained from hedonic regressions into classic Laspeyres, Paasche and Fischer price index formulas.

2.2 Selected Problems in Hedonic Modelling

Hedonic regression models face all the classical problems of econometric modelling of cross-sectional data, such as heteroskedasticity of the error term, collinearity of the independent variables or spatial autocorrelation (cf. Greene 2011; Wooldridge 2010). In this section two crucial issues for hedonic modelling will be briefly addressed: model specification and data collection.

2.2.1 Specification Problems

The specification problems in hedonic modeling comprise the following three aspects:

1. specifying correctly the class of good,
2. correctly determining the set of characteristics,
3. getting the right functional form.

In the theory of hedonic model the class of good signifies all the variants of the given commodity which may be correctly described by a common hedonic function. In practice it is not always easy to correctly define the class of the commodity. For example, it is very probable that the group of commodities specified as "apartments in Wrocław" in its majority is comprised by flats of size 50–100 m^2 and price 150–500 thousands of PLN. However, it is almost certain that in that group much bigger (e.g. 300 m^2), and much more luxurious (e.g. 1,500 thousands of PLN) units occur as well. Such heterogeneity of goods might be difficult to capture by the same hedonic function, causing the indispensability of a compromise between the accuracy of the estimated hedonic model and the generality of obtained results.

The number and type of characteristics which comprise the vector X depend on the nature of a given good and its technical properties. The determination of significant attributes may be based on technological information concerning production processes, marketing data on the needs and preferences of the consumers, as well as the use of statistical information obtained from the dataset. Frequently, however, the choice of independent variables is limited by the availability of data. Therefore, the possibility biases caused by omitted variables must be taken into account.

The theory of hedonic methods gives little suggestions as to how the relationship between the price of a good and its characteristics should be specified. Many studies present either the a priori assumption of the functional form of hedonic regression, or the approach of using the functional form which fits the data best (for the survey of hedonic empirical research on real estate market see, e.g., Herath and Maier (2010) . The commonly used functional forms of hedonic regressions are linear ($P = \beta_0 + \sum_{j=1}^{m} \beta_j X_j + \epsilon$), exponential (log-lin) ($\ln P = \beta_0 + \sum_{j=1}^{m} \beta_j X_j + \epsilon$), double-log (log-log) ($\ln P = \beta_0 + \sum_{j=1}^{m} \beta_j \ln X_j + \epsilon$) and logarithmic (lin-log) ($P = \beta_0 + \sum_{j=1}^{m} \beta_j \ln X_j + \epsilon$). A convenient method for choosing functional form of hedonic regression is Box–Cox transformation (cf. Box and Cox 1964) of the dependent variable (a similar transformation may be applied for the independent variables of the model leading to further extensions of the set of the potential functional forms):

$$B\left(P_i, \lambda\right) = \begin{cases} \frac{P_i^\lambda - 1}{\lambda} & \text{for } \lambda \neq 0 \\ \ln P_i & \text{for } \lambda = 0 \end{cases} \tag{3}$$

The Box–Cox method consists in comparing the whole spectrum of potential functional forms of the model for various values of parameter λ and choosing the one which yields the highest value of likelihood function. This procedure allows to consider a large family of functional forms, including linear (for $\lambda = 1$) and exponential (for $\lambda = 0$) approaches.

2.2.2 Data Requirements

The discussed research problem is not new, however it has not been sufficiently fathomed, especially from the Polish perspective. The main reasons and possible explanations of such unsatisfactory level of research in the hedonic method area are, on the one hand, the complexity and difficulty of the related issues, and on the other shortage of suitable databases. The demands on the data for the purposes of hedonic analysis are numerous. The data should consist of large number of observations on

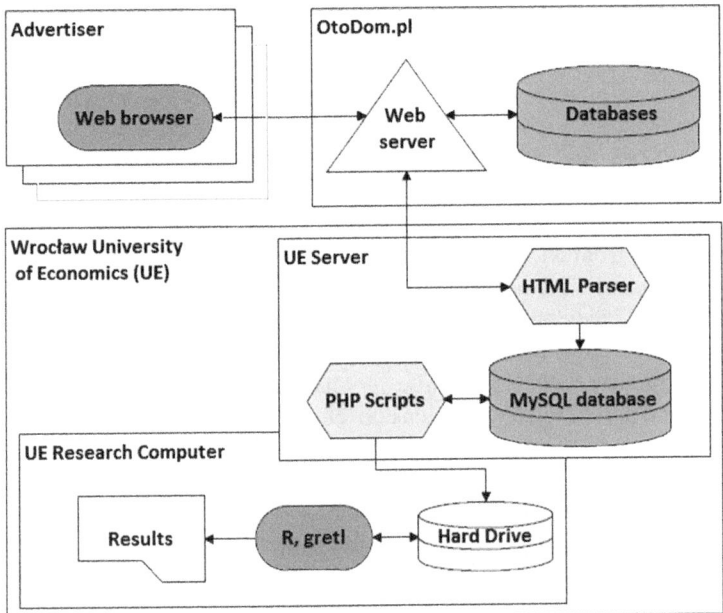

Fig. 1 Infrastructure for data gathering and analysis

prices of real estates and sufficiently extensive set of characteristics. Since such datasets are difficult to obtain, for this research special infrastructure designed for data collection was developed. The schematic diagram of the infrastructure is presented in Fig. 1.

In essence, the collected data came from Internet site www.OtoDom.pl, which is one of the biggest Polish advertising site for both individual and institutional real estate sellers. Each advertiser for the price of small fee can place sale offer, which is then stored in www.OtoDom.pl databases and visible for the potential buyers on the website. Since the direct access to the databases is impossible, special tool (called HTML Parser) was designed. The tool itself is a set of mutually related PHP scripts integrated with SQL database, and is using PHP framework *Simple HTML DOM Parser* written by S.C. Chen (http://simplehtmldom.sourceforge.net/). In order to gather the data the user has to determine the specifics of the chosen commodity. In case of real estates the user defines the type of the real estate (e.g. houses, apartments, parcels), the location (e.g. region, city, district) and the type of the market (secondary or primary). Basing on the information provided by the user the programme from all the offers published on www.OtoDom.pl website chooses only the suitable advertisements. Subsequently necessary for the research data is collected and stored in the SQL database. Afterwards, another PHP script transfers the data to the research computer in the format prefered by the user. The main disadvantage of such data source is the fact that the collected prices are offer prices, and not transaction prices.

3 Description of the Dataset

Using the above-mentioned infrastructure for data collection extensive dataset on apartments from the secondary real estate market in Wrocław has been collected. The database consists of 13,920 offers of apartments for sale advertised in the year 2011, including 3,340 offers from Fabryczna district (D1), 1,525 from Psie Pole district (D2), 2,349 from Śródmieście district (D3), 1,768 from Stare Miasto district (D4) and 4,938 from Krzyki district (D5). The information gathered included such variables as price, size, address (from which, using geolocation, distance to the city centre was calculated), number of rooms, building age, number of floors in the building as well as the floor on which the apartment was located. Descriptions and basic summary statistics for those variables are given in Table 1.

The second part of the information collected concerned additional features of the apartments, such as: balcony, garage, terrace, separate kitchen, garden, basement, whether the apartment is located on the ground floor, whether there is a lift in the building, etc. Such characteristics may be incorporated into the model in the form of dummy variables, taking the value of 1 if an apartment possessess given feature, and 0 otherwise. Table 2 presents only some of the collected dummy variables, namely those which were the most common in the dataset.

Table 1 Variables description and summary statistics (NA—the percentage of unavailable observations)

Variable name	Description	Mean	Min	Max	NA (%)
PRICE	Apartment price [PLN]	366,620	110, 000	676, 000	0.00
SIZE	Apartment size [m^2]	56.32	16.00	105.20	0.00
DISTANCE	Distance to city centre [km]	3.87	0.2	13.23	21.12
NOROOM	No. of rooms	2,51	0	5	0.00
YEAR	Year of the building	1990.5	1,867	2,012	61.38
FLOOR	Apartment floor	2.74	0	14	0.00
NOFLOOR	No. of floors in building	4.50	0	14	0.00

Table 2 Dummy variables description and percentages of their occurrence in dataset

Variable name	Description	Percentage (%)
GRFLOOR	Dummy for ground floor	14.50
BALCONY	Dummy for balcony	39.40
GARAGE	Dummy for garage	7.15
BASEMENT	Dummy for basement	19.19
GARDEN	Dummy for garden	0.91
TERRACE	Dummy for terrace	1.83
LIFT	Dummy for lift	5.77
TLEVELS	Dummy for two levels	0.92
SKITCH	Dummy for separate kitchen	23.38

The described characteristics may be divided into two groups: structure-related (such as size, number of rooms, building age, number of floors in the building, additional features of the apartment), and location-related (distance to city centre, city district).

4 Research Steps and Results

In order to determine the correct functional relationship f for the model (2), the Box–Cox transformation of the dependent variable was performed. The results are presented in Fig. 2, where the values of log-likelihood function for various values of parameter λ are compared.

The highest value of log-likelihood function equal to $-147{,}395$ was found for $\lambda = -0.03788$. However, the 95 % confidence interval included $\lambda = 0$ as well, which indicates the correctness of the exponential (log-lin) function (to be more specific, a mixed approach was applied, as one of the independent variables–*SIZE*— was transformed using the logarithm function as well).

The model $\ln P = \beta_0 + \sum_{j=1}^{m} \beta_j X_j + \epsilon$ was estimated using Ordinary Least Squares method, and since the presence of heteroskedasticity was detected (White's test statistics $LM = 796.92$) re-estimated using White's Weighted Least Squares method. The procedure involves OLS estimation of the model of interest, followed by an auxiliary regression to generate an estimate of the error variance, then finally weighted least squares, using as weight the reciprocal of the estimated variance (cf. Whhite 1980). Afterwards, using the results of t-test, the insignificant variables were rejected, and the best models were chosen with the help of AIC and SC information criteria, when necessary. For both models the Variance Inflation Factors for each variable did not exceed 3, indicating lack of serious collinearity problems. The highest correlation coefficients were observed for the pairs (*DISTANCE, D1_FAB*) about 0.41, and (*DISTANCE, D4_SM*) about -0.4. The results of OLS estimation (model (1) W_OLS) and WLS estimation (model (2) W_WLS) are presented in Table 3.

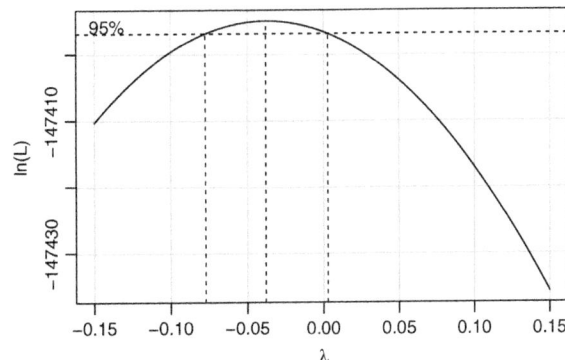

Fig. 2 The values of log-likelihood function for various λ

Table 3 Estimation results for models of Wrocław apartments' prices (t-ratios in parenthesis; *** indicates significance on the level 0,01; ** on the level 0,05; * on the level 0,1)

ln *PRICE*	(1)	(2)
	W_OLS	W_WLS
n	9565	9565
Adj. R^2	0.7543	0.7820
constant	9.971***	10.02***
	(447.8)	(408.8)
ln *SIZE*	0.7392***	0.7243***
	(122.8)	(101.7)
DISTANCE	−0.004947***	−0.006605***
	(−6.315)	(−9.444)
NOROOM	0.007440***	0.01267***
	(4.311)	(4.986)
FLOOR	−0.002482***	−0.001170*
	(−2.979)	(−1.819)
GRFLOOR	−0.01300***	−0.01363***
	(−2.694)	(−2.98)
NOFLOOR	−0.006149***	−0.007044***
	(−9.515)	(−13.63)
BALCONY	0.03519***	0.03228***
	(11.38)	(11.63)
GARAGE	0.08477***	0.08279***
	(14.87)	(14.35)
TERRACE	0.1019***	0.09540***
	(9.423)	(10.07)
TLEVELS	0.04822***	0.05756***
	(3.292)	(3.911)
SKITCH	−0.008729**	−0.009316***
	(−2.553)	(−3.091)
D1_FAB	−0.1694***	−0.1607***
	(−26.66)	(−23.01)
D2_PP	−0.1871***	−0.1791***
	(−24.49)	(−22.41)
D3_SROD	−0.1525***	−0.1509***
	(−26.24)	(−20.74)
D5_KRZ	−0.1158***	−0.1178***
	(−21.43)	(−18.07)

Both models were estimated using 9,565 observations (some parts of the initial data were discarded as outliers, and some were missing). The differences in parameters estimates are in most cases small, however in view of the presence of heteroskedasticity the results of model (2) will be further analysed. All the variables are statistically significant, and the signs of the parameters are consistent with the expectations. The prices of the apartments in Wrocław are influenced by their size: 1 % increase in the size of the apartment result in a 0.72 % increase

Fig. 3 Differences in apartments' prices in Wrocław districts: Fabryczna (D1), Psie pole (D2), Śródmieście (D3) and Krzyki (D5) in comparison with Stare miasto (D4)

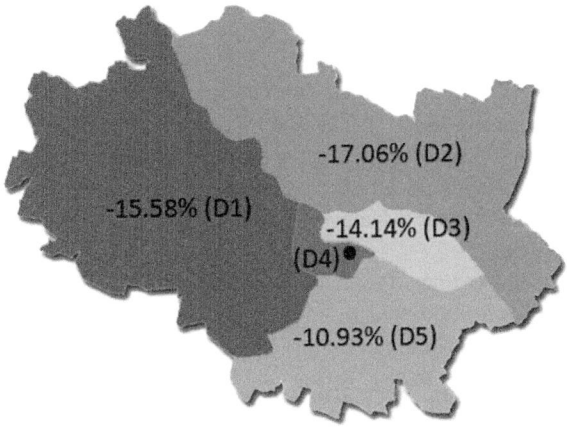

in the apartment's price, ceteris paribus (moreover the parameter for the variable *NOROOM* is positive and significant). Another factor is the location of the apartment within the building. The more floors the building has and the higher the apartment is located (except the ground floor), the lower the price. Furthermore, the apartments located on the ground floor are significantly cheaper. The presence of additional features significantly increases the apartment price. The price of a flat endowed with garage is about 8.6 % higher (($\exp(\beta) - 1) \cdot 100\%$); a balcony and terrace increase the price by 3.3 % and 10 %, respectively. The last factor is the location of the apartment in the city—the apartments closer to the city centre are more expensive than similar apartments in the city outskirts.

One of the research hypotheses was that the central districts of Wrocław (namely D4 and D3) along with the southern Krzyki district (D5) are more expensive locations than the remaining two districts. The obtained results support this hypothesis. The location of Wrocław's districts from the most to the least expensive is presented in Fig. 3. The apartments in the city centre (Stare miasto (D4)) are the most costly and the succession of the other districts is as follows: (D5) Krzyki district (similar in size and other characteristics apartment located in Krzyki is cheaper by almost 11 %), (D3) Śródmieście district (cheaper by over 14 %), (D1) Fabryczna district (cheaper by about 15.5 %) and (D2) Psie Pole district (cheaper by approximately 17 %).

The fit of the model is relatively high with adjusted R^2 equal to about 78 %. In order to additionally evaluate the model performance, 1 % of the initial dataset was not used in the estimation process and was intended for the out-of-sample testing. Only 3 % of the obtained forecasts were outside of the 95 % confidence intervals, and the following values of error measures were reported RMSE = 51,632, MAE = 37,500, MAPE = 9.90 %.

5 Final Remarks

The study attempted the estimation of a hedonic model for the secondary apartment market in Wrocław. The obtained results provide the means for pricing apartments on the basis of their most significant characteristics. Moreover, hedonic prices for particular characteristics were estimated, including the prices which are not directly observable on the market (such as distance to the city centre, location in a given district).

The estimated model is relatively well behaved and yields acceptable forecasts, however additional research could provide improvements in its predictive power. Further research directions should include:

- estimating the models with more location and neighbourhood-related variables (e.g. distance to green areas, communication facilities),
- missing data handling (imputation mechanisms),
- investigating in greater detail the problem of outliers and influential observations (e.g. DFFITS (cf. Belsley et al. 1980), robust methods),
- researching usefulness of obtained results in creating quality-adjusted price indices.

References

Belsley, D. A., Kuh, E., & Welsh, R. E. (1980). *Regression Diagnostics: Identifying Influential Data and Sources of Collinearity. Wiley Series in Probability and Mathematical Statistics.* New York: Wiley.

Berndt, E. R., Griliches, Z., & Rappaport, N. J. (1995). Econometric estimates of prices indexes for personal computers in the 1990s. *Journal of Econometrics, 68*(1), 243–268.

Boskin, M. J., Dulberger, E. R., & Griliches, Z. (1996). *Toward a more accurate measure of the cost of living: final report.* Washington: Diane Publishing Company.

Box, G. E. P., & Cox, D. R. (1964). An analysis of transformations. *Journal of the Royal Statistical Society. Series B (Methodological), 26*(2), 211–252.

Brachinger, H. W. (2002) *Statistical Theory of Hedonic Price Indices.* DQE Working Papers, 1, Department of Quantitative Economics. Switzerland: University of Freiburg/Fribourg.

Can, A. (1992). Specification and estimation of hedonic housing price models. *Regional Science and Urban Economics, 22*(3), 453–474.

Court, A. (1939). Hedonic Price Indexes with Automotive Examples. *The dynamics of automobile demand.* New York: General Motors Corporation (pp. 99–117).

Dziechciarz, J. (2004). Regresja Hedoniczna. Próba Wskazania Obszarów Stosowalności. In A. Zelias (Ed.) *Przestrzenno-czasowe Modelowanie i Prognozowanie Zjawisk Gospodarczych.* Kraków: Wydawnictwo Akademii Ekonomicznej w Krakowie (pp. 163–175).

Greene, W. H. (2011). *Econometric analysis,* 7th edn. New Jersey: Prentice Hall.

Haas, G. C. (1922). *Sale Prices as a Basis for Farm Land Appraisal.* Technical Bulletin 9, University of Minnesota, Agricultural Experiment Station.

Herath, S., Maier, G. (2010). *The Hedonic Price Method in Real Estate and Housing Market Research. A Review of the Literature.* SRE-Discussion Papers, 2010/03, WU Vienna University of Economics and Business, Vienna.

Sheppard, S. (1999). Hedonic Analysis of Housing Markets. In: P.C. Cheshire & E.S. Mills (Eds.) *Handbook of regional and urban economics* Vol. 3, (pp.1595–1635). Amsterdam: Elsevier.

Triplet, T. J. (2006). *Handbook on hedonic indexes and quality adjustments in price indexes.* OECD Directorate for science, technology and industry. Paris: OECD Publishing.

Triplett, J. (1986). The economic interpretation of hedonic methods. *Survey of Current Business, 36*(1), 36–40.

Wallace, H. A. (1926). Comparative farmland values in iowa. *The Journal of Land and Public Utility Economics, 2*(4), 385–392.

Waugh, F. V. (1929). *Quality as a determinant of vegetable prices.* New York: Columbia University Press,

White, H. (1980). A heteroskedasticity-consistent covariance matrix estimator and a direct test for heteroskedasticity. *Econometrica, 48*(4), 817–838.

Wooldridge, J. M. (2010). *Econometric analysis of cross section and panel data,* 2nd edn. Cambridge: The MIT Press.

Smart Growth Path as the Basis for the European Union Countries Typology

Elżbieta Sobczak and Beata Bal-Domańska

Abstract The concept of smart growth integrates activities in the area of smart specialization, creativity and innovation influencing development opportunities of particular European countries. The objective of the paper is to classify the EU countries with regard to smart growth paths by means of multivariate statistical analysis methods. The concept of smart growth path was defined considering the direction and intensity of changes occurring in the area of smart specialization, creativity and innovation. These paths became the basis for the European Union member states classification carried out using cluster analysis methods. The presented analysis is of dynamic nature and allows for the smart growth patterns typology.

1 Introduction

In 2010 the European Union accepted Europe 2020 development strategy defining objectives aimed at providing support for the EU member countries in overcoming economic crisis and ensuring smart, sustainable and facilitating social inclusion growth (see Europe 2010).

Smart growth defined in the strategy is to be achieved through the development of knowledge-based economy and innovation. Three pillars of smart growth were identified: smart specialization, creativity and innovation. Additionally, the indicators facilitating their quantification were also specified.

The concept of smart growth refers to numerous former theoretical concepts and models of regional growth, among which the dominating role is played by the regional innovation systems (Cooke et al. 1997), innovation environments—*milieu innovateur* (analysed by GREMI research group *Groupe de Recherche Européen sur les Milieux Innovateurs*; French for the European Research Group into Innovative Milieu), learning regions (Florida 1995; Morgan 1997) and innovation clusters (Porter 1998). Smart growth is based on investments in education and research. It results in the establishment of modern economic structures, the development of

E. Sobczak (✉) • B. Bal-Domańska
Wrocław University of Economics, Nowowiejska 3, 58-500 Jelenia Góra, Poland
e-mail: elzbieta.sobczak@ue.wroc.pl; beata.bal-domanska@ue.wroc.pl

© Springer-Verlag Berlin Heidelberg 2015
B. Lausen et al. (eds.), *Data Science, Learning by Latent Structures, and Knowledge Discovery*, Studies in Classification, Data Analysis, and Knowledge Organization, DOI 10.1007/978-3-662-44983-7_45

human capital and innovation. For the purposes of the study it was accepted that smart growth is a complex phenomenon based on three pillars defined as smart specialization (SS), creativity (C) and innovation (I).

The objective of the conducted research is to perform the classification and typology of the European Union member states with regard to smart growth paths in the period 2002–2011 and present the models of changes in the distinguished smart growth pillars in 2011 comparing to 2002.

2 The Information Basis and the Stages of Research Procedure

The statistical information, indispensable for the empirical research, was obtained based on Eurostat database. The study included 27 EU member states and the time range of research covered the period of 2002–2011.

The following research procedure was applied:

1. The selection of indicators for smart growth pillars.
2. The construction of smart growth aggregate models for the EU countries.
3. The construction of smart growth paths by identifying:

   • the individual benchmarks of smart growth,
   • the leading pillars of smart growth.

4. The smart growth paths classification of EU countries in the period of 2002–2011.
5. The classification of EU countries regarding the model of changes in the pillars of smart growth in 2011 comparing to 2002.

Stage 1.   The selection of indicators for smart growth pillars

Table 1 presents indicators which were used for the level assessment of smart specialization, creativity and innovation growth in the EU states.

Stage 2.   The construction of smart growth aggregate models for the EU countries

Aggregate measures are applied for the purposes of quantifying and measuring smart specialization, creativity and innovation in the EU countries. Normalization with zero minimum (see Kukuła 2000), expressed by the formula (1) for stimulant characteristics, was applied in the normalization of smart growth identifiers:

$$z_{itj} = \frac{x_{itj} - \min_i x_{itj}}{\max_i x_{itj} - \min_i x_{itj}}, \tag{1}$$

where: z_{itj}—normalized value of j-th identifier (variable) in i-th object (country) and in t-th period; x_{itj}—value of j-th identifier (variable) in i-th object (country) and in t-th period; $i = 1, 2, \ldots, N$—object number, $j = 1, 2, \ldots, M$—identifier number, $t = 1, 2, \ldots, T$—period number.

Table 1 The set of diagnostic indicators for smart growth pillars

Smart growth		
Pillar I	Pillar II	Pillar III
–Smart specialization	–Creativity	–Innovation
KIS—employment in knowledge intensive services as the percentage of total employment (%)	TETR—share of tertiary education employment in total employment (%)	R&De—research and development expenditure in enterprise sector (% of GDP)
HTMS—employment in high and medium high-technology manufacturing as the percentage of total employment (%)	HRST—human capital in science and technology as the percentage of active population (%)	R&Dgov—research and development expenditure in government sector (% of GDP)
	LLL—participation in education and training, the age range of 25–64 as the percentage of total population (%)	EPO—patent applications to the EPO per million of labour force

Source: Author's compilation based on: European regional space classification in the perspective of smart growth concept—dynamic approach (NCN grant no. 2011/01/B/HS4/04743)

The method of averaged standardized sum was used as the aggregating function for normalized values of diagnostic characteristics describing the particular pillars of smart growth:

$$AM_{it}^k = \frac{1}{m^k} \sum_{j=1}^{m^k} z_{itj}, \tag{2}$$

where: AM_{it}^k—the aggregate measure of smart growth for k-th pillar ($k = SS, C, I$), m^k—the number of diagnostic characteristics describing k-th pillar.

The aggregate model of smart growth of i-th object in t-th period takes the following form:

$$SG_{it} = \left[AM_{it}^{SS}, AM_{it}^C, AM_{it}^I \right], \tag{3}$$

where: AM_{it}^{SS}—aggregate measure of smart specialization in i-th object and t-th period, AM_{it}^C—aggregate measure of creativity in i-th measure in t-th period, AM_{it}^I—aggregate innovation measure for i-th object and t-th period.

Stage 3.   The construction of smart growth path

a. The identification of the individual benchmarks of smart growth

The smart growth identifiers normalized by applying the normalization with zero minimum method take values in the range of [0, 1] and therefore aggregate measures for particular pillars of smart growth are characterized by the following property:

$$AM_{it}^{SS}, AM_{it}^C, AM_{it}^I \in [0, 1]. \tag{4}$$

Therefore, a global benchmark of smart growth and the leading element on the growth path of the studied objects is represented by the following point

$$P_o^1 = [1, 1, 1]. \tag{5}$$

The path of smart growth is illustrated by a straight line crossing the following points:

$$P_o^0 = [0, 0, 0]$$
$$P_o^1 = [1, 1, 1]. \tag{6}$$

Placing an i-th object on the path of smart growth consists in specifying the individual benchmark of smart growth in line with the following formula:

$$IBSG_{it} = \max(AM_{it}^{SS}, AM_{it}^{C}, AM_{it}^{I}). \tag{7}$$

b. The identification of the leading pillars of smart growth

Smart growth pillar characterized by the maximum aggregate measure value is the leading one in a given object.

Stage 4.   The smart growth path classification of EU countries in the period of 2002–2011

The below observation matrix is the basis for the EU countries classification of individual benchmark of smart growth (IBSG):

$$[IBSG_{it}]_{N \times T}. \tag{8}$$

In order to classify countries according to smart growth path in the period of 2002–2011 the following procedure was performed:

- specifying diversification between the studied countries by means of squared Euclidean distance,
- hierarchical classification of countries into homogenous groups applying Ward method. The review of information on distance measures and classification methods possible to apply may be found, among others, in the studies by: Hellwig (1968), Grabiński et al. (1989), Pociecha et al. (1988), Walesiak (2006),
- presentation of initial classification results on a dendrogram,
- classification of countries by means of k-means method,
- typology of the obtained classes.

Stage 5.   The classification of EU countries regarding the model of changes in the pillars of smart growth in 2011 comparing to 2002

The model of changes in the pillars of smart growth as the basis for the EU countries classification is as follows:

$$(P_o^1 - SG_{i1}) - (P_o^1 - SG_{iT}), \tag{9}$$

where: $P_o^1 = [1, 1, 1]$—global benchmark of smart growth, SG_{i1}, SG_{iT}—the aggregate model of smart growth of i-th object and 1-st and T-th period (formula 3).

In order to classify countries according to the model of changes in smart growth pillars, in 2011 comparing to 2002, the same procedure was performed as in stage 4.

3 Empirical Analysis Results

Figure 1 presents the distribution of aggregate measure values of smart specialization, creativity and innovation of the EU countries in the period of 2002–2011. The levels of smart specialization and innovation in the studied countries do not change significantly in the analyzed years. A significant increase in the aggregate measure values of growth (for the minimum, median and maximum value) was observed for creativity.

The next stage of the research procedure was to construct paths of smart growth. Table 2 presents values of individual benchmarks for smart growth regarding the studied countries in 2002 and 2011 (further referred to as *IBSG*). Putting a country on the path of smart growth means the identification of such benchmarks. Table 2 shows EU countries ranked by the decreasing values of *IBSG* and the leading pillar

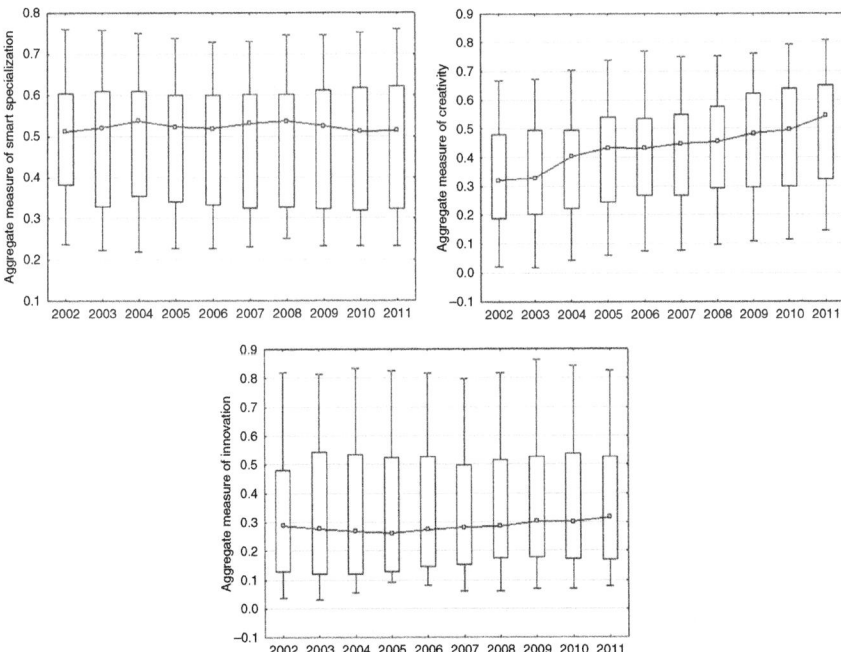

Fig. 1 Aggregate measure values of smart specialization, creativity and innovation growth of the EU countries in the period of 2002–2011. Source: Author's compilation in STATISTICA program

Table 2 The leading pillars of smart growth in 2002 and 2011 listed by the decreasing values of individual benchmarks in smart growth in the EU countries

2002					2011			
No.	Country	IBSG	Leading pillar		No.	Country	IBSG	Leading pillar
1.	Finland	0.8216	Innovation		1.	Germany	0.8279	Innovation
2.	Germany	0.7705	Innovation		2.	Finland	0.8121	Innovation
3.	Sweden	0.7604	Smart specialization		3.	Denmark	0.7968	Creativity
4.	Denmark	0.7279	Smart specialization		4.	Luxembourg	0.7735	Creativity
5.	United Kingdom	0.6604	Smart specialization		5.	Sweden	0.7663	Creativity
6.	Malta	0.6158	Smart specialization		6.	United Kingdom	0.7074	Creativity
7.	France	0.6116	Innovation		7.	Czech Rep.	0.6952	Smart specialization
8.	Italy	0.6040	Smart specialization		8.	Netherlands	0.6529	Creativity
9.	Belgium	0.6007	Smart specialization		9.	Ireland	0.6518	Creativity
10.	Ireland	0.5505	Smart specialization		10.	Estonia	0.6438	Creativity
11.	Hungary	0.5375	Smart specialization		11.	Belgium	0.6391	Creativity
12.	Netherlands	0.5287	Innovation		12.	Slovakia	0.6386	Smart specialization
13.	Slovenia	0.5237	Smart specialization		13.	Cyprus	0.6204	Creativity
14.	Czech Rep.	0.5115	Smart specialization		14.	Slovenia	0.5901	Smart specialization
15.	Estonia	0.5111	Smart specialization		15.	France	0.5880	Smart specialization

(continued)

Table 2 (continued)

2002				2011			
No.	Country	IBSG	Leading pillar	No.	Country	IBSG	Leading pillar
16.	Austria	0.5096	Smart specialization	16.	Lithuania	0.5825	Creativity
17.	Slovakia	0.5032	Smart specialization	17.	Hungary	0.5824	Smart specialization
18.	Cyprus	0.4802	Creativity	18.	Spain	0.5779	Creativity
19.	Luxembourg	0.4801	Innovation	19.	Italy	0.5457	Smart specialization
20.	Romania	0.4102	Smart specialization	20.	Austria	0.5211	Innovation
21.	Poland	0.3941	Smart specialization	21.	Malta	0.4797	Smart specialization
22.	Spain	0.3814	Smart specialization	22.	Latvia	0.4461	Creativity
23.	Bulgaria	0.3413	Smart specialization	23.	Poland	0.4014	Smart specialization
24.	Latvia	0.3281	Creativity	24.	Greece	0.3499	Creativity
25.	Lithuania	0.3196	creativity	25.	Portugal	0.3230	Smart specialization
26.	Portugal	0.2436	Smart specialization	26.	Bulgaria	0.3153	Smart specialization
27.	Greece	0.2374	Smart specialization	27.	Romania	0.2325	Smart specialization

Source: Author's compilation

for each country. As it can be noticed, in case of the vast majority of EU countries (19) smart specialization represented the leading pillar, with innovation (5) and creativity (3) to follow.

In 2002 Finland and Germany achieved the highest values of individual benchmarks in smart growth (respectively 0.8216 and 0.7705). Innovation was the leading pillar in these countries. The lowest level, on the smart growth path in 2002, was occupied by Portugal and Greece, with smart specialization as the leading pillar in these countries. The individual smart growth models represented the following values 0.2436 and 0.2374.

Innovation was of dominating significance in the smart growth of five countries (Finland, Germany, France, the Netherlands and Luxembourg), while creativity only in case of three (Cyprus, Latvia, Lithuania).

It is noticeable that in 2011 the number of countries, in which smart specialization was the leading pillar, has decreased (from 19 to 11), while the number of countries characterized by the creative pillar, as the dominating one, has rapidly grown (from 3 to 13); primarily as the result of increase of both employment education level and the size of human resources in science and technology. In 2011 the highest positions, on the smart growth path, were occupied by Germany and Finland with innovation as the leading pillar, while the lowest by Bulgaria and Romania in case of which smart specialization was identified as the leading pillar.

The subsequent step of research procedure was to classify smart growth paths of the EU countries in the period of 2002–2011. The classification was based on the matrix of individual benchmarks regarding smart growth (see (9)). Based on the dendrogram connection, using Ward method, the number of classes was determined. The division of EU countries into four classes was accepted. Figure 2

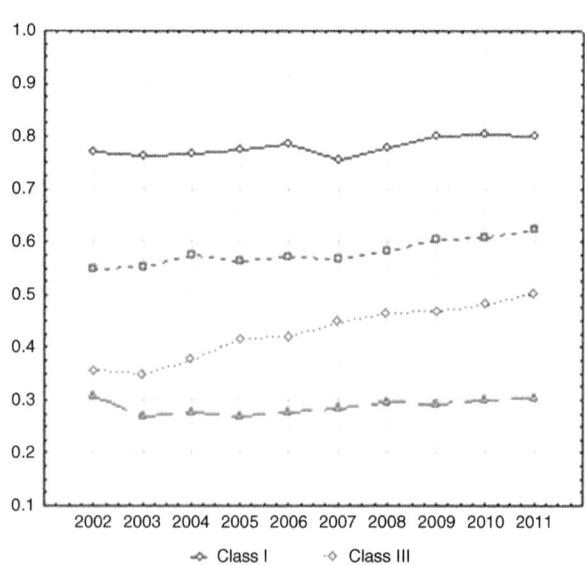

Fig. 2 The diagram presents mean values of individual smart growth benchmarks in the selected classes of the EU countries in the period 2002–2011. Source: Author's compilation in STATISTICA program

Table 3 Smart growth path classification of the EU countries in the period of 2002–2011

Classes	Mean values of *IBSG*	Changes of smart growth path	Countries	Number of countries
I	High	Stagnation	Denmark, Germany, Finland, Sweden	4
II	Mid-high	Slight stagnation	Belgium, Czech Rep., Estonia, Ireland, France, Italy, Cyprus, Luxembourg, Hungary, Malta, Netherlands, Austria, Slovenia, Slovakia, United Kingdom	15
III	Mid-low	Dynamic improvement	Spain, Latvia, Lithuania, Poland	4
IV	Low	Stagnation	Bulgaria, Greece, Portugal, Romania	4

Source: Author's compilation

presents the mean values of *IBSG* in the four classes of EU states in the period of 2002–2011 obtained as the result of k-means method application. Table 3 presents the composition of the distinguished classes.

Class II turned out to be the most numerous and covers 15 countries featuring mid-high level of individual smart growth benchmarks and moderate improvement of the growth path. According to this classification the first class covers the states characterized by the high level of *IBSG* and the stagnation in the growth path. The key feature of the states included in the third class is the mid-low level of *IBSG* but, at the same time, they present a dynamic improvement of the growth path. Finally, the fourth class lists the states featuring low level of *IBSG* and stagnation in the growth path.

The last stage of the research was the classification of EU countries regarding the model of changes on the path of smart growth. The conducted analysis allowed for indicating changes which occurred in 2011 comparing to 2002 as distant from the global growth pattern. The results show the model of changes on the growth path as the basis for the EU countries classification. Based on the tree diagram connections using Ward method the number of classes was determined. Similarly to the previous classification the division of EU states into four classes was adopted.

Figure 3 presents the mean values for distance changes from the global growth benchmark in the distinguished classes of EU countries in 2011 comparing to 2002 obtained as the results of k-means method application.

Table 4 presents the composition of classes. The most common model of countries is represented by class I. It consisted in an increasing creativity accompanied by insignificant changes in the remaining pillars. The first class covers countries featuring improvement only in the area of creativity (the smallest of all the four classes).The second class includes the states characterized by the improvement in

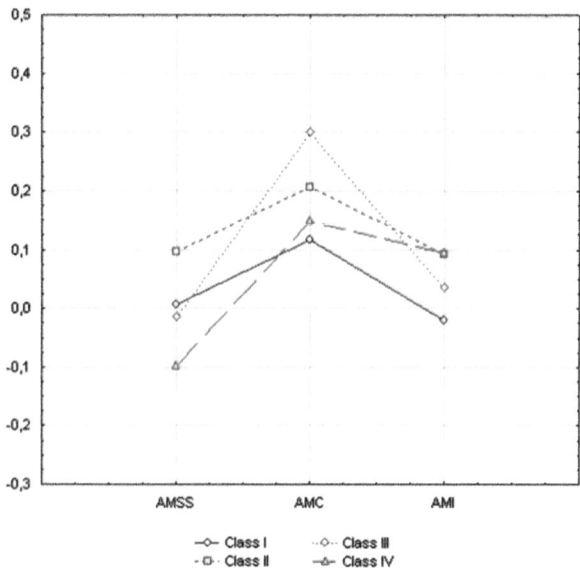

Fig. 3 The diagram of mean values changes in the pillars of smart growth in 2011 comparing to 2002. Source: Author's compilation in STATISTICA program

Table 4 The classification of EU countries regarding the model of changes in the pillars of smart growth in 2011 comparing to 2002

Classes	The model of changes	Countries in smart growth pillars	Number of countries
I	AMC growth	Belgium, Bulgaria, Germany, Greece, France, Italy, Cyprus, Hungary, Netherlands, Finland, United Kingdom	11
II	AMSS, AMC, AMI growth	Czech Rep., Spain, Portugal, Slovenia, Slovakia	5
III	AMC, AMI growth	Ireland, Lithuania, Luxemburg, Poland, Sweden	5
IV	AMSS drop, AMC, AMI growth	Denmark, Estonia, Latvia, Malta, Austria, Romania	6

all smart growth pillars, the most significant one being the creativity pillar. The third class features the states showing the most dynamic improvement in the creativity pillar, out of all the four classes, and a moderate improvement in the innovation pillar.

Finally, the fourth class is characterized by the worsening situation in the smart specialization pillar and, at the same time, by an average improvement of creativity and innovation.

4   Conclusions

The conducted analysis and assessment of smart growth in the European Union countries allow for the following conclusions:

1. In 2002 smart specialization was the leading pillar of smart growth in the majority of EU countries (19), whereas in 2011 the position of the leading pillar was taken over by creativity (13).
2. Germany and Finland were the closest to the global benchmark of smart growth in 2011 (the leading pillar—innovation, *IBSG* for these countries was, respectively, 0.8279 and 0.8121) while Romania and Bulgaria were the most distant countries from the global benchmark of smart growth in 2011 (the leading pillar—smart specialization, *IBSG* for these countries was, respectively, 0.2325 and 0.3153).
3. As far as the criterion of smart growth paths is concerned, in the period of 2002–2011, two classes of countries appear particularly interesting:

   - the countries featuring high and relatively stable values of individual benchmarks smart growth benchmarks: Denmark, Germany, Finland and Sweden,
   - the countries characterized by the dynamic improvement and, at the same time, the relatively low level of individual smart growth benchmarks: Spain, Latvia, Lithuania and Poland.

4. Regarding the criterion of the changes model in the growth path pillars, in 2011 comparing to 2002, two other classes, out of four, appear to be particularly interesting:

   - the countries featuring the most dynamic creativity growth: Ireland, Lithuania, Luxembourg, Poland and Sweden,
   - the countries featuring the decreasing smart specialization level: Denmark, Estonia, Latvia, Malta, Austria and Romania.

Acknowledgements The study was conducted within the framework of research grant NCN no. 2011/01/B/HS4/04743 entitled: European regional space classification in the perspective of smart growth concept—dynamic approach.

References

Cooke, P., et al. (1997). Regional innovation systems: institutional and organizational dimensions. *Research Policy, 26*, 475–491.

Florida, R. (1995). Towards the learning region. *Futures, 27*, 527–536.

Grabiński, S., et al. (1989). *Metody taksonomii numerycznej w modelowaniu zjawisk społeczno–gospodarczych [Numerical taxonomy methods in social and economic phenomena modelling].* Warsaw: PWN Publishers.

Hellwig, Z. (1968). Zastosowanie metody taksonomicznej do typologicznego podziału krajów ze względu na poziom ich rozwoju oraz zasoby i strukturę wykwalifikowanych kadr [The

Application of Taxonomic Method for Typological Division of Countries Regarding Their Development Level as Well as the Resources and Structure of Qualified Personnel]. *Przegląd Statystyczny [Statistical Review]*, *4*, 307–327.

Kukuła, K. (2000). *Metoda unitaryzacji zerowej [Normalization with zero minimum method]*. Warsaw: PWN Publishers.

Morgan, K. (1997). The Learning region: institutions, innovations and regional renewal. *Regional Studies*, *31*, 491–503.

Pociecha, J. et al. (1988). *Metody taksonomiczne w badaniach społeczno–ekonomicznych [Taxonomic Methods in Social and Economic Studies]*. PWN Publishers: Warszawa.

Porter, M. E. (1998). *The competitive advantage of nations*. London: Macmillan.

Walesiak, M. (2006). *Uogólniona miara odległości w statystycznej analizie wielowymiarowej [Generalised distance measure in statistical multivariate analysis]*. Wrocław: Wrocław University of Economics Publishing House.

The Influence of Upper Level NUTS on Lower Level Classification of EU Regions

Andrzej Sokołowski, Małgorzata Markowska, Danuta Strahl, and Marek Sobolewski

Abstract The Nomenclature of Territorial Units for Statistics or Nomenclature of Units for Territorial Statistics (NUTS) is a geocode standard for referencing the subdivision of countries for statistical purposes. It covers the member states of the European Union. For each EU member country, a hierarchy of three levels is established by Eurostat. In 27 EU countries we have 97 regions at NUTS1, 271 regions at NUTS2 and 1,303 regions at NUTS3. They are subject of many statistical analysis involving clustering methods. Having a partition of units on a given level, we can ask the question, whether this partition has been influenced by the upper level division of Europe. For example, after finding groups of homogeneous levels of NUTS 2 regions we would like to know if the partition has been influenced by differences between countries. In the paper we propose a procedure for testing the statistical significance of influence of upper level units on a given partition. If there is no such influence, we can expect that the number of between-groups borders which are also country borders should have a proper probability distribution. A simulation procedure for finding this distribution and its critical values for testing significance is proposed in this paper. The real data analysis shown as an example deals with the innovativeness of German districts and the influence of government regions on innovation processes.

Project has been financed by the Polish National Centre for Science, decision DEC-2013/09/B/HS4/0509.

A. Sokołowski (✉)
Cracow University of Economics, Kraków, Poland
e-mail: sokolows@uek.krakow.pl

D. Strahl • M. Markowska
Wrocław University of Economics, Wrocław, Poland
e-mail: dstrahl@ae.jgora.pl; mmarkowska@ae.jgora.pl

M. Sobolewski
Rzeszow University of Technology, Rzeszów, Poland
e-mail: msobolew@prz.edu.pl

© Springer-Verlag Berlin Heidelberg 2015   525
B. Lausen et al. (eds.), *Data Science, Learning by Latent Structures, and Knowledge Discovery*, Studies in Classification, Data Analysis, and Knowledge Organization, DOI 10.1007/978-3-662-44983-7_46

1   The Idea for the Influence Test

The analysed subject is loosely related to geographic boundary analysis (Jacquez et al. 2000). We are not looking for borders separating regions with different levels of a given variable(s) (Womble 1951), they are given by the administrative division of countries. Problems relatively similar to ours are subboundary significance (Oden et al. 1993) and boundary overlapping (Jacquez 1995).

The Nomenclature of Territorial Units for Statistics or Nomenclature of Units for Territorial Statistics (NUTS) is a geocode standard for referencing the subdivision of countries for statistical purposes. It covers member states of European Union. For each EU member country, a hierarchy of three levels is established by Eurostat. In 27 EU countries we have 97 regions at NUTS1, 271 regions at NUTS2 and 1,303 regions at NUTS3 (since July 1, 2013 Croatia joined European Union with 2 NUTS2 units and 21 NUTS3 units). They are subject of many statistical analysis involving clustering methods. Having a partition of units on a given level, we can ask the question, whether this partition has been influenced by the upper level division of Europe. For example, after finding groups of homogeneous levels of NUTS2 regions we would like to know if the partition has been influenced by differences between countries.

The general idea of the proposed procedure can be explained as follows:

- Lower level NUTS units (e.g. NUTS2) are clustered and we obtain non-overlapping groups of these units.
- The border between units is called active if these units belong to different groups.
- The border is a lower level border if units belong to the same upper level NUTS unit.
- The border is an upper level border if units do not belong to the same upper lever NUTS unit.
- In a given country we have N borders between lower level NUTS. M of them are also the upper level borders.
- The partition obtained by clustering activate some of the lower level borders.

The aim of this paper is to study the distribution of the number of upper level borders among active ones, for a given number of groups. We are testing the following hypotheses:

H0:   Upper level NUTS does not influence the partition of lower level NUTS units.
H1:   Upper level NUTS influences the partition of lower level NUTS units.

We use the test statistics

$$U = \frac{\text{Number of upper level active borders}}{\text{Total number of active borders}}. \tag{1}$$

The proposed test is right-sided, since big number of active upper level borders indicates the influence of decisions made on the upper level on the situation in the lower level units. The distribution of the U statistic depends on the upper level number of units, number of borders and geographical structure of both upper and lower units. Thus this distribution cannot be derived analytically in general application, but must be studied separately for each country, by a Monte Carlo approach.[1]

2 Simulation Results

For partition simulations we use the simplified version (with a given number of groups) of the random partition generator proposed by Sokołowski (1979):

- Set k (number of groups)
- Assign random number from uniform distribution to each object.
- Order objects according to values of this random variable.
- Now we have $(n - 1)$ potential borders between objects.
- Assign random number from uniform distribution to each potential border.
- Make active first k borders with the biggest values of these random numbers.

We studied the distribution of the U statistics for Germany's and their NUTS1 (upper level) and NUTS2 (lower level) units. We considered partitions of Germany's NUTS2 units from 2 to 10 groups. With 1,000 simulation runs we found that the distribution of the U statistics can be approximated by a normal distribution. Empirical distribution for $k = 2$ is presented in Fig. 1.

From simulation results we found that the critical values for $\alpha = 0.10$ and $\alpha = 0.05$ can be easily smoothed by straight line and they depend on k (Figs. 2 and 3). Smoothed critical values are given in Table 1.

---

[1]Practically all statistics and methods within geographic boundary analysis are evaluated in this way (see: Oden et al. 1993; Fortin 1994; Jacquez 1995; Jacquez et al. 2000)

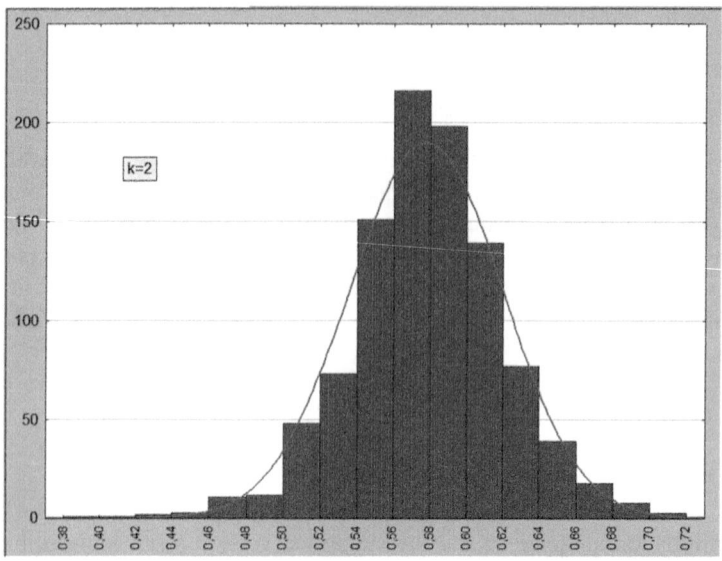

Fig. 1 Simulated distribution of U statistic for $k = 2$

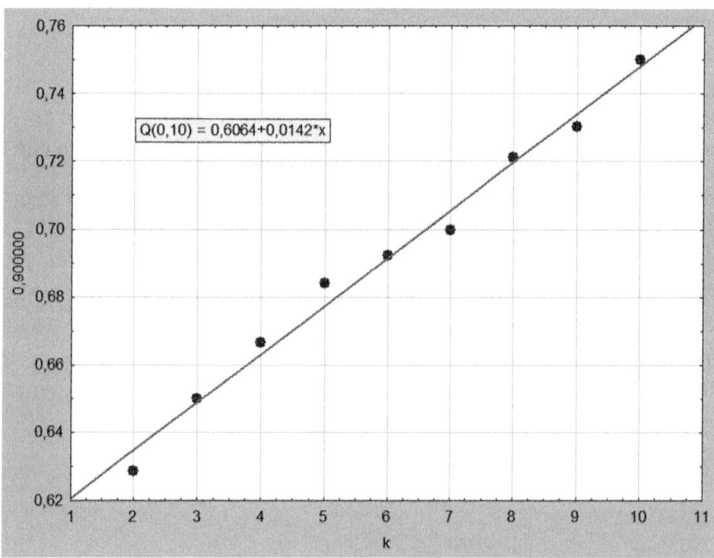

Fig. 2 Critical values of U statistic for $\alpha = 0.10$ for Germany NUTS1 and 2 comparisons

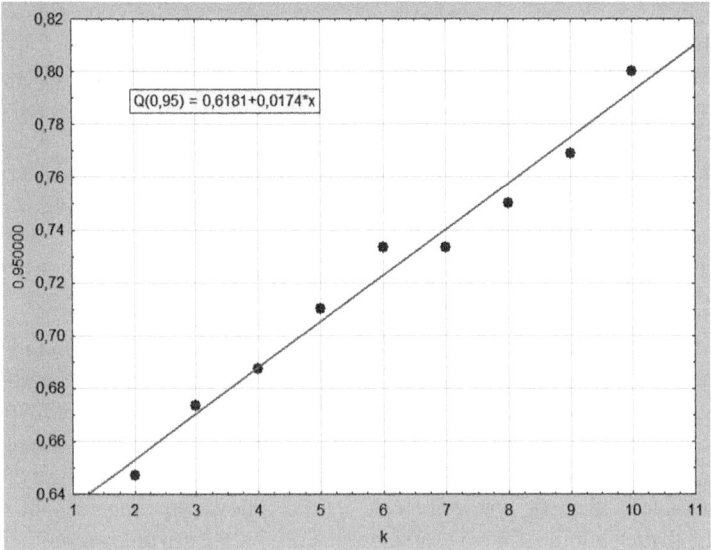

Fig. 3 Critical values of U statistic for $\alpha = 0.05$ for Germany NUTS1 and 2 comparisons

Table 1 Smoothed critical
U-values for testing the
influence of NUTS1 on
NUTS2 units in Germany

Number of groups	$\alpha = 0.10$	$\alpha = 0.05$
2	0.6348	0.6529
3	0.6490	0.6703
4	0.6632	0.6877
5	0.6774	0.7051
6	0.6916	0.7225
7	0.7058	0.7399
8	0.7200	0.7573
9	0.7342	0.7747
10	0.7484	0.7921

3 Example

We analysed the innovativeness of the German economy on the basis of 2008
data available in Eurostat. The following variables characterized German NUTS2
units:

- percentage share of tertiary education graduates in total workforce number in a
  region,
- percentage share of population aged 25–64 participating in lifelong learning in a
  region,
- workforce employed in knowledge-intensive services as the percentage share of
  total workforce,

- workforce employed in knowledge-intensive services as the percentage share of total workforce in services,
- human resources for science and technology, i.e. total number of workforce actually employed in S&T professions, in relation to professionally active population,
- workforce employed in high and mid-tech industry (as % of total workforce), number of patents registered in a given year in the European Patent Office (EPO) per one million of workforce,
- percentage share of workforce employed in high and mid-tech industry in the total workforce number employed in industry.

Wards method (Fig. 4) suggested four groups. Their content is given in Table 2.

In the obtained partition we have 39 NUTS1 active borders and 16 NUTS2 active borders. With that $U = 39/(39 + 16) = 0.7091$ which gives $p = 0.0296$ (which was calculated using estimated standard deviation and normal approximation, the approach applied also by Fortin 1994). So we reject the null hypothesis. NUTS1 regions (lands) play an important role in the innovation processes in German economy. Upper administrative level significantly influences innovations in NUTS2 regions.

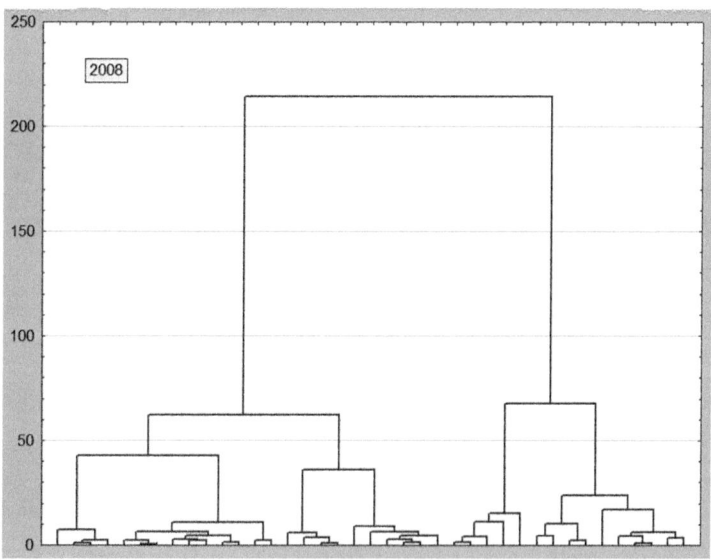

Fig. 4 Wards dendrogram for the innovativeness of German NUTS2 regions

Table 2 Groups of German NUTS2 regions

Group	NUTS2 regions
1	Stuttgart, Karlsruhe, Freiburg, Tübingen, Mittelfranken, Unterfranken, Rheinhessen-Pfalz
2	Oberbayern, Berlin, Hamburg, Köln
3	Bremen, Gießen, Braunschweig
4	Niederbayern, Oberpfalz, Oberfranken, Brandenburg-Nordost, Brandenburg-Südwest, Kassel, Mecklemburg-Vorpommern, Hannover, Lüneburg, Weser-Ems, Düsseldorf, Münster, Detmold, Arnsberg, Koblenz, Trier, Saarland, Chemnitz, Dresden, Leipzig, Sachnen-Anhalt, Schleswig-Holstein, Thüringen

References

Fortin, M.-J. (1994). Edge detection algorithm for two-dimensional ecological data. *Ecology, 75,* 956–965.

Jacquez, G. M. (1995). The map comparison problem: Tests for the overlap of geographic boundaries. *Statistics in Medicine, 14,* 2343–2361.

Jacquez, G. M., Maruca, S., & Fortin, M.-J. (2000). From fields to objects: A review of geographic boundary analysis. *Journal of Geographical Systems, 2,* 221–241.

Oden, N. L., Sokal, R. R., Fortin, M.-J., & Goebl, H. (1993). Categorical wombling: Detecting regions of significant change in spatially located categorical variables. *Geographical Analysis, 25,* 315–336.

Sokołowski, A. (1979). Generowanie losowego podziału zbioru skończonego. *Prace Naukowe Akademii Ekonomicznej we Wrocławiu, 160*(182), 413–415.

Womble, W. H. (1951). Differential systematics. *Science, 114,* 315–322.

Part X
Data Analysis in Library Science

Multilingual Subject Retrieval: Bibliotheca Alexandrina's Subject Authority File and Linked Subject Data

Magda El-Sherbini

1 Introduction

The Bibliotheca Alexandrina (BA) has been developing its own authority file since September 2006. The file includes subject headings, personal names, corporate bodies, series, and uniform titles. The BA authority file is unique in that it is based on the actual collection of materials in Arabic, French, and English.

Items in the BA collection are cataloged in the language of publication. A catalog record for an English language book is created in English, while an Arabic language book is cataloged in Arabic. It is important to note that catalog records for Arabic materials are created in that script. This enables the users to search for Arabic language items by typing their queries in the Arabic script. Unlike other catalogs, the BA does not Romanize its Arabic language records.

The interface of the BA on-line catalog allows the user to search the catalog in English, French, and Arabic. While searching for a subject term in English, search results will display the English term and its equivalents in French and Arabic, if the library owns materials in those languages in that subject category. This is accomplished through linking data in the subject catalog. BA catalogers assign subject headings for their Arabic materials and link them to equivalent terms from the Library of Congress Subject Headings (LCSHs) and Les notices d'autorité de BN-OPALE PLUS (Rameau).

This study of the BA authority file is the result of a research project conducted at the BA in 2011.

The first part of my research dealt with the Arabic name and corporate body authority file. An article on this subject was published in 2013 by El-Sherbini

M. El-Sherbini (✉)
The Ohio State University Libraries, Columbus, OH 43210, USA
e-mail: el-sherbini.1@osu.edu

© Springer-Verlag Berlin Heidelberg 2015   535
B. Lausen et al. (eds.), *Data Science, Learning by Latent Structures,*
and Knowledge Discovery, Studies in Classification, Data Analysis,
and Knowledge Organization, DOI 10.1007/978-3-662-44983-7_47

(2013). That study included a detailed description of the policies and processes involved in creating the BA authority file; a brief historical background about the BA; the library's organizational structure; the BA's on-line catalog (Virginia Tech Library System integrated library system known as "Virtua"); discussion of issues related to creating authority records for classical Arabic names; consequences of using various reference sources to identify elements of classical Arabic names; problems with Arabic scripts; reasons for creating a local BA authority file; local policies for creating BA authority records for classical and modern Arabic names and corporate names; authority work processing for Arabic materials at the BA; and the contribution of authority records to the Virtual International Authority File (VIAF).

The present study describes the subject authority file that is being developed at the BA. The BA has created a tri-lingual database of subject terms that describes their collections of materials in Arabic, English, and French. The main focus of this subject authority file is on the civilization of the Arab and Islamic World and its interaction with the outside world. The BA authority file is based on authorized subject headings and standard thesauri, such as LCSH and the Qa'imat ru'us al-mawdu'at al-'Arabiyah al-qiyasiyah lil-maktabat wa-marakiz al-malumat wa-qawaid al-bayanat" by Sha'ban Khalifah.

In preparation for this project, the author prepared a set of questions to be addressed to the librarians at the Bibliotheca Alexandrina.

• What are the BA workflows for creating subject authority records?
• What are the steps of creating the subject authority records?
• What is the linking process?
• Observations

2   Literature Review

The importance of providing multilingual access to library resources has been acknowledged by a growing number of institutions around the world. The topic has been discussed at international conferences since the early 1990s. Hamilton's annotated bibliography (2008) reviews important contributions related to multilingual access published prior to 2009. A number of projects provided crosswalks or links between thesauri in multiple languages. One such initiative, the Multilingual Subject Access (MACS) project that aims at providing links between subject headings in English, French, and German was described by Landry (2004).

Unfortunately, there are no published documents on the BA's subject authority file. The only information that is available comes from PowerPoint presentations by Iman Khairy, former authority control librarian at the BA. In her *Dewey in Authority Records: Bibliotheca Alexandrina's Experience* PowerPoint presentation, Khairy (2010) outlined briefly the structure of the Bibliotheca Alexandrina Subject Headings (BASHs) database and the tools used to formulate the Arabic subject headings. The Western African Manuscript Database project (1987) has developed Arabic subject headings for a collection of Arabic manuscripts that have been collected in West Africa. The database of subject terms is searchable in Arabic and English.

A study by El-Sherbini and Chen (2011) indicated that while most libraries in North America use the LCSH for assigning subject headings for non-Latin script materials, many end users and librarians would prefer a system that is more open to multilingual, multi-script subject headings. Although the survey respondents did not indicate a strong preference for adding a tagging feature to the library catalog, they highlighted areas of opportunity for libraries to make significant improvements to the catalog.

3   The BA Authority Processing Workflows for Latin Script and the Arabic Materials

The BA cataloging workflow is divided into parts, each part being performed by a different group of librarians and staff. During the initial phase, library materials are processed by catalogers who perform copy and original cataloging without checking and verifying subject headings. Authority control processing is conducted by another group of librarians during the second phase of processing. These two functions are performed independently of each other, by two different groups of staff.

For Latin-script materials, the BA relies on the OCLC database as a source of catalog records. The BA staff search the OCLC database. If a copy is found, the record is downloaded into the local system. If no copy is identified, a new record is created in the local online catalog through a template. The BA staff do not, however, contribute newly created bibliographic or authority records directly to the OCLC database.

During this process, the BA catalogers do not check the access points for the downloaded record. However, when creating original records, they consult the LC/OCLC authority file to verify access points. This includes personal names and subject headings. If the heading is found in the LC/OCLC authority file, the cataloger cuts this heading and pastes it in their local original cataloging template. If the heading is not found, access points are created locally, based on AACR2 and their local policy.

The workflow for Arabic materials is significantly different from the Latin-script materials workflow. In the Arabic materials workflow, the BA staff create the bibliographic records from scratch. All cataloging is original and the description is in the Arabic script. If the need arises, the LC catalog and authority file are consulted to help in determining the main entry and access points in the bibliographic record.

All the authority work, whether for Arabic or Latin-script materials, is performed after the initial bibliographic records have been created in the BA database. The daily report of new access points created is generated every day. This list includes subjects, personal names, corporate bodies, and series. Authority processing staff

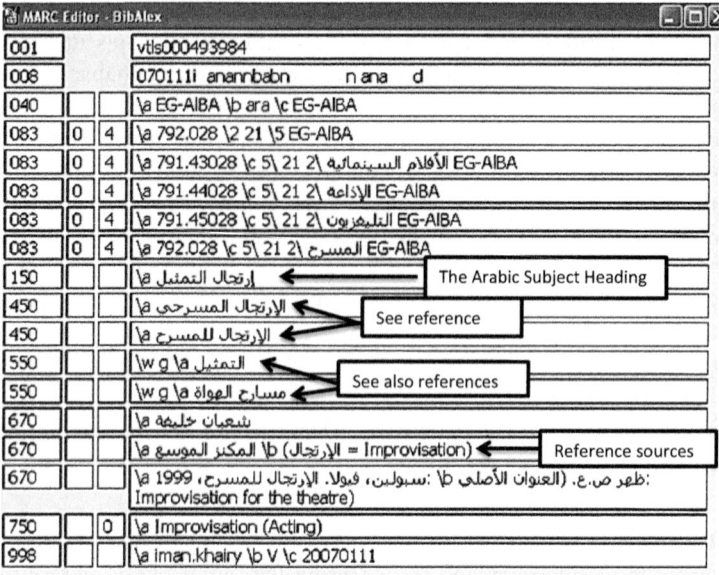

Fig. 1 Arabic authority record created

can recognize the type of heading by a MARC 21 tag that accompanies the heading. Because catalogers do not check the BA authority file during cataloging, a single heading could appear more than once in the daily report.

The authority control staff search the daily report of authority headings for all the new access points that enter the system. They select from the daily report the new headings and search these headings in the local VIRTUA database to check for duplication. If a heading is not found in the local system and the heading is in Arabic, they create a new authority record (Fig. 1). If the heading is in Latin-script, they search the LC/OCLC authority file and download the authority record. If the authority record is not found, a new authority record will be created locally.

The authority control staff link the equivalent subject headings in other languages in authority records through MARC 21 tag 750, the second indicator 7 (Fig. 2). This is done through a search in the BA database to determine if there are titles under subject in other languages that are equivalent. For example, if the downloaded subject authority records are for the subject "Molecular biology," a search in the BA database shows that there are records in Arabic and French languages that have subject headings equivalent to the English subject heading. In this case, the subject heading in French and Arabic will be added to the authority records and will be linked to the bibliographic record in Arabic and French. The downloaded records are automatically loaded into the database and merged with the bibliographic records and all updates in the authority file are changed in the bibliographic records automatically. The process of linking the data will be addressed in a separate section in more detail.

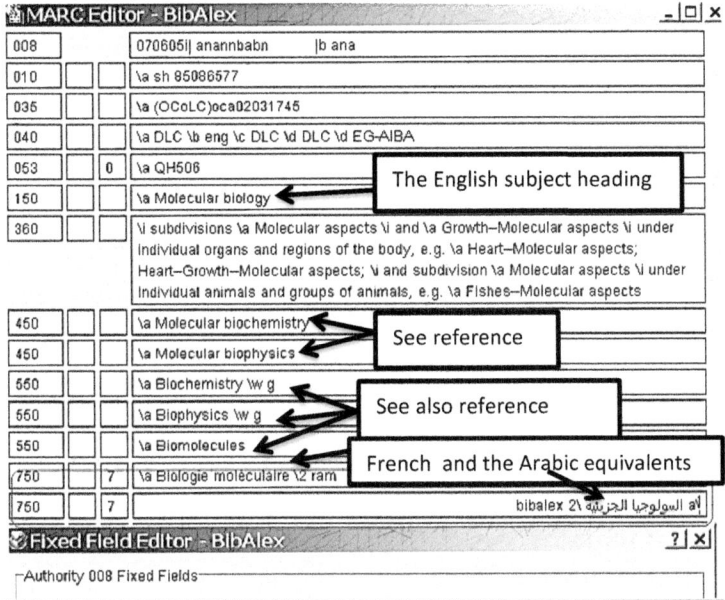

Fig. 2 An authority record downloaded from LC/OCLC for headings in English

4 Steps in Creating a Subject Authority Record

The authority processing staff check the VTLS utilities database reports daily to review the new headings that have been created. To access the new headings, the authority cataloger will open the VTLS utilities menu and select "Authority Messages" from the "Cataloging" option. This will display the various reports, including a report that indicates that new headings have been created in MARC 21 1xx, 7xx, or 6xx. Records can be retrieved by a specific date or date range, and can be limited by the cataloger's initials. The cataloger will then click on the new subject headings in MARC 21 tag 6xx that needs an authority record (Fig. 3).

The first step in the process of creating an authority record is to search the BA authority file for the heading "Media literacy." If the subject is found, the bibliographic record is checked for MARC 21 tag 650, second indicator "0." When the second indicator is coded "0," it means that the subject heading comes from the LCSH. If the subject is from the LCSH, the next step is to open the OCLC Authority file to download the full authority record.

The BA downloads or creates authority records only for the LC subject headings as indicated in MARC 21 tag 650 second indicator "0." If the 650 tag has a different indictor, the authority cataloger leaves the subject without creating or downloading an authority record for it. In our example, the subject "Media literacy" is found in the

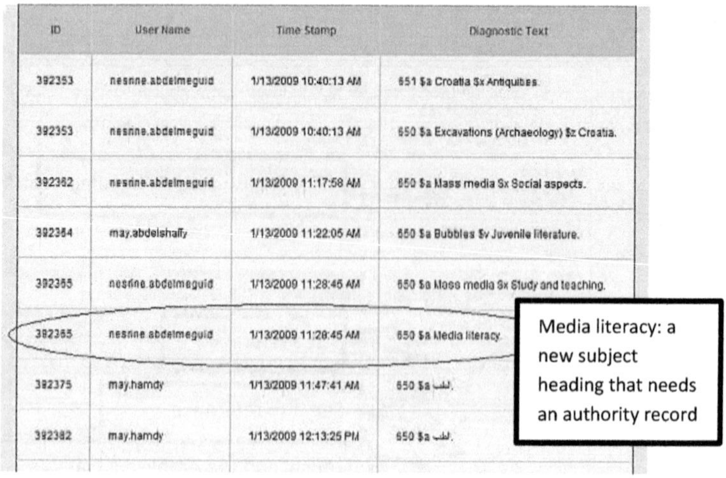

Fig. 3 A list of new headings in the daily report

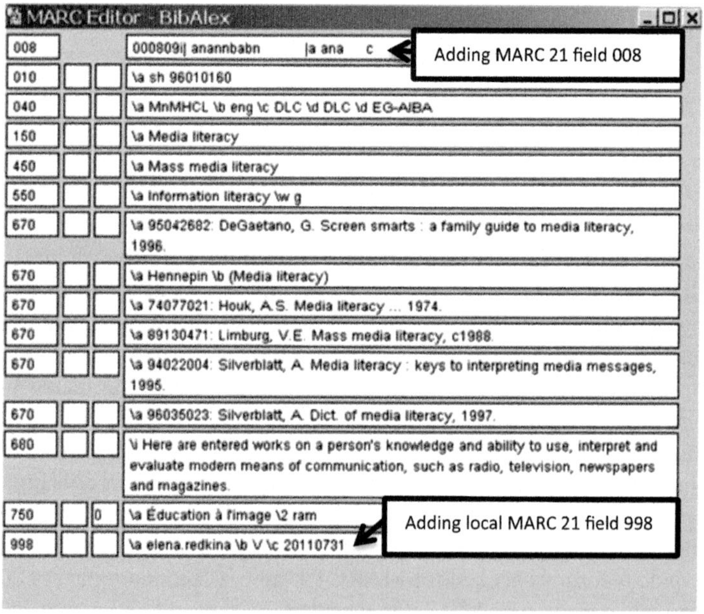

Fig. 4 Edit the authority record by adding the 008 and 998 fields

LCSH and the authority cataloger searched the OCLC Authority file and followed specific steps to export the record into the BA authority file and edit it (Fig. 4). The authority record is created, as shown in Fig. 5.

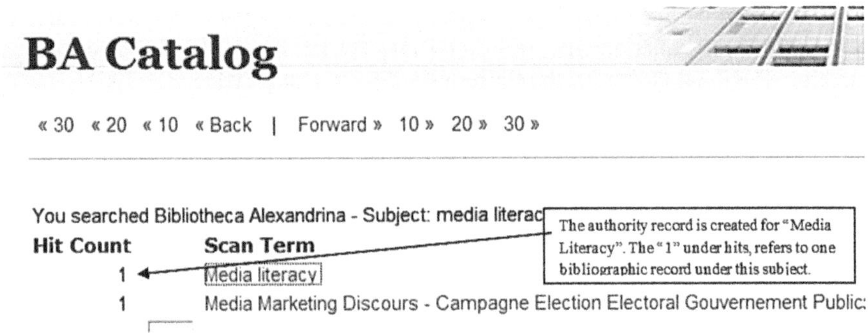

BA Catalog

« 30   « 20   « 10   « Back | Forward »   10 »   20 »   30 »

You searched Bibliotheca Alexandrina - Subject: media literac[y]

Hit Count	Scan Term
1	Media literacy
1	Media Marketing Discours - Campagne Election Electoral Gouvernement Public:

The authority record is created for "Media Literacy". The "1" under hits, refers to one bibliographic record under this subject.

Fig. 5 An English language authority record is created

5 Subject Headings Proposals

Since the BA collections represent several languages and cataloging traditions, no single subject thesaurus can accommodate these diverse collections. This is particularly true when providing subject access points for the Arabic collection. The Latin-script thesauri are limited in their coverage of the Arabic culture and civilization, especially in the religion aspects of this culture. The BA makes use of these thesauri, but also relies on the Arabic thesauri.

Qa'imat Ru'us al-Mawdu'at al-Arabiyah al-Qiyasiyah lil-Maktabat wa-Marakiz al-Ma'lumat wa-Qawa'id al-Bayanat (List of Standard Arabic Subject Headings for Libraries, Information Centers and Databases (QRMAK) by Sha'ban 'Abd al-'Aziz Khalifah and Muhammad 'Awad 'Ayidi (2001)

> QRMAK is the main list used for assigning Arabic subject headings. It is a general subject headings list that includes free-floating subdivisions and MARC 21 code. The subject headings are listed in Arabic. There is no electronic version of this thesaurus, making it very difficult to update or revise the headings. It does not identify broader, narrower or related subjects.

al-Faysal: Maknaz Arabi shamil fi ulum al-hadith: qism ulum al-din al-Islaami (1994)

> The subjects are in Arabic and cover Islamic subjects and other subjects like economics, language, etc. It provides broader, narrower as well as related subjects.

The Expanded Thesaurus (CD-Rom) produced by Abdul Hameed Shoman Foundation, Dubai Municipality, Juma Al-Majed Centre for Culture & Heritage, collected & edited by Mahmoud Itayem.

> It is a general thesaurus that covers all subjects with 50,000 subjects and 25,000 descriptors. The subjects are in Arabic, English, and French. The English and French equivalent subjects are translations of the Arabic subjects and are not mapped to an authorized subject list such as the LCSH and Rameau. It does not include free-floating subdivision.

Qaimat Ru'us al-Mawduat al-Tibbiyah = Medical subject headings (MeSH), National Library of Medicine, translated into Arabic by the World Health Organization, Regional Office for the Eastern Mediterranean (2001)

> It covers medical and scientific subjects in Arabic and English. The Arabic subjects are equivalent to English subjects from MeSH. It provides broader, narrower and related subjects. It is difficult to search the list since the subjects are arranged alphabetically in Arabic, and there is no index available for searching the subjects in English.

The BA staff are using the LCSH to assign subject headings to materials in English and Rameau to assign French subject headings for the French language materials.

If a subject does not exist in any of the above lists, the BA staff create a proposal for a new subject and identify the equivalents from LCSH and Rameau. The proposed new subject is coded in MARC 21 tag 6xx with the second indicator 4 to indicate that the source of this subject is not specified.

The process of proposing an Arabic subject heading is complex and depends on the skill of the authority control staff. The process of proposing new Arabic subject headings is as follows.

- The authority control staff check if the proposed Arabic subject already exists in the BA authority file under a related term. If the term is found, the cataloger will be informed to use this subject instead. If the term is not found, the authority control staff will check "al-Maknaz al-Muwassa (The Expanded Thesaurus)" which is a general thesaurus and any other specialized thesauri that will help in formulating the appropriate subject heading.
- If the term is not found in any of the Arabic thesauri, the authority control staff will undertake various methods to establish this heading. They might search "Google" to find if there are any equivalents or more useful terms related to that heading.
- If the term has other expressions, they use the most frequently used term and make cross-references to the other terms.
- If they find any broader, narrower or related terms, they will use these as well.
- In formulating the heading, the authority control staff have to consider the display and the indexing in the online authority file and the catalog.
- Some headings have not been accepted as authorized subject headings. In such instances, they may use the heading as a temporary heading, subject to future revision.
- New Arabic subject heading proposals are added to the BA authority file after receiving confirmation from the authority control team that the heading is valid.

For the Latin-script collection, the catalogers do not propose a new subject heading. The BA is not a direct member of the Program for Cooperative Cataloging, Subject Authority Cooperative Program (PCC/SACO). If the heading is not in the LC/OCLC authority file, the authority control processing staff either leave the heading alone or submit a proposal for the new heading through the Middle East Librarian's Association's (MELA) SACO funnel submissions.

6 Linked Subject Headings in Arabic French, and English as Equivalents

Linking data allows the users to navigate from one language to another and to link the subjects in multilingual collections. The BA follows MARC 21's "Model B" of Appendix C for multilingual/multi-script authority records (2001). The MARC 21 7xx tags (2012) in the authority records are used for linking subject headings in Arabic, English, and French.

During cataloging, the BA catalogers use English, Arabic, and French subject lists to assign subject headings for materials in these languages. If the book is in Arabic, the subject heading is selected from the Arabic Subject Headings lists, and if it is in English, the subject heading will be selected from the LCSH. For French language materials, the Rameau subject list will be used to select the subject heading. Each subject displays titles under its respective language.

The authority processing staff are responsible for linking the subject heading equivalents. This step is performed after the bibliographic record is created. During the creation of a new subject authority record, the authority processing staff will search the online catalog by the equivalent subjects to locate records in another language that might be under these equivalent subjects. If an equivalent subject is found, the authority staff will retrieve the authority record for each subject heading in its own respective language and add links to the equivalent in the 750 tag, second indicator 7. In the following example, the authority record "Database management" was retrieved and updated to add the equivalent subject headings in Arabic and French in MARC 21 field 750 tag, second indicator 7 (Fig. 6).

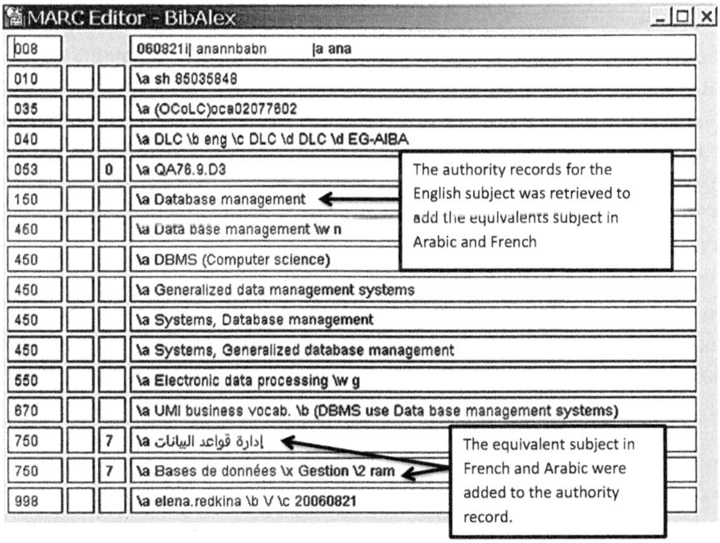

Fig. 6 Adding the French and Arabic equivalent subjects to the English language authority record

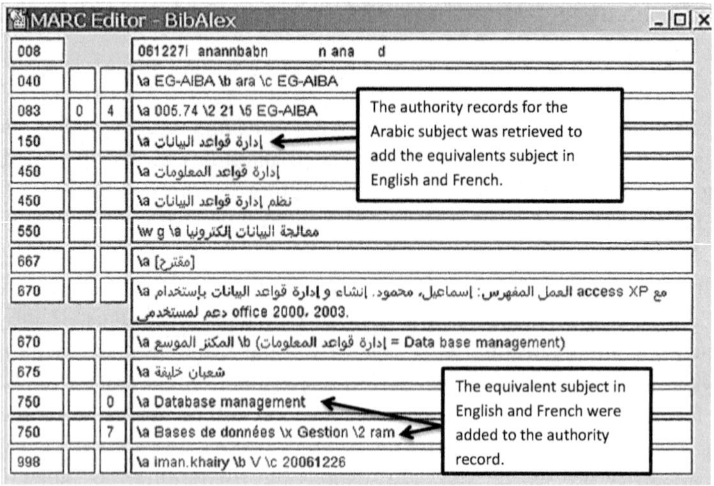

MARC Editor – BibAlex _ □ x

008			061227i anannbabn n ana d
040			\a EG-AIBA \b ara \c EG-AIBA
083	0	4	\a 005.74 \2 21 \5 EG-AIBA
150			إدارة قواعد البيانات a\
460			إدارة قواعد المعلومات a\
450			نظم إدارة قواعد البيانات a\
550			معالجة البيانات إلكترونيا a\ g \w
667			[مقترح] a\
670			مع access XP العمل المفهرس: إسماعيل، محمود. إنشاء وإدارة قواعد البيانات بإستخدام a\ دعم لمستخدمي office 2000، 2003.
670			a\ (إدارة قواعد المعلومات = Data base management) \b المكنز الموسع a\
675			شعبان خليفة a\
750		0	\a Database management
750		7	\a Bases de données \x Gestion \2 ram
998			\a iman.khairy \b V \c 20061226

The authority records for the Arabic subject was retrieved to add the equivalents subject in English and French.

The equivalent subject in English and French were added to the authority record.

Fig. 7 Adding the English and French equivalent subjects to the Arabic authority record

The Arabic authority record will be retrieved and the English and French equivalent subjects will be added to the Arabic authority record in MARC 21 field 750 second indicator "7" (Fig. 7). Next, the French authority record will be retrieved to add the equivalent subjects in English and Arabic.

7 Search Results

Searching by the English subject heading "Database management," for example, will retrieve this subject and will indicate the number of titles under this subject. The plus sign in front of this subject indicates that there are other titles under this subject heading in different languages (Fig. 8). By clicking the plus sign, the user will be prompted to the authority record that shows that there are other titles in Arabic and French (Fig. 9). This is a result of linking the Arabic and the French subject to the English subject in the authority record.

If the search is conducted for the French subject term, the authority record will point to equivalent subjects in Arabic and/or English, if the library holds titles in these languages. If searches are done using the Arabic subject, the search results will show equivalent subjects in French and/or English, if there are titles in these languages in the collection.

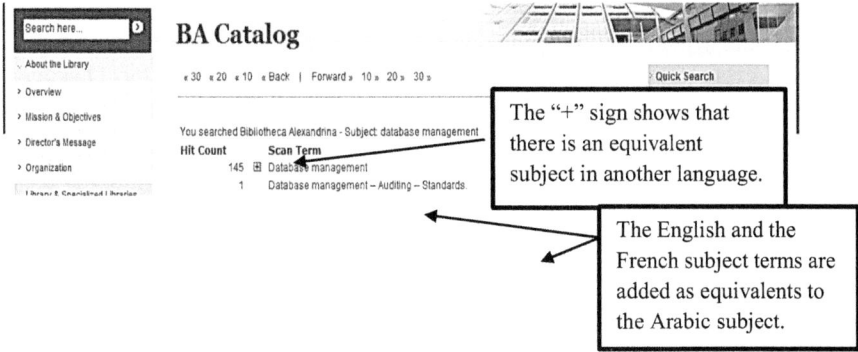

Fig. 8 Searching under the English subject heading

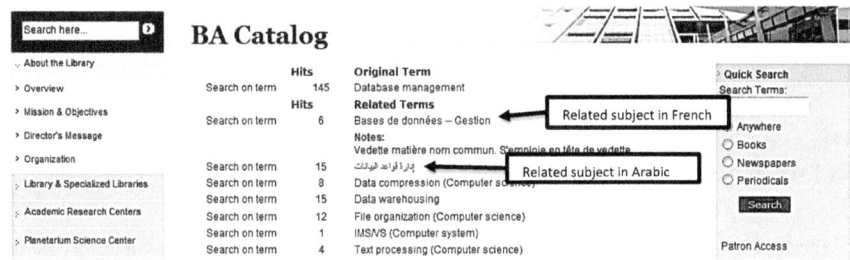

Fig. 9 Shows the "see also" references in the authority record

8 General Observations

The structure of the BA authority file is similar to the LC/OCLC authority file in that it conforms to the accepted cataloging rules. The BA has established procedures for creating a standard form for the subject heading. It differs from the LC practice in that the authority records are created based on the language and script of the item cataloged. The BA does not use Romanization when creating catalog records for Arabic script materials.

The process of linking data manually helps eliminate unintended errors in the database, as the cataloger studies the context in which the terms are used in the original language and establishes exact links with equivalent terms in the other languages. This process is time consuming and requires a high level of subject expertise.

The current process of assigning subject headings results in substantial duplication. The BA staff download bibliographic records into their database. The system harvests the access points and generates a daily report. When the authority control cataloger receives the list, he/she needs to check each subject term from the daily report in the authority database to avoid duplication. This process could be simplified by creating programming that would check the database and flag those headings that need to be linked or created.

The BA relies on the print Arabic thesauri for assigning Arabic subject headings. Many of them are highly specialized in dealing with complex fields such as Islamic religion and culture. Until recently, these subject terms existed only in difficult to obtain print editions, or even in a manuscript form. These books are not updated and the BA staff often have to assign new terms in their local database.

One of the benefits of the BA authority file is the addition of the Arabic subject terms to the database.

Introducing them into a subject authority database and linking them with their English or French equivalents from LCSH and Rameau is an important contribution of the BA file. The advantage of creating the BA their local authority file is that it opens the door to the possibility of creating an authority center that would serve the Arab world through cooperative arrangements. Some of the solutions developed at the BA can lead to future developments in resource sharing and cooperation on a global scale.

References

El-Sherbini, M. (2013). Bibliotheca Alexandrina's model for Arabic name authority control. *Library Resources & Technical Services, 5*(1), 4–17.

El-Sherbini, M., & Chen, S. (2011). An assessment of the need to provide non-Roman subject access to the library online catalog. *Cataloging & Classification Quarterly, 49*(6), 457–483.

Hamilton, M. B. (2008). An annotated bibliography of resources related to multilingual library cataloging and access. http://www.meaningliberation.info/804bibliography.

Khairy, I. (2010). Dewey in authority records: Bibliotheca Alexandrina's experience. http://www.slainte.org.uk/edug/docs/2010/Khairy-EDUG2010.ppt.

Landry, P. (2004). Multilingual subject access: The linking approach of MACS. *Cataloging & Classification Quarterly, 37*(3–4), 177–191.

Library of Congress, MARC 21 Authorities. (2012). 7xx—Heading linking entries-general information. http://www.loc.gov/marc/authority/ad7xx.html.

Library of Congress, MARC 21 Authority. (2001). Appendix: Multiscript records, model B. www.loc.gov/marc/authority/ecadmulti.html#modelb.

MELA: Committee on Cataloging. SACO funnel submissions. https://sites.google.com/site/melacataloging/resources/saco.

VIAF: Virtual International Authority file. http://viaf.org.

Western African Arabic Manuscript Database. (1987). http://westafricanmanuscripts.org/history.htm.

The VuFind Based "MT-Katalog": A Customized Music Library Service at the University of Music and Drama Leipzig

Anke Hofmann and Barbara Wiermann

Abstract For some time, large academic libraries have been offering discovery systems in order to allow access not only to their library holdings but also to their licensed electronic materials. Most of these libraries integrate huge commercial indices into their discovery systems. But it is only now that special libraries are starting to discuss whether those indices meet their demands, too.

As a part of a cooperative project, the library of the University of Music and Drama in Leipzig, Germany, installed the open source system *VuFind* (see http://www.vufind.org). It was accompanied by discussions about how to develop a discovery system that is transparent and suitable to the users' needs.

The following paper will show the reflections on that matter that finally led to the new *MT-Katalog*, which offers higher comfort of use and broader scope of search than our previous catalogue. We will take into account which additional musical and music-related e-resources can be found, selected and integrated. The paper will also provide ideas for improved use and enhancement of metadata.

1 Introduction

Since autumn 2011, a project called *finc* (*find in catalogue*) is running at Leipzig University Library. The goal of this project is to introduce a next-generation catalogue to 11 Saxon university libraries. *Finc* is financed by the European Regional Development Fund (ERDF) and runs for 3 years. A team of about 7 IT specialists and librarians develops 11 platforms in cooperation with the library and IT staff of each involved library. These platforms are meant to meet the individual

A. Hofmann
Hochschule für Musik und Theater "Felix Mendelssohn Bartholdy" Leipzig,
Grassistr. 8, 04107 Leipzig, Germany
e-mail: anke.hofmann@hmt-leipzig.de

B. Wiermann (✉)
Sächsische Landesbibliothek- Staats- und Universitätsbibliothek Dresden (SLUB), 01054 Dresden
e-mail: barbara.wiermann@slub-dresden.de

B. Lausen et al. (eds.), *Data Science, Learning by Latent Structures, and Knowledge Discovery*, Studies in Classification, Data Analysis, and Knowledge Organization, DOI 10.1007/978-3-662-44983-7_48

needs of the different institutions and their patrons. The library of the University of Music and Drama Leipzig is 1 of the 11 project partners. In summer 2012, we were the second library of the consortium to publish our new catalogue, the so-called *MT-Katalog*.

In this paper, (1) we want to start with some general remarks on the project *finc*, the software and architecture used in the consortium and our so-called *MT-Katalog*. (2) We want to show how we use the discovery system to improve the search options for our patrons, making better use of the already existing metadata and further enhancing them. And (3) we are going to discuss which external resources we have integrated into the system and which problems we have encountered dealing with metadata from different projects of varying origins.

2   The Project *Finc*: Software and Architecture

The Saxon consortium of 11 university libraries uses the open source software *VuFind* for its next-generation catalogue. *VuFind* was developed at the University of Villanova and is by now disseminated almost worldwide.[1] The key decision to use *VuFind* was taken on the basis of a proof-of-concept undertaken in summer 2010, which showed the possibility (1) to integrate authority files within *VuFind* and (2) to connect *VuFind* with the ILS *Libero* by Lib-IT to get real-time availability data and patrons' information.[2] These were the preset criteria that the target software had to meet.

Figure 1 shows the architecture of the consortium. The metadata of the traditional library materials are taken from the Union Catalogue of the *Südwestdeutsche Bibliotheksverbund* (Southwestern German Library Network, *SWB*), which is responsible for the libraries of the Federal States Baden-Wuerttemberg and Saxony. They comprise title data enhanced with authority files. These metadata are combined with the metadata of additional external resources, which together build up the consortial index for all 11 libraries. The different resources receive a tag that specifies for which library they are relevant. In order to get information such as the availability status or details from the user accounts in real time, the discovery system is connected to our local ILS. In addition, as we will see later on, a few local title fields are taken over.

For many libraries, the key feature of a discovery system is a mega-index, which comes along with many commercial discovery systems but is not part of the open source software *VuFind*. In the Saxon *finc* consortium, the larger universities decided to combine *VuFind* with *Primo Central,* the mega-index of Exlibris.[3]

---

[1] See http://www.vufind.org.

[2] The ILS Libero is used by 10 of the 11 project partners. See http://www.lib-it.de.

[3] See http://www.exlibrisgroup.com.

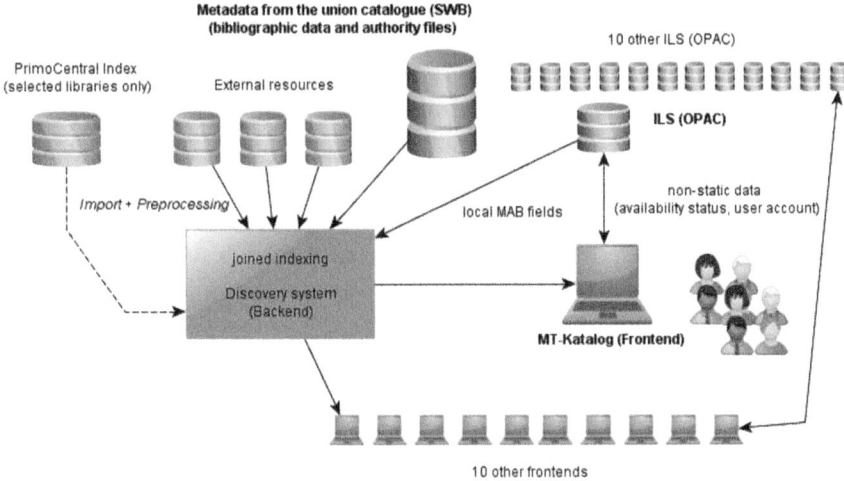

Fig. 1 Architecture of the Saxon *finc* consortium (simplified presentation)

According to the company, the index comprises hundreds of millions data sets for scholarly e-resources, journals and e-books from all subject areas.

Our library, however, opted against a mega-index for several reasons, one of which was of course a cost–benefit consideration. In addition, we were reluctant with regard to the content of the mega-index and its compatibility to the profile of the university and the library.[4] The following questions occurred to us: How much relevant literature on music and drama is included? How is the ratio between English- and German-language titles, which is especially important considering the fact that musicology is still a discipline with a strong German focus? How far is it reasonable and appropriate to promote research publications through a mega-index, when the students and professors at our university are just as interested in printed music and audio-visual materials, which are not included in a mega-index?

After turning down the mega-index, we had to define our own goals for the discovery system. Instead of improving the search situation through *Primo Central*, we now had to aggregate selected music and theatre resources on our own, choosing those that exactly fit the needs of our patrons. Furthermore, we tried to optimize the ease of search by building additional facets, creating new access points and enhancing the metadata.

---

[4]The University of Music and Drama Leipzig has about 950 students working towards Bachelor or Master Degrees in classical and popular music, in music pedagogy, musicology and drama. The library holds about 125,000 music scores, 50,000 books and 20,000 audio-visual materials.

3   Improving the Search Options by an Optimized Use of Metadata

In comparison with Google, which uses a full text search, traditional library catalogues stick to a metadata search. Using search engine technology of a next-generation catalogue nevertheless offers many new options, even when not using a mega-index with full texts. Since *VuFind* is based on *MARC* data processing, we had to take a closer look at the original *MAB* data[5] and *PICA* data[6] and deal with concordances and similar issues. This advanced some ideas how to better explore and enhance the existing data. Besides, in comparison to traditional ILS the discovery system is much more flexible concerning issues such as how many different fields can be indexed and how many and which facets can be built.

3.1   Customized Facets

At this time we have established seven facets (see Table 1), which take into account the special needs of our patrons.

In the following, we would like to have a closer look at two examples to show their relevance.

3.1.1   A Film Facet

Students and professors of the University of Music and Drama Leipzig are highly interested in the DVD collection of our library, which by lack of space is in the

Table 1 Facets of the *MT-Katalog*

Zugang	Access
Medientyp	Media type
Datensammlung	Collection
Ausgabeform	Musical representation
Besetzung	Media of performance/instrumentation
Filmgattung	Film genre
Person/institution	Author/corporate author

---

[5]*MAB* (Maschinelles Austauschformat für Bibliotheken) has long been the standard metadata format used by German libraries. In 2006 the German national library stopped any further development of *MAB*. It slowly gets replaced by *MARC 21*. See http://www.dnb.de/EN/Standardisierung/Formate/MAB/mab_node.html.

[6]The so-called *PICA* format has its origins in the catalogue systems of the Pica Foundation (*Stichting Pica,* Leiden/Netherland). See http://en.wikipedia.org/wiki/OCLC_PICA.

closed stacks. This means that browsing DVDs at the shelves is not possible. They only can be searched for in the catalogue. It is a matter of fact that by a mere author and title search, there is no way to really explore a collection.

A closer look at the metadata for DVDs nevertheless opens new perspectives. According to the cataloguing rules for the *SWB* Union catalogue, librarians are asked to record the genre of films such as Spielfilm (motion pictures), Openin-szenierung (opera production), Literaturverfilmung (screen adaption of literature), Dokumentarfilm (documentary).[7] In most library catalogues this information serves no other purpose than a mere descriptive one. Even in the *SWB* Union catalogue itself, the information is neither searchable nor used as a facet. In our *MT-Katalog*, the relevant field gets indexed now. This allows us to build a film facet, which offers the opportunity to explore the DVD collection by genre.

This simple example points to a typical mismatch in the library world. We produce many standardized data and we often do not use it in our systems to create relevant and comfortable search options for our patrons.

3.1.2   A Musical Presentation Facet

One of the biggest challenges in the search of music prints is the musical presentation form; that is, to differentiate between full scores, vocal scores, performance materials, piano reductions, and so on. The German music cataloguing rules *RAK-Musik*[8] do not lay down standardized information on musical presentation. Instead, information on musical presentation is entered in the fields for title statement or statement of responsibility and optionally in the edition statement. In addition, these entries are in most cases not standardized but taken as given on the music print. The wording therefore shows a great variety in form and language (Fig. 2). The content of these fields thus has only a descriptive character and cannot be used in an acceptable and systematic way as a search term.

Since musical presentation information is of high importance for musicians, for the *MT-Katalog* we decided to create a standardized vocabulary for it that could be used as search terms and facet. This meant (1) to define this standardized terminology, (2) to choose a field where to incorporate the information, (3) to look for technical routines of adding the information retrospectively for previous holdings, (4) to add musical presentation from now on for new items (although it is not part of the cataloguing rules *RAK-Musik*).

We opted for five terms (Table 2) that are closely related to the experiences of our everyday library work.

---

[7]For a list of the standardized terms, see http://pollux2.bsz-bw.de/cgi-bin/help.pl?cmd=kat&val=4205.

[8]http://d-nb.info/970364628/34.

Fig. 2 Variety of terms used for the musical presentation "Klavierauszug"

Table 2 Standardized terms for musical presentation in the *MT-Katalog*

Aufführungsmaterial	Performance material
Partitur	Full score
Stimmen	Parts
Klavierauszug	Piano score
Quelle	Source material

Multiple use of the terms is possible, which is especially important for "Quelle" (source material). This is generally given for facsimiles, microforms, digitized materials of music manuscripts and early music prints.

For now, the new standardized terms for the musical presentation are only catalogued at a local field in our ILS, from where they are imported and indexed in the discovery system. In the medium-term, we are working on a solution to integrate these data within the library network.[9]

The real challenge was to generate the information retrospectively, and to add a musical presentation field with a standardized entry point for all the media items that were already part of our system. To achieve this, we first collected information from all the fields that might be relevant to retrieval of musical presentation data. We then aggregated this information and extracted the most likely musical presentation form for each item. In the many cases where this was not possible automatically, we performed manual checking and verification. Finally, we imported the extracted information back into the catalog, placing it in a new field of its own.

---

[9]This will probably be connected with the implementation of *RDA* in the German-speaking countries.

Having fulfilled this challenging task, it is now quite straightforward to add musical presentation as a standard field for new items. The facet for musical presentation is the feature of the new catalogue for which we get the most positive feedback from our patrons.

3.2 Verbal Representation of the RVK Classification

For subject indexing, the library of the University of Music and Drama Leipzig uses the classification system *RVK* (*Regensburger Verbundklassifikation*).[10] In the traditional OPAC, the alphanumerical notations were hyperlinked to get from one book to other books on the same topic. Since alphanumerical notations are not self-explanatory, the users never really became familiar with this feature. The discovery system offers the opportunity to improve the feature in at least two ways.

When getting the metadata of the *SWB*, we now import also the verbal representation of the classification system and display it with a mouse-over effect. Like this, the users are made aware of the information behind the alphanumerical system. Furthermore, we do not hyperlink the notation anymore, but add a separate link called "Ähnliche Treffer finden" (Find similar items), which is much more explicit to our patrons.

Moreover, we are currently running a test to index the verbal representations of the *RVK* to make them searchable as additional keywords. We would like to figure out whether these terms (that are partly very broad) are of relevance to our users when conducting a thematic search (Fig. 3).

Korea : Einführung in die Musiktradition Koreas

Beteiligte:	Burde, Wolfgang [Hrsg.]
Verfasserangabe:	hrsg. von Wolfgang Burde
Format:	Buch
Sprache:	Deutsch
veröffentlicht:	Mainz [u.a.] : Schott, 1985
Umfang:	192 S. : zahlr. Ill., Notenbeispiele.
Gesamtaufnahme:	Weltmusik
Schlagworte:	Korea > Musik
RVK-Notation:	LS 42000 ⊕ Ähnliche Treffer finden
Tags:	Keine Tags, Fügen Sie den ersten Tag hi

Allgemeines (einschl. Geschichte, Theorie)
Musikwissenschaft
I. Musica theoretica
Musikalische Volks- und Völkerkunde: Musica theoretica
Musik des Orients (und Asiens insgesamt) - Theoretica
Musikkulturen Ostasiens
Koreanische Musik
Allgemeines (einschl. Geschichte, Theorie)

Fig. 3 Verbal representation of the *RVK* classification system

[10]See http://rvk.uni-regensburg.de.

4 External Resources

As mentioned above, the University of Music and Drama Leipzig opted against a mega-index and instead decided for a selection of well-chosen external resources that closely fit the needs of our patrons. As until today, we integrated 14 external resources of different origins that focus on music and drama.

Figure 4 shows the ratio between the different groups of online materials now in the catalogue[11]: Licensed content, e-papers, e-books and e-scores (5,500 data sets) are a minority in our field. There are also relatively few digital-born publications (e-papers and e-books) in open-access repositories (11,721). Most of them are articles or books on pedagogy, not on music or drama. The catalogue has a real focus on retro-digitized material (51,921). Many libraries have special projects for digitizing music manuscripts and prints, which are of high importance for musicians (especially those involved with historically informed performance practice) and also for musicologists. With the *Naxos Music Library*, we have a comparatively high amount of licensed audio files (38,907). Finally, there is a special segment with data from the *International Music Score Library Project* (*IMSLP*),[12] a wiki most heavily used in the music world (51,722 works). The data are displayed in the graph separately, since (1) these are a mix of digitized music scores and digital-born music scores, and (2) the numbers are not comparable to the other projects because each data set describes a work and not an edition. Behind each work there can be many different editions in different formats. The number of scores in *IMSLP* is about 250,000 by today. (In other words, this segment would even be much larger if editions rather than works were indexed.)

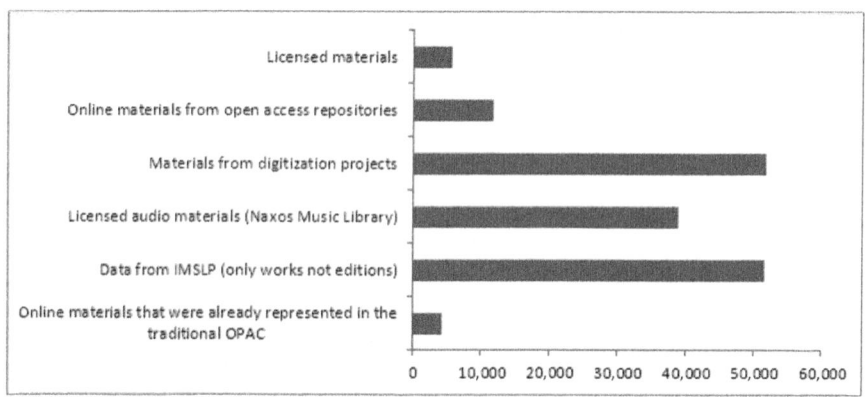

Fig. 4 Different types of electronic resources and their absolute numbers in the *MT-Katalog*

---

[11] All numbers as from October 15, 2013.

[12] See http://www.imslp.org.

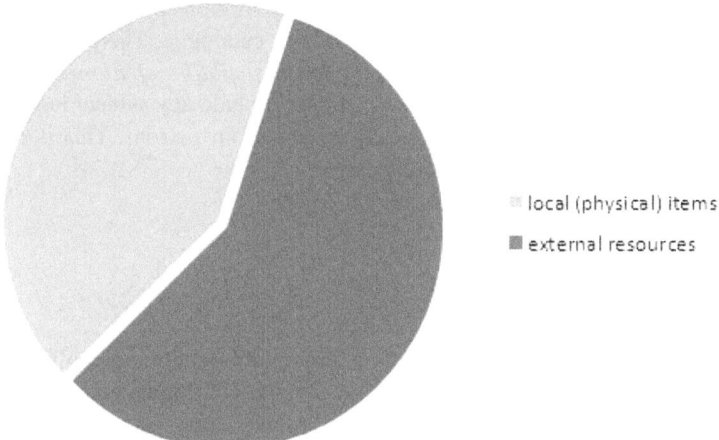

Fig. 5 Relative proportions of local (physical) and external (electronic) items in the *MT-Katalog*

Altogether, we succeeded in more than doubling the title data that the library catalogue offers to our patrons (Fig. 5).

When having a look at the chart in Fig. 4, our make-up may seem kind of arbitrary, and it is therefore appropriate to ask for our selection criteria. Besides the thematic relevance, external resources have to fulfill the following requirements: (1) the projects we are interested in have to offer data sets and images as part of a license or open access; (2) an appropriate interface is needed to download the data; (3) when dealing with larger data compilations comprising materials of different subjects areas, the data sets must include criteria that allow to narrow down the compilations thematically.

Collections like *Naxos Music Library*, *Music Online*, *Early Music Online* and *Music Treasures Consortium* were integrated into the *MT-Katalog* in their entirety. Metadata from *Dissonline*, the database for German online dissertations, were selected by Dewey decimal classification.[13] The relevant music data from the digital collections of the Bavarian State Library Munich were filtered by a special indication for the "Sondersammelgebiet Musik."[14] Unfortunately, the data sets do not offer a consistent criterion to limit them to the subject area of drama. For *Gallica*, the *Bibliothèque nationale de France Paris* provides a separate OAI set[15] for music scores, which is a very convenient service—yet it causes us to miss all digitized literature on music.

---

[13]We used the following notations music (780), film (791), theatre and dance (792) and pedagogy (370).

[14]"Sondersammelgebiete" is a German special collection system funded by the *Deutsche Forschungsgemeinschaft*.

[15]See http://www.openarchives.org/.

We are highly interested to integrate online dissertations not only from Germany, but also from other European countries into our catalogue. Therefore, we looked at *DART*—the Europe E-theses Portal of the *Association of European Research Libraries*.[16] However, since the data sets do not include any subject indication, we cannot filter the data for titles that are relevant for our patrons. That is one of the reasons, why hitherto we set the project aside.

5 Summary

In our above reflections, one may note a huge contrast: On the one hand, we are scrutinizing our own metadata as described in part II to improve the search options of our users. On the other hand, we are integrating external resources with metadata that are based on different cataloguing rules, are provided in different languages, or even come from outside the library world (such as *IMSLP* data). We have definitely lost the homogenous quality of the data in our catalogue, which used to guarantee a certain reliability for searches. In addition, we have lost such real achievements of the library world as, for example, a consistent use of authority files and uniform titles. Even for most of the library projects, one has to assert that the metadata are delivered without authority files or even without the relevant identifiers (*VIAF* or *GND*).[17] For the moment, we consider this an acceptable development as it comes with huge improvements in the amount and variety of information that is available to our patrons.

In summary, we have shown how discovery systems offer many opportunities to improve the search comfort by their search engine technology, supporting a better use of the existing metadata, a metadata enhancement or even integrating new resources. Nevertheless, it seems to be one of the major challenges in the context of discovery systems to keep the special quality of library data when exchanging and reusing it. Using search engine technology offers new possibilities of integrating data from different origins, but for an optimized search we still need standardizations as well as a data mapping to make searches as reliable as possible for our patrons. This is particularly true for systems specialized on selected subject areas such as the *MT-Katalog*. These fundamental challenges can only be addressed with enhanced national and international collaboration.

---

[16]See http://www.dart-europe.eu.

[17]See http://www.dnb.de/EN/Standardisierung/GND/gnd_node.html and http://viaf.org.

Index

© Springer-Verlag Berlin Heidelberg 2015
B. Lausen et al. (eds.), *Data Science, Learning by Latent Structures, and Knowledge Discovery*, Studies in Classification, Data Analysis, and Knowledge Organization, DOI 10.1007/978-3-662-44983-7

Printed by Printforce, the Netherlands